Statistical Inference as Severe Testing

How to Get Beyond the Statistics Wars

Mounting failures of replication in the social and biological sciences give a new urgency to critically appraising proposed reforms. This book pulls back the cover on disagreements between experts charged with restoring integrity to science. It denies two pervasive views of the role of probability in inference: to assign degrees of belief, and to control error rates in a long run. If statistical consumers are unaware of assumptions behind rival evidence reforms, they can't scrutinize the consequences that affect them (in personalized medicine, psychology, and so on). The book sets sail with a simple tool: If little has been done to rule out flaws in inferring a claim, then it has not passed a severe test. Many methods advocated by data experts do not stand up to severe scrutiny, and are even in tension with successful strategies for blocking or accounting for cherry picking and selective reporting. Through a series of excursions, tours, and exhibits, the philosophy and history of inductive inference come alive, while philosophical tools are put to work to solve problems about science and pseudoscience, induction and falsification.

Deborah G. Mayo is Professor Emerita in the Department of Philosophy at Virginia Tech and is a visiting professor at the London School of Economics and Political Science, Centre for the Philosophy of Natural and Social Science. She is the author of *Error and the Growth of Experimental Knowledge* (1996), which won the 1998 Lakatos Prize awarded to the most outstanding contribution to the philosophy of science during the previous six years. She co-edited *Error and Inference: Recent Exchanges on Experimental Reasoning, Reliability, and the Objectivity and Rationality of Science* (2010, Cambridge University Press) with Aris Spanos, and has published widely in the philosophy of science, statistics, and experimental inference.

"In this lively, witty, and intellectually engaging book, Deborah Mayo returns to first principles to make sense of statistics. She takes us beyond statistical formalism and recipes, and asks us to think philosophically about the enterprise of statistical inference itself. Her contribution will be a welcomed addition to statistical learning. Mayo's timely book will shrink enlarged posteriors and overinflated significance, by focusing on whether our inferences have been severely tested, which is where we should be focused."

– Nathan A. Schachtman, Lecturer in Law, Columbia Law School

"Whether or not you agree with her basic stance on statistical inference, if you are interested in the subject – and all scientists ought to be – Deborah Mayo's writings are a must. Her views on inference are all the more valuable for being contrary to much current consensus. Her latest book will delight many and infuriate others but force all who are serious about these issues to think. Her capacity to jolt the complacent is second to none."

– Stephen Senn, author of *Dicing with Death*

"Deborah Mayo's insights into the philosophical dimensions of these problems are unsurpassed in their originality, their importance, and the breadth of understanding on which they are based. Here she combines perspectives from philosophy of science and the foundations of statistics to eliminate mirages produced by misunderstandings both philosophical and statistical, while putting into focus the ways in which her error-statistical approach is relevant to current problems of scientific inquiry in various disciplines."

– Kent Staley, Saint Louis University

"This book by Deborah Mayo is a timely examination of the use of statistics in science. Her severity requirement demands that the scientist provide a sharp question and related data. Absent that, the observer should withhold judgment or outright reject. It is time to get tough. Funding agencies should take note."

– S. Stanley Young, Ph.D., FASA FAAAS

Statistical Inference as Severe Testing

How to Get Beyond the Statistics Wars

Deborah G. Mayo

Virginia Tech

CAMBRIDGE
UNIVERSITY PRESS

University Printing House, Cambridge CB2 8BS, United Kingdom

One Liberty Plaza, 20th Floor, New York, NY 10006, USA

477 Williamstown Road, Port Melbourne, VIC 3207, Australia

314–321, 3rd Floor, Plot 3, Splendor Forum, Jasola District Centre, New Delhi – 110025, India

79 Anson Road, #06–04/06, Singapore 079906

Cambridge University Press is part of the University of Cambridge.

It furthers the University's mission by disseminating knowledge in the pursuit of education, learning, and research at the highest international levels of excellence.

www.cambridge.org
Information on this title: www.cambridge.org/9781107054134
DOI: 10.1017/9781107286184

First published 2018
3rd printing 2019

Printed in the United Kingdom by TJ International Ltd. Padstow, Cornwall

A catalog record for this publication is available from the British Library.

Library of Congress Cataloging-in-Publication Data
Names: Mayo, Deborah G., author.
Title: Statistical inference as severe testing : how to get beyond the statistics wars / Deborah G. Mayo (Virginia Tech).
Description: Cambridge : Cambridge University Press, 2018. | Includes bibliographical references and index.
Identifiers: LCCN 2018014718 | ISBN 9781107054134 (alk. paper)
Subjects: LCSH: Mathematical statistics. | Inference. | Error analysis (Mathematics) | Fallacies (Logic) | Deviation (Mathematics)
Classification: LCC QA276 .M3755 2018 | DDC 519.5/4–dc23
LC record available at https://lccn.loc.gov/2018014718

ISBN 978-1-107-05413-4 Hardback
ISBN 978-1-107-66464-7 Paperback

Additional resources for this publication at www.cambridge.org/mayo

To George W. Chatfield
for his magnificent support

Itinerary

Preface

The Statistics Wars

Today's "statistics wars" are fascinating: They are at once ancient and up to the minute. They reflect disagreements on one of the deepest, oldest, philosophical questions: How do humans learn about the world despite threats of error due to incomplete and variable data? At the same time, they are the engine behind current controversies surrounding high-profile failures of replication in the social and biological sciences. How should the integrity of science be restored? Experts do not agree. This book pulls back the curtain on why.

Methods of statistical inference become relevant primarily when effects are neither totally swamped by noise, nor so clear cut that formal assessment of errors is relatively unimportant. Should probability enter to capture degrees of belief about claims? To measure variability? Or to ensure we won't reach mistaken interpretations of data too often in the long run of experience? Modern statistical methods grew out of attempts to systematize doing all of these. The field has been marked by disagreements between competing tribes of frequentists and Bayesians that have been so contentious – likened in some quarters to religious and political debates – that everyone wants to believe we are long past them. We now enjoy unifications and reconciliations between rival schools, it will be said, and practitioners are eclectic, prepared to use whatever method "works." The truth is, long-standing battles still simmer below the surface in questions about scientific trustworthiness and the relationships between Big Data-driven models and theory. The reconciliations and unifications have been revealed to have serious problems, and there's little agreement on which to use or how to interpret them. As for eclecticism, it's often not clear what is even meant by "works." The presumption that all we need is an agreement on numbers – never mind if they're measuring different things – leads to pandemonium. Let's brush the dust off the pivotal debates, walk into the museums where we can see and hear such founders as Fisher, Neyman, Pearson, Savage, and many others. This is to simultaneously zero in on the arguments between metaresearchers – those doing research on research – charged with statistical reforms.

Statistical Inference as Severe Testing

Why are some arguing in today's world of high-powered computer searches that statistical findings are mostly false? The problem is that high-powered methods can make it easy to uncover impressive-looking findings even if they are false: spurious correlations and other errors have not been severely probed. We set sail with a simple tool: If little or nothing has been done to rule out flaws in inferring a claim, then it has not passed a *severe test*. In the severe testing view, probability arises in scientific contexts to assess and control how capable methods are at uncovering and avoiding erroneous interpretations of data. That's what it means to *view statistical inference as severe testing*. A claim is severely tested to the extent it has been subjected to and passes a test that probably would have found flaws, were they present. You may be surprised to learn that many methods advocated by experts do not stand up to severe scrutiny, are even in tension with successful strategies for blocking or accounting for cherry picking and selective reporting!

The severe testing perspective substantiates, using modern statistics, the idea Karl Popper promoted, but never cashed out. The goal of *highly well-tested* claims differs sufficiently from *highly probable* ones that you can have your cake and eat it too: retaining both for different contexts. Claims may be "probable" (in whatever sense you choose) but terribly tested by these data. In saying we may view statistical inference as severe testing, I'm not saying statistical inference is always about formal statistical testing. The testing metaphor grows out of the idea that before we have evidence for a claim, it must have passed an analysis that could have found it flawed. The probability that a method commits an erroneous interpretation of data is an *error probability*. Statistical methods based on error probabilities I call *error statistics*. The value of error probabilities, I argue, is not merely to control error in the long run, but because of what they teach us about the source of the data in front of us. The concept of severe testing is sufficiently general to apply to any of the methods now in use, whether for exploration, estimation, or prediction.

Getting Beyond the Statistics Wars

Thomas Kuhn's remark that only in the face of crisis "do scientists behave like philosophers" (1970), holds some truth in the current statistical crisis in science. Leaders of today's programs to restore scientific integrity have their own preconceptions about the nature of evidence and inference, and about "what we really want" in learning from data. Philosophy of science can also alleviate such conceptual discomforts. Fortunately, you needn't accept the severe testing view in order to employ it as a tool for bringing into focus the crux of all these issues. It's a tool for excavation, and for keeping us afloat in the marshes and quicksand that often mark today's controversies. Nevertheless, important consequences will follow once this tool is used. First there will be a reformulation of existing tools (tests, confidence intervals, and others) so as to avoid misinterpretations and abuses. The debates on statistical inference generally concern inference after a statistical model and data statements are in place, when in fact the most interesting work involves the local inferences needed to get to that point. A primary asset of error statistical methods is their contributions to designing, collecting, modeling, and learning from data. The severe testing view provides the much-needed link between a test's error probabilities and what's required for a warranted inference in the case at hand. Second, instead of rehearsing the same criticisms over and over again, challengers on all sides should now begin by grappling with the arguments we trace within. Kneejerk assumptions about the superiority of one or another method will not do. Although we'll be excavating the actual history, it's the methods themselves that matter; they're too important to be limited by what someone 50, 60, or 90 years ago thought, or to what today's discussants *think* they thought.

Who is the Reader of This Book?

This book is intended for a wide-ranging audience of readers. It's directed to consumers and practitioners of statistics and data science, and anyone interested in the methodology, philosophy, or history of statistical inference, or the controversies surrounding widely used statistical methods across the physical, social, and biological sciences. You might be a researcher or science writer befuddled by the competing recommendations offered by large groups ("megateams") of researchers (should P-values be set at 0.05 or 0.005, or not set at all?). By viewing a contentious battle in terms of a difference in goals – finding highly probable versus highly well-probed hypotheses – readers can see why leaders of rival tribes often talk right past each other. A fair-minded assessment may finally be possible. You may have a skeptical bent, keen to hold the experts accountable. Without awareness of the assumptions behind proposed reforms you can't scrutinize consequences that will affect you, be it in medical advice, economics, or psychology.

Your interest may be in improving statistical pedagogy, which requires, to begin with, recognizing that no matter how sophisticated the technology has become, the nature and meaning of basic statistical concepts are more unsettled than ever. You could be teaching a methods course in psychology wishing to intersperse philosophy of science in a way that is both serious and connected to immediate issues of practice. You might be an introspective statistician, focused on applications, but wanting your arguments to be on surer philosophical grounds.

Viewing statistical inference as severe testing will offer philosophers of science new avenues to employ statistical ideas to solve philosophical problems of induction, falsification, and demarcating science from pseudoscience. Philosophers of experiment should find insight into how statistical modeling bridges gaps between scientific theories and data. Scientists often question the relevance of philosophy of science to scientific practice. Through a series of excursions, tours, and exhibits, tools from the philosophy and history of statistics will be put directly to work to illuminate and solve problems of practice. I hope to galvanize philosophers of science and experimental philosophers to further engage with the burgeoning field of data science and reproducibility research.

Fittingly, the deepest debates over statistical foundations revolve around very simple examples, and I stick to those. This allows getting to the nitty-gritty logical issues with minimal technical complexity. If there's disagreement even there, there's little hope with more complex problems. (I try to use the notation of discussants, leading to some variation.) The book would serve as a one-semester course, or as a companion to courses on research methodology, philosophy of science, or interdisciplinary research in science and society. Each tour gives a small set of central works from statistics or philosophy, but since the field is immense, I reserve many important references for further reading on the CUP-hosted webpage for this book, www.cambridge.org/mayo.

Relation to Previous Work

While (1) philosophy of science provides important resources to tackle foundational problems of statistical practice, at the same time, (2) the statistical method offers tools for solving philosophical problems of evidence and inference. My earlier work, such as *Error and the Growth of Experimental Knowledge* (1996), falls under the umbrella of (2), using statistical science for philosophy of science: to model scientific inference, solve problems about evidence (problem of induction), and evaluate methodological rules (does more weight accrue to a hypothesis if it is prespecified?). *Error and Inference* (2010), with its joint work and exchanges with philosophers and statisticians, aimed to bridge the two-way street of (1) and (2). This work, by contrast, falls under goal (1): tackling foundational problems of statistical practice. While doing so will constantly find us entwined with philosophical problems of inference, it is the arguments and debates currently engaging practitioners that take the lead for our journey.

Join me, then, on a series of six excursions and 16 tours, during which we will visit three leading museums of statistical science and philosophy of science, and engage with a host of tribes marked by family quarrels, peace treaties, and shifting alliances.[1]

---

[1] A bit of travel trivia for those who not only read to the end of prefaces, but check its footnotes: two museums will be visited twice, one excursion will have no museums. With one exception, we engage current work through interaction with tribes, not museums. There's no extra cost for the 26 souvenirs: A–Z.

Acknowledgments

I am deeply grateful to my colleague and frequent co-author, Aris Spanos. More than anyone else, he is to be credited for encouraging the connection between the error statistical philosophy of Mayo (1996) and statistical practice, and for developing an error statistical account of misspecification testing. I thank him for rescuing me time and again from being stalled by one or another obstacle. He has given me massive help with the technical aspects of this book, and with revisions to countless drafts of the entire manuscript.

My ideas were importantly influenced by Sir David Cox. My debt is to his considerable work on statistical principles, his involvement in conferences in 2006 (at Virginia Tech) and 2010 (at the London School of Economics), and through writing joint papers (2006,[2] 2010[3]). I thank him for his steadfast confidence in this project, and for discussions leading to my identifying the unsoundness in arguments for the (strong) Likelihood Principle (Mayo 2014b) – an important backdrop to the evidential interpretation of error probabilities that figures importantly within.

I have several people to thank for valuable ideas on many of the topics in this book through extensive blog comments (errorstatistics.com), and/or discussions on portions of this work: John Byrd, Nancy Cartwright, Robert Cousins, Andrew Gelman, Gerd Gigerenzer, Richard Gill, Prakash Gorroochurn, Sander Greenland, Brian Haig, David Hand, Christian Hennig, Thomas Kepler, Daniël Lakens, Michael Lew, Oliver Maclaren, Steven McKinney, Richard Morey, Keith O'Rourke, Caitlin Parker, Christian Roberts, Nathan Schachtman, Stephen Senn, Cosma Shalizi, Kent Staley, Niels Waller, Larry Wasserman, Corey Yanofsky, and Stanley Young.

Older debts recalled are to discussions and correspondence with George Barnard, Ronald Giere, Erich Lehmann, Paul Meehl, and Karl Popper.

Key ideas in this work grew out of exchanges with Peter Achinstein, Alan Chalmers, Clark Glymour, Larry Laudan, Alan Musgrave, and John Worrall, published in Mayo and Spanos (2010). I am grateful for the stimulating conversations on aspects of this research during seminars and conferences at the London School of Economics in 2008, 2010, 2012, and 2016, Centre for Philosophy of Natural and Social Sciences. I thank Virginia Tech and Doug Lind, chair of the philosophy department, for support and professional accommodations which were essential to this project. I obtained valuable feedback

---

[2] *First Symposium on Philosophy, History, and Methodology of Error (ERROR 06)*, Virginia Tech.
[3] *Statistical Science & Philosophy of Science: Where Do/Should They Meet in 2010 (and Beyond)?* The London School of Economics and Political Science, Centre for the Philosophy of Natural and Social Science.

from graduate students of a 2014 seminar (with A. Spanos) on Statistical Inference and Modeling at Virginia Tech.

I owe special thanks to Diana Gillooly and Cambridge University Press for supporting this project even when it existed only as a ten-page summary, and for her immense help throughout. I thank Esther Migueliz, Margaret Patterson, and Adam Kratoska for assistance in the production and preparation of this manuscript. For the figures in this work, I'm very appreciative for all Marcos Jiménez' work. I am grateful to Mickey Mayo for graphics for the online component. I thank Madeleine Avirov, Mary Cato, Michael Fay, Nicole Jinn, Caitlin Parker, and Ellen Woodall for help with the copy-editing. For insightful comments and a scrupulous review of this manuscript, copy-editing, library, and indexing work, I owe mammoth thanks to Jean Anne Miller. For other essential support, I am indebted to Melodie Givens and William Hendricks.

I am grateful to my son, Isaac Chatfield, for technical assistance, proofing, and being the main person to cook real food. My deepest debt is to my husband, George W. Chatfield, for his magnificent support of me and the study of E.R.R.O.R.S.[4] I dedicate this book to him.

---

[4] Experimental Reasoning, Reliability, and the Objectivity, and Rationality of Science.

Excursion 1 How to Tell What's True about Statistical Inference

Itinerary

Tour I Beyond Probabilism and Performance

I'm talking about a specific, extra type of integrity that is [beyond] not lying, but bending over backwards to show how you're maybe wrong, that you ought to have when acting as a scientist. (Feynman 1974/1985, p. 387)

It is easy to lie with statistics. Or so the cliché goes. It is also very difficult to uncover these lies without statistical methods – at least of the right kind. Self-correcting statistical methods are needed, and, with minimal technical fanfare, that's what I aim to illuminate. Since Darrell Huff wrote *How to Lie with Statistics* in 1954, ways of lying with statistics are so well worn as to have emerged in reverberating slogans:

- Association is not causation.
- Statistical significance is not substantive significance.
- No evidence of risk is not evidence of no risk.
- If you torture the data enough, they will confess.

Exposés of fallacies and foibles ranging from professional manuals and task forces to more popularized debunking treatises are legion. New evidence has piled up showing lack of replication and all manner of selection and publication biases. Even expanded "evidence-based" practices, whose very rationale is to emulate experimental controls, are not immune from allegations of illicit cherry picking, significance seeking, *P*-hacking, and assorted modes of extraordinary rendition of data. Attempts to restore credibility have gone far beyond the cottage industries of just a few years ago, to entirely new research programs: statistical fraud-busting, statistical forensics, technical activism, and widespread reproducibility studies. There are proposed methodological reforms – many are generally welcome (preregistration of experiments, transparency about data collection, discouraging mechanical uses of statistics), some are quite radical. If we are to appraise these evidence policy reforms, a much better grasp of some central statistical problems is needed.

Getting Philosophical

Are philosophies about science, evidence, and inference relevant here? Because the problems involve questions about uncertain evidence, probabilistic models, science, and pseudoscience – all of which are intertwined with technical statistical concepts and presuppositions – they certainly ought to be. Even in an open-access world in which we have become increasingly fearless about taking on scientific complexities, a certain trepidation and groupthink take over when it comes to philosophically tinged notions such as inductive reasoning, objectivity, rationality, and science versus pseudoscience. The general area of philosophy that deals with knowledge, evidence, inference, and rationality is called *epistemology*. The epistemological standpoints of leaders, be they philosophers or scientists, are too readily taken as canon by others. We want to understand what's true about some of the popular memes: "All models are false," "Everything is equally subjective and objective," "*P*-values exaggerate evidence," and "[M]ost published research findings are false" (Ioannidis 2005) – at least if you publish a single statistically significant result after data finagling. (Do people do that? Shame on them.) Yet R. A. Fisher, founder of modern statistical tests, denied that an isolated statistically significant result counts.

[W]e need, not an isolated record, but a reliable method of procedure. In relation to the test of significance, we may say that a phenomenon is experimentally demonstrable when we know how to conduct an experiment which will rarely fail to give us a statistically significant result. (Fisher 1935b/1947, p. 14)

Satisfying this requirement depends on the proper use of background knowledge and deliberate design and modeling.

This opening excursion will launch us into the main themes we will encounter. You mustn't suppose, by its title, that I will be talking about how to tell the truth using statistics. Although I expect to make some progress there, my goal is to tell what's true about statistical methods themselves! There are so many misrepresentations of those methods that telling what is true about them is no mean feat. It may be thought that the basic statistical concepts are well understood. But I show that this is simply not true.

Nor can you just open a statistical text or advice manual for the goal at hand. The issues run deeper. Here's where I come in. Having long had one foot in philosophy of science and the other in foundations of statistics, I will zero in on the central philosophical issues that lie below the surface of today's raging debates. "Getting philosophical" is not about articulating rarified concepts divorced from statistical practice. It is to provide tools to avoid obfuscating the terms and issues being bandied about. Readers should be empowered to understand the core presuppositions on which rival positions are based – and on which they depend.

Do I hear a protest? "There is nothing philosophical about our criticism of statistical significance tests" (someone might say). The problem is that a small *P*-value is invariably, and erroneously, interpreted as giving a small probability to the null hypothesis." Really? *P*-values are not intended to be used this way; presupposing they ought to be so interpreted grows out of a specific conception of the role of probability in statistical inference. *That conception is philosophical.* Methods characterized through the lens of over-simple epistemological orthodoxies are methods misapplied and mischaracterized. This may lead one to lie, however unwittingly, about the nature and goals of statistical inference, when what we want is to tell what's true about them.

1.1 Severity Requirement: Bad Evidence, No Test (BENT)

Fisher observed long ago, "[t]he political principle that anything can be proved by statistics arises from the practice of presenting only a selected subset of the data available" (Fisher 1955, p. 75). If you report results selectively, it becomes easy to prejudge hypotheses: yes, the data may accord amazingly well with a hypothesis *H*, but such a method is practically guaranteed to issue so good a fit even if *H* is false and not warranted by the evidence. If it is predetermined that a way will be found to either obtain or interpret data as evidence for *H*, then data are not being taken seriously in appraising *H*. *H* is essentially immune to having its flaws uncovered by the data. *H* might be said to have "passed" the test, but it is a test that lacks stringency or severity. Everyone understands that this is bad evidence, or no test at all. I call this the *severity requirement*. In its weakest form it supplies a *minimal requirement* for evidence:

> *Severity Requirement (weak): One does not have evidence for a claim if nothing has been done to rule out ways the claim may be false.* If data *x* agree with a claim *C* but the method used is practically guaranteed to find such agreement, and had little or no capability of finding flaws with *C* even if they exist, then we have bad evidence, no test (BENT).

The "practically guaranteed" acknowledges that even if the method had some slim chance of producing a disagreement when *C* is false, we still regard the evidence as lousy. Little if anything has been done to rule out erroneous construals of data. We'll need many different ways to state this minimal principle of evidence, depending on context.

A Scandal Involving Personalized Medicine

A recent scandal offers an example. Over 100 patients signed up for the chance to participate in the Duke University (2007–10) clinical trials that promised a custom-tailored cancer treatment. A cutting-edge prediction model developed by Anil Potti and Joseph Nevins purported to predict your response to one or another chemotherapy based on large data sets correlating properties of various tumors and positive responses to different regimens (Potti et al. 2006). Gross errors and data manipulation eventually forced the trials to be halted. It was revealed in 2014 that a whistleblower – a student – had expressed concerns that

... in developing the model, only those samples which fit the model best in cross validation were included. Over half of the original samples were removed. . . . This was an incredibly biased approach. (Perez 2015)

In order to avoid the overly rosy predictions that ensue from a model built to fit the data (called the training set), a portion of the data (called the test set) is to be held out to "cross validate" the model. If any unwelcome test data are simply excluded, the technique has obviously not done its job. Unsurprisingly, when researchers at a different cancer center, Baggerly and Coombes, set out to avail themselves of this prediction model, they were badly disappointed: "When we apply the same methods but maintain the separation of training and test sets, predictions are poor" (Coombes et al. 2007, p. 1277). Predicting which treatment would work was no better than chance.

You might be surprised to learn that Potti dismissed their failed replication on grounds that they didn't use his method (Potti and Nevins 2007)! But his technique had little or no ability to reveal the unreliability of the model, and thus failed utterly as a cross check. By contrast, Baggerly and Coombes' approach informed about what it *would be like* to apply the model to brand new patients – the intended function of the cross validation. Medical journals were reluctant to publish Baggerly and Coombes' failed replications and report of critical flaws. (It eventually appeared in a statistics journal, *Annals of Applied Statistics* 2009, thanks to editor Brad Efron.) The clinical trials – yes on patients – were only shut down when it was discovered Potti had exaggerated his honors in his CV! The bottom line is, tactics that stand in the way of discovering weak spots, whether for prediction or explanation, create obstacles to the severity requirement; it would be puzzling if accounts of statistical inference failed to place this requirement, or something akin to it, right at the center – or even worse, permitted loopholes to enable such moves. Wouldn't it?

Do We Always Want to Find Things Out?

The severity requirement gives a minimal principle based on the fact that highly insevere tests yield bad evidence, no tests (BENT). We can all agree on this much, I think. We will explore how much mileage we can get from it. It applies at a number of junctures in collecting and modeling data, in linking data to statistical inference, and to substantive questions and claims. This will be our linchpin for understanding what's true about statistical inference. In addition to our minimal principle for evidence, one more thing is needed, at least during the time we are engaged in this project: *the goal of finding things out.*

The desire to find things out is an obvious goal; yet most of the time it is not what drives us. We typically may be uninterested in, if not quite resistant to, finding flaws or incongruencies with ideas we like. Often it is entirely proper to gather information to make your case, and ignore anything that fails to support it. Only if you really desire to find out something, or to challenge so-and-so's ("trust me") assurances, will you be prepared to stick your (or their) neck out to conduct a genuine "conjecture and refutation" exercise. Because you want to learn, you will be prepared to risk the possibility that the conjecture is found flawed.

We hear that "motivated reasoning has interacted with tribalism and new media technologies since the 1990s in unfortunate ways" (Haidt and Iyer 2016). Not only do we see things through the tunnel of our tribe, social media and web searches enable us to live in the echo chamber of our tribe more than ever. We might think we're trying to find things out but we're not. Since craving truth is rare (unless your life depends on it) and the "perverse incentives" of publishing novel results so shiny, the wise will invite methods that make uncovering errors and biases as quick and painless as possible. Methods of inference that fail to satisfy the minimal severity requirement fail us in an essential way.

With the rise of Big Data, data analytics, machine learning, and bioinformatics, statistics has been undergoing a good deal of introspection. Exciting results are often being turned out by researchers without a traditional statistics background; biostatistician Jeff Leek (2016) explains: "There is a structural reason for this: data was sparse when they were trained and there wasn't any reason for them to learn statistics." The problem goes beyond turf battles. It's discovering that many data analytic applications are missing key ingredients of statistical thinking. Brown and Kass (2009) crystalize its essence. "Statistical thinking uses probabilistic descriptions of variability in (1) inductive reasoning and (2) analysis of procedures for data collection, prediction, and scientific inference" (p. 107). A word on each.

(1) Types of statistical inference are too varied to neatly encompass. Typically we employ data to learn something about the process or mechanism producing the data. The claims inferred are not specific events, but statistical generalizations, parameters in theories and models, causal claims, and general predictions. Statistical inference goes beyond the data – by definition that

makes it an *inductive* inference. The risk of error is to be expected. There is no need to be reckless. The secret is controlling and learning from error. Ideally we take precautions in advance: *pre-data*, we devise methods that make it hard for claims to pass muster unless they are approximately true or adequately solve our problem. With data in hand, *post-data*, we scrutinize what, if anything, can be inferred.

What's the essence of analyzing procedures in (2)? Brown and Kass don't specifically say, but the gist can be gleaned from what vexes them; namely, ad hoc data analytic algorithms where researchers "have done nothing to indicate that it performs well" (p. 107). Minimally, statistical thinking means never ignoring the fact that there are alternative methods: Why is this one a good tool for the job? Statistical thinking requires stepping back and examining a method's capabilities, whether it's designing or choosing a method, or scrutinizing the results.

A Philosophical Excursion

Taking the severity principle then, along with the aim that we desire to find things out without being obstructed in this goal, let's set sail on a philosophical excursion to illuminate statistical inference. Envision yourself embarking on a special interest cruise featuring "exceptional itineraries to popular destinations worldwide as well as unique routes" (Smithsonian Journeys). What our cruise lacks in glamour will be more than made up for in our ability to travel back in time to hear what Fisher, Neyman, Pearson, Popper, Savage, and many others were saying and thinking, and then zoom forward to current debates. There will be exhibits, a blend of statistics, philosophy, and history, and even a bit of theater. Our standpoint will be pragmatic in this sense: my interest is not in some ideal form of knowledge or rational agency, no omniscience or God's-eye view – although we'll start and end surveying the landscape from a hot-air balloon. I'm interested in the problem of how we get the kind of knowledge we do manage to obtain – and how we can get more of it. Statistical methods should not be seen as tools for what philosophers call "rational reconstruction" of a piece of reasoning. Rather, they are forward-looking tools to find something out faster and more efficiently, and to discriminate how good or poor a job others have done.

The job of the philosopher is to clarify but also to provoke reflection and scrutiny precisely in those areas that go unchallenged in ordinary practice. My focus will be on the issues having the most influence, and being most liable to obfuscation. Fortunately, that doesn't require an abundance of technicalities, but you can opt out of any daytrip that appears too technical: an idea not caught in one place should be illuminated in another. Our philosophical excursion may well land us in positions that are provocative to all existing sides of the debate about probability and statistics in scientific inquiry.

Methodology and Meta-methodology

We are studying statistical methods from various schools. What shall we call methods for doing so? Borrowing a term from philosophy of science, we may call it our meta-methodology – it's one level removed.[1] To put my cards on the table: A severity scrutiny is going to be a key method of our meta-methodology. It is fairly obvious that we want to scrutinize how capable a statistical method is at detecting and avoiding erroneous interpretations of data. So when it comes to the role of probability as a pedagogical tool for our purposes, severity – its assessment and control – will be at the center. The term "severity" is Popper's, though he never adequately defined it. It's not part of any statistical methodology as of yet. Viewing statistical inference as severe testing lets us stand one level removed from existing accounts, where the air is a bit clearer.

Our intuitive, minimal, requirement for evidence connects readily to formal statistics. The probabilities that a statistical method lands in erroneous interpretations of data are often called its *error probabilities*. So an account that revolves around control of error probabilities I call an *error statistical account*. But "error probability" has been used in different ways. Most familiar are those in relation to hypotheses tests (Type I and II errors), significance levels, confidence levels, and power – all of which we will explore in detail. It has occasionally been used in relation to the proportion of false hypotheses among those now in circulation, which is different. For now it suffices to say that none of the formal notions directly give severity assessments. There isn't even a statistical school or tribe that has explicitly endorsed this goal. I find this perplexing. That will not preclude our immersion into the mindset of a futuristic tribe whose members use error probabilities for assessing severity; it's just the ticket for our task: understanding and getting beyond the statistics wars. We may call this tribe the *severe testers*.

We can keep to testing language. See it as part of the meta-language we use to talk about formal statistical methods, where the latter include estimation, exploration, prediction, and data analysis. I will use the term "hypothesis," or just "claim," for any conjecture we wish to entertain; it need not be one set out in advance of data. Even predesignating hypotheses, by the way, doesn't

---

[1] This contrasts with the use of "metaresearch" to describe work on methodological reforms by non-philosophers. This is not to say they don't tread on philosophical territory often: they do.

preclude bias: that view is a holdover from a crude empiricism that assumes data are unproblematically "given," rather than selected and interpreted. Conversely, using the same data to arrive at and test a claim can, in some cases, be accomplished with stringency.

As we embark on statistical foundations, we must avoid blurring formal terms such as probability and likelihood with their ordinary English meanings. Actually, "probability" comes from the Latin *probare*, meaning to try, test, or prove. "Proof" in "The proof is in the pudding" refers to how you put something to the test. You must show or demonstrate, not just believe strongly. Ironically, using probability this way would bring it very close to the idea of measuring well-testedness (or how well shown). But it's not our current, informal English sense of probability, as varied as that can be. To see this, consider "improbable." Calling a claim improbable, in ordinary English, can mean a host of things: I bet it's not so; all things considered, given what I know, it's implausible; and other things besides. Describing a claim as *poorly tested* generally means something quite different: little has been done to probe whether the claim holds or not, the method used was highly unreliable, or things of that nature. In short, our informal notion of poorly tested comes rather close to the lack of severity in statistics. There's a difference between finding H poorly tested by data x, and finding x renders H improbable – in any of the many senses the latter takes on. The existence of a Higgs particle was thought to be probable if not necessary before it was regarded as well tested around 2012. Physicists had to show or demonstrate its existence for it to be well tested. It follows that you are free to pursue our testing goal without implying there are no other statistical goals. One other thing on language: I will have to retain the terms currently used in exploring them. That doesn't mean I'm in favor of them; in fact, I will jettison some of them by the end of the journey.

To sum up this first tour so far, statistical inference uses data to reach claims about aspects of processes and mechanisms producing them, accompanied by an assessment of the properties of the inference methods: their capabilities to control and alert us to erroneous interpretations. We need to report if the method has satisfied the most minimal requirement for solving such a problem. Has anything been tested with a modicum of severity, or not? The severe tester also requires reporting of what has been poorly probed, and highlights the need to "bend over backwards," as Feynman puts it, to admit where weaknesses lie. In formal statistical testing, the crude dichotomy of "pass/fail" or "significant or not" will scarcely do. We must determine the magnitudes (and directions) of any statistical discrepancies warranted, and the limits to any substantive claims you may be entitled to infer from the statistical ones. Using just our minimal principle of evidence, and a sturdy pair of shoes, join me on a tour of statistical inference, back to the leading museums of statistics, and forward to current offshoots and statistical tribes.

Why We Must Get Beyond the Statistics Wars

Some readers may be surprised to learn that the field of statistics, arid and staid as it seems, has a fascinating and colorful history of philosophical debate, marked by unusual heights of passion, personality, and controversy for at least a century. Others know them all too well and regard supporting any one side largely as proselytizing. I've heard some refer to statistical debates as "theological." I do not want to rehash the "statistics wars" that have raged in every decade, although the significance test controversy is still hotly debated among practitioners, and even though each generation fights these wars anew – with task forces set up to stem reflexive, recipe-like statistics that have long been deplored.

The time is ripe for a fair-minded engagement in the debates about statistical foundations; more than that, it is becoming of pressing importance. Not only because

> (i)   these issues are increasingly being brought to bear on some very public controversies;

nor because

> (ii)   the "statistics wars" have presented new twists and turns that cry out for fresh analysis

– as important as those facets are – but because what is at stake is a critical standpoint that we may be in danger of losing. Without it, we forfeit the ability to communicate with, and hold accountable, the "experts," the agencies, the quants, and all those data handlers increasingly exerting power over our lives. Understanding the nature and basis of statistical inference must not be considered as all about mathematical details; it is at the heart of what it means to reason scientifically and with integrity about any field whatever. Robert Kass (2011) puts it this way:

We care about our philosophy of statistics, first and foremost, because statistical inference sheds light on an important part of human existence, inductive reasoning, and we want to understand it. (p. 19)

Isolating out a particular conception of statistical inference as severe testing is a way of telling what's true about the statistics wars, and getting beyond them.

Chutzpah, No Proselytizing

Our task is twofold: not only must we analyze statistical methods; we must also scrutinize the jousting on various sides of the debates. Our meta-level standpoint will let us rise above much of the cacophony; but the excursion will involve a dose of chutzpah that is out of the ordinary in professional discussions. You will need to critically evaluate the texts and the teams of critics, including brilliant leaders, high priests, maybe even royalty. Are they asking the most unbiased questions in examining methods, or are they like admen touting their brand, dragging out howlers to make their favorite method look good? (I am not sparing any of the statistical tribes here.) There are those who are earnest but brainwashed, or are stuck holding banners from an earlier battle now over; some are wedded to what they've learned, to what's in fashion, to what pays the rent.

Some are so jaundiced about the abuses of statistics as to wonder at my admittedly herculean task. I have a considerable degree of sympathy with them. But, I do not sympathize with those who ask: "why bother to clarify statistical concepts if they are invariably misinterpreted?" and then proceed to misinterpret them. Anyone is free to dismiss statistical notions as irrelevant to them, but then why set out a shingle as a "statistical reformer"? You may even be shilling for one of the proffered reforms, thinking it the road to restoring credibility, when it will do nothing of the kind.

You might say, since rival statistical methods turn on issues of philosophy and on rival conceptions of scientific learning, that it's impossible to say anything "true" about them. You just did. It's precisely these interpretative and philosophical issues that I plan to discuss. Understanding the issues is different from settling them, but it's of value nonetheless. Although statistical disagreements involve philosophy, statistical practitioners and not philosophers are the ones leading today's discussions of foundations. Is it possible to pursue our task in a way that will be seen as neither too philosophical nor not philosophical enough? Too statistical or not statistically sophisticated enough? Probably not, I expect grievances from both sides.

Finally, I will not be proselytizing for a given statistical school, so you can relax. Frankly, they all have shortcomings, insofar as one can even glean a clear statement of a given statistical "school." What we have is more like a jumble with tribal members often speaking right past each other. View the severity requirement as a heuristic tool for telling what's true about statistical controversies. Whether you resist some of the ports of call we arrive at is unimportant; it suffices that visiting them provides a key to unlock current mysteries that are leaving many consumers and students of statistics in the dark about a crucial portion of science.

1.2 Probabilism, Performance, and Probativeness

> I shall be concerned with the foundations of the subject. But in case it should be thought that this means I am not here strongly concerned with practical applications, let me say right away that confusion about the foundations of the subject is responsible, in my opinion, for much of the misuse of the statistics that one meets in fields of application such as medicine, psychology, sociology, economics, and so forth. (George Barnard 1985, p. 2)

While statistical science (as with other sciences) generally goes about its business without attending to its own foundations, implicit in every statistical methodology are core ideas that direct its principles, methods, and interpretations. I will call this its *statistical philosophy*. To tell what's true about statistical inference, understanding the associated philosophy (or philosophies) is essential. Discussions of statistical foundations tend to focus on how to interpret probability, and much less on the overarching question of how probability ought to be used in inference. Assumptions about the latter lurk implicitly behind debates, but rarely get the limelight. If we put the spotlight on them, we see that there are two main philosophies about the roles of probability in statistical inference: We may dub them *performance* (in the long run) and *probabilism*.

The performance philosophy sees the key function of statistical method as controlling the relative frequency of erroneous inferences in the long run of applications. For example, a frequentist statistical test, in its naked form, can be seen as a rule: whenever your outcome exceeds some value (say, $X > x^*$), reject a hypothesis H_0 and infer H_1. The value of the rule, according to its performance-oriented defenders, is that it can ensure that, regardless of which hypothesis is true, there is both a low probability of erroneously rejecting H_0 (rejecting H_0 when it is true) as well as erroneously accepting H_0 (failing to reject H_0 when it is false).

The second philosophy, probabilism, views probability as a way to assign degrees of belief, support, or plausibility to hypotheses. Many keep to a comparative report, for example that H_0 is more believable than is H_1 given data x; others strive to say H_0 is less believable given data x than before, and offer a quantitative report of the difference.

What happened to the goal of scrutinizing BENT science by the severity criterion? Neither "probabilism" nor "performance" directly captures that demand. To take these goals at face value, it's easy to see why they come up short. Potti and Nevins' strong belief in the reliability of their prediction model for cancer therapy scarcely made up for the shoddy testing. Neither is good long-run performance a sufficient condition. Most obviously, there may be no long-run repetitions, and our interest in science is often just the particular statistical inference before us. Crude long-run requirements may be met by silly methods. Most importantly, good performance alone fails to get at *why* methods work when they do; namely – I claim – to let us assess and control the stringency of tests. This is the key to answering a burning question that has caused major headaches in statistical foundations: why should a low relative frequency of error matter to the appraisal of the inference at hand? It is not probabilism or performance we seek to quantify, but *probativeness*.

I do not mean to disparage the long-run performance goal – there are plenty of tasks in inquiry where performance is absolutely key. Examples are screening in high-throughput data analysis, and methods for deciding which of tens of millions of collisions in high-energy physics to capture and analyze. New applications of machine learning may lead some to say that only low rates of prediction or classification errors matter. Even with prediction, "black-box" modeling, and non-probabilistic inquiries, there is concern with solving a problem. We want to know if a good job has been done in the case at hand.

Severity (Strong): Argument from Coincidence

The weakest version of the severity requirement (Section 1.1), in the sense of easiest to justify, is negative, warning us when BENT data are at hand, and a surprising amount of mileage may be had from that negative principle alone. It is when we recognize how poorly certain claims are warranted that we get ideas for improved inquiries. In fact, if you wish to stop at the negative requirement, you can still go pretty far along with me. I also advocate the positive counterpart:

> *Severity (strong): We have evidence for a claim C just to the extent it survives a stringent scrutiny.* If C passes a test that was highly capable of finding flaws or discrepancies from C, and yet none or few are found, then the passing result, *x*, is evidence for C.

One way this can be achieved is by an *argument from coincidence*. The most vivid cases occur outside formal statistics.

Some of my strongest examples tend to revolve around my weight. Before leaving the USA for the UK, I record my weight on two scales at home, one digital, one not, and the big medical scale at my doctor's office. Suppose they are well calibrated and nearly identical in their readings, and they also all pick up on the extra 3 pounds when I'm weighed carrying three copies of my 1-pound book, *Error and the Growth of Experimental Knowledge* (EGEK). Returning from the UK, to my astonishment, not one but all three scales show anywhere from a 4–5 pound gain. There's no difference when I place the three books on the scales, so I must conclude, unfortunately, that I've gained around 4 pounds. Even for me, that's a lot. I've surely falsified the supposition that I lost weight! From this informal example, we may make two rather obvious points that will serve for less obvious cases. First, there's the idea I call lift-off.

> *Lift-off: An overall inference can be more reliable and precise than its premises individually.*

Each scale, by itself, has some possibility of error, and limited precision. But the fact that all of them have me at an over 4-pound gain, while none show any difference in the weights of EGEK, pretty well seals it. Were one scale off balance, it would be discovered by another, and would show up in the weighing of books. They cannot all be systematically misleading just when it comes to objects of unknown weight, can they? Rejecting a conspiracy of the scales, I conclude I've gained weight, at least 4 pounds. We may call this an *argument from coincidence*, and by its means we can attain lift-off. Lift-off runs directly counter to a seemingly obvious claim of drag-down.

> *Drag-down: An overall inference is only as reliable/precise as is its weakest premise.*

The drag-down assumption is common among empiricist philosophers: As they like to say, "It's turtles all the way down." Sometimes our inferences do stand as a kind of tower built on linked stones – if even one stone fails they all come tumbling down. Call that a *linked* argument.

Our most prized scientific inferences would be in a very bad way if piling on assumptions invariably leads to weakened conclusions. Fortunately we also can build what may be called *convergent* arguments, where lift-off is attained. This seemingly banal point suffices to combat some of the most well entrenched skepticisms in philosophy of science. And statistics happens to be the science par excellence for demonstrating lift-off!

Now consider what justifies my weight conclusion, based, as we are supposing it is, on a strong argument from coincidence. No one would say: "I can be assured that by following such a procedure, in the long run I would rarely report weight gains erroneously, but I can tell nothing from these readings about my weight now." To justify my conclusion by long-run performance would be absurd. Instead we say that the procedure had enormous capacity to reveal if any of the scales were wrong, and from this I argue about the source of the readings: *H*: I've gained weight. Simple as that. It would be a preposterous coincidence if none of the scales registered even slight weight shifts when weighing objects of known weight, and yet were systematically misleading when applied to my weight. You see where I'm going with this. This is the key – granted with a homely example – that can fill a very important gap in frequentist foundations: Just because an account is touted as having a long-run rationale, it does not mean it lacks a short run rationale, or even one relevant for the particular case at hand.

Nor is it merely the improbability of all the results were *H* false; it is rather like denying an evil demon has read my mind just in the cases where I do not know the weight of an object, and deliberately deceived me. The argument to "weight gain" is an example of an argument from coincidence to the absence of an error, what I call:

> *Arguing from Error*: There is evidence an error is absent to the extent that a procedure with a very high capability of signaling the error, if and only if it is present, nevertheless detects no error.

I am using "signaling" and "detecting" synonymously: It is important to keep in mind that we don't know if the test output is correct, only that it gives a signal or alert, like sounding a bell. Methods that enable strong arguments to the absence (or presence) of an error I call *strong error probes*. Our ability to develop strong arguments from coincidence, I will argue, is the basis for solving the "problem of induction."

Glaring Demonstrations of Deception

Intelligence is indicated by a capacity for deliberate deviousness. Such deviousness becomes self-conscious in inquiry: An example is the use of a placebo to find out what it would be like if the drug has no effect. What impressed me the most in my first statistics class was the demonstration of how apparently impressive results are readily produced when nothing's going on, i.e., "by chance alone." Once you see how it is done, and done easily, there is no going back. The toy hypotheses used in statistical testing are nearly always overly simple as scientific hypotheses. But when it comes to framing rather blatant deceptions, they are just the ticket!

When Fisher offered Muriel Bristol-Roach a cup of tea back in the 1920s, she refused it because he had put the milk in first. What difference could it make? Her husband and Fisher thought it would be fun to put her to the test (1935a). Say she doesn't claim to get it right all the time but does claim that she has some genuine discerning ability. Suppose Fisher subjects her to 16 trials and she gets 9 of them right. Should I be impressed or not? By a simple experiment of randomly assigning milk first/tea first Fisher sought to answer this stringently. But don't be fooled: a great deal of work goes into controlling biases and confounders before the experimental design can work. The main point just now is this: so long as lacking ability is sufficiently like the canonical "coin tossing" (Bernoulli) model (with the probability of success at each trial of 0.5), we can learn from the test procedure. In the Bernoulli model, we record success or failure, assume a fixed probability of success θ on each trial, and that trials are independent. If the probability of getting even more successes than she got, merely by guessing, is fairly high, there's little indication of special tasting ability. The probability of at least 9 of 16 successes, even if $\theta = 0.5$, is 0.4. To abbreviate, Pr(at least 9 of 16 successes; H_0: $\theta = 0.5$) = 0.4. This is the P-value of the observed difference; an unimpressive 0.4. You'd expect as many or even more "successes" 40% of the time merely by guessing. It's also the *significance level attained* by the result. (I often use P-value as it's shorter.) Muriel Bristol-Roach pledges that if her performance may be regarded as scarcely better than guessing, then she hasn't shown her ability. Typically, a small value such as 0.05, 0.025, or 0.01 is required.

Such artificial and simplistic statistical hypotheses play valuable roles at stages of inquiry where what is needed are blatant standards of "nothing's going on." There is no presumption of a metaphysical chance agency, just that there is expected variability – otherwise one test would suffice – and that probability models from games of chance can be used to distinguish genuine from spurious effects. Although the goal of inquiry is to find things out, the hypotheses erected to this end are generally approximations and may be deliberately false. To present statistical hypotheses as identical to substantive scientific claims is to mischaracterize them. We want to tell what's true about statistical inference. Among the most notable of these truths is:

> P-values can be readily invalidated due to how the data (or hypotheses!) are generated or selected for testing.

If you fool around with the results afterwards, reporting only successful guesses, your report will be invalid. You may claim it's very difficult to get such an impressive result due to chance, when in fact it's very easy to do so, with selective reporting. Another way to put this: your *computed* P-value is small, but the *actual* P-value is high! Concern with spurious findings, while an ancient problem, is considered sufficiently serious to have motivated the American Statistical Association to issue a guide on how not to interpret P-values (Wasserstein and Lazar 2016); hereafter, ASA 2016 Guide. It may seem that if a statistical account is free to ignore such fooling around then the problem disappears! It doesn't.

Incidentally, Bristol-Roach got all the cases correct, and thereby taught her husband a lesson about putting her claims to the test.

Peirce

The philosopher and astronomer C. S. Peirce, writing in the late nineteenth century, is acknowledged to have anticipated many modern statistical ideas (including randomization and confidence intervals). Peirce describes how "so accomplished a reasoner" as Dr. Playfair deceives himself by a technique we know all too well – scouring the data for impressive regularities (2.738). Looking at the specific gravities of three forms of carbon, Playfair seeks and discovers a formula that holds for all of them (each is a root of the atomic weight of carbon, which is 12). Can this regularity be expected to hold in general for metalloids? It turns out that half of the cases required Playfair to modify the formula after the fact. If one limits the successful instances to ones where the formula was predesignated, and not altered later on, only half satisfy Playfair's formula. Peirce asks, how often would such good agreement be found due to chance? Again, should we be impressed?

Peirce introduces a mechanism to arbitrarily pair the specific gravity of a set of elements with the atomic weight of another. By design, such agreements could only be due to the chance pairing. Lo and behold, Peirce finds about the same number of cases that satisfy Playfair's formula. "It thus appears that there is no more frequent agreement with Playfair's proposed law than what is due to chance" (2.738).

At first Peirce's demonstration seems strange. He introduces an accidental pairing just to simulate the ease of obtaining so many agreements in an entirely imaginary situation. Yet that suffices to show Playfair's evidence is BENT. The popular inductive accounts of his time, Peirce argues, do not prohibit adjusting the formula to fit the data, and, because of that, they would persist in Playfair's error. The same debate occurs today, as when Anil Potti (of the Duke scandal) dismissed the whistleblower Perez thus: "we likely disagree with what constitutes validation" (Nevins and Potti 2015). Erasing genomic data that failed to fit his predictive model was justified, Potti claimed, by the fact that other data points fit (Perez 2015)! Peirce's strategy, as that of Coombes et al., is to introduce a blatant standard to put the method through its paces, without bogus agreements. If the agreement is no better than bogus agreement, we deny there is evidence for a genuine regularity or valid prediction. Playfair's formula may be true, or probably true, but Peirce's little demonstration is enough to show his method did a lousy job of testing it.

Texas Marksman

Take an even simpler and more blatant argument of deception. It is my favorite: the Texas Marksman. A Texan wants to demonstrate his shooting prowess. He shoots all his bullets any old way into the side of a barn and then paints a bull's-eye in spots where the bullet holes are clustered. This fails utterly to severely test his marksmanship ability. When some visitors come to town and notice the incredible number of bull's-eyes, they ask to meet this marksman and are introduced to a little kid. How'd you do so well, they ask? Easy, I just drew the bull's-eye around the most tightly clustered shots. There is impressive "agreement" with shooting ability, he might even compute how improbably so many bull's-eyes would occur by chance. Yet his ability to shoot was not tested in the least by this little exercise. There's a real effect all right, but it's not caused by his marksmanship! It serves as a potent analogy for a cluster of formal statistical fallacies from data-dependent findings of "exceptional" patterns.

The term "apophenia" refers to a tendency to zero in on an apparent regularity or cluster within a vast sea of data and claim a genuine regularity. One of our fundamental problems (and skills) is that we're apopheniacs. Some investment funds, none that we actually know, are alleged to produce several portfolios by random selection of stocks and send out only the one that did best. Call it the Pickrite method. They want you to infer that it would be a preposterous coincidence to get so great a portfolio if the Pickrite method were like guessing. So their methods are genuinely wonderful, or so you are to infer. If this had been their only portfolio, the probability of doing so well by luck is low. But the probability of at least one of many portfolios doing so well (even if each is generated by chance) is high, if not guaranteed.

Let's review the rogues' gallery of glaring arguments from deception. The lady tasting tea showed how a statistical model of "no effect" could be used to amplify our ordinary capacities to discern if something really unusual is going on. The P-value is the probability of at least as high a success rate as observed, assuming the test or null hypothesis, the probability of success is 0.5. Since even more successes than she got is fairly frequent through guessing alone (the P-value is moderate), there's poor evidence of a genuine ability. The Playfair and Texas sharpshooter examples, while quasi-formal or informal, demonstrate how to invalidate reports of significant effects. They show how gambits of post-data adjustments or selection can render a method highly capable of spewing out impressive looking fits even when it's just random noise.

We appeal to the same statistical reasoning to show the problematic cases as to show genuine arguments from coincidence.

So am I proposing that a key role for statistical inference is to identify ways to spot egregious deceptions (BENT cases) and create strong arguments from coincidence? Yes, I am.

Spurious P-values and Auditing

In many cases you read about you'd be right to suspect that someone has gone circling shots on the side of a barn. Confronted with the statistical news flash of the day, your first question is: Are the results due to selective reporting, cherry picking, or any number of other similar ruses? This is a central part of what we'll call *auditing* a significance level.

A key point too rarely appreciated: Statistical facts about P-values themselves demonstrate how data finagling can yield spurious significance. This is true for all error probabilities. That's what a self-correcting inference account should do. Ben Goldacre, in *Bad Pharma* (2012), sums it up this way: the gambits give researchers an abundance of chances to find something when the tools assume you have had just one chance. Scouring different subgroups and otherwise "trying and trying again" are classic ways to blow up the actual probability of obtaining an impressive, but spurious, finding – and that remains so even if you ditch P-values and never compute them. FDA rules are designed to outlaw such gambits. To spot the cheating or questionable research practices (QRPs) responsible for a finding may not be easy. New research tools are being developed to detect them. Unsurprisingly, P-value analysis is relied on to discern spurious P-values (e.g., by lack of replication, or, in analyzing a group of tests, finding too many P-values in a given range). Ultimately, a qualitative severity scrutiny is necessary to get beyond merely raising doubts to falsifying purported findings.

Association Is Not Causation: Hormone Replacement Therapy (HRT)

Replicable results from high-quality research are sound, except for the sin that replicability fails to uncover: systematic bias.[2] Gaps between what is actually producing the statistical effect and what is inferred open the door by which biases creep in. Stand-in or proxy variables in statistical models may have little to do with the phenomenon of interest.

---

[2] This is the traditional use of "bias" as a systematic error. Ioannidis (2005) alludes to biasing as behaviors that result in a reported significance level differing from the value it actually has or ought to have (e.g., post-data endpoints, selective reporting). I will call those biasing selection effects.

So strong was the consensus-based medical judgment that hormone replacement therapy helps prevent heart disease that many doctors deemed it "unethical to ask women to accept the possibility that they might be randomized to a placebo" (The National Women's Health Network (NWHN) 2002, p. 180). Post-menopausal women who wanted to retain the attractions of being "Feminine Forever," as in the title of an influential tract (Wilson 1971), were routinely given HRT. Nevertheless, when a large randomized controlled trial (RCT) was finally done, it revealed statistically significant increased risks of heart disease, breast cancer, and other diseases that HRT was to have helped prevent. The observational studies on HRT, despite reproducibly showing a benefit, had little capacity to unearth biases due to "the healthy women's syndrome." There were confounding factors separately correlated with the beneficial outcomes enjoyed by women given HRT: they were healthier, better educated, and less obese than women not taking HRT. (That certain subgroups are now thought to benefit is a separate matter.)

Big Data scientists are discovering there may be something in the data collection that results in the bias being "hard-wired" into the data, and therefore even into successful replications. So replication is not enough. Beyond biased data, there's the worry that lab experiments may be only loosely connected to research claims. Experimental economics, for instance, is replete with replicable effects that economist Robert Sugden calls "exhibits." "An exhibit is an experimental design which reliably induces a surprising regularity" with at best an informal hypothesis as to its underlying cause (Sugden 2005, p. 291). Competing interpretations remain. (In our museum travels, "exhibit" will be used in the ordinary way.) In analyzing a test's capability to control erroneous interpretations, we must consider the porousness at multiple steps from data, to statistical inference, to substantive claims.

Souvenir A: Postcard to Send

The gift shop has a postcard listing the four slogans from the start of this Tour. Much of today's handwringing about statistical inference is unified by a call to block these fallacies. In some realms, trafficking in too-easy claims for evidence, if not criminal offenses, are "bad statistics"; in others, notably some social sciences, they are accepted cavalierly – much to the despair of panels on research integrity. We are more sophisticated than ever about the ways researchers can repress unwanted, and magnify wanted, results. Fraud-busting is everywhere, and the most important grain of truth is this: all the fraud-busting is based on error statistical reasoning (if only on the meta-level). The minimal requirement to avoid BENT isn't met. It's hard to see how one can grant the criticisms while denying the critical logic.

We should oust mechanical, recipe-like uses of statistical methods that have long been lampooned, and are doubtless made easier by Big Data mining. They should be supplemented with tools to report magnitudes of effects that have and have not been warranted with severity. But simple significance tests have their uses, and shouldn't be ousted simply because some people are liable to violate Fisher's warning and report isolated results. They should be seen as a part of a conglomeration of error statistical tools for distinguishing genuine and spurious effects. They offer assets that are essential to our task: they have the means by which to register formally the fallacies in the postcard list. The failed statistical assumptions, the selection effects from trying and trying again, all alter a test's error-probing capacities. This sets off important alarm bells, and we want to hear them. Don't throw out the error-control baby with the bad statistics bathwater.

The slogans about lying with statistics? View them, not as a litany of embarrassments, but as announcing what any responsible method must register, if not control or avoid. Criticisms of statistical tests, where valid, boil down to problems with the critical alert function. Far from the high capacity to warn, "Curb your enthusiasm!" as correct uses of tests do, there are practices that make sending out spurious enthusiasm as easy as pie. This is a failure for sure, but don't trade them in for methods that cannot detect failure at all. If you're shopping for a statistical account, or appraising a statistical reform, your number one question should be: does it embody trigger warnings of spurious effects? Of bias? Of cherry picking and multiple tries? If the response is: "No problem; if you use our method, those practices require no change in statistical assessment!" all I can say is, if it sounds too good to be true, you might wish to hold off buying it.

We shouldn't be hamstrung by the limitations of any formal methodology. Background considerations, usually absent from typical frequentist expositions, must be made more explicit; taboos and conventions that encourage "mindless statistics" (Gigerenzer 2004) eradicated. The severity demand is what we naturally insist on as consumers. We want methods that are highly capable of finding flaws just when they're present, and we specify worst case scenarios. With the data in hand, we custom tailor our assessments depending on how severely (or inseverely) claims hold up. Here's an informal statement of the severity requirements (weak and strong):

Severity Requirement (weak): If data *x* agree with a claim *C* but the method was practically incapable of finding flaws with *C* even if they exist, then *x* is poor evidence for *C*.

Severity (strong): If *C* passes a test that was highly capable of finding flaws or discrepancies from *C*, and yet none or few are found, then the passing result, *x*, is an indication of, or evidence for, *C*.

You might aver that we are too weak to fight off the lures of retaining the status quo – the carrots are too enticing, given that the sticks aren't usually too painful. I've heard some people say that evoking traditional mantras for promoting reliability, now that science has become so crooked, only makes things worse. Really? Yes there is gaming, but if we are not to become utter skeptics of good science, we should understand how the protections can work. In either case, I'd rather have rules to hold the "experts" accountable than live in a lawless wild west. I, for one, would be skeptical of entering clinical trials based on some of the methods now standard. There will always be cheaters, but give me an account that has eyes with which to spot them, and the means by which to hold cheaters accountable. That is, in brief, my basic statistical philosophy. The stakes couldn't be higher in today's world. Feynman said to take on an "extra type of integrity" that is not merely the avoidance of lying but striving "to check how you're maybe wrong." I couldn't agree more. But we laywomen are still going to have to proceed with a cattle prod.

1.3 The Current State of Play in Statistical Foundations: A View From a Hot-Air Balloon

> How can a discipline, central to science and to critical thinking, have two methodologies, two logics, two approaches that frequently give substantively different answers to the same problems? ... Is complacency in the face of contradiction acceptable for a central discipline of science? (Donald Fraser 2011, p. 329)

> We [statisticians] are not blameless ... we have not made a concerted professional effort to provide the scientific world with a unified testing methodology. (J. Berger 2003, p. 4)

From the aerial perspective of a hot-air balloon, we may see contemporary statistics as a place of happy multiplicity: the wealth of computational ability allows for the application of countless methods, with little handwringing about foundations. Doesn't this show we may have reached "the end of statistical foundations"? One might have thought so. Yet, descending close to a marshy wetland, and especially scratching a bit below the surface, reveals unease on all sides. The false dilemma between probabilism and long-run performance lets us get a handle on it. In fact, the Bayesian versus frequentist dispute arises as a dispute between probabilism and performance. This gets to my second reason for why the time is right to jump back into these debates: the "statistics wars" present new twists and turns. Rival tribes are more likely to live closer and in mixed neighborhoods since around the turn of the century. Yet, to the beginning student, it can appear as a jungle.

Statistics Debates: Bayesian versus Frequentist

> These days there is less distance between Bayesians and frequentists, especially with the rise of objective [default] Bayesianism, and we may even be heading toward a coalition government. (Efron 2013, p. 145)

A central way to formally capture probabilism is by means of the formula for conditional probability, where $\Pr(x) > 0$:

$$\Pr(H|x) = \frac{\Pr(H \text{ and } x)}{\Pr(x)}.$$

Since $\Pr(H \text{ and } x) = \Pr(x|H)\Pr(H)$ and $\Pr(x) = \Pr(x|H)\Pr(H) + \Pr(x|\sim H)\Pr(\sim H)$, we get:

$$\Pr(H|x) = \frac{\Pr(x|H)\Pr(H)}{\Pr(x|H)\Pr(H) + \Pr(x|\sim H)\Pr(\sim H)},$$

where $\sim H$ is the denial of H. It would be cashed out in terms of all rivals to H within a frame of reference. Some call it Bayes' Rule or inverse probability. Leaving probability uninterpreted for now, if the data are very improbable given H, then our probability in H after seeing x, the *posterior* probability $\Pr(H|x)$, may be lower than the probability in H prior to x, the *prior* probability $\Pr(H)$. Bayes' Theorem is just a theorem stemming from the definition of conditional probability; it is only when statistical inference is thought to be encompassed by it that it becomes a statistical philosophy. Using Bayes' Theorem doesn't make you a Bayesian.

Larry Wasserman, a statistician and master of brevity, boils it down to a contrast of goals. According to him (2012b):

The Goal of Frequentist Inference: Construct procedure with frequentist guarantees [i.e., low error rates].

The Goal of Bayesian Inference: Quantify and manipulate your degrees of beliefs. In other words, Bayesian inference is the Analysis of Beliefs.

At times he suggests we use $B(H)$ for belief and $F(H)$ for frequencies. The distinctions in goals are too crude, but they give a feel for what is often regarded as the Bayesian-frequentist controversy. However, they present us with the false dilemma (performance or probabilism) I've said we need to get beyond.

Today's Bayesian–frequentist debates clearly differ from those of some years ago. In fact, many of the same discussants, who only a decade ago were arguing for the irreconcilability of frequentist P-values and Bayesian measures, are now smoking the peace pipe, calling for ways to unify and marry the two. I want to show you what really drew me back into the Bayesian–frequentist debates sometime around 2000. If you lean over the edge of the gondola, you can hear some Bayesian family feuds starting around then or a bit after. Principles that had long been part of the Bayesian hard core are being questioned or even abandoned by members of the Bayesian family. Suddenly sparks are flying, mostly kept shrouded within Bayesian walls, but nothing can long be kept secret even there. Spontaneous combustion looms. Hard core subjectivists are accusing the increasingly popular "objective (non-subjective)" and "reference" Bayesians of practicing in bad faith; the new frequentist–Bayesian unificationists are taking pains to show they are not subjective; and some are calling the new Bayesian kids on the block "pseudo Bayesian." Then there are the Bayesians camping somewhere in the middle (or perhaps out in left field) who, though they still use the Bayesian umbrella, are flatly denying the very idea that Bayesian updating fits anything they actually do in statistics. Obeisance to Bayesian reasoning remains, but on some kind of a priori philosophical grounds. Let's start with the unifications.

While subjective Bayesianism offers an algorithm for coherently updating prior degrees of belief in possible hypotheses $H_1, H_2, \ldots, H_n$, these unifications fall under the umbrella of non-subjective Bayesian paradigms. Here the prior probabilities in hypotheses are not taken to express degrees of belief but are given by various formal assignments, ideally to have minimal impact on the posterior probability. I will call such Bayesian priors *default*. Advocates of unifications are keen to show that (i) default Bayesian methods have good performance in a long series of repetitions – so probabilism may yield performance; or alternatively, (ii) frequentist quantities are similar to Bayesian ones (at least in certain cases) – so performance may yield probabilist numbers. Why is this not bliss? Why are so many from all sides dissatisfied?

True blue subjective Bayesians are understandably unhappy with non-subjective priors. Rather than quantify prior beliefs, non-subjective priors are viewed as primitives or conventions for obtaining posterior probabilities. Take Jay Kadane (2008):

The growth in use and popularity of Bayesian methods has stunned many of us who were involved in exploring their implications decades ago. The result . . . is that there are users of these methods who do not understand the *philosophical basis of the methods they are using*, and hence may misinterpret or badly use the results . . . No doubt helping people to use Bayesian methods more appropriately is an important task of our time. (p. 457, emphasis added)

I have some sympathy here: Many modern Bayesians aren't aware of the traditional philosophy behind the methods they're buying into. Yet there is not just one philosophical basis for a given set of methods. This takes us to one of the most dramatic shifts in contemporary statistical foundations. It had long been assumed that only subjective or personalistic Bayesianism had a shot at providing genuine philosophical foundations, but you'll notice that groups holding this position, while they still dot the landscape in 2018, have been gradually shrinking. Some Bayesians have come to question whether the widespread use of methods under the Bayesian umbrella, however useful, indicates support for subjective Bayesianism as a foundation.

Marriages of Convenience?

The current frequentist–Bayesian unifications are often marriages of convenience; statisticians rationalize them less on philosophical than on practical grounds. For one thing, some are concerned that methodological conflicts are bad for the profession. For another, frequentist tribes, contrary to expectation, have not disappeared. Ensuring that accounts can control their error probabilities remains a desideratum that scientists are unwilling to forgo. Frequentists have an incentive to marry as well. Lacking a suitable epistemic interpretation of error probabilities – significance levels, power, and confidence levels – frequentists are constantly put on the defensive. Jim Berger (2003) proposes a construal of significance tests on which the tribes of Fisher, Jeffreys, and Neyman could agree, yet none of the chiefs of those tribes concur (Mayo 2003b). The success stories are based on agreements on numbers that are not obviously true to any of the three philosophies. Beneath the surface – while it's not often said in polite company – the most serious disputes live on. I plan to lay them bare.

If it's assumed an evidential assessment of hypothesis H should take the form of a posterior probability of H – a form of probabilism – then P-values and confidence levels are applicable only through misinterpretation and mistranslation. Resigned to live with P-values, some are keen to show that construing them as posterior probabilities is not so bad (e.g., Greenland and Poole 2013). Others focus on long-run error control, but cede territory wherein probability captures the epistemological ground of statistical inference. Why assume significance levels and confidence levels lack an authentic epistemological function? I say they do: to secure and evaluate how well probed and how severely tested claims are.

Eclecticism and Ecumenism

If you look carefully between dense forest trees, you can distinguish unification country from lands of eclecticism (Cox 1978) and ecumenism (Box 1983), where tools first constructed by rival tribes are separate, and more or less equal (for different aims). Current-day eclecticisms have a long history – the dabbling in tools from competing statistical tribes has not been thought to pose serious challenges. For example, frequentist methods have long been employed to check or calibrate Bayesian methods (e.g., Box 1983); you might test your statistical model using a simple significance test, say, and then proceed to Bayesian updating. Others suggest scrutinizing a posterior probability or a likelihood ratio from an error probability standpoint. What this boils down to will depend on the notion of probability used. If a procedure frequently gives high probability for *claim C* even if *C* is false, severe testers deny convincing evidence has been provided, and never mind about the meaning of probability.

One argument is that throwing different methods at a problem is all to the good, that it increases the chances that at least one will get it right. This may be so, provided one understands how to interpret competing answers. Using multiple methods is valuable when a shortcoming of one is rescued by a strength in another. For example, when randomized studies are used to expose the failure to replicate observational studies, there is a presumption that the former is capable of discerning problems with the latter. But what happens if one procedure fosters a goal that is not recognized or is even opposed by another? Members of rival tribes are free to sneak ammunition from a rival's arsenal – but what if at the same time they denounce the rival method as useless or ineffective?

Decoupling. On the horizon is the idea that statistical methods may be decoupled from the philosophies in which they are traditionally couched. In an attempted meeting of the minds (Bayesian and error statistical), Andrew Gelman and Cosma Shalizi (2013) claim that "implicit in the best Bayesian practice is a stance that has much in common with the error-statistical approach of Mayo" (p. 10). In particular, Bayesian model checking, they say, uses statistics to satisfy Popperian criteria for *severe tests*. The idea of error statistical foundations for Bayesian tools is not as preposterous as it may seem. The concept of severe testing is sufficiently general to apply to any of the methods now in use.

On the face of it, any inference, whether to the adequacy of a model or to a posterior probability, can be said to be warranted just to the extent that it has withstood severe testing. Where this will land us is still futuristic.

Why Our Journey?

> We have all, or nearly all, moved past these old [Bayesian-frequentist] debates, yet our textbook explanations have not caught up with the eclecticism of statistical practice. (Kass 2011, p. 1)

When Kass proffers "a philosophy that matches contemporary attitudes," he finds resistance to his big tent. Being hesitant to reopen wounds from old battles does not heal them. Distilling them in inoffensive terms just leads to the marshy swamp. Textbooks can't "catch-up" by soft-peddling competing statistical accounts. They show up in the current problems of scientific integrity, irreproducibility, questionable research practices, and in the swirl of methodological reforms and guidelines that spin their way down from journals and reports.

From an elevated altitude we see how it occurs. Once high-profile failures of replication spread to biomedicine, and other "hard" sciences, the problem took on a new seriousness. Where does the new scrutiny look? By and large, it collects from the earlier social science "significance test controversy" and the traditional philosophies coupled to Bayesian and frequentist accounts, along with the newer Bayesian–frequentist unifications we just surveyed. This jungle has never been disentangled. No wonder leading reforms and semi-popular guidebooks contain misleading views about all these tools. No wonder we see the same fallacies that earlier reforms were designed to avoid, and even brand new ones. Let me be clear, I'm not speaking about flat-out howlers such as interpreting a P-value as a posterior probability. By and large, they are more subtle; you'll want to reach your own position on them. It's not a matter of switching your tribe, but excavating the roots of tribal warfare. To tell what's true about them. I don't mean understand them at the socio-psychological levels, although there's a good story there (and I'll leak some of the juicy parts during our travels).

How can we make progress when it is difficult even to tell what is true about the different methods of statistics? We must start afresh, taking responsibility to offer a new standpoint from which to interpret the cluster of tools around which there has been so much controversy. Only then can we alter and extend their limits. I admit that the statistical philosophy that girds our explorations is not out there ready-made; if it was, there would be no need for our holiday cruise. While there are plenty of giant shoulders on which we stand, we won't be restricted by the pronouncements of any of the high and low priests, as sagacious as many of their words have been. In fact, we'll brazenly question some of their most entrenched mantras. Grab on to the gondola, our balloon's about to land.

In Tour II, I'll give you a glimpse of the core behind statistics battles, with a firm promise to retrace the steps more slowly in later trips.

Tour II Error Probing Tools versus Logics of Evidence

1.4   The Law of Likelihood and Error Statistics

If you want to understand what's true about statistical inference, you should begin with what has long been a holy grail – to use probability to arrive at a type of logic of evidential support – and in the first instance you should look not at full-blown Bayesian probabilism, but at comparative accounts that sidestep prior probabilities in hypotheses. An intuitively plausible logic of comparative support was given by the philosopher Ian Hacking (1965) – the Law of Likelihood. Fortunately, the Museum of Statistics is organized by theme, and the Law of Likelihood and the related Likelihood Principle is a big one.

> *Law of Likelihood (LL):* Data x are better evidence for hypothesis H_1 than for H_0 if x is more probable under H_1 than under H_0: $\Pr(x; H_1) > \Pr(x; H_0)$, that is, the *likelihood ratio (LR)* of H_1 over H_0 exceeds 1.

H_0 and H_1 are statistical hypotheses that assign probabilities to values of the random variable X. A fixed value of X is written x_0, but we often want to generalize about this value, in which case, following others, I use x. The *likelihood of the hypothesis H*, given data x, is the probability of observing x, under the assumption that H is true or adequate in some sense. Typically, the ratio of the likelihood of H_1 over H_0 also supplies the quantitative measure of comparative support. Note, when X is continuous, the probability is assigned over a small interval around X, to avoid probability 0.

Does the Law of Likelihood Obey the Minimal Requirement for Severity?

Likelihoods are vital to all statistical accounts, but they are often misunderstood because the data are fixed and the hypothesis varies. Likelihoods of hypotheses should not be confused with their probabilities. Two ways to see this. First, suppose you discover all of the stocks in Pickrite's promotional letter went up in value (x) – all winners. A hypothesis H to explain this is that their method always succeeds in picking winners. *H entails x*, so the likelihood of H given x is 1. Yet we wouldn't say H is therefore highly probable, especially without reason to put to rest that they culled the winners post hoc. For a second way, at any time, the same phenomenon may be perfectly predicted or explained by two rival theories; so both theories are equally likely on the data, even though they cannot both be true.

Suppose Bristol-Roach, in our Bernoulli tea tasting example, got two correct guesses followed by one failure. The observed data can be represented as $x_0 = \langle 1,1,0 \rangle$. Let the hypotheses be different values for θ, the probability of success on each independent trial. The likelihood of the hypothesis $H_0 : \theta = 0.5$, given x_0, which we may write as Lik(0.5), equals $(1/2)(1/2)(1/2) = 1/8$. Strictly speaking, we should write Lik($\theta;x_0$), because it's always computed given data x_0; I will do so later on. The likelihood of the hypothesis $\theta = 0.2$ is Lik(0.2) $= (0.2)(0.2)(0.8) = 0.032$. In general, the likelihood in the case of Bernoulli independent and identically distributed trials takes the form: Lik(θ)$= \theta^s(1 - \theta)^f$, $0 < \theta < 1$, where s is the number of successes and f the number of failures. Infinitely many values for θ between 0 and 1 yield positive likelihoods; clearly then, likelihoods do not sum to 1, or any number in particular. Likelihoods do not obey the probability calculus.

The Law of Likelihood (LL) will immediately be seen to fail our minimal severity requirement – at least if it is taken as an account of inference. Why? There is no onus on the Likelihoodist to predesignate the rival hypotheses – you are free to search, hunt, and post-designate a more likely, or even maximally likely, rival to a test hypothesis H_0.

Consider the hypothesis that $\theta = 1$ on trials one and two and 0 on trial three. That makes the probability of x maximal. For another example, hypothesize that the observed pattern would always recur in three-trials of the experiment (I. J. Good said in his cryptoanalysis work these were called "kinkera"). Hunting for an impressive fit, or trying and trying again, one is sure to find a rival hypothesis H_1 much better "supported" than H_0 even when H_0 is true. As George Barnard puts it, "there *always* is such a rival hypothesis, viz. that things just had to turn out the way they actually did" (1972, p. 129).

Note that for any outcome of n Bernoulli trials, the likelihood of $H_0 : \theta = 0.5$ is $(0.5)^n$, so is quite small. The likelihood ratio (LR) of a best-supported alternative compared to H_0 would be quite high. Since one could always erect such an alternative,

(*) Pr(LR in favor of H_1 over H_0; H_0) = maximal.

Thus the LL permits BENT evidence. The severity for H_1 is minimal, though the particular H_1 is not formulated until the data are in hand. I call such maximally fitting, but minimally severely tested, hypotheses *Gellerized*, since Uri Geller was apt to erect a way to explain his results in ESP trials. Our Texas sharpshooter is analogous because he can always draw a circle around a cluster of bullet holes, or around each single hole. One needn't go to such an extreme rival, but it suffices to show that the LL does not control the probability of erroneous interpretations.

What do we do to compute (*)? We look beyond the specific observed data to the behavior of the general rule or method, here the LL. The output is always a comparison of likelihoods. We observe one outcome, but we can consider that for any outcome, unless it makes H_0 maximally likely, we can find an H_1 that is more likely. This lets us compute the relevant properties of the method: its inability to block erroneous interpretations of data. As always, a severity assessment is one level removed: you give me the rule, and I consider its latitude for erroneous outputs. We're actually looking at the probability distribution of the rule, over outcomes in the sample space. This distribution is called a *sampling distribution*. It's not a very apt term, but nothing has arisen to replace it. For those who embrace the LL, once the data are given, it's irrelevant what other outcomes could have been observed but were not. Likelihoodists say that such considerations make sense only if the concern is the performance of a rule over repetitions, but not for inference from the data. Likelihoodists hold to "the irrelevance of the sample space" (once the data are given). This is the key contrast between accounts based on error probabilities (error statistical accounts) and logics of statistical inference.

Hacking "There is No Such Thing as a Logic of Statistical Inference"

Hacking's (1965) book was so ahead of its time that by the time philosophers of science started to get serious about philosophy of statistics, he had already broken the law he had earlier advanced. Hacking (1972, 1980) admits to having been caught up in the "logicist" mindset wherein we assume a logical relationship exists between any data and hypothesis; and even denies (1980, p. 145) there is any such thing.

In his review of A. F. Edwards' (1972) book *Likelihood*, Hacking (1972) gives his main reasons for rejecting the LL:

We capture enemy tanks at random and note the serial numbers on their engines. We know the serial numbers start at 0001. We capture a tank number 2176. How many did the enemy make? On the likelihood analysis, the best-supported guess is: 2176. Now one can defend this remarkable result by saying that it does not follow that we should estimate the actual number as 2176 only that comparing individual numbers, 2176 is better supported than any larger figure. My worry is deeper. Let us compare the relative likelihood of the two hypotheses, 2176 and 3000. Now pass to a situation where we are measuring, say, widths of a grating in which error has a normal distribution with known variance; we can devise data and a pair of hypotheses about the mean which will have the same log-likelihood ratio. I have no inclination to say that the relative support in the tank case is 'exactly the same as' that in the normal distribution case, even though the likelihood ratios are the same. (pp. 136–7)

Likelihoodists will insist that the law may be upheld by appropriately invoking background information, and by drawing distinctions between evidence, belief, and action.

Royall's Road to Statistical Evidence

Statistician Richard Royall, a longtime leader of Likelihoodist tribes, has had a deep impact on current statistical foundations. His views are directly tied to recent statistical reforms – even if those reformers go Bayesian rather than stopping, like Royall, with comparative likelihoods. He provides what many consider a neat proposal for settling disagreements about statistical philosophy. He distinguishes three questions: belief, action, and evidence:

1. What do I believe, now that I have this observation?
2. What should I do, now that I have this observation?
3. How should I interpret this observation as evidence regarding [H_0] versus [H_1]? (Royall 1997, p. 4)

Can we line up these three goals to my probabilism, performance, and probativeness (Section 1.2)? No. Probativeness gets no pigeonhole. According to Royall, what to believe is captured by Bayesian posteriors, how to act is captured by a frequentist performance (in some cases he will add costs). What's his answer to the evidence question? The Law of Likelihood.

Let's use one of Royall's first examples, appealing to Bernoulli distributions again – independent, dichotomous trials, "success" or "failure":

Medical researchers are interested in the success probability, θ, associated with a new treatment. They are particularly interested in how θ relates to the old treatment's success probability, believed to be about 0.2. They have reason to hope that θ is considerably greater, perhaps 0.8 or even greater. (Royall 1997, p. 19)

There is a set of possible outcomes, a sample space, S, and a set of possible parameter values, a parameter space Ω. He considers two hypotheses:

$\theta = 0.2$ and $\theta = 0.8$.

These are *simple* or *point* hypotheses. To illustrate take a miniature example with only $n = 4$ trials where each can be a "success" $\{X = 1\}$ or a "failure" $\{X = 0\}$. A possible result might be $x_0 = \langle 1,1,0,1 \rangle$. Since $\Pr(X = 1) = \theta$ and $\Pr(X = 0) = (1 - \theta)$, the probability of x_0 is $(\theta)(\theta)(1 - \theta)(\theta)$. Given independent trials, they multiply. Under the two hypotheses, given $\langle 1,1,0,1 \rangle$, the likelihoods are

$\text{Lik}(H_0) = (0.2)(0.2)(0.8)(0.2) = 0.0064,$

$\text{Lik}(H_1) = (0.8)(0.8)(0.2)(0.8) = 0.1024.$

A hypothesis that would make the data most probable would be that $\theta = 1$, on the three trials that yield successes, and 0 where it yields failure.

We typically denigrate "just so" stories, purposely erected to fit the data, as "unlikely." Yet they are *most* likely in the technical sense! So in hearing likelihood used formally, you must continually keep this swap of meanings in mind. (We call them Gellerized only if they pass with minimal severity.) If θ is to be constant on each trial, as in the Bernoulli model, the maximum likely hypothesis equates θ with the relative frequency of success, 0.75. [Exercise for reader: find Lik (0.75)]

Exhibit (i): Law of Likelihood Compared to a Significance Test. Here Royall contrasts his handling of the medical example to the standard significance test:

A standard statistical analysis of their observations would use a *Bernoulli*(θ) statistical model and test the composite hypotheses H_0: $\theta \leq 0.2$ versus H_1: $\theta > 0.2$. That analysis would show that H_0 can be rejected in favor of H_1 at any significance level greater than 0.003, a result that is conventionally taken to mean that the observations are very strong evidence supporting H_1 over H_0. (Royall 1997, p. 19; substituting H_0 and H_1 for H_1 and H_2.)

So the significance tester looks at the composite hypotheses H_0: $\theta \leq 0.2$ vs. H_1: $\theta > 0.2$, rather than his point hypotheses $\theta = 0.2$ and $\theta = 0.8$. Here, she would look at how much larger the mean success rate is in the sample $(X_1 + X_2 + \ldots X_{17})/17$, which we abbreviate as $\bar{x} = 9/17 = 0.53$, compared to what is expected under H_0, put in standard deviation units. Using Royall's numbers, the observed success rate is

$$\bar{x} = 9/17 = .53;$$

$$\sigma = \sqrt{[\theta\,(1-\theta)]}, \text{ which, under the null, is } \sqrt{[0.2\,(0.8)]} = 0.4.$$

The *test statistic* d(**X**) is $\sqrt{17}(\bar{X} - 0.2)/\sigma$; it gets larger and larger the more the data deviate from what is expected under H_0 – as is sensible for a good test statistic. Its value is

$$d(\boldsymbol{x}_0) = \sqrt{17}\,(0.53 - 0.2)/\,0.4 \simeq 3.3.$$

The significance level associated with d($\boldsymbol{x}_0$) is

$$\text{Pr}(d(\boldsymbol{X}) \geq d(\boldsymbol{x}); H_0) \simeq 0.003.$$

This is read, "the probability $d(X)$ would be at least as large as the particular value $d(x_0)$, under the supposition that H_0 adequately describes the data generation procedure" (see Souvenir C). It's not strictly a conditional probability – a subtle point that won't detain us here. We continue to follow Royall's treatment, though we'd want to distinguish the mere *indication* of an isolated significant result from strong *evidence*. We'd also have to audit for model assumptions and selection effects, but we assume these check out; after all, Royall's likelihood account also depends on the model holding.

We'd argue along the following lines: were H_0 a reasonable description of the process, then with very high probability you would not be able to regularly produce $d(x)$ values as large as this:

$$\Pr(d(X) < d(x); H_0) \simeq 0.997.$$

So if you manage to get such a large difference, I may infer that x *indicates* a genuine effect. Let's go back to Royall's contrast, because he's very unhappy with this.

Why Does the LL Reject Composite Hypotheses?

Royall tells us that his account is unable to handle composite hypotheses, even this one (for which there is a uniformly most powerful [UMP] test over all points in H_0). He does not conclude that his test comes up short. He and other Likelihoodists maintain that any genuine test or "rule of rejection" should be restricted to comparing the likelihood of H versus some point alternative H' *relative to* fixed data x (Royall 1997, pp. 19–20). It is a virtue. No wonder the Likelihoodist disagrees with the significance tester. In their view, a simple significance test is not a "real" testing account because it is not a comparative appraisal. Elliott Sober, a well-known philosopher of science, echoes Royall: "The fact that significance tests don't contrast the null with alternatives suffices to show that they do not provide a good rule for rejection" (Sober 2008, p. 56). Now, Royall's significance test *has* an alternative $H_1: \theta > 0.2$! It's just not a point alternative but is compound or composite (including all values greater than 0.2). The form of inference, admittedly, is not of the comparative ("evidence favoring") variety. In this discussion, H_0 and H_1 replace his H_1 and H_2.

What untoward consequences occur if we consider composite hypotheses (according to the Likelihoodist)? The problem is that even though the likelihood of $\theta = 0.2$ is small, there are values within alternative $H_1: \theta > 0.2$ that are even less likely on the data $\bar{x} = 0.53$. For instance consider $\theta = 0.9$.

[B]ecause H_0 contains some simple hypotheses that are better supported than some hypotheses in H_1 (e.g., $\theta = 0.2$ is better supported than $\theta = 0.9$ by a likelihood ratio of

LR = $(0.2/0.9)^9(0.8/0.1)^8$ = 22.2), the law of likelihood does not allow the characterization of these observations as strong evidence for H_1 over H_0. (Royall 1997, p. 20)

For Royall, rejecting H_0: $\theta \leq 0.2$ and inferring H_1: $\theta > 0.2$ is to assert *every* parameter point within H_1 is more likely than every point in H_0. That seems an idiosyncratic meaning to attach to "infer evidence of $\theta > 0.2$"; but it explains this particular battle. It still doesn't explain the alleged problem for the significance tester who just takes it to mean what it says:

To reject H_0: $\theta \leq 0.2$ is to infer *some* positive discrepancy from 0.2.

We readily agree with Royall that there's a problem with taking a rejection of H_0: $\theta \leq 0.2$, with $\bar{x} = 0.53$, as evidence of a discrepancy as large as $\theta = 0.9$. It's terrible evidence even that θ is as large as 0.7 or 0.8. Here's how a tester articulates this terrible evidence.

Consider the test rule: infer evidence of a discrepancy from 0.2 as large as 0.9, based on observing $\bar{x} = 0.53$. The data differ from 0.2 in the direction of H_1, but to take that difference as indicating an underlying $\theta > 0.9$ would be wrong with probability ~1. Since the standard error of the mean, $\sigma_{\bar{x}}$, is 0.1, alternative 0.9 is more than $3\sigma_{\bar{x}}$ greater than 0.53. ($\sigma_{\bar{x}} = \sigma/\sqrt{n}$) The inference gets low severity.

We'll be touring significance tests and confidence bounds in detail later. We're trying now to extract some core contrasts between error statistical methods and logics of evidence such as the LL. According to the LL, so long as there is a point within H_1 that is less likely given x than is H_0, the data are "evidence *in favor* of the null hypothesis, not evidence *against* it" (Sober 2008, pp. 55–6). He should add "as compared to" some less likely alternative. We never infer a statistical hypothesis according to the LL, but rather a likelihood ratio of two hypotheses, neither of which might be likely. The significance tester and the comparativist hold very different images of statistical inference.

Can an account restricted to comparisons answer the questions: is x good evidence for H? Or is it a case of bad evidence, no test? Royall says no. He declares that all attempts to say whether x is good evidence for H, or even if x is better evidence for H than is y, are utterly futile. Similarly, "What *does* the [LL] say when one hypothesis attaches the same probability to two different observations? It says absolutely nothing . . . [it] applies when two different hypotheses attach probabilities to the same observation" (Royall 2004, p. 148). That cuts short important tasks of inferential scrutiny. Since model checking concerns the adequacy of a single model, the Likelihoodist either forgoes such checks or must go beyond the paradigm.

Still, if the model can be taken as adequate, and the Likelihoodist gives a sufficiently long list of comparisons, the differences between us don't seem so marked. Take Royall:

One statement that we can make is that the observations are only weak evidence in favor of $\theta = 0.8$ versus $\theta = 0.2$ (LR = 4) ... and at least moderately strong evidence for $\theta = 0.5$ over any value $\theta > 0.8$ (LR) > 22). (1977, p. 20)

Nonetheless, we'd want to ask: what do these numbers mean? Is 22 a lot? Is 4 small? We're back to Hacking's attempt to compare tank cars with widths of a grating. How do we calibrate them? Neyman and Pearson's answer, we'll see, is to look at the probability of so large a likelihood ratio, under various hypotheses, as in (*).

LRs and Posteriors. Royall is loath to add prior probabilities to the assessment of the import of the evidence. This, he says, allows the LR to be "a precise and objective numerical measure of the strength of evidence" in comparing hypotheses (2004, p. 123). At the same time, Royall argues, the LL "constitutes the essential core of the Bayesian account of evidence ... the Bayesian who rejects the [LL] undermines his own position" (ibid., p. 146). The LR, after all, is the factor by which the ratio of posterior probabilities is changed by the data. Consider just two hypotheses, switching from the ";" in the significance test to conditional probability "|":[1]

$$\Pr(H_0|x) = \frac{\Pr(x|H_0)\,\Pr(H_0)}{\Pr(x|H_0)\,\Pr(H_0) \,+\, \Pr(x|H_1)\,\Pr(H_1)}.$$

Likewise:

$$\Pr(H_1|x) = \frac{\Pr(x|H_1)\,\Pr(H_1)}{\Pr(x|H_1)\,\Pr(H_1) \,+\, \Pr(x|H_0)\,\Pr(H_0)}.$$

The denominators equal $\Pr(x)$, so they cancel in the LR:

$$\frac{\Pr(H_1|x)}{\Pr(H_0|x)} = \frac{\Pr(x|\,H_1)\Pr(H_1)}{\Pr(x|\,H_0)\Pr(H_0)}.$$

All of this assumes the likelihoods and the model are deemed adequate.

---

[1] Divide the numerator and the denominator by $\Pr(x|H_0)\Pr(H_0)$. Then

$$\Pr(H_0|x) = \frac{1}{1 + \frac{\Pr(x|H_1)\Pr(H_1)}{\Pr(x|H_0)\Pr(H_0)}}$$

Data Dredging: Royall Bites the Bullet

Return now to our most serious problem: The Law of Likelihood permits finding evidence in favor of a hypothesis deliberately arrived at using the data, even in the extreme case that it is Gellerized. Allan Birnbaum, who had started out as a Likelihoodist, concludes, "the likelihood concept cannot be construed so as to allow useful appraisal, and thereby possible control, of probabilities of erroneous interpretations" (Birnbaum 1969, p. 128). But Royall has a clever response. Royall thinks control of error probabilities arises only in answering his second question about action, not evidence. He is prepared to bite the bullet. He himself gives the example of a "trick deck." You've shuffled a deck of ordinary-looking playing cards; you turn over the top card and find an ace of diamonds:

According to the law of likelihood, the hypothesis that the deck consists of 52 aces of diamonds (H_1) is better supported than the hypothesis that the deck is normal (H_N) [by the factor 52] . . . Some find this disturbing. (Royall 1997, pp. 13–14)

Royall does not. He admits:

. . . it seems unfair; no matter what card is drawn, the law implies that the corresponding trick-deck hypothesis (52 cards just like the one drawn) is better supported than the normal-deck hypothesis. Thus even if the deck is normal we will always claim to have found strong evidence that it is not. (ibid.)

What he is admitting then is, given any card:

Pr(LR favors trick deck hypothesis; normal deck) = 1.

Even though different trick deck hypotheses would be formed for different outcomes, we may compute the sampling distribution (*). The severity for "trick deck" would be 0. It need not be this extreme to have BENT results, but you get the idea.

What's Royall's way out? At the level of a report on comparative likelihoods, Royall argues, there's no need for a way out. To Royall, it only shows a confusion between evidence and belief.[2] If you're not convinced the deck has 52 aces of diamonds rather than being a normal deck "it does not mean that the observation is not strong evidence in favor of H_1 versus H_N" where H_N is a normal deck (ibid., p. 14). It just wasn't strong enough to overcome your prior beliefs. If you regard the maximally likely alternative as unpalatable, you should have given it a suitably low prior degree of probability. The more likely hypothesis is still favored on grounds of evidence, but your posterior belief

---

[2] He notes that the comparative evidence for a trick versus a normal deck is not evidence against a normal deck alone (pp. 14–15).

may be low. Don't confuse evidence with belief! For the question of evidence, your beliefs have nothing to do with it, according to Royall's Likelihoodist.

What if we grant the Likelihoodist this position? What do we do to tackle the essential challenge to the credibility of statistical inference today, when it's all about Texas Marksmen, hunters, snoopers, and cherry pickers? These moves, which play havoc with a test's ability to control erroneous interpretations, do not alter the evidence at all, say Likelihoodists. The fairest reading of Royall's position might be this: the data indicate only the various LRs. If they are the same, it matters not whether hypotheses arose through data dredging – at least, so long as you are in the category of "what the data say." As soon as you're troubled, you slip into the category of belief. What if we're troubled by the ease of exaggerating findings when you're allowed to rummage around? What if we wish to clobber the Texas sharpshooter method, never mind my beliefs in the particular claims they infer. You might aver, we should never be considering trick deck hypotheses, but this is the example Royall gives, and he is a, if not the, leading Likelihoodist.

To him, appealing to error probabilities is relevant only pre-data, which wouldn't trouble the severe tester so much if Likelihoodists didn't regard them as relevant only for a performance goal, not inference. Given that frequentists have silently assented to the performance use of error probabilities, it's perhaps not surprising that others accept this. The problem with cherry picking is not about long runs, it's that a poor job has been done in the case at hand. The severity requirement reflects this intuition. By contrast, Likelihoodists hold that likelihood ratios, and unadjusted P-values, still convey what the data say, even with claims arrived at through data dredging. It's true you can explore, arrive at H, then test H on other data; but isn't the reason there's a need to test on new data that your assessment will otherwise fail to convey how well tested H is?

Downsides to the "Appeal to Beliefs" Solution to Inseverity

What's wrong with Royall's appeal to prior beliefs to withhold support to a "just so" hypothesis? It may get you out of a jam in some cases. Here's why the severe tester objects. First, she insists on distinguishing the *evidential* warrant for one and the same hypothesis H in two cases: one where it was constructed post hoc, cherry picked, and so on, a second where it was predesignated. A cherry-picked hypothesis H could well be believable, but we'd still want to distinguish the evidential credit H deserves in the two cases. Appealing to priors can't help, since here there's one and the same H.

Perhaps someone wants to argue that the mode of testing alters the degree of belief in *H*, but this would be non-standard (violating the Likelihood Principle to be discussed shortly). Philosopher Roger Rosenkrantz puts it thus: The LL entails the irrelevance "of whether the theory was formulated in advance or suggested by the observations themselves" (Rosenkrantz 1977, p. 121). For Rosenkrantz, a default Bayesian last I checked, this irrelevance of predesignation is altogether proper. By contrast, he admits, "Orthodox (non-Bayesian) statisticians have found this to be strong medicine indeed!" (ibid.). Many might say instead that it is bad medicine. Take, for instance, something called the CONSORT, the Consolidated Standards of Reporting Trials from RCTs in medicine:

Selective reporting of outcomes is widely regarded as misleading. It undermines the validity of findings, particularly when driven by statistical significance or the direction of the effect [4], and has memorably been described in the New England Journal of Medicine as "Data Torturing" [5]. (COMpare Team 2015)

This gets to a second problem with relying on beliefs to block data-dredged hypotheses. Post-data explanations, even if it took a bit of data torture, are often incredibly convincing, and you don't have to be a sleaze to really believe them. Goldacre (2016) expresses shock that medical journals continue to report outcomes that were altered post-data – he calls this *outcome-switching*. Worse, he finds, some journals defend the practice because they are convinced that their very good judgment entitles them to determine when to treat post-designated hypotheses as if they were predesignated. Unlike the LL, the CONSORT and many other best practice guides view these concerns as an essential part of reporting what the data say. Now you might say this is just semantics, as long as, in the end, they report that outcome-switching occurred. Maybe so, provided the report mentions why it would be misleading to hide the information. At least people have stopped referring to frequentist statistics as "Orthodox."

There is a third reason to be unhappy with supposing the only way to block evidence for "just so" stories is by the *deus ex machina* of a low prior degree of belief: it misidentifies what the problem really *is*. The influence of the biased selection is not on the believability of *H* but rather on the capability of the test to have unearthed errors. The error probing capability of the testing procedure is being diminished. If you engage in cherry picking, you are not "sincerely trying," as Popper puts it, to find flaws with claims, but instead you are finding evidence in favor of a well-fitting hypothesis that you deliberately construct – barred only if your intuitions say it's unbelievable. The job that was supposed to be accomplished by an account of statistics now has to be performed by *you*. Yet you are the one most likely to follow your preconceived opinions, biases, and pet theories. If an account of statistical inference or evidence doesn't supply self-critical tools, it comes up short in an *essential* way. So says the severe tester.

Souvenir B: Likelihood versus Error Statistical

Like pamphlets from competing political parties, the gift shop from this tour proffers pamphlets from these two perspectives.

To the Likelihoodist, points in favor of the LL are:

- The LR offers "a precise and objective numerical measure of the strength of statistical evidence" for one hypotheses over another; it is a frequentist account and does not use prior probabilities (Royall 2004, p. 123).
- The LR is fundamentally related to Bayesian inference: the LR is the factor by which the ratio of posterior probabilities is changed by the data.
- A Likelihoodist account does not consider outcomes other than the one observed, unlike *P*-values, and Type I and II errors. (Irrelevance of the sample space.)
- Fishing for maximally fitting hypotheses and other gambits that alter error probabilities do not affect the assessment of evidence; they may be blocked by moving to the "belief" category.

To the error statistician, problems with the LL include:

- LRs do not convey the same evidential appraisal in different contexts.
- The LL denies it makes sense to speak of how well or poorly tested a single hypothesis is on evidence, essential for model checking; it is inapplicable to composite hypothesis tests.
- A Likelihoodist account does not consider outcomes other than the one observed, unlike *P*-values, and Type I and II errors. (Irrelevance of the sample space.)
- Fishing for maximally fitting hypotheses and other gambits that alter error probabilities do not affect the assessment of evidence; they may be blocked by moving to the "belief" category.

Notice, the last two points are identical for both. What's a selling point for a Likelihoodist is a problem for an error statistician.

1.5 Trying and Trying Again: The Likelihood Principle

> The likelihood principle emphasized in Bayesian statistics implies, among other things, that the rules governing when data collection stops are irrelevant to data interpretation. (Edwards, Lindman, and Savage 1963, p. 193)

Several well-known gambits make it altogether easy to find evidence in support of favored claims, even when they are unwarranted. A responsible statistical inference report requires information about whether the method used is capable of controlling such erroneous interpretations of data or not. Now we see that adopting a statistical inference account is also to buy into principles for processing data, hence criteria for "what the data say," hence grounds for charging an inference as illegitimate, questionable, or even outright cheating. The best way to survey the landscape of statistical debates is to hone in on some pivotal points of controversy – saving caveats and nuances for later on.

Consider for example the gambit of "trying and trying again" to achieve statistical significance, stopping the experiment only when reaching a nominally significant result. Kosher, or not? Suppose somebody reports data showing a statistically significant effect, say at the 0.05 level. Would it matter to your appraisal of the evidence if you found out that each time they failed to find significance, they went on to collect more data, until finally they did? A rule for when to stop sampling is called a *stopping rule*.

The question is generally put by considering a random sample X that is Normally distributed with mean μ and standard deviation $\sigma = 1$, and we are testing the hypotheses:

$$H_0: \mu = 0 \text{ against } H_1: \mu \neq 0.$$

This is a two-sided test: a discrepancy in either direction is sought. (The details of testing are in Excursions 3 and thereafter.) To ensure a significance level of 0.05, H_0 is rejected whenever the sample mean differs from 0 by more than $1.96\sigma/\sqrt{n}$, and, since $\sigma = 1$, the rule is: Declare x is statistically significant at the 0.05 level whenever $|\overline{X}| > 1.96/\sqrt{n}$. However, instead of fixing the sample size in advance, n is determined by the optional stopping rule:

Optional stopping rule: keep sampling until $|\overline{X}| \geq (1.96/\sqrt{n})$.

Equivalently, since the test statistic $d(X) = (\overline{X} - 0)/\sqrt{n}$:

Keep sampling until $|d(X)| \geq 1.96$.

Our question was: would it be relevant to your evaluation of the evidence if you learned she'd planned to keep running trials until reaching 1.96? Having failed to rack up a 1.96 difference after, say, 10 trials, she goes on to 20, and failing yet again, she goes to 30 and on and on until finally, say, on trial 169 she gets a 1.96 difference. Then she stops and declares the statistical significance is ~0.05.

This is an example of what's called a *proper stopping rule*: the probability it will stop in a finite number of trials is 1, regardless of the true value of μ. Thus, in one of the most seminal papers in statistical foundations, by Ward Edwards,

Harold Lindman, and Leonard (Jimmie) Savage (E, L, & S) tell us, "if an experimenter uses this procedure, then with probability 1 he will eventually reject any sharp null hypothesis, even though it be true" (1963, p. 239). Understandably, they observe, the significance tester frowns on optional stopping, or at least requires the auditing of the P-value to require an adjustment. Had n been fixed, the significance level would be 0.05, but with optional stopping it increases.

Imagine instead if an account advertised itself as ignoring stopping rules. What if an account declared:

In general, suppose that you collect data of any kind whatsoever – not necessarily Bernoullian, nor identically distributed, nor independent of each other. . . – stopping only when the data thus far collected satisfy some criterion of a sort that is sure to be satisfied sooner or later, then the import of the sequence of n data actually observed will be exactly the same as it would be had you planned to take exactly n observations in the first place. (ibid., pp. 238–9)

I've been teasing you, because these same authors who warn that to ignore stopping rules is to guarantee rejecting the null hypothesis even if it's true are the individuals who tout the irrelevance of stopping rules in the above citation – E, L, & S. They call it the *Stopping Rule Principle*. Are they contradicting themselves?

No. It is just that what looks to be, and indeed is, cheating from the significance testing perspective is not cheating from these authors' Bayesian perspective. "[F]requentist test results actually depend not only on what x was observed, but on how the experiment was stopped" (Carlin and Louis 2008, p. 8). Yes, but shouldn't they? Take a look at Table 1.1: by the time one reaches 50 trials, the probability of attaining a nominally significant 0.05 result is not 0.05 but 0.32. The actual or overall significance level is the probability of finding a 0.05 nominally significant result at some stopping point *or other*, up to the point it stops. The actual significance level accumulates.

Well-known statistical critics from psychology, Joseph Simmons, Leif Nelson, and Uri Simonsohn, place at the top of their list of requirements the need to block flexible stopping: "Researchers often decide when to stop data collection on the basis of interim data analysis . . . many believe this practice exerts no more than a trivial influence on false-positive rates" (Simmons et al. 2011, p. 1361). "Contradicting this intuition" they show the probability of erroneous rejections balloons. "A researcher who starts with 10 observations per condition and then tests for significance after every new . . . observation finds a significant effect 22% of the time" erroneously (ibid., p. 1362). Yet the followers of the Stopping Rule Principle deny it makes a difference to evidence. On their account, it *doesn't*. It's easy to see why there's disagreement.

Table 1.1 The effect of repeated significance tests (the "try and try again" method)

Number of trials n	Probability of rejecting H_0 with a result nominally significant at the 0.05 level at or before n trials, given H_0 is true
1	0.05
2	0.083
10	0.193
20	0.238
30	0.280
40	0.303
50	0.320
60	0.334
80	0.357
100	0.375
200	0.425
500	0.487
750	0.512
1000	0.531
Infinity	1.000

The Likelihood Principle

By what magic can such considerations disappear? One way to see the vanishing act is to hold, with Royall, that "what the data have to say" is encompassed in likelihood ratios. This is the gist of a very important principle of evidence, the *Likelihood Principle* (LP). Bayesian inference requires likelihoods plus prior probabilities in hypotheses; but the LP has long been regarded as a crucial part of their foundation: to violate it is to be *incoherent* Bayesianly. Disagreement about the LP is a pivot point around which much philosophical debate between frequentists and Bayesians has turned. Here is a statement of the LP:

According to Bayes's Theorem, $\Pr(x|\mu)$ ... constitutes the entire evidence of the experiment, that is it tells all that the experiment has to tell. More fully and more precisely, if y is the datum of some other experiment, and if it happens that $\Pr(x|\mu)$ and $\Pr(y|\mu)$ are proportional functions of μ (that is constant multiples of each other), then each of the two data x and y have exactly the same thing to say about the value of μ ... (Savage 1962, p. 17; replace λ with μ)

Some go further and claim that if x and y give the same likelihood, "they should give the same inference, analysis, conclusion, decision, action or anything else" (Pratt et al. 1995, p. 542). Does the LP entail the LL? No. Bayesians, for example, generally hold to the LP, but would insist on priors that go beyond the LL. Even the converse may be denied (according to Hacking) but this is not of concern to us.

Weak Repeated Sampling Principle. For sampling theorists (my error statisticians), by contrast, this example "taken in the context of examining consistency with $\theta = 0$, is enough to refute the strong likelihood principle" (Cox 1978, p. 54), since, with probability 1, it will stop with a "nominally" significant result even though $\theta = 0$. It contradicts what Cox and Hinkley call "the weak repeated sampling principle" (Cox and Hinkley 1974, p. 51). "[W]e should not follow procedures which for some possible parameter values would give, in hypothetical repetitions, misleading conclusions most of the time" (ibid., pp. 45–6).

For Cox and Hinkley, to report a 1.96 standard deviation difference from optional stopping just the same as if the sample size had been fixed, is to discard relevant information for inferring inconsistency with the null, while "according to any approach that is in accord with the strong likelihood principle, the fact that this particular stopping rule has been used is irrelevant" (ibid., p. 51). What they call the "strong" likelihood principle will just be called the LP here. (A weaker form boils down to sufficiency, see Excursion 3.)

Exhibit (ii): How Stopping Rules Drop Out. Our question remains: by what magic can such considerations disappear? Formally, the answer is straightforward. Consider two versions of the above experiment: In the first, 1.96 is reached via fixed sample size ($n = 169$); in the second, by means of optional stopping that ended at 169. While $d(x) = d(y)$, because of the stopping rule, the likelihood of y differs from that of x by a constant k, that is,

$$\Pr(x|H_i) = k\Pr(y|H_i) \text{ for constant } k.$$

Given that likelihoods enter as ratios, such proportional likelihoods are often said to be the "same." Now suppose inference is by Bayes' Theorem. Since likelihoods enter as ratios, the constant k drops out. This is easily shown. I follow E, L, & S; p. 237.

For simplicity, suppose the possible hypotheses are exhausted by two, H_0 and H_1, neither with probability of 0.

To show $\Pr(H_0|y) = \Pr(H_0|x)$:

(1) We are given the proportionality of likelihoods, for an arbitrary value of k:

$$\Pr(y|H_0) = k\Pr(x|H_0),$$

$$\Pr(y|H_1) = k\Pr(x|H_1).$$

(2) By definition:

$$\Pr(H_0|y) = \frac{\Pr(y|H_0)\Pr(H_0)}{\Pr(y)}.$$

The denominator $\Pr(y) = \Pr(y|H_0) \Pr(H_0) + \Pr(y|H_1) \Pr(H_1)$.

Now substitute for each term in (2) the proportionality claims in (1). That is, replace $\Pr(y|H_0)$ with $k\Pr(x|H_0)$ and $\Pr(y|H_1)$ with $k\Pr(x|H_1)$.

(3) The result is

$$\Pr(H_0|y) = \frac{k\Pr(x|H_0) \Pr(H_0)}{k\Pr(x)} = \Pr(H_0|x).$$

The posterior probabilities are the same whether the 1.96 result emerged from optional stopping, Y, or fixed sample size, X.

This essentially derives the LP from inference by Bayes' Theorem, and shows the equivalence for the particular case of interest, optional stopping. As always, when showing a Bayesian computation I use the conditional probability "|" rather than the ";" of the frequentist.[3]

The 1959 Savage Forum: What Counts as Cheating?

My colleague, well-known Bayesian I. J. Good, would state it as a "paradox":

[I]f a Fisherian is prepared to use optional stopping (which usually he is not) he can be sure of rejecting a true null hypothesis provided that he is prepared to go on sampling for a long time. The way I usually express this 'paradox' is that a Fisherian [but not a Bayesian] can cheat by pretending he has a plane to catch like a gambler who leaves the table when he is ahead. (Good 1983, p. 135)

The lesson about who is allowed to cheat depends on your statistical philosophy. Error statisticians require that the overall and not the "computed" significance level be reported. To them, cheating would be to report the significance level you got after trying and trying again in just *the same way* as if the test had a fixed sample size (Mayo 1996, p. 351). Viewing statistical methods as tools for severe tests, rather than as probabilistic logics of evidence, makes a deep difference to the tools we seek. Already we find ourselves thrust into some of the knottiest and most intriguing foundational issues.

This is Jimmie Savage's message at a 1959 forum deemed sufficiently important to occupy a large gallery of the Museum of Statistics (hereafter "The Savage Forum" (Savage 1962)). Attendees include Armitage, Barnard,

---

[3] $\Pr(x) = \Pr(x \ \& \ H_0) + \Pr(x \ \& \ H_1)$, where H_0 and H_1 are exhaustive.

Bartlett, Cox, Good, Jenkins, Lindley, Pearson, Rubin, and Smith. Savage announces to this eminent group of statisticians that if adjustments in significance levels are required for optional stopping, which they are, then the fault must be with significance levels. Not all agreed. Needling Savage on this issue, was Peter Armitage:

I feel that if a man deliberately stopped an investigation when he had departed sufficiently far from his particular hypothesis, then 'Thou shalt be misled if thou dost not know that.' If so, prior probability methods seem to appear in a less attractive light than frequency methods where one can take into account the method of sampling. (Armitage 1962, p. 72)

Armitage, an expert in sequential trials in medicine, is fully in favor of them, but he thinks stopping rules should be reflected in overall inferences. He goes further:

[Savage] remarked that, using conventional significance tests, if you go on long enough you can be sure of achieving any level of significance; does not the same sort of result happen with Bayesian methods? (ibid., p. 72)

He has in mind using a type of uniform prior probability for μ, wherein the posterior for the null hypothesis matches the significance level. (We return to this in Excursion 6. For $\sigma = 1$, its distribution is Normal($\bar{x}$, $1/n$).)

Not all cases of trying and trying again injure error probabilities. Think of trying and trying again until you find a key that fits a lock. When you stop, there's no probability of being wrong. (We return to this in Excursion 4.)

Savage's Sleight of Hand

Responding to Armitage, Savage engages in a bit of sleight of hand. Moving from the problematic example to one of two predesignated point hypotheses, H_0: $\mu = \mu_0$, and H_1: $\mu = \mu_1$, he shows that the error probabilities are controlled in that case. In particular, the probability of obtaining a result that makes H_1 r times more likely than H_0 is less than $1/r$: $\Pr(\text{LR} > r; H_0) < 1/r$. But, that wasn't Armitage's example; nor does Savage return to it. Now, it is open to Likelihoodists to resist being saddled "with ideas that are alien to them" (Sober 2008, p. 77). Since the Likelihoodist keeps to this type of comparative appraisal, they can set bounds to the probabilities of error. However, the bounds are no longer impressively small as we add hypotheses, even if they are predesignated[4] (Mayo and Kruse 2001).

---

[4] A general result, stated in Kerridge (1963, p. 1109), is that with k simple hypotheses, where H_0 is true and $H_1, \ldots, H_{k-1}$ are false, and equal priors, "*the frequency with which, at the termination of sampling the posterior probability of the true hypothesis is* p *or less cannot exceed* $(k - 1)p/(1 - p)$." Such bounds depend on having countably additive probability, while the uniform prior in Armitage's example imposes finite additivity.

Something more revealing is going on when the Likelihoodist sets pre-data bounds. Why the sudden concern with showing the rule for comparative evidence would very improbably find evidence in favor of the wrong hypothesis? This is an error probability. So it appears they also care about error probabilities – at least before-trial – or they are noting, for those of us who do, that they also have error control in the simple case of predesignated point hypotheses. The severe tester asks: If you want to retain these pre-data safeguards, why allow them to be spoiled by data-dependent hypotheses and stopping rules?

Some have said: the evidence is the same, but you take into account things like stopping rules and data-dependent selections *afterwards*. When making an inference, this *is* afterwards, and we need an epistemological rationale to pick up on their influences *now*. Perhaps knowing someone uses optional stopping warrants a high belief he's trying to deceive you, leading to a high enough prior belief in the null. Maybe so, but this is to let priors reflect methods in a non-standard way. Besides, Savage (1961, p. 583) claimed optional stopping "is no sin," so why should it impute deception? So far as I know, subjective Bayesians have resisted the idea that rules for stopping alter the prior. Couldn't you pack the concern in some background *B*? You could, but you would need another account to justify doing so, thereby only pushing back the issue. I've discussed an assortment of attempts elsewhere: Mayo (1996), Mayo and Kruse (2001), Mayo (2014b). Others have too, discussed here and elsewhere; please see our online sources (preface).

Arguments from Intentions: All in Your Head?

A funny thing happened at the Savage Forum: George Barnard announces he no longer holds the LP for the two-sided test under discussion, only for the predesignated point alternatives. Savage is shocked to hear it:

I learned the stopping rule principle from Professor Barnard, in conversation in the summer of 1952. Frankly, I then thought it a scandal that anyone in the profession could advance an idea so patently wrong, even as today I can scarcely believe that some people resist an idea so patently right. (Savage 1962, p. 76)

The argument Barnard gave him was that the plan for when to stop was a matter of the researchers' intentions, all wrapped up in their heads. While Savage denies he was ever sold on the argument from intentions, it's a main complaint you will hear about taking account, not just of stopping rules, but of error probabilities in general. Take the subjective Bayesian philosophers Howson and Urbach (1993):

A significance test inference, therefore, depends not only on the outcome that a trial produced, but also on the outcomes that it could have produced but did not. And the latter are determined by certain private intentions of the experimenters, embodying their stopping rule. It seems to us that this fact precludes a significance test delivering any kind of judgment about empirical support. (p. 212)

The truth is, whether they're hidden or not turns on your methodology being able to pick up on them. So the deeper question is: *ought* your account pick up on them?

The answer isn't a matter of mathematics, it depends on your goals and perspective – yes on your philosophy of statistics. Ask yourself: What features lead you to worry about cherry picking, and selective reporting? Why do the CONSORT and myriad other best practice manuals care? Looking just at the data and hypotheses – as a "logic" of evidence would – you will not see the machinations. Nevertheless, these machinations influence the capabilities of the tools. Much of the handwringing about irreproducibility is the result of wearing blinders as to the construction and selection of both hypotheses and data. In one sense, all test specifications are determined by a researcher's intentions; that doesn't make them private or invisible to us. They're visible to accounts with antennae to pick up on them!

You might try to deflect the criticism of stopping rules by pointing out that some stopping rules do alter priors. Armitage wasn't ignoring that, nor are we. These are called informative stopping rules, and examples are rather contrived. For instance, "a man who wanted to know how frequently lions watered at a certain pool was chased away by lions" (E, L, & S 1963, p. 239). They add, "we would not give a facetious example had we been able to think of a serious one." In any event, this is irrelevant for the Armitage example, which is non-informative.

Error Probabilities Violate the LP

> [I]t seems very strange that a frequentist could not analyze a given set of data, such as $(x_1, \ldots, x_n)$ if the stopping rule is not given . . . [D]ata should be able to speak for itself. (Berger and Wolpert 1988, p. 78)

Inference by Bayes' Theorem satisfies this intuition, which sounds appealing; but for our severe tester, data no more speak for themselves in the case of stopping rules than with cherry picking, hunting for significance, and the like. We may grant to the Bayesian that

[The] irrelevance of stopping rules to statistical inference restores a simplicity and freedom to experimental design that had been lost by classical emphasis on significance levels (in the sense of Neyman and Pearson). (E, L, & S 1963, p. 239)

The question is whether this latitude is desirable. If you are keen to use statistical methods critically, as our severe tester, you'll be suspicious of a simplicity and freedom to mislead.

Admittedly, this should have been more clearly spelled out by Neyman and Pearson. They rightly note:

In order to fix a limit between 'small' and 'large' values of [the likelihood ratio] we must know how often such values appear when we deal with a true hypothesis. (Pearson and Neyman 1930, p. 106)

That's true, but putting it in terms of the desire "to control the error involved in rejecting a true hypothesis" it is easy to dismiss it as an affliction of a frequentist concerned only with long-run performance. Bayesians and Likelihoodists are free of this affliction. Pearson and Neyman should have said: ignoring the information as to how readily true hypotheses are rejected, we cannot determine if there really is evidence of inconsistency with them.

Our minimal requirement for evidence insists that data only provide genuine or reliable evidence for H if H survives a severe test – a test H would probably have failed if false. Here the hypothesis H of interest is the non-null of Armitage's example: the existence of a genuine effect. A warranted inference to H depends on the test's ability to find H false when it is, i.e., when the null hypothesis is true. The severity conception of tests provides the link between a test's error probabilities and what's required for a warranted inference.

The error probability computations in significance levels, confidence levels, power, all depend on violating the LP! Aside from a concern with "intentions," you will find two other terms used in describing the use of error probabilities: a concern with (i) outcomes other than the one observed, or (ii) the sample space. Recall Souvenir B, where Royall, who obeys the LP, speaks of "the irrelevance of the sample space" once the data are in hand. It's not so obvious what's meant. To explain, consider Jay Kadane: "Significance testing violates the Likelihood Principle, which states that, having observed the data, inference must rely only on what happened, and not on what might have happened but did not" (Kadane 2011, p. 439). According to Kadane, the probability statement: $\Pr(|d(X)| > 1.96) = 0.05$ "is a statement about d(X) before it is observed. After it is observed, the event $\{d(X) > 1.96\}$ either happened or did not happen and hence has probability either one or zero" (ibid.).

Knowing d(x) = 1.96, Kadane is saying there's no more uncertainty about it. But would he really give it probability 1? That's generally thought to invite the problem of "known (or old) evidence" made famous by Clark Glymour (1980). If the probability of the data x is 1, Glymour argues, then Pr($x|H$) also is 1, but then Pr($H|x$) = Pr(H)Pr($x|H$)/Pr(x) = Pr(H), so there is no boost in probability given x. So does that mean known data don't supply evidence? Surely not. Subjective Bayesians try different solutions: either they abstract to a context prior to knowing x, or view the known data as an instance of a general type, in relation to a sample space of outcomes. Put this to one side for now in order to continue the discussion.[5]

Kadane is emphasizing that Bayesian inference is *conditional* on the particular outcome. So once x is known and fixed, other possible outcomes that could have occurred but didn't are irrelevant. Recall finding that Pickrite's procedure was to build k different portfolios and report just the one that did best. It's as if Kadane is asking: "Why are you considering other portfolios that you might have been sent but were not, to reason from the one that you got?" Your answer is: "Because that's how I figure out whether your boast about Pickrite is warranted." With the "search through k portfolios" procedure, the possible outcomes are the success rates of the k different attempted portfolios, each with its own null hypothesis. The actual or "audited" P-value is rather high, so the severity for H: Pickrite has a reliable strategy, is low (1 − p). For the holder of the LP to say that, once x is known, we're not allowed to consider the other chances they gave themselves to find an impressive portfolio, is to put the kibosh on a crucial way to scrutinize the testing process.

Interestingly, nowadays, non-subjective or default Bayesians concede they "have to live with some violations of the likelihood and stopping rule principles" (Ghosh, Delampady, and Samanta 2010, p. 148) since their prior probability distributions are influenced by the sampling distribution. Is it because ignoring stopping rules can wreak havoc with the well-testedness of inferences? If that is their aim, too, then that is very welcome. Stay tuned.

---

[5] Colin Howson, a long-time subjective Bayesian, has recently switched to being a non-subjective Bayesian at least in part because of the known evidence problem (Howson 2017, p. 670).

Souvenir C: A Severe Tester's Translation Guide

Just as in ordinary museum shops, our souvenir literature often probes treasures that you didn't get to visit at all. Here's an example of that, and you'll need it going forward. There's a confusion about what's being done when the significance tester considers the set of all of the outcomes leading to a $d(x)$ greater than or equal to 1.96, i.e., $\{x: d(x) \geq 1.96\}$, or just $d(x) \geq 1.96$. This is generally viewed as throwing away the particular x, and lumping all these outcomes together. What's really happening, according to the severe tester, is quite different. What's actually being signified is that we are interested in the method, not just the particular outcome. Those who embrace the LP make it very plain that data-dependent selections and stopping rules drop out. To get them to drop in, we signal an interest in what the test procedure *would have* yielded. This is a counterfactual and is altogether essential in expressing the properties of the method, in particular, the probability it would have yielded some nominally significant outcome *or other*.

When you see $\Pr(d(X) \geq d(x_0); H_0)$, or $\Pr(d(X) \geq d(x_0); H_1)$, for any particular alternative of interest, insert:

"the test procedure would have yielded"

just before the $d(X)$. In other words, this expression, with its inequality, is a signal of interest in, and an abbreviation for, the error probabilities associated with a test.

Applying the Severity Translation. In Exhibit (i), Royall described a significance test with a Bernoulli(θ) model, testing H_0: $\theta \leq 0.2$ vs. H_1: $\theta > 0.2$. We blocked an inference from observed difference $d(x) = 3.3$ to $\theta = 0.8$ as follows. (Recall that $\bar{x} = 0.53$ and $d(x_0) \simeq 3.3$.)

> *We computed* $\Pr(d(X) > 3.3; \theta = 0.8) \simeq 1$.
>
> *We translate it as* $\Pr$(The test would yield $d(X) > 3.3; \theta = 0.8) \simeq 1$.

We then reason as follows:

> *Statistical inference*: If $\theta = 0.8$, then the method would virtually always give a difference larger than what we observed. Therefore, the data indicate $\theta < 0.8$.

(This follows for rejecting H_0 in general.) When we ask: "How often would your test have found such a significant effect even if H_0 is approximately true?" we are asking about the properties of the experiment that *did* happen.

The counterfactual "would have" refers to how the procedure would behave in general, not just with these data, but with other possible data sets in the sample space.

Exhibit (iii). Analogous situations to the optional stopping example occur even without optional stopping, as with selecting a data-dependent, maximally likely, alternative. Here's an example from Cox and Hinkley (1974, 2.4.1, pp. 51–2), attributed to Allan Birnbaum (1969).

A single observation is made on X, which can take values 1, 2, . . ., 100. "There are 101 possible distributions conveniently indexed by a parameter θ taking values 0, 1,..., 100" (ibid.). We are not told what θ is, but there are 101 possible point hypotheses about the value of θ: from 0 to 100. If X is observed to be r, written $X = r$ $(r \neq 0)$, then the most likely hypothesis is $\theta = r$: in fact, $\Pr(X = r;$ $\theta = r) = 1$. By contrast, $\Pr(X = r; \theta = 0) = 0.01$. Whatever value r that is observed, hypothesis $\theta = r$ is 100 times as likely as is $\theta = 0$. Say you observe $X = 50$, then H: $\theta = 50$ is 100 times as likely as is $\theta = 0$. So "even if in fact $\theta = 0$, we are certain to find evidence apparently pointing strongly against $\theta = 0$, if we allow comparisons of likelihoods chosen in the light of the data" (Cox and Hinkley 1974, p. 52). This does not happen if the test is restricted to two preselected values. In fact, if $\theta = 0$ the probability of a ratio of 100 in favor of the false hypothesis is 0.01.[6]

Allan Birnbaum gets the prize for inventing chestnuts that deeply challenge both those who do, and those who do not, hold the Likelihood Principle!

Souvenir D: Why We Are So New

What's Old? You will hear critics say that the reason to overturn frequentist, sampling theory methods – all of which fall under our error statistical umbrella – is that, well, they've been around a long, long time. First, they are scarcely stuck in a time warp. They have developed with, and have often been the source of, the latest in modeling, resampling, simulation, Big Data, and machine learning techniques. Second, all the methods have roots in long-ago ideas. Do you know what is really up-to-the-minute in this time of massive, computer algorithmic methods and "trust me" science? A new vigilance about retaining hard-won error control techniques. Some thought that, with enough data, experimental design

---

[6] From Cox and Hinkley 1974, p. 51. The likelihood function corresponds to the normal distribution of $\overline{X}$ around μ with SE $\sigma/\sqrt{n}$. The likelihood at $\mu = 0$ is $\exp(-0.5k^2)$ times that at $\mu = \overline{x}$. One can choose k to make the ratio small. "That is, even if in fact $\mu = 0$, there always appears to be strong evidence against $\mu = 0$, at least if we allow comparison of the likelihood at $\mu = 0$ against any value of μ and hence in particular against the value of μ giving maximum likelihood". However, if we confine ourselves to comparing the likelihood at $\mu = 0$ with that at some fixed $\mu = \mu'$, this difficulty does not arise.

could be ignored, so we have a decade of wasted microarray experiments. To view outcomes other than what you observed as irrelevant to what x_0 says is also at odds with cures for irreproducible results. When it comes to cutting-edge fraud-busting, the ancient techniques (e.g., of Fisher) are called in, refurbished with simulation.

What's really old and past its prime is the idea of a logic of inductive inference. Yet core discussions of statistical foundations today revolve around a small cluster of (very old) arguments based on that vision. Tour II took us to the crux of those arguments. Logics of induction focus on the relationships between given data and hypotheses – so outcomes other than the one observed drop out. This is captured in the Likelihood Principle (LP). According to the LP, trying and trying again makes no difference to the probabilist: it is what someone intended to do, locked up in their heads.

It is interesting that frequentist analyses often need to be adjusted to account for these 'looks at the data,'... That Bayesian analysis claims no need to adjust for this 'look elsewhere' effect – called the *stopping rule principle* – has long been a controversial and difficult issue... (J. Berger 2008, p. 15)

The irrelevance of optional stopping is an asset for holders of the LP. For the task of criticizing and debunking, this puts us in a straightjacket. The warring sides talk past each other. We need a new perspective on the role of probability in statistical inference that will illuminate, and let us get beyond, this battle.

New Role of Probability for Assessing What's Learned. A passage to locate our approach within current thinking is from Reid and Cox (2015):

Statistical theory continues to focus on the interplay between the roles of probability as representing physical haphazard variability ... and as encapsulating in some way, directly or indirectly, aspects of the uncertainty of knowledge, often referred to as epistemic. (p. 294)

We may avoid the need for a different version of probability by appeal to a notion of calibration, as measured by the behavior of a procedure under hypothetical repetition. That is, we study assessing uncertainty, as with other measuring devices, by assessing the performance of proposed methods under hypothetical repetition. Within this scheme of repetition, probability is defined as a hypothetical frequency. (p. 295)

This is an ingenious idea. Our meta-level appraisal of methods proceeds this way too, but with one important difference. A key question for us is the proper epistemic role for probability. It is standardly taken as providing a probabilism, as an assignment of degree of actual or rational belief in a claim, absolute or comparative. We reject this. We proffer an alternative theory: a severity assessment. An account of what is warranted and unwarranted to infer – a normative epistemology – is not a matter of using probability to assign rational beliefs, but to control and assess how well probed claims are.

If we keep the presumption that the epistemic role of probability is a degree of belief of some sort, then we can "avoid the need for a different version of probability" by supposing that good/poor performance of a method warrants high/low belief in the method's output. Clearly, poor performance is a problem, but I say a more nuanced construal is called for. The idea that partial or imperfect knowledge is all about degrees of belief is handed down by philosophers. Let's be philosophical enough to challenge it.

New Name? An error statistician assesses inference by means of the error probabilities of the method by which the inference is reached. As these stem from the sampling distribution, the conglomeration of such methods is often called "sampling theory." However, sampling theory, like classical statistics, Fisherian, Neyman–Pearsonian, or frequentism are too much associated with hardline or mish-mashed views. Our job is to clarify them, but in a new way. Where it's apt for taking up discussions, we'll use "frequentist" interchangeably with "error statistician." However, frequentist error statisticians tend to embrace the long-run performance role of probability that I find too restrictive for science. In an attempt to remedy this, Birnbaum put forward the "confidence concept" (Conf), which he called the "one rock in a shifting scene" in statistical thinking and practice. This "one rock," he says, takes from the Neyman–Pearson (N-P) approach "techniques for systematically appraising and bounding the probabil- ities (under respective hypotheses) of seriously misleading interpretations of data" (Birnbaum 1970, p.1033). Extending his notion to a composite alternative:

> Conf: An adequate concept of statistical evidence should find strong evidence against H_0 (for $\sim H_0$) with small probability α when H_0 is true, and with much larger probability $(1 - \beta)$ when H_0 is false, increasing as discrepancies from H_0 increase.

This is an entirely right-headed pre-data performance requirement, but I agree with Birnbaum that it requires a reinterpretation for evidence post-data (Birnbaum 1977). Despite hints and examples, no such evidential interpreta- tion has been given. The switch that I'm hinting at as to what's required for an evidential or epistemological assessment is key. Whether one uses a frequentist or a propensity interpretation of error probabilities (as Birnbaum did) is not essential. *What we want is an error statistical approach that controls and assesses a test's stringency or severity.* That's not much of a label. For short, we call someone who embraces such an approach a severe tester. For now I will just venture that a severity scrutiny illuminates all statistical approaches currently on offer.

Excursion 2 Taboos of Induction and Falsification

Itinerary

Tour I Induction and Confirmation

Cox: [I]n some fields foundations do not seem very important, but we both think that foundations of statistical inference are important; why do you think that is?

Mayo: I think because they ask about fundamental questions of evidence, inference, and probability ... we invariably cross into philosophical questions about empirical knowledge and inductive inference. (Cox and Mayo 2011, p. 103)

Contemporary philosophy of science presents us with some taboos: Thou shalt not try to find solutions to problems of induction, falsification, and demarcating science from pseudoscience. It's impossible to understand rival statistical accounts, let alone get beyond the statistics wars, without first exploring how these came to be "lost causes." I am not talking of ancient history here: these problems were alive and well when I set out to do philosophy in the 1980s. I think we gave up on them too easily, and by the end of Excursion 2 you'll see why. Excursion 2 takes us into the land of "Statistical Science and Philosophy of Science" (StatSci/PhilSci). Our Museum Guide gives a terse thumbnail sketch of Tour I. Here's a useful excerpt:

> Once the Problem of Induction was deemed to admit of no satisfactory, non-circular solutions (~1970s), philosophers of science turned to building formal logics of induction using the deductive calculus of probabilities, often called Confirmation Logics or Theories. A leader of this Confirmation Theory movement was Rudolf Carnap. A distinct program, led by Karl Popper, denies there is a logic of induction, and focuses on Testing and Falsification of theories by data. At best a theory may be accepted or corroborated if it fails to be falsified by a severe test. The two programs have analogues to distinct methodologies in statistics: Confirmation theory is to Bayesianism as Testing and Falsification are to Fisher and Neyman–Pearson.

Tour I begins with the traditional Problem of Induction, then moves to Carnapian confirmation and takes a brief look at contemporary formal epistemology. Tour II visits Popper, falsification, and demarcation, moving into Fisherian tests and the replication crisis. Redolent of Frank Lloyd Wright's Guggenheim Museum in New York City, the StatSci/PhilSci Museum is arranged in concentric sloping oval floors that narrow as you go up. It's as if we're in a three-dimensional Normal curve. We begin in a large exposition on the ground floor. Those who start on the upper floors forfeit a central Rosetta Stone to decipher today's statistical debates.

2.1   The Traditional Problem of Induction

Start with the *asymmetry of falsification and confirmation*. One black swan falsifies the universal claim that *C*: all swans are white. Observing a single white swan, while a *positive instance* of *C*, wouldn't allow inferring generalization *C*, unless there was only one swan in the entire population. If the generalization refers to an infinite number of cases, as most people would say about scientific theories and laws, then no matter how many positive instances observed, you couldn't infer it with certainty. It's always possible there's a black swan out there, a *negative instance*, and it would only take one to falsify *C*. But surely we think enough positive instances of the right kind might warrant an argument for inferring *C*. Enter the problem of induction. First, a bit about arguments.

Soundness versus Validity

An *argument* is a group of statements, one of which is said to follow from or be supported by the others. The others are premises, the one inferred, the conclusion. A deductively *valid* argument is one where if its premises are all true, then its conclusion must be true. Falsification of "all swans are white" follows a deductively valid argument. Let ~*C* be the denial of claim *C*.

> (1) *C*: All swans are white.
>   *x* is a swan but is black.
>   Therefore, ~*C*.

We can also infer, validly, what follows if a generalization *C* is true.

> (2) *C*: All swans are white.
>   *x* is a swan.
>   Therefore, *x* is white.

However, validity is not the same thing as *soundness*. Here's a case of argument form (2):

> (3) All philosophers can fly.
>   Mayo is a philosopher.
>   Therefore, Mayo can fly.

Validity is a matter of form. Since (3) has a valid form, it is a valid argument. But its conclusion is false! That's because it is *unsound*: at least one of its premises is false (the first). No one can stop you from applying deductively valid arguments, regardless of your statistical account. Don't assume you will get truth thereby. Bayes' Theorem can occur in a valid argument, within a formal system of probability:

> (4) If $Pr(H_1)$, $\ldots$, $Pr(H_n)$ are the prior probabilities of an exhaustive set of hypotheses, and $Pr(x|H_i)$ the corresponding likelihoods.
> Data x are given, and $Pr(H_1|x)$ is defined.
> Therefore, $Pr(H_1|x) = p$.[1]

The conclusion is the posterior probability $Pr(H_1|x)$. It can be inferred only if the argument is *sound*: all the givens must hold (at least approximately). To deny that all of statistical inference is reducible to Bayes' Theorem is not to preclude your using this or any other deductive argument. What you need to be concerned about is their soundness. So, you will still need a way to vouchsafe the premises.

Now to the traditional philosophical problem of induction. What is it? Why has confusion about induction and the threat of the traditional or "logical" problem of induction made some people afraid to dare use the "I" word? The traditional problem of induction seeks to justify a type of argument: one taking a form of *enumerative induction* (EI) (or the *straight rule* of induction). Infer from past cases of A's that were B's to all or most A's will be B's:

> EI: All observed A_1, A_2, $\ldots$, A_n have been B's.
> Therefore, H: all A's are B's.

It is not a deductively valid argument, because clearly its premises can all be true while its conclusion false. It's *invalid*, as is so for any inductive argument. As Hume (1739) notes, nothing changes if we place the word "probably" in front of the conclusion: it is justified to infer from past A's being B's that, *probably*, all or most A's will be B's. To "rationally" justify induction is to supply a reasoned argument for using EI. The traditional problem of induction, then, involves trying to find an argument to justify a type of argument!

Exhibit (i): Justifying Induction Is Circular. In other words, the traditional problem of induction is to justify the conclusion:

---

[1] i.e., $p = \dfrac{Pr(x|H_1)Pr(H_1)}{Pr(x|H_1)Pr(H_1) + \cdots + Pr(x|H_n)Pr(H_n)}$

Conclusion: EI is rationally justified, it's a reliable rule.

We need an argument for concluding EI is reliable. Using an inductive argument to justify induction lands us in a circle. We'd be using the method we're trying to justify, or begging the question. What about a deductively valid argument? The premises would have to be things we know to be true, otherwise the argument would not be sound. We might try:

Premise 1: EI has been reliable in a set of observed cases.

Trouble is, this premise can't be used to deductively infer EI will be reliable *in general*: the known cases only refer to the past and present, not the future. Suppose we add a premise:

Premise 2: Methods that have worked in past cases will work in future cases.

Yet to assume Premise 2 is true is to use EI, and thus, again, to beg the question.

Another idea for the additional premise is in terms of assuming nature is uniform. We do not escape: to assume the *uniformity of nature* is to assume EI is a reliable method. Therefore, induction cannot be rationally justified. It is called the *logical* problem of induction because logical argument alone does not appear able to solve it. All attempts to justify EI assume past successes of a rule justify its general reliability, which is to assume EI – what we're trying to show.

I'm skimming past the rest of a large exhibition on brilliant attempts to solve induction in this form. Some argue that although an attempted justification is circular it is not *viciously* circular. (An excellent source is Skyrms 1986.)

But wait. Is inductive enumeration a rule that has been reliable even in the past? No. It is reasonable to expect that unobserved or future cases will be very different from the past, that apparent patterns are spurious, and that observed associations are not generalizable. We would only want to justify inferences of that form if we had done a good job ruling out the many ways we know we can be misled by such an inference. That's not the way confirmation theorists see it, or at least, saw it.

Exhibit (ii): Probabilistic (Statistical) Affirming the Consequent. Enter logics of confirmation. Conceding that we cannot justify the inductive method (EI), philosophers sought logics that represented apparently plausible inductive reasoning. The thinking is this: never mind trying to convince a skeptic of the inductive method, we give up on that. But we know what we mean. We need only to make sense of the habit of applying EI. True to the logical positivist spirit of the 1930s–1960s, they sought evidential relationships between statements of evidence and conclusions. I sometimes call them evidential-relation (E-R) logics. They didn't renounce enumerative induction, they sought logics that embodied it. Begin by fleshing out the full argument behind EI:

> If H: all A's are B's, then all observed A's (A$_1$, A$_2$, . . ., A$_n$) are B's.
> All observed A's (A$_1$, A$_2$, . . ., A$_n$) are B's.
> Therefore, H: all A's are B's.

The premise that we added, the first, is obviously true; the problem is that the second premise can be true while the conclusion false. The argument is deductively *invalid* – it even has a name: *affirming the consequent*. However, its probabilistic version is weaker. *Probabilistic affirming the consequent* says only that the conclusion is probable or gets a boost in confirmation or probability – a *B-boost*. It's in this sense that Bayes' Theorem is often taken to ground a plausible confirmation theory. It probabilistically justifies EI in that it embodies probabilistic affirming the consequent.

How do we obtain the probabilities? Rudolf Carnap's audacious program (1962) had been to assign probabilities of hypotheses or statements by deducing them from the logical structure of a particular (first order) language. These were called *logical probabilities*. The language would have a list of properties (e.g., "is a swan," "is white") and individuals or names (e.g., i, j, k). The task was to assign equal initial probability assignments to states of this mini world, from which we could deduce the probabilities of truth functional combinations. The degree of probability, usually understood as a rational degree of belief, would hold between two statements, one expressing a hypothesis and the other the data. $C(H,x)$ symbolizes "the confirmation of H, given x." Once you have chosen the initial assignments to core states of the world, calculating degrees of confirmation is a formal or syntactical matter, much like deductive logic. The goal was to somehow measure the *degree of implication* or confirmation that x affords H. Carnap imagined the scientist coming to the inductive logician to have the rational degree of confirmation in H evaluated, given her evidence. (I'm serious.) Putting aside the difficulty of listing all properties of scientific interest, from where do the initial assignments come?

Carnap's first attempt at a C-function resulted in no learning! For a toy illustration, take a universe with three items, i, j, k, and a single property B. "Bk" expresses "k has property B." There are eight possibilities, each called a *state description*. Here's one: {Bi, ~Bj, ~Bk}. If each is given initial probability of $\frac{1}{8}$, we have what Carnap called the logic $c^{\dagger}$. The degree of confirmation that j will be black given that i was white = $\frac{1}{2}$, which is the same as the initial confirmation of Bi (since it occurs in four of eight state descriptions). Nothing has been learned: $c^\dagger$ is scrapped. By apportioning initial probabilities more coarsely, one could learn, but there was an infinite continuum of inductive logics characterized by choosing the value of a parameter he called λ (λ continuum). λ in effect determines how much uniformity and regularity to expect. To restrict the field, Carnap had to postulate what he called "inductive intuitions." As a logic student, I too found these attempts tantalizing – until I walked into my first statistics class. I was also persuaded by philosopher Wesley Salmon:

Carnap has stated that the ultimate justification of the axioms is inductive intuition. I do not consider this answer an adequate basis for a concept of rationality. Indeed, I think that *every* attempt, including those by Jaakko Hintikka and his students, to ground the concept of rational degree of belief in logical probability suffers from the same unacceptable *apriorism*. (Salmon 1988, p. 13).

This program, still in its heyday in the 1980s, was part of a general logical positivist attempt to reduce science to observables plus logic (no metaphysics). Had this reductionist goal been realized, which it wasn't, the idea of scientific inference being reduced to particular predicted observations might have succeeded. Even with that observable restriction, the worry remained: what does a highly probable claim, according to a particular inductive logic, have to do with the real world? How can it provide "a guide to life?" (e.g., Kyburg 2003, Salmon 1966). The epistemology is restricted to inner coherence and consistency. However much contemporary philosophers have gotten beyond logical positivism, the hankering for an inductive logic remains. You could say it's behind the appeal of the default (non-subjective) Bayesianism of Harold Jeffreys, and other attempts to view probability theory as extending deductive logic.

Exhibit (iii): A Faulty Analogy Between Deduction and Induction. When we heard Hacking announce (Section 1.4): "there is no such thing as a logic of statistical inference" (1980, p. 145), it wasn't only the failed attempts to build one, but the recognition that the project is "founded on a false analogy with deductive logic" (ibid.). The issue here is subtle, and we'll revisit it through our journey. I agree with Hacking, who is agreeing with C. S. Peirce:

In the case of analytic [deductive] inference we know the probability of our conclusion (if the premises are true), but in the case of synthetic [inductive] inferences we only know the degree of trustworthiness of our proceeding. (Peirce 2.693)

In getting new knowledge, in ampliative or inductive reasoning, the conclusion should go beyond the premises; probability enters to qualify the overall "trustworthiness" of the method. Hacking not only retracts his Law of Likelihood (LL), but also his earlier denial that Neyman–Pearson statistics is inferential. "I now believe that Neyman, Peirce, and Braithwaite were on the right lines to follow in the analysis of inductive arguments" (Hacking 1980, p. 141). Let's adapt some of Hacking's excellent discussion.

When we speak of an inference, it could mean the entire argument including premises and conclusion. Or it could mean just the conclusion, or statement inferred. Let's use "inference" to mean the latter – the claim detached from the premises or data. A statistical procedure of *inferring* refers to a method for reaching a statistical inference about some aspect of the source of the data, together with its probabilistic properties: in particular, its capacities to avoid erroneous (and ensure non-erroneous) interpretations of data. These are the method's error probabilities. My argument from coincidence to weight gain (Section 1.3) inferred H: I've gained at least 4 pounds. The inference is qualified by the detailed data (group of weighings), and information on how capable the method is at blocking erroneous pronouncements of my weight. I argue that, very probably, my scales would not produce the weight data they do (e.g., on objects with known weight) were H false. What is being qualified probabilistically is the inferring or testing process.

By contrast, in a probability or confirmation logic, what is generally detached is the probability of H, given data. It is a *probabilism*. Hacking's diagnosis in 1980 is that this grows out of an abiding logical positivism, with which he admits to having been afflicted. There's this much analogy with deduction: In a deductively valid argument: if the premises are true then, necessarily, the conclusion is true. But we don't attach the "necessarily" to the conclusion. Instead it qualifies the entire argument. So mimicking deduction, why isn't the inductive task to qualify the method in some sense, for example, report that it would probably lead to true or approximately true conclusions? That would be to show the reliable performance of an inference method. If that's what an inductive method requires, then Neyman–Pearson tests, which afford good performance, are inductive.

My main difference from Hacking here is that I don't argue, as he seems to, that the warrant for the inference is that it stems from a method that very probably gets it right (so I may hope it is right this time). It's not that the method's reliability "rubs off" on this particular claim. I say inference C may be detached as *indicated* or *warranted*, having passed a severe test (a test that C probably would have failed, if false in a specified manner). This is the central point of Souvenir D. The logician's "semantic entailment" symbol, the double turnstile: "$|=$", could be used to abbreviate "entails severely":

Data + capacities of scales $|=_{SEV}$ I've gained at least k pounds.

(The premises are on the left side of $|=$.) However, I won't use this notation.

Keeping to a deductive logic of probability, we never detach an inference. This is in sync with a probabilist such as Bruno de Finetti:

The calculus of probability can say absolutely nothing about reality . . . As with the logic of certainty, the logic of the probable adds nothing of its own: it merely helps one to see the implications contained in what has gone before. (de Finetti 1974, p. 215)

These are some of the first clues we'll be collecting on a wide difference between statistical inference as a deductive logic of probability, and an inductive testing account sought by the error statistician. When it comes to inductive learning, we want our inferences to go beyond the data: we want lift-off. To my knowledge, Fisher is the only other writer on statistical inference, aside from Peirce, to emphasize this distinction.

In deductive reasoning all knowledge obtainable is already latent in the postulates. Rigour is needed to prevent the successive inferences growing less and less accurate as we proceed. The conclusions are never more accurate than the data. In inductive reasoning we are performing part of the process by which new knowledge is created. The conclusions normally grow more and more accurate as more data are included. It should never be true, though it is still often *said*, that the conclusions are no more accurate than the data on which they are based. (Fisher 1935b, p. 54)

2.2   Is Probability a Good Measure of Confirmation?

> It is often assumed that the degree of confirmation of x by y must be the same as the (relative) probability of x given y, i.e., that $C(x, y) = \Pr(x, y)$. My first task is to show the inadequacy of this view. (Popper 1959, p. 396; substituting Pr for P)

If your suitcase rings the alarm at an airport, this might slightly increase the probability of its containing a weapon, and slightly decrease the probability that it's clean. But the probability it contains a weapon is so small that the probability it's clean remains high, even if it makes the alarm go off. These facts illustrate a tension between two ways a probabilist might use probability to measure confirmation. A test of a philosophical confirmation theory is whether it elucidates or is even in sync with intuitive methodological principles about evidence or testing. Which, if either, fits with intuitions?

The most familiar interpretation is that H is confirmed by x if x gives a boost to the probability of H, *incremental* confirmation. The components of $C(H,x)$ are allowed to be any statements, and, in identifying C with Pr, no reference to a probability model is required. There is typically a background variable k, so that x confirms H relative to k: to the extent that $\Pr(H|x \text{ and } k) > \Pr(H \text{ and } k)$. However, for readability, I will drop the explicit inclusion of k. More generally, if H entails x, then assuming $\Pr(x) \neq 1$ and $\Pr(H) \neq 0$, we have $\Pr(H|x) > \Pr(H)$.

This is an instance of probabilistic affirming the consequent. (Note: if $\Pr(H|x)$ > $\Pr(H)$ then $\Pr(x|H)$ > $\Pr(x)$.)

> (1) *Incremental* (B-boost): H is confirmed by x iff $\Pr(H|x) > \Pr(H)$,
> H is disconfirmed iff $\Pr(H|x) < \Pr(H)$.

("iff" denotes if and only if.) Also plausible is an *absolute* interpretation:

> (2) *Absolute*: H is confirmed by x iff $\Pr(H|x)$ is high, at least greater than $\Pr(\sim H|x)$.

Since $\Pr(\sim H|x) = 1 - \Pr(H|x)$, (2) is the same as defining x confirms H: $\Pr(H|x)$ > 0.5. From (1), x (the alarm) *dis*confirms the hypothesis H: the bag is clean, because its probability has gone down, however slightly. Yet from (2) x confirms H: bag is clean, because $\Pr(H)$ is high to begin with.

There's a conflict. Thus, if (1) seems plausible, then probability, $\Pr(H|x)$, isn't a satisfactory way to define confirmation. At the very least, we must distinguish between an incremental and an absolute measure of confirmation for H. No surprise there. From the start Carnap recognized that "the verb 'to confirm' is ambiguous"; Carnap and most others choose the "making firmer" or incremental connotation as better capturing what is meant than that of "making firm" (Carnap 1962, p. xviii). Incremental confirmation is generally used in current Bayesian epistemology. Confirmation is a B-boost.

The first point Popper's making in the epigraph is this: to identify confirmation and probability "$C = \Pr$" leads to this type of conflict. His example is a single toss of a homogeneous die: The data x: an even number occurs; hypothesis H: a 6 will occur. It's given that $\Pr(H) = 1/6$, $\Pr(x) = 1/2$. The probability of H is increased by data x, while $\sim H$ is undermined by x (its probability goes from 5/6 to 4/6). If we identify probability with degree of confirmation, x confirms H and disconfirms $\sim H$. However, $\Pr(H|x) < \Pr(\sim H|x)$. So H is less well confirmed given x than is $\sim H$, in the sense of (2). Here's how Popper puts it, addressing Carnap: How can we say H is confirmed by x, while $\sim H$ is not; but at the same time $\sim H$ is confirmed to a higher degree with x than is H? (Popper 1959, p. 390).[2]

---

[2] Let HJ be (H & J). To show: If there is a case where x confirms HJ more than x confirms J, then degree of probability cannot equal degree of confirmation.

   (i) $C(HJ, x) > C(J, x)$ is given.
   (ii) $J = \sim HJ$ or HJ by logical equivalence.
   (iii) $C(HJ, x) > C(\sim HJ$ or $HJ, x)$ by substituting (ii) in line (i).

Since $\sim HJ$ and HJ are mutually exclusive, we have from the special addition rule for probability:

   (iv) $\Pr(HJ, x) \leq \Pr(\sim HJ$ or $HJ, x)$.

So if $\Pr = C$, (iii) and (iv) yield a contradiction. (Adapting Popper 1959, p. 391)

Moreover, Popper continues, confirmation theorists don't use $Pr(H|x)$ alone (as they would if $C = Pr$), but myriad functions of probability to capture how much x has firmed up H. A number of measures offer themselves for the job. A simple B-boost would report the ratio R: $Pr(H|x)/Pr(H)$, which in Popper's example is 2. Or we can use the likelihood ratio of H compared to ~H. Since I used LR in Excursion 1, where the two hypotheses are not exhaustive, let's write [LR] to denote

$$[LR]: \frac{Pr(x|H)}{Pr(x|\sim H)} = (1/0.4) = 2.5.$$

Many other ways of measuring the increase in confirmation that x affords H could do as well. (For some excellent lists see Popper 1959 and Fitelson 2002.)

What shall we say about the numbers like 2, 2.5? Do they mean the same thing in different contexts? Then there's the question of computing $Pr(x|\sim H)$, the *catchall factor*. It doesn't offer problems in this case because ~H, the *catchall hypothesis*, is just an event statement. It's far more problematic once we move to genuine statistical hypotheses. Recall how Royall's Likelihoodist avoids the composite catchall factor by restricting his likelihood ratios to two simple statistical hypotheses.

Popper's second point is that "the probability of a statement . . . simply does not express an appraisal of the severity of the tests a theory has passed, or of the manner in which it has passed these tests" (pp. 394–5). Ultimately, Popper denies that severity can be completely formalized by any C function. Is there nothing in between a pure formal-syntactical approach and leaving terms at a vague level? I say there is.

Consider for a moment philosopher Peter Achinstein – a Carnap student. Achinstein (2000, 2001) declared that scientists should not take seriously philosophical accounts of confirmation because they make it too easy to confirm. Furthermore, scientists look to empirical grounds for confirmation, whereas philosophical accounts give us formal (non-empirical) a priori measures. (I call it Achinstein's "Dean's problem" because he made the confession to a Dean asking about the relevance of philosophy – not usually the best way to keep funding for philosophy.) Achinstein rejects confirmation as increased firmness, denying it is either necessary or sufficient for evidence (rejects (1)).[3] He requires for H to be confirmed by x that the posterior of H given x be rather high, a version of (2): $Pr(H|x) \gg Pr(\sim H|x)$, but that's not all. He requires that, before we apply

---

[3] Why is a B-boost not necessary for Achinstein? Suppose you know x: the newspaper says Harry won, and it's never wrong. Then a radio, also assumed 100% reliable, announces y: Harry won. Statement y, Achinstein thinks, should still count as evidence for H: he won. I agree.

confirmation measures, the components have an appropriate explanatory relationship to each other. Yet this requires an adequate way to make explanatory inferences before getting started. It's not clear how the formalism helped. He still considers himself a Bayesian epistemologist – a term that has replaced confirmation theorist – but the probabilistic representation threatens to be mostly a kind of bookkeeping for inferential work done in some other way.

Achinstein is right to object that (1) incremental confirmation makes it too easy to have evidence. After all, *J*: Mike drowns in the Pacific Ocean, entails *x*: there is a Pacific Ocean; yet *x* does not seem to be evidence for *J*. Still the generally favored position is to view confirmation as (1) a B-boost.

Exhibit (iv): Paradox of Irrelevant Conjunctions. Consider a famous argument due to Glymour (1980). If we allow that *x* confirms *H* so long as $\Pr(H|x) > \Pr(H)$, it seems everything confirms everything, so long as one thing is confirmed!

The first piece of the argument is the problem of irrelevant conjunctions – also called the "tacking paradox." If *x* confirms *H*, then *x* also confirms (*H* & *J*), even if hypothesis *J* is just "tacked on" to *H*. As with most of these chestnuts, there is a long history (e.g., Earman 1992, Rosenkrantz 1977) but I consider a leading contemporary representative, Branden Fitelson. Fitelson (2002) and Hawthorne and Fitelson (2004) define the statement "*J* is an *irrelevant conjunct* to *H*, with respect to evidence *x*" as meaning $\Pr(x|J) = \Pr(x|J \ \& \ H)$. For instance, *x* might be radioastronomic data in support of

> *H*: the General Theory of Relativity (GTR) deflection of light effect is 1.75″ and
>
> *J*: the radioactivity of the Fukushima water being dumped in the Pacific Ocean is within acceptable levels.

(A) If *x* confirms *H*, then *x* confirms (*H* & *J*), where $\Pr(x|H \ \& \ J) = \Pr(x|H)$ for any *J* consistent with *H*.

The reasoning is as follows:

> (i) $\Pr(x|H)/\Pr(x) > 1$ (*x Bayesian-confirms H*).
> (ii) $\Pr(x|H \ \& \ J) = \Pr(x|H)$ (*J*'s irrelevance is given).

Substituting (ii) into (i) gives $\Pr(x|H \ \& \ J)/\Pr(x) > 1$.

> Therefore *x* Bayesian-confirms (*H* & *J*).[4]

---

[4] To expand the reasoning, first observe that $\Pr(H|x)/\Pr(H) = \Pr(x|H)/\Pr(x)$ and $\Pr(H \ \& \ J|x)/\Pr(H \ \& \ J) = \Pr(x|H \ \& \ J)/\Pr(x)$, both by Bayes' Theorem. So, when $\Pr(H|x)/\Pr(H) > 1$, we also have $\Pr(x|H)/\Pr(x) > 1$. This, together with $\Pr(x|H \ \& \ J) = \Pr(x|H)$ (given), yields $\Pr(x|H \ \& \ J)/\Pr(x) > 1$. Thus, we also have $\Pr(H \ \& \ J|x)/\Pr(H \ \& \ J) > 1$.

However, it is also plausible to hold what philosophers call the "special consequence" condition: If x confirms a claim W, and W entails J, then x confirms J. In particular:

(B) If x confirms $(H \ \& \ J)$, then x confirms J.

(B) gives the second piece of the argument. From (A) and (B) we have, if x confirms H, then x confirms J for any irrelevant J consistent with H (neither H nor J have probabilities 0 or 1).

It follows that if x confirms any H, then x confirms any J.

This absurd result, however, assumed (B) (special consequence) and most Bayesian epistemologists reject it. This is the gist of Fitelson's solution to tacking, updated in Hawthorne and Fitelson (2004). It is granted that x confirms the conjunction $(H \ \& \ J)$, while denying x confirms the irrelevant conjunct J. Aren't they uncomfortable with (A), allowing $(H \ \& \ J)$ to be confirmed by x?

I'm inclined to agree with Glymour that we are not too happy with an account of evidence that tells us deflection of light data confirms the conjunction of the GTR deflection and the radioactivity of the Fukushima water is within acceptable levels, while assuring us that x does not confirm the conjunct, that the Fukushima water has acceptable levels of radiation (1980, p. 31). Moreover, suppose we measure the confirmation boost by

R: $\Pr(H|x)/\Pr(x)$.

Then, Fitelson points out, the conjunction $(H \ \& \ J)$ is just as well confirmed by x as is H!

However, granting confirmation is an incremental B-boost doesn't commit you to measuring it by R. The conjunction $(H \ \& \ J)$ gets less of a confirmation boost than does H if we use, instead of R, the likelihood ratio [LR] of H against $\sim H$:

[LR]: $\Pr(x|H)/\Pr(x|\sim H)$.[5]

This avoids the counterintuitive result, or so it is claimed. (Note: $\Pr(H|x) > \Pr(H)$ iff $\Pr(x|H) > \Pr(x)$, but measuring the boost by R differs from measuring it with [LR].)

---

[5] Recall that Royall restricts the likelihood ratio to non-composite hypotheses, whereas here $\sim H$ is the Bayesian catchall.

What Does the Severity Account Say?

Our account of inference disembarked way back at (1): that x confirms H so long as $\Pr(H|x) > \Pr(H)$. That is, we reject probabilistic affirming the consequent. In the simplest case, H entails x, and x is observed. (We assume the probabilities are well defined, and H doesn't already have probability 1.) H gets a B-boost, but there are many other "explanations" of x. It's the same reason we reject the Law of Likelihood (LL). Unless stringent probing has occurred, finding an H that fits x is not difficult to achieve even if H is false. H hasn't passed severely. Now severely passing is obviously stronger than merely finding some evidence for H, and the confirmation theorist is only saying a B-boost suffices for some evidence. To us, to have *any* evidence, or even the weaker notion of an "indication," requires a minimal threshold of severity be met.

How about tacking? As always, the error statistician needs to know the relevant properties of the test procedure or rule, and just handing me the H's, x's, and relative probabilities will not suffice. The process of tacking, at least one form, is this – once you have an incrementally confirmed H with data x, tack on any consistent J and announce "x confirms $(H \& J)$." Let's allow that $(H \& J)$ fits or accords with x (since GTR entails or renders probable the deflection data x). However, the very claim: "$(H \& J)$ is confirmed by x" has been subjected to a radically non-risky test. Nothing has been done to measure the radioactivity of the Fukushima water being dumped into the ocean. B-boosters might reply, "We're admitting J is irrelevant and gets no confirmation," but our testing intuitions tell us then it's crazy to regard $(H \& J)$ as confirmed. They will point out other examples where this doesn't seem crazy. But what matters is that it's being permitted in general.

We should *punish* a claim to have evidence for H with a tacked-on J, when nothing has been done to refute J. Imagine the chaos. Are we to allow positive trial data on diabetes patients given drug D to confirm the claim that D improves survival of diabetes patients *and* Roche's artificial knee is effective, when there's only evidence for one? If the confirmation theorist simply stipulates that (1) defines confirmation, then it's within your rights to deny it captures ordinary notions of evidence. On the other hand, if you do accept (1), then why are you bothered at all by tacking? Many are not.

Patrick Maher (2004) argues that if B-boosting is confirmation, then there is nothing counterintuitive about data confirming irrelevant conjunctions; Fitelson should not even be conceding "he bites the bullet." It makes sense that $(H \& J)$ increases the probability assignment to x just as much as does H, for J the irrelevant conjunct. The supposition that this is problematic and that therefore one must move away from R: $\Pr(x|H)/\Pr(x)$ sits uneasily with the fact that R > 1 is just what confirmation boost means. Rather than "solve" the problem by saying we can measure boost so that (H & J) gets less confirmation than H, using [LR], why not see it as what's meant by an irrelevant conjunct J: J doesn't improve the ability to predict *x*. Other philosophers working in this arena, Crupi and Tentori (2010), notice that [LR] is not without problems. In particular, if *x* *dis*confirms hypothesis Q, then (Q & J) isn't as badly disconfirmed as Q is, for irrelevant conjunct J. Just as (H & J) gets less of a B-boost than does H, (Q & J) gets less disconfirmation in the case where *x* disconfirms J. This too makes sense on the [LR] measure, though I will spare the details. Their intuitions are that this is worse than the irrelevant conjunction case, and is not solved by the use of [LR]. Interesting new measures are offered. Again, this seems to our tester to reflect the tension between Bayes boosts and good tests.

What They Call Confirmation We Call Mere "Fit" or "Accordance"

> In opposition to [the] inductivist attitude, I assert that C(H,*x*) must not be interpreted as the degree of corroboration of H by *x*, unless *x* reports the results of *our sincere efforts to overthrow H*. The requirement of sincerity cannot be formalized – no more than the inductivist requirement that *x* must represent our total observational knowledge. (Popper 1959, p. 418, substituting H for h; *x* for e)

Sincerity! Popper never held that severe tests turned on a psychological notion, but he was at a loss to formalize severity. A fuller passage from Popper (1959) is worth reading if you get a chance.[6] All the measures of confirmation, be it R or LR, or one of the others, count merely as "fit" or "accordance" measures to Popper and to the severe tester. They may each be relevant for different problems – that there are different dimensions for fit is to be expected. These measures do not capture what's needed to determine if much (or anything) has been done to find H is flawed. What we need to add are the associated error probabilities. Error probabilities do not enter into these standard confirmation theories – which isn't to say they couldn't. If R is used and observed to be *r*, we want to compute Pr(R > *r*; ~(H & J)). Here, the probability of getting R > 1 is maximal (since (H & J) entails *x*), even if ~(H & J) is true. So *x* is "bad evidence,

---

[6] "I must insist that C(h, e) can be interpreted as degree of corroboration only if e is *a report on the severest tests we have been able to design*. It is this point that marks the difference between the attitude of the inductivist, or verificationist, and my own attitude. The inductivist or verificationist wants *affirmation* for his hypothesis. He hopes to make it *firmer* by his evidence e and he looks out for *'firmness'* – for *'confirmation.'* ... Yet if e is *not* a report about the results of our sincere attempts to overthrow h, then we shall simply deceive ourselves if we think we can interpret C(h,e) as degree of corroboration, or anything like it." (Popper 1959, p. 418).

no test" (BENT) for the conjunction.[7] It's not a psychological "sincerity" being captured; nor is it purely context free. Popper couldn't capture it as he never made the error probability turn.

Time prevents us from entering multiple other rooms displaying paradoxes of confirmation theory, where we'd meet up with such wonderful zombies as the white shoe confirming all ravens are black, and the "grue" paradox, which my editor banished from my 1996 book. (See Skyrms 1986.) Enough tears have been shed. Yet they shouldn't be dismissed too readily; they very often contain a puzzle of deep relevance for statistical practice. There are two reasons the tacking paradox above is of relevance to us. The first concerns a problem that arises for both Popperians and Bayesians. There is a large-scale theory T that predicts x, and we want to discern which portion of T to credit. Severity says: do not credit those portions that could not have been found false, even if they're false. They are poorly tested. This may not be evident until long after the experiment. We don't want to say there is evidence for a large-scale theory such as GTR just because one part was well tested. On the other hand, it may well be that all relativistic theories with certain properties have passed severely.

Second, the question of whether to measure support with a Bayes boost or with posterior probability arises in Bayesian statistical inference as well. When you hear that what you want is some version of probabilism, be sure to ask if it's a boost (and if so which kind) or a posterior probability, a likelihood ratio, or something else. Now statisticians might rightly say, we don't go around tacking on hypotheses like this. True, the Bayesian epistemologist invites trouble by not clearly spelling out corresponding statistical models. They seek a formal logic, holding for statements about radiation, deflection, fish, or whatnot. I think this is a mistake. That doesn't preclude a general account for statistical inference; it just won't be purely formal.

Statistical Foundations Need Philosophers of Statistics

> The idea of putting probabilities over hypotheses delivered to philosophy a godsend, an entire package of superficiality. (Glymour 2010, p. 334)

Given a formal epistemology, the next step is to use it to represent or justify intuitive principles of evidence. The problem to which Glymour is alluding is this: you can start with the principle you want your confirmation logic to reflect, and then *reconstruct* it using probability. The task, for the formal epistemologist, becomes the problem of assigning priors and likelihoods that mesh with the principle you want to defend. Here's an example. Some think

---

[7] The real problem is that $\Pr(x; H \& J) = \Pr(x; H \& {\sim}J)$.

that GTR got more confirmation than a rival theory (e.g., Brans-Dicke theory) because the latter is made to fit the data thanks to adjustable parameters (Jefferys and Berger 1992). Others think the fact it had adjustable parameters does not alter the confirmation (Earman 1992). They too can reconstruct the episode so that Brans-Dicke pays no penalty. The historical episode can be "rationally reconstructed" to accord with either philosophical standpoint.

Although the problem of statistical inference is only a small part of what today goes under the umbrella of formal epistemology, progress in the statistics wars would advance more surely if philosophers regularly adopted the language of statistics. Not only would we be better at the job of clarifying the conceptual discomforts among practitioners of statistics and modeling, some of the classic problems of confirmation could be scotched using the language of random variables and their distributions.[8] Philosophy of statistics had long been ahead of its time, in the sense of involving genuinely interdisciplinary work with statisticians, scientists, and philosophers of science. We need to return to that. There are many exceptions, of course; yet to try to list them would surely make me guilty of leaving several out.

---

[8] For a discussion and justification of the use of "random variables," see Mayo (1996).

Tour II Falsification, Pseudoscience, Induction

We'll move from the philosophical ground floor to connecting themes from other levels, from Popperian falsification to significance tests, and from Popper's demarcation to current-day problems of pseudoscience and irreplication. An excerpt from our Museum Guide gives a broad-brush sketch of the first few sections of Tour II:

> Karl Popper had a brilliant way to "solve" the problem of induction: Hume was right that enumerative induction is unjustified, but science is a matter of deductive falsification. Science was to be demarcated from pseudoscience according to whether its theories were testable and falsifiable. A hypothesis is deemed severely tested if it survives a stringent attempt to falsify it. Popper's critics denied he could sustain this and still be a deductivist . . .
>
> Popperian falsification is often seen as akin to Fisher's view that "every experiment may be said to exist only in order to give the facts a chance of disproving the null hypothesis" (1935a, p. 16). Though scientists often appeal to Popper, some critics of significance tests argue that they are used in decidedly non-Popperian ways. Tour II explores this controversy.

While Popper didn't make good on his most winning slogans, he gives us many seminal launching-off points for improved accounts of falsification, corroboration, science versus pseudoscience, and the role of novel evidence and predesignation. These will let you revisit some thorny issues in today's statistical crisis in science.

2.3 Popper, Severity, and Methodological Probability

Here's Popper's summary (drawing from Popper, *Conjectures and Refutations*, 1962, p. 53):

- [Enumerative] induction . . . is a myth. It is neither a psychological fact . . . nor one of scientific procedure.
- The actual procedure of science is to operate with conjectures. . .
- Repeated observation and experiments function in science as tests of our conjectures or hypotheses, i.e., as attempted refutations.

- [It is wrongly believed that using the inductive method can] *serve as a criterion of demarcation between science and pseudoscience. . . .* None of this is altered in the least if we say that induction makes theories only probable.

There are four key, interrelated themes:

(1) **Science and Pseudoscience.** Redefining scientific method gave Popper a new basis for demarcating genuine science from questionable science or pseudoscience. Flexible theories that are easy to confirm – theories of Marx, Freud, and Adler were his exemplars – where you open your eyes and find confirmations everywhere, are low on the scientific totem pole (ibid., p. 35). For a theory to be scientific it must be testable and falsifiable.

(2) **Conjecture and Refutation.** The problem of induction is a problem only if it depends on an unjustifiable procedure such as enumerative induction. Popper shocked everyone by denying scientists were in the habit of inductively enumerating. It doesn't even hold up on logical grounds. To talk of "another instance of an A that is a B" assumes a conceptual classification scheme. How else do we recognize it as another item under the umbrellas A and B? (ibid., p. 44). You can't just observe, you need an interest, a point of view, a problem.

The actual procedure for learning in science is to operate with conjectures in which we then try to find weak spots and flaws. Deductive logic is needed to draw out the remote logical consequences that we actually have a shot at testing (ibid., p. 51). From the scientist down to the amoeba, says Popper, we learn by trial and error: conjecture and refutation (ibid., p. 52). The crucial difference is the extent to which we constructively learn how to reorient ourselves after clashes.

Without waiting, passively, for repetitions to impress or impose regularities upon us, we actively try to impose regularities upon the world. . . These may have to be discarded later, should observation show that they are wrong. (ibid., p. 46)

(3) **Observations Are Not Given.** Popper rejected the time-honored empiricist assumption that observations are known relatively unproblematically. If they are at the "foundation," it is only because there are apt methods for testing their validity. We dub claims observable *because* or to the extent that they are open to stringent checks. (Popper: "anyone who has learned the relevant technique can test it" (1959, p. 99).) Accounts of hypothesis appraisal that start with "evidence *x*," as in confirmation logics, vastly oversimplify the role of data in learning.

(4) **Corroboration Not Confirmation, Severity Not Probabilism.** Last, there is his radical view on the role of probability in scientific inference. Rejecting probabilism, Popper not only rejects Carnap-style logics of confirmation, he denies scientists are interested in highly probable hypotheses (in any sense). They seek bold, informative, interesting conjectures and ingenious and severe attempts to refute them. If one uses a logical notion of probability, as philosophers (including Popper) did at the time, the high content theories are highly improbable; in fact, Popper said universal theories have 0 probability. (Popper also talked of statistical probabilities as propensities.)

These themes are in the spirit of the error statistician. Considerable spadework is required to see what to keep and what to revise, so bring along your archeological shovels.

Demarcation and Investigating Bad Science

There is a reason that statisticians and scientists often refer back to Popper; his basic philosophy – at least his most winning slogans – are in sync with ordinary intuitions about good scientific practice. Even people divorced from Popper's full philosophy wind up going back to him when they need to demarcate science from pseudoscience. Popper's right that if using enumerative induction makes you scientific then anyone from an astrologer to one who blithely moves from observed associations to full blown theories is scientific. Yet the criterion of testability and falsifiability – as typically understood – is nearly as bad. It is both too strong and too weak. Any crazy theory found false would be scientific, and our most impressive theories aren't deductively falsifiable. Larry Laudan's famous (1983) "The Demise of the Demarcation Problem" declared the problem taboo. This is a highly unsatisfactory situation for philosophers of science. Now Laudan and I generally see eye to eye, perhaps our disagreement here is just semantics. I share his view that what really matters is determining if a hypothesis is warranted or not, rather than whether the theory is "scientific," but surely Popper didn't mean logical falsifiability sufficed. Popper is clear that many unscientific theories (e.g., Marxism, astrology) are falsifiable. It's clinging to falsified theories that leads to unscientific practices. (Note: The use of a strictly falsified theory for prediction, or because nothing better is available, isn't unscientific.) I say that, with a bit of fine-tuning, we can retain the essence of Popper to capture what makes an inquiry, if not an entire domain, scientific.

Following Laudan, philosophers tend to shy away from saying anything general about science versus pseudoscience – the predominant view is that there is no such thing. Some say that there's at most a kind of "family resemblance" amongst domains people tend to consider scientific (Dupré 1993, Pigliucci 2010, 2013). One gets the impression that the demarcation task is being left to committees investigating allegations of poor science or fraud. They are forced to articulate what to count as fraud, as bad statistics, or as mere questionable research practices (QRPs). People's careers depend on their ruling: they have "skin in the game," as Nassim Nicholas Taleb might say (2018). The best one I know – the committee investigating fraudster Diederik Stapel – advises making philosophy of science a requirement for researchers (Levelt Committee, Noort Committee, and Drenth Committee 2012). So let's not tell them philosophers haven given up on it.

Diederik Stapel. A prominent social psychologist "was found to have committed a serious infringement of scientific integrity by using fictitious data in his publications" (Levelt Committee 2012, p. 7). He was required to retract 58 papers, relinquish his university degree and much else. The authors of the report describe walking into a culture of confirmation and verification bias. They could scarcely believe their ears when people they interviewed "defended the serious and less serious violations of proper scientific method with the words: that is what I have learned in practice; everyone in my research environment does the same, and so does everyone we talk to at international conferences" (ibid., p. 48). Free of the qualms that give philosophers of science cold feet, they advance some obvious yet crucially important rules with Popperian echoes:

> One of the most fundamental rules of scientific research is that an investigation must be designed in such a way that facts that might refute the research hypotheses are given at least an equal chance of emerging as do facts that confirm the research hypotheses. Violations of this fundamental rule, such as continuing to repeat an experiment until it works as desired, or excluding unwelcome experimental subjects or results, inevitably tend to confirm the researcher's research hypotheses, and essentially render the hypotheses immune to the facts. (ibid., p. 48)

Exactly! This is our minimal requirement for evidence: If it's so easy to find agreement with a pet theory or claim, such agreement is bad evidence, no test, BENT. To scrutinize the scientific credentials of an inquiry is to determine if there was a serious attempt to detect and report errors and biasing selection effects. We'll meet Stapel again when we reach the temporary installation on the upper level: The Replication Revolution in Psychology.

The issue of demarcation (point (1)) is closely related to Popper's conjecture and refutation (point (2)). While he regards a degree of dogmatism to be necessary before giving theories up too readily, the trial and error methodology "gives us a chance to survive the elimination of an inadequate hypothesis –

when a more dogmatic attitude would eliminate it by eliminating us" (Popper 1962, p. 52). Despite giving lip service to testing and falsification, many popular accounts of statistical inference do not embody falsification – even of a statistical sort.

Nearly everyone, however, now accepts point (3), that observations are not just "given" – knocking out a crucial pillar on which naïve empiricism stood. To the question: What came first, hypothesis or observation? Popper answers, another hypothesis, only lower down or more local. Do we get an infinite regress? No, because we may go back to increasingly primitive theories and even, Popper thinks, to an inborn propensity to search for and find regularities (ibid., p. 47). I've read about studies appearing to show that babies are aware of what is statistically unusual. In one, babies were shown a box with a large majority of red versus white balls (Xu and Garcia 2008, Gopnik 2009). When a succession of white balls are drawn, one after another, with the contents of the box covered with a screen, the babies looked longer than when the more common red balls were drawn. I don't find this far-fetched. Anyone familiar with preschool computer games knows how far toddlers can get in solving problems without a single word, just by trial and error.

Greater Content, Greater Severity. The position people are most likely to take a pass on is (4), his view of the role of probability. Yet Popper's central intuition is correct: if we wanted highly probable claims, scientists would stick to low-level observables and not seek generalizations, much less theories with high explanatory content. In this day of fascination with Big Data's ability to predict what book I'll buy next, a healthy Popperian reminder is due: humans also want to understand and to explain. We want bold "improbable" theories. I'm a little puzzled when I hear leading machine learners praise Popper, a realist, while proclaiming themselves fervid instrumentalists. That is, they hold the view that theories, rather than aiming at truth, are just instruments for organizing and predicting observable facts. It follows from the success of machine learning, Vladimir Cherkassky avers, that "realism is not possible." This is very quick philosophy! ". . . [I]n machine learning we are given a set of [random] data samples, and the goal is to select the best model (function, hypothesis) from a given set of possible models" (Cherkassky 2012). Fine, but is the background knowledge required for this setup itself reducible to a prediction–classification problem? I say no, as would Popper. Even if Cherkassky's problem is relatively theory free, it wouldn't follow this is true for all of science. Some of the most impressive "deep learning" results in AI have been criticized for lacking the ability to generalize beyond observed "training" samples, or to solve open-ended problems (Gary Marcus 2018).

A valuable idea to take from Popper is that probability in learning attaches to a method of conjecture and refutation, that is to testing: it is *methodological probability*. An error probability is a special case of a methodological probability. We want methods with a high probability of teaching us (and machines) how to distinguish approximately correct and incorrect interpretations of data, even leaving murky cases in the middle, and how to advance knowledge of detectable, while strictly unobservable, effects.

The choices for probability that we are commonly offered are stark: "in here" (beliefs ascertained by introspection) or "out there" (frequencies in long runs, or chance mechanisms). This is the "epistemology" versus "variability" shoehorn we reject (Souvenir D). To qualify the method by which H was tested, frequentist performance is necessary, but it's not sufficient. The assessment must be relevant to ensuring that claims have been put to severe tests. You can talk of a test having a type of *propensity* or capability to have discerned flaws, as Popper did at times. A highly explanatory, high-content theory, with interconnected tentacles, has a higher probability of having flaws discerned than low-content theories that do not rule out as much. Thus, when the bolder, higher content, theory stands up to testing, it may earn higher overall severity than the one with measly content. That a theory is plausible is of little interest, in and of itself; what matters is that it is *im*plausible for it to have passed these tests were it false or incapable of adequately solving its set of problems. It is the fuller, unifying, theory developed in the course of solving interconnected problems that enables severe tests.

Methodological probability is not to quantify my beliefs, but neither is it about a world I came across without considerable effort to beat nature into line. Let alone is it about a world-in-itself which, by definition, can't be accessed by us. Deliberate effort and ingenuity are what allow me to ensure I shall come up against a brick wall, and be forced to reorient myself, at least with reasonable probability, when I test a flawed conjecture. The capabilities of my tools to uncover mistaken claims (its error probabilities) are real properties of the tools. Still, they are *my* tools, specially and imaginatively designed. If people say they've made so many judgment calls in building the inferential apparatus that what's learned cannot be objective, I suggest they go back and work some more at their experimental design, or develop better theories.

Falsification Is Rarely Deductive. It is rare for any interesting scientific hypotheses to be logically falsifiable. This might seem surprising given all the applause heaped on falsifiability. For a scientific hypothesis H to be deductively falsified, it would be required that some observable result taken together with H yields a logical contradiction (A & ~A). But the only theories that deductively prohibit observations are of the sort one mainly finds in philosophy books: All swans are white is falsified by a single non-white swan. There are some statistical claims and contexts, I will argue, where it's possible to achieve or come close to deductive falsification: claims such as, these data are independent and identically distributed (IID). Going beyond a mere denial to replacing them requires more work.

However, interesting claims about mechanisms and causal generalizations require numerous assumptions (substantive and statistical) and are rarely open to deductive falsification. How then can good science be all about falsifiability? The answer is that we can erect reliable rules for falsifying claims with severity. We corroborate their denials. If your statistical account denies we can reliably falsify interesting theories, it is irrelevant to real-world knowledge. Let me draw your attention to an exhibit on a strange disease, kuru, and how it falsified a fundamental dogma of biology.

Exhibit (v): Kuru. Kuru (which means "shaking") was widespread among the Fore people of New Guinea in the 1960s. In around 3–6 months, Kuru victims go from having difficulty walking, to outbursts of laughter, to inability to swallow and death. Kuru, and (what we now know to be) related diseases, e.g., mad cow, Creutzfeldt–Jakob, and scrapie, are "spongiform" diseases, causing brains to appear spongy. Kuru clustered in families, in particular among Fore women and their children, or elderly parents. They began to suspect transmission was through mortuary cannibalism. Consuming the brains of loved ones, a way of honoring the dead, was also a main source of meat permitted to women. Some say men got first dibs on the muscle; others deny men partook in these funerary practices. What we know is that ending these cannibalistic practices all but eradicated the disease. No one expected at the time that understanding kuru's cause would falsify an established theory that only viruses and bacteria could be infectious. This "central dogma of biology" says:

H: All infectious agents have nucleic acid.

Any infectious agent free of nucleic acid would be *anomalous* for H – meaning it goes against what H claims. A separate step is required to decide when H's anomalies should count as falsifying H. There needn't be a cut-off so much as a standpoint as to when continuing to defend H becomes bad science. Prion researchers weren't looking to test the central dogma of biology, but to understand kuru and related diseases. The anomaly erupted only because kuru appeared to be transmitted by a protein alone, by changing a normal protein shape into an abnormal fold. Stanley Prusiner called the infectious protein a prion – for which he received much grief. He thought, at first, he'd made a mistake "and was puzzled when the data kept telling me that our preparations contained protein but not nucleic acid" (Prusiner 1997). The anomalous results would not go away and, eventually, were demonstrated via experimental transmission to animals. The discovery of prions led to a "revolution" in molecular biology, and Prusiner received a Nobel prize in 1997. It is *logically* possible that nucleic acid is somehow involved. But continuing to block the falsification of *H* (i.e., block the "protein only" hypothesis) precludes learning more about prion diseases, which now include Alzheimer's. (See Mayo 2014a.)

Insofar as we falsify general scientific claims, we are all methodological falsificationists. Some people say, "I know my models are false, so I'm done with the job of falsifying before I even begin." Really? That's not falsifying. Let's look at your method: always infer that *H* is false, fails to solve its intended problem. Then you're bound to infer this even when this is erroneous. Your method fails the minimal severity requirement.

Do Probabilists Falsify? It isn't obvious a probabilist desires to falsify, rather than supply a probability measure indicating disconfirmation, the opposite of a B-boost (a B-bust?), or a low posterior. Members of some probabilist tribes propose that Popper is subsumed under a Bayesian account by taking a low value of $\Pr(x|H)$ to falsify *H*. That could not work. Individual outcomes described in detail will easily have very small probabilities under *H* without being genuine anomalies for *H*. To the severe tester, this as an attempt to distract from the inability of probabilists to falsify, insofar as they remain probabilists. What about comparative accounts (Likelihoodists or Bayes factor accounts), which I also place under probabilism? Reporting that one hypothesis is more likely than the other is not to falsify anything. Royall is clear that it's wrong to even take the comparative report as evidence against one of the two hypotheses: they are not exhaustive. (Nothing turns on whether you prefer to put Likelihoodism under its own category.) Must all such accounts abandon the ability to falsify? No, they can *indirectly* falsify hypotheses by adding a methodological falsification rule. A natural candidate is to falsify *H* if its posterior probability is sufficiently low (or, perhaps, sufficiently disconfirmed). Of course, they'd need to justify the rule, ensuring it wasn't often mistaken.

The Popperian (Methodological) Falsificationist Is an Error Statistician

When is a statistical hypothesis to count as falsified? Although extremely rare events may occur, Popper notes:

such occurrences would not be physical effects, because, on account of their immense improbability, *they are not reproducible at will* . . . If, however, we find *reproducible* deviations from a macro effect . . . deduced from a probability estimate . . . then we must assume that the probability estimate is *falsified.* (Popper 1959, p. 203)

In the same vein, we heard Fisher deny that an "isolated record" of statistically significant results suffices to warrant a reproducible or genuine effect (Fisher 1935a, p. 14). Early on, Popper (1959) bases his statistical falsifying rules on Fisher, though citations are rare. Even where a scientific hypothesis is thought to be deterministic, inaccuracies and knowledge gaps involve error-laden predictions; so our methodological rules typically involve inferring a statistical hypothesis. Popper calls it a *falsifying hypothesis.* It's a hypothesis inferred in order to falsify some other claim. A first step is often to infer an anomaly is real, by falsifying a "due to chance" hypothesis.

The recognition that we need methodological rules to warrant falsification led Popperian Imre Lakatos to dub Popper's philosophy "methodological falsificationism" (Lakatos 1970, p. 106). If you look at this footnote, where Lakatos often buried gems, you read about "the philosophical basis of some of the most interesting developments in modern statistics. The Neyman–Pearson approach rests completely on methodological falsificationism" (ibid., p. 109, note 6). Still, neither he nor Popper made explicit use of N-P tests. Statistical hypotheses are the perfect tool for "falsifying hypotheses." However, this means you can't be a falsificationist and remain a strict deductivist. When statisticians (e.g., Gelman 2011) claim they are deductivists like Popper, I take it they mean they favor a testing account like Popper, rather than inductively building up probabilities. The falsifying hypotheses that are integral for Popper also necessitate an evidence-transcending (inductive) statistical inference.

This is hugely problematic for Popper because being a strict Popperian means never having to justify a claim as true or a method as reliable. After all, this was part of Popper's escape from induction. The problem is this: Popper's account rests on severe tests, tests that would probably falsify claims if false, but he cannot warrant saying a method is probative or severe, because that would mean it was reliable, which makes Popperians squeamish. It would appear to concede to his critics that Popper has a "whiff of induction" after all. But it's not inductive enumeration. Error statistical methods (whether from statistics or informal) can supply the severe tests Popper sought. This leads us to Pierre Duhem, physicist and philosopher of science.

Duhemian Problems of Falsification

Consider the simplest form of deductive falsification: If H entails observation O, and we observe $\sim O$, then we infer $\sim H$. To infer $\sim H$ is to infer H is false, or there is some discrepancy in what H claims about the phenomenon in question. As with any argument, in order to *detach* its conclusion (without which there is no *inference*), the premises must be true or approximately true. But O is derived only with the help of various additional claims. In statistical contexts, we may group these under two umbrellas: auxiliary factors linking substantive and statistical claims, A_1 & $\cdots$ & A_n, and assumptions of the statistical model E_1 & $\cdots$ & E_k. You are to imagine a great big long conjunction of factors, in the following argument:

1. If H & $(A_1$ & $\cdots$ & $A_n)$ & $(E_1$ & $\cdots$ & $E_k)$, then O.
2. $\sim O$.
3. Therefore, either $\sim H$ or $\sim A_1$ or ... or $\sim A_n$ or $\sim E_1$ or ... or $\sim E_k$.

This is an instance of deductively valid *modus tollens*. The catchall $\sim H$ itself is an exhaustive list of alternatives. This is too ugly for words. Philosophers, ever appealing to logic, often take this as the entity facing scientists who are left to fight their way through a great big disjunction: either H or one (or more) of the assumptions used in deriving observation claim O is to blame for anomaly $\sim O$.

When we are faced with an anomaly for H, Duhem argues, "The only thing the experiment teaches us is . . . there is at least one error; but where this error lies is just what it does not tell us" (Duhem 1954, p. 185). *Duhem's problem* is the problem of pinpointing what is warranted to blame for an observed anomaly with a claim H.

Bayesian philosophers deal with Duhem's problem by assigning each of the elements used to derive a prediction a prior probability. Whether H itself, or one of the A_i or E_k, is blamed is a matter of their posterior probabilities. Even if a failed prediction lowers the probability of hypothesis H, its posterior probability may still remain high, while the probability in A_{16}, say, drops down. The trouble is that one is free to tinker around with these assignments so that an auxiliary is blamed, and a main hypothesis H retained, or the other way around. Duhem's problem is what's really responsible for the anomaly (Mayo 1997a) – what's *warranted* to blame. On the other hand, the Bayesian approach is an excellent way to formally reconstruct Duhem's position. In his view, different researchers may choose to restore consistency according to their beliefs or to what Duhem called good sense, "bon sens." Popper was allergic to such a thing.

How can Popper, if he is really a deductivist, solve Duhem in order to falsify? At best he'd subject each of the conjuncts to as stringent a test as possible, and falsify accordingly. This still leaves, Popper admits, a disjunction of non-falsified hypotheses (he thought infinitely many)! Popperian philosophers of science advise you to choose a suitable overall package of hypotheses, assumptions, auxiliaries, on a set of criteria: simplicity, explanatory power, unification and so on. There's no agreement on which, nor how to define them. On this view, you can't really solve Duhem, you accept or "prefer" (as Popper said) the large-scale research program or paradigm as a whole. It's intended to be an advance over *bon sens* in blocking certain types of tinkering (see Section 2.4). There's a remark in the Popper museum display I only recently came across:

[W]e can be reasonably successful in attributing our refutations to definite portions of the theoretical maze. (For we *are* reasonably successful in this – a fact which must remain inexplicable for one who adopts Duhem's and Quine's view on the matter.) (1962, p. 243)

That doesn't mean he supplied an account for such attributions. He should have, but did not. There is a tendency to suppose Duhem's problem, like demarcation and induction, is insoluble and that it's taboo to claim to solve it. Our journey breaks with these taboos.

We should reject these formulations of Duhem's problem, starting with the great big conjunction in the antecedent of the conditional. It is vintage "rational reconstruction" of science, a very linear but straight-jacketed way to view the problem. Falsifying the central dogma of biology (infection requires nucleic acid) involved no series of conjunctions from H down to observations, but moving from *the bottom up*, as it were. The first clues that no nucleic acids were involved came from the fact that prions are not eradicated with techniques known to kill viruses and bacteria (e.g., UV irradiation, boiling, hospital disinfectants, hydrogen peroxide, and much else). If it were a mistake to regard prions as having no nucleic acid, then at least one of these known agents would have eradicated it. Further, prions are deactivated with substances known to kill proteins. Post-positive philosophers of science, many of them, are right to say philosophers need to pay more attention to experiments (a trend I call the New Experimentalism), but this must be combined with an account of statistical inference.

Frequentist statistics "allows interesting parts of a complicated problem to be broken off and solved separately" (Efron 1986, p. 4). We invent methods that take account of the effect of as many unknowns as possible, perhaps randomizing the rest. I never had to affirm that each and every one of my scales worked in my weighing example, the strong argument from coincidence lets me rule out, with severity, the possibility that accidental errors were producing precisely the same artifact in each case. Duhem famously compared the physicist to the doctor, as opposed to the watchmaker who can pull things apart. But the doctor may determine what it would be like if such and such were operative and *distinguish* the effects of different sources. The effect of violating an assumption of a constant mean looks very different from a changing variance; despite all the causes of a sore throat, strep tests are quite reliable. Good research should at least be able to embark on inquiries to solve their Duhemian problems.

Popper Comes Up Short. Popper's account rests on severe tests, tests that would probably have falsified a claim if false, but he cannot warrant saying any such thing. High corroboration, Popper freely admits, is at most a report on past successes with little warrant for future reliability.

Although Popper's work is full of exhortations to put hypotheses through the wringer, to make them "suffer in our stead in the struggle for the survival of the fittest" (Popper 1962, p. 52), the tests Popper sets out are white-glove affairs of logical analysis . . . it is little wonder that they seem to tell us only that there is an error somewhere and that they are silent about its source. We have to become shrewd inquisitors of errors, interact with them, simulate them (with models and computers), amplify them: we have to learn to make them talk. (Mayo 1996, p. 4)

Even to falsify non-trivial claims – as Popper grants – requires grounds for inferring a reliable effect. Singular observation statements will not do. We need "lift-off." Popper never saw how to solve the problem of "drag down" wherein empirical claims are only as reliable as the data involved in reaching them (Excursion 1). We cannot just pick up his or any other past account. Yet there's no reason to be hamstrung by the limits of the logical positivist or empiricist era. Scattered measurements are not of much use, but with adequate data massaging and averaging we can estimate a quantity of interest far more accurately than individual measurements. Recall Fisher's "it should never be true" in Exhibit (iii), Section 2.1. Fisher and Neyman–Pearson were ahead of Popper here (as was Peirce). When Popper wrote me "I regret not studying statistics," my thought was "not as much as I do."

Souvenir E: An Array of Questions, Problems, Models

It is a fundamental contribution of modern mathematical statistics to have recognized the explicit need of a model in analyzing the significance of experimental data. (Suppes 1969, p. 33)

Our framework cannot abide by oversimplifications of accounts that blur statistical hypotheses and research claims, that ignore assumptions of data or limit the entry of background information to any one portal or any one form. So what do we do if we're trying to set out the problems of statistical inference? I appeal to a general account (Mayo 1996) that builds on Patrick Suppes' (1969) idea of a hierarchy of models between models of data, experiment, and theory. Trying to cash out a full-blown picture of inquiry that purports to represent all

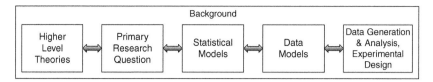

Figure 2.1 Array of questions, problems, models.

contexts of inquiry is a fool's errand. Or so I discovered after many years of trying. If one is not to land in a Rube Goldberg mess of arrows and boxes, only to discover it's not pertinent to every inquiry, it's best to settle for pigeonholes roomy enough to organize the interconnected pieces of a given inquiry as in Figure 2.1.

Loosely, there's an inferential move from the data model to the primary claim or question via the statistical test or inference model. Secondary questions include a variety of inferences involved in generating and probing conjectured answers to the primary question. A sample: How might we break down a problem into one or more local questions that can be probed with reasonable severity? How should we generate and model raw data, put them in canonical form, and check their assumptions? Remember, we are using "tests" to encompass probing any claim, including estimates. It's standard to distinguish "confirmatory" and "exploratory" contexts, but each is still an inferential or learning problem, although criteria for judging the solutions differ. In explorations, we may simply wish to infer that a model is worth developing further, that another is wildly off target.

Souvenir F: Getting Free of Popperian Constraints on Language

Popper allows that anyone who wants to define induction as the procedure of corroborating by severe testing is free to do so; and I do. Free of the bogeyman that induction must take the form of a probabilism, let's get rid of some linguistic peculiarities inherited by current-day Popperians (critical rationalists). They say things such as: it is *warranted* to infer (prefer or believe) H (because H has passed a severe test), but there is no *justification* for H (because "justifying" H would mean H was true or highly probable). In our language, if H passes a severe test, you can say it is warranted, corroborated, justified – along with whatever qualification is appropriate. I tend to use "warranted." The Popperian "hypothesis H is corroborated by data x" is such a tidy abbreviation of "H has passed a severe test with x" that we may use the two interchangeably. I've already co-opted Popper's description of science as *problem solving*. A hypothesis can be seen as a potential solution to a problem (Laudan 1978). For example, the theory of protein folding purports to solve the problem of how pathological prions are transmitted. The problem might be to explain, to predict, to unify, to suggest new problems, etc. When we severely probe, it's not for falsity per se, but to investigate if a problem has been adequately solved by a model, method, or theory.

In rejecting probabilism, there is nothing to stop us from speaking of believing in *H*. It's not the direct output of a statistical inference. A post-statistical inference might be to believe a severely tested claim; disbelieve a falsified one. There are many different grounds for believing something. We may be tenacious in our beliefs in the face of given evidence; they may have other grounds, or be prudential. By the same token, talk of deciding to conclude, infer, prefer, or act can be fully epistemic in the sense of assessing evidence, warrant, and well-testedness. Popper, like Neyman and Pearson, employs such language because it allows talking about inference distinct from assigning probabilities to hypotheses. Failing to recognize this has created unnecessary combat.

Live Exhibit (vi): Revisiting Popper's Demarcation of Science. Here's an experiment: try shifting what Popper says about theories to a related claim about inquiries to find something out. To see what I have in mind, let's listen to an exchange between two fellow travelers over coffee at Statbucks.

TRAVELER 1:   If mere logical falsifiability suffices for a theory to be scientific, then, we can't properly oust astrology from the scientific pantheon. Plenty of nutty theories have been falsified, so by definition they're scientific. Moreover, scientists aren't always looking to subject well-corroborated theories to "grave risk" of falsification.

TRAVELER 2:   I've been thinking about this. On your first point, Popper confuses things by making it sound as if he's asking: *When is a theory unscientific?* What he is actually asking or should be asking is: *When is an inquiry into a theory, or an appraisal of claim H, unscientific?* We want to distinguish meritorious modes of inquiry from those that are BENT. If the test methods enable ad hoc maneuvering, sneaky face-saving devices, then the inquiry – the handling and use of data – is unscientific. Despite being logically falsifiable, theories can be rendered immune from falsification by means of cavalier methods for their testing. Adhering to a falsified theory no matter what is poor science. Some areas have so much noise and/or flexibility that they can't or won't distinguish warranted from unwarranted explanations of failed predictions. Rivals may find flaws in one another's inquiry or model, but the criticism is not constrained by what's actually responsible. This is another way inquiries can become unscientific.[1]

---

[1] For example, astronomy, but not astrology, can reliably solve its Duhemian puzzles. Chapter 2, Mayo (1996), following my reading of Kuhn (1970) on "normal science."

She continues:

> On your second point, it's true that Popper talked of wanting to subject theories to grave risk of falsification. I suggest that it's really our *inquiries* into, or tests of, the theories that we want to subject to grave risk. The onus is on interpreters of data to show how they are countering the charge of a poorly run test. I admit this is a modification of Popper. One could reframe the entire demarcation problem as one of the characters of an inquiry or test.

She makes a good point. In addition to blocking inferences that fail the minimal requirement for severity:

> *A scientific inquiry or test: must be able to embark on a reliable probe to pinpoint blame for anomalies (and use the results to replace falsified claims and build a repertoire of errors).*

The parenthetical remark isn't absolutely required, but is a feature that greatly strengthens scientific credentials. Without solving, not merely embarking on, some Duhemian problems there are no interesting falsifications. The ability or inability to pin down the source of failed replications – a familiar occupation these days – speaks to the scientific credentials of an inquiry. At any given time, even in good sciences there are anomalies whose sources haven't been traced – unsolved Duhemian problems – generally at "higher" levels of the theory-data array. Embarking on solving these is the impetus for new conjectures. Checking test assumptions is part of working through the Duhemian maze. The reliability requirement is: infer claims just to the extent that they pass severe tests. There's no sharp line for demarcation, but when these requirements are absent, an inquiry veers into the realm of questionable science or pseudoscience. Some physicists worry that highly theoretical realms can't be expected to be constrained by empirical data. Theoretical constraints are also important. We'll flesh out these ideas in future tours.

2.4 Novelty and Severity

> When you have put a lot of ideas together to make an elaborate theory, you want to make sure, when explaining what it fits, that those things it fits are not just the things that gave you the idea for the theory; but that the finished theory makes something else come out right, in addition. (Feynman 1974, p. 385)

This "something else that must come out right" is often called a "novel" predictive success. Whether or not novel predictive success is required is a very old battle that parallels debates between frequentists and inductive logicians, in both statistics and philosophy of science, for example, between Mill and Peirce and Popper and Keynes. Walking up the ramp from the ground floor to the gallery of Statistics, Science, and Pseudoscience, the novelty debate is used to intermix Popper and statistical testing.

When Popper denied we can capture severity formally, he was reflecting an astute insight: there is a tension between the drive for a logic of confirmation and our strictures against practices that lead to poor tests and ad hoc hypotheses. Adhering to the former downplays or blocks the ability to capture the latter, which demands we go beyond the data and hypotheses. Imre Lakatos would say we need to know something about the *history* of the hypothesis: how was it developed? Was it the result of deliberate and ad hoc attempts to spare one's theory from refutation? Did the researcher continue to adjust her theory in the face of an anomaly or apparent discorroborating result? (He called these "exception incorporations".) By contrast, the confirmation theorist asks: why should it matter how the hypothesis inferred was arrived at, or whether data-dependent selection effects were operative? When holders of the Likelihood Principle (LP) wonder why data can't speak for themselves, they're echoing the logical empiricist (Section 1.4). Here's Popperian philosopher Alan Musgrave:

> According to modern logical empiricist orthodoxy, in deciding whether hypothesis *h* is confirmed by evidence *e*, ... we must consider only the statements *h* and *e*, and the logical relations between them. It is quite irrelevant whether *e* was known first and *h* proposed to explain it, or whether *e* resulted from testing predictions drawn from *h*. (Musgrave 1974, p. 2)

John Maynard Keynes likewise held that the "... question as to whether a particular hypothesis happens to be propounded before or after examination of [its instances] is quite irrelevant (Keynes 1921/1952, p. 305). Logics of confirmation ran into problems because they insisted on purely formal or syntactical criteria of confirmation that, like deductive logic, "should contain no reference to the specific subject-matter" (Hempel 1945, p. 9) in question. The Popper–Lakatos school attempts to avoid these shortcomings by means of novelty requirements:

> *Novelty Requirement*: For data to warrant a hypothesis *H* requires not just that (i) *H* agree with the data, but also (ii) the data should be novel or surprising or the like.

For decades Popperians squabbled over how to define novel predictive success. There's (1) *temporal novelty* – the data were not already available before the hypothesis was erected (Popper, early); (2) *theoretical novelty* – the data were not already predicted by an existing hypothesis (Popper, Lakatos); and (3) *use-*

novelty – the data were not used to construct or select the hypothesis. Defining novel success is intimately linked to defining Popperian severity.

Temporal novelty is untenable: known data (e.g., the perihelion of Mercury, anomalous for Newton) are often strong evidence for theories (e.g., GTR). Popper ultimately favored theoretical novelty: *H* passes a severe test with *x*, when *H* entails *x*, and *x* is theoretically novel – according to a letter he sent me. That, of course, freed me to consider my own notion as distinct. (We replace "entails" with something like "accords with.") However, as philosopher John Worrall (1978, pp. 330–1) shows, to require theoretical novelty prevents passing *H* with severity, so long as there's already a hypothesis that predicts the data or phenomenon *x* (it's not clear which). Why should the first hypothesis that explains *x* be better tested?

I take the most promising notion of novelty to be a version of use-novelty: *H* passes a test with data *x* severely, so long as *x* was not used to construct *H* (Worrall 1989). Data can be known, so long as they weren't used in building *H*, presumably to ensure *H* accords with *x*. While the idea is in sync with the error statistical admonishment against "peeking at the data" and finding your hypothesis in the data – it's far too vague as it stands. Watching this debate unfold in philosophy, I realized none of the notions of novelty were either sufficient or necessary for a good test (Mayo 1991).

You will notice that statistical researchers go out of their way to state a prediction at the start of a paper, presenting it as temporally novel, and if *H* is temporally novel, it also satisfies use-novelty. If *H* came first, the data could not have been used to arrive at *H*. This stricture is desirable, but to suppose it suffices for a good test grows out of a discredited empiricist account where the data are *given* rather than the product of much massaging and interpretation. There is as much opportunity for bias to arise in interpreting or selectively reporting results, with a known hypothesis, as there is in starting with data and artfully creating a hypothesis. Nor is violating use-novelty a matter of the implausibility of *H*. On the contrary, popular psychology thrives by seeking to explain results by means of hypotheses expected to meet with approval, at least in a given political tribe. Preregistration of the detailed protocol is supposed to cure this. We come back to this.

Should use-novelty be *necessary* for a good test? Is it ever okay to use data to arrive at a hypothesis *H* as well as to test *H* – even if the data use ensures agreement or disagreement with *H*? The answers, I say, are no and yes, respectively. Violations of use-novelty need not be pejorative. A trivial example: count all the people in the room and use it to fix the parameter of the number in the room. Or, less trivially, think of confidence intervals: we use the data to form the interval estimate. The estimate is really a hypothesis about the value of the parameter. The same data warrant the hypothesis constructed! Likewise, using the same data to arrive at and test assumptions of statistical models can be entirely reliable. What matters is not novelty, in any of the senses, but severity in the error statistical sense. Even where our intuition is to prohibit use-novelty violations, the requirement is murky. We should instead consider specific ways that severity can be violated. Let's define:

> *Biasing Selection Effects*: when data or hypotheses are selected or generated (or a test criterion is specified), in such a way that the minimal severity requirement is violated, seriously altered, or incapable of being assessed.[2]

Despite using this subdued label, it's too irresistible to banish entirely a cluster of colorful terms for related gambits – double-counting, cherry picking, fishing, hunting, significance seeking, searching for the pony, trying and trying again, data dredging, monster barring, look elsewhere effect, and many others besides – unless we're rushing. New terms such as *P*-hacking are popular, but don't forget that these crooked practices are very old.[3]

To follow the Popper–Lakatos school (although entailment is too strong):

> *Severity Requirement:* for data to warrant a hypothesis *H* requires not just that
> (S-1) *H* agrees with the data (*H* passes the test), but also
> (S-2) with high probability, *H* would not have passed the test so well, were *H* false.

This describes corroborating a claim, it is "strong" severity. Weak severity denies *H* is warranted if the test method would probably have passed *H* even if false. While severity got its start in this Popperian context, in future excursions, we will need more specifics to describe both clauses (S-1) and (S-2).

2.5 Fallacies of Rejection and an Animal Called NHST

One of Popper's prime examples of non-falsifiable sciences was Freudian and Adlerian psychology, which gave psychologist Paul Meehl conniptions

---

[2] As noted earlier, I follow Ioannidis in using bias this way, in speaking of selections.
[3] For a discussion of novelty and severity in philosophy of science, see Chapter 8 of Mayo (1996). Worrall and I have engaged in a battle over this in numerous places (Mayo 2010d, Worrall 1989, 2010). Related exchanges include Mayo 2008, Hitchcock and Sober 2004.

because he was a Freudian as well as a Popperian. Meehl castigates Fisherian significance tests for providing a sciency aura to experimental psychology, when they seem to violate Popperian strictures: "[T]he almost universal reliance on merely refuting the null hypothesis as the standard method for corroborating substantive theories in the soft areas . . . is basically unsound, poor scientific strategy . . ." (Meehl 1978, p. 817). Reading Meehl, Lakatos wrote, "one wonders whether the function of statistical techniques in the social sciences is not primarily to provide a machinery for producing phoney corroborations and . . . an increase in pseudo-intellectual garbage" (Lakatos 1978, pp. 88–9, note 4).

Now Meehl is a giant when it comes to criticizing statistical practice in psychology, and a good deal of what contemporary critics are on about was said long ago by him. He's wrong, though, to pin the blame on "Sir Ronald" (Fisher). Corroborating substantive theories merely by means of refuting the null? Meehl may be describing what is taught and permitted in the "soft sciences," but the practice of moving from statistical to substantive theory violates the testing methodologies of both Fisher and Neyman–Pearson. I am glad to see Gerd Gigerenzer setting the record straight on this point, given how hard he, too, often is on Fisher:

It should be recognized that, according to Fisher, rejecting the null hypothesis is not equivalent to accepting the efficacy of the cause in question. The latter cannot be established on the basis of one single experiment, but requires obtaining more significant results when the experiment, or an improvement of it, is repeated at other laboratories or under other conditions. (Gigerenzer et al. 1989, pp. 95–6)

According to Gigerenzer et al., "careless writing on Fisher's part, combined with selective reading of his early writings has led to the identification of the two, and has encouraged the practice of demonstrating a phenomenon on the basis of a single statistically significant result" (ibid., p. 96). I don't think Fisher can be accused of carelessness here; he made two crucial clarifications, and the museum display case bears me out. The first is that "[W]e need, not an isolated record, but a reliable method of procedure" (Fisher 1935a, p. 14), from Excursion 1. The second is Fisher's requirement that even a genuine statistical effect H fails to warrant a substantive research hypothesis H^*. Using "$\not\Rightarrow$" to abbreviate "does not imply": $H \not\Rightarrow H^*$.

Here's David Cox defining significance tests over 40 years ago:

. . . we mean by a significance test a procedure for measuring the consistency of data with a null hypothesis . . . there is a function $d = d(y)$ of the observations, called a test statistic, and such that the larger is $d(y)$ the stronger is the inconsistency of y with H_0, in the respect under study . . . we need to be able to compute, at least approximately,

$$p_{obs} = \Pr(d \geq d(y_{obs}); H_0)$$

called the observed level of significance [or P-value].

Application of the significance test consists of computing approximately the value of p_{obs} and using it as a summary measure of the degree of consistency with H_0, in the respect under study. (Cox 1977, p. 50; replacing t with d)

Statistical test requirements follow non-statistical tests, Cox emphasizes, though at most H_0 entails some results with high probability. Say 99% of the time the test would yield $\{d < d_0\}$, if H_0 adequately describes the data-generating mechanism where d_0 abbreviates $d(x_0)$. Observing $\{d \geq d_0\}$ indicates inconsistency with H_0 in the respect tested. (Implicit alternatives, Cox says, "lurk in the undergrowth," given by the test statistic.) So significance tests reflect statistical *modus tollens*, and its reasoning follows severe testing – BUT, an isolated low P-value won't suffice to infer a genuine effect, let alone a research claim. Here's a list of *fallacies of rejection*.

1. The reported (nominal) statistical significance result is *spurious* (it's not even an actual P-value). This can happen in two ways: biasing selection effects, or violated assumptions of the model.
2. The reported statistically significant result is genuine, but it's an isolated effect not yet indicative of a genuine experimental phenomenon. (Isolated low P-value $\not\Rightarrow$ H: statistical effect.)
3. There's evidence of a genuine statistical phenomenon but either (i) the magnitude of the effect is less than purported, call this a *magnitude error*,[4] or (ii) the substantive interpretation is unwarranted ($H \not\Rightarrow H^*$).

I will call an *audit* of a P-value, a check of any of these concerns, generally in order, depending on the inference. That's why I place the background information for auditing throughout our "series of models" representation (Figure 2.1). Until audits are passed, the relevant statistical inference is to be reported as "unaudited." Until 2 is ruled out, it's a mere "indication," perhaps, in some settings, grounds to get more data.

Meehl's criticism is to a violation described in 3(ii). Like many criticisms of significance tests these days, it's based on an animal that goes by the acronym NHST (null hypothesis significance testing). What's wrong with NHST in relation to Fisherian significance tests? The museum label says it for me:

---

[4] This is the term used by Andrew Gelman.

If NHST permits going from a single small P-value to a genuine effect, it is illicit; and if it permits going directly to a substantive research claim it is doubly illicit!

We can add: if it permits biasing selection effects it's triply guilty. Too often NHST refers to a monster describing highly fallacious uses of Fisherian tests, introduced in certain social sciences. I now think it's best to drop the term NHST. Statistical tests will do, although our journey requires we employ the terms used in today's battles.

Shall we blame the wise and sagacious Meehl with selective reading of Fisher? I don't know. Meehl gave me the impression that he was irked that using significance tests seemed to place shallow areas of psychology on a firm falsification footing; whereas, more interesting, deep psycho-analytic theories were stuck in pseudoscientific limbo. He and Niels Waller give me the honor of being referred to in the same breath as Popper and Salmon:

For the corroboration to be strong, we have to have 'Popperian risk', … 'severe test' [as in Mayo], or what philosopher Wesley Salmon called a *highly improbable coincidence* ["damn strange coincidence"]. (Meehl and Waller 2002, p. 284)

Yet we mustn't blur an argument from coincidence merely to a real effect, and one that underwrites arguing from coincidence to research hypothesis H^*. Meehl worried that, by increasing the sample size, trivial discrepancies can lead to a low P-value, and using NHST, evidence for H^* too readily attained. Yes, if you plan to perform an illicit inference, then whatever makes the inference easier (increasing sample size) is even more illicit. Since proper statistical tests block such interpretations, there's nothing anti-Popperian about them.

The fact that selective reporting leads to unreplicable results is an *asset* of significance tests: If you obtained your apparently impressive result by violating Fisherian strictures, preregistered tests will give you a much deserved hard time when it comes to replication. On the other hand, evidence of a statistical effect H does give a B-boost to H^*, since if H^* is true, a statistical effect follows (statistical affirming the consequent).

Meehl's critiques rarely mention the methodological falsificationism of Neyman and Pearson. Why is the field that cares about power – which is defined in terms of N-P tests – so hung up on simple significance tests? We'll disinter the answer later on. With N-P tests, the statistical alternative to the null hypothesis is made explicit: the null and alternative exhaust the possibilities. There can be no illicit jumping of levels from statistical to causal (from H to H^*). Fisher didn't allow jumping either, but he was less explicit. Statistically significant increased yields in Fisher's controlled trials on fertilizers, as Gigerenzer notes, are intimately linked to a causal alternative. If the fertilizer does not increase yield (H^* is false, so $\sim H^*$ is true), then no statistical increase is expected, if the test is run well.[5] Thus, finding statistical increases (rejecting H_0) is grounds to falsify $\sim H^*$ and find evidence of H^*. Unlike the typical psychology experiment, here rejecting a statistical null very nearly warrants a statistical causal claim. If you want a statistically significant effect to (statistically) warrant H^* show:

> If $\sim H^*$ is true (research claim H^* is false), then H_0 won't be rejected as inconsistent with data, at least not regularly.

Psychology should move to an enlightened reformulation of N-P and Fisher (see Section 3.3). To emphasize the Fisherian (null hypothesis only) variety, we follow the literature in calling them "simple" significance tests. They are extremely important in their own right: They are the basis for testing assumptions without which statistical methods fail scientific requirements. View them as just one member of a panoply of error statistical methods.

Statistics Can't Fix Intrinsic Latitude. The problem Popper found with Freudian and Adlerian psychology is that any observed behavior could be readily interpreted through the tunnel of either theory. Whether a man jumped in the water to save a child, or if he failed to save her, you can invoke Adlerian inferiority complexes, or Freudian theories of sublimation or Oedipal complexes (Popper 1962, p. 35). Both Freudian and Adlerian theories can explain whatever happens. This latitude has nothing to do with statistics. As we learned from Exhibit (vi), Section 2.3, we should really speak of the latitude offered by the overall inquiry: research question, auxiliaries, and interpretive rules. If it has self-sealing facets to account for any data, then it fails to probe with severity. Statistical methods cannot fix this. Applying statistical methods is just window dressing. Notice that Freud/Adler, as Popper describes them, are amongst the few cases where the latitude really is part of the theory or terminology. It's not obvious that Popper's theoretical novelty bars this, unless one of Freud/Adler is deemed first. We've arrived at the special topical installation on:

---

[5] Gigerenzer calls such a "no increase" hypothesis the substantive null hypothesis.

2.6   The Reproducibility Revolution (Crisis) in Psychology

> I was alone in my tastefully furnished office at the University of Groningen. . . . I opened the file with the data that I had entered and changed an unexpected 2 into a 4; then, a little further along, I changed a 3 into a 5. . . . When the results are just not quite what you'd so badly hoped for; when you know that that hope is based on a thorough analysis of the literature; . . . then, surely, you're entitled to adjust the results just a little? . . . I looked at the array of data and made a few mouse clicks to tell the computer to run the statistical analyses. When I saw the results, the world had become logical again. (Stapel 2014, p. 103)

This is Diederik Stapel describing his "first time" – when he was still collecting data and not inventing them. After the Stapel affair (2011), psychologist Daniel Kahneman warned that he "saw a train wreck looming" for social psychology and called for a "daisy chain" of replication to restore credibility to some of the hardest hit areas such as priming studies (Kahneman 2012). Priming theory holds that exposure to an experience can unconsciously affect subsequent behavior. Kahneman (2012) wrote: "right or wrong, your field is now the poster child for doubts about the integrity of psychological research." One of the outgrowths of this call was the 2011–2015 Reproducibility Project, part of the Center for Open Science Initiative at the University of Virginia. In a nutshell: This is a crowd-sourced effort to systematically subject published statistically significant findings to checks of reproducibility. In 2011, 100 articles from leading psychology journals from 2008 were chosen; in August of 2015, it was announced only around 33% could be replicated (depending on how that was defined). Whatever you think of the results, it's hard not to be impressed that a field could organize such a self-critical project, obtain the resources, and galvanize serious-minded professionals to carry it out.

First, on the terminology: The American Statistical Association (2017, p. 1) calls a study "reproducible if you can take the original data and the computer code used . . . and reproduce all of the numerical findings . . ." In the case of Anil Potti, they couldn't reproduce his numbers. By contrast, replicability refers to "the act of repeating an entire study, independently of the original investigator without the use of the original data (but generally using the same methods)" (ibid.).[6] The Reproducibility Project, however, is really a replication project (as the ASA defines it). These points of terminology shouldn't affect

---

[6] This is a "direct replication," whereas a "conceptual replication" probes the same hypothesis but through a different phenomenon.

our discussion. The Reproducibility Project is appealing to what most people have in mind in saying a key feature of science is reproducibility, namely replicability.

So how does the Reproducibility Project proceed? A team of (self-selected) knowledgeable replicators, using a protocol that is ideally to be approved by the initial researchers, reruns the study on new subjects. A failed replication occurs when there's a non-statistically significant or *negative* result, that is, a *P*-value that is not small (say >0.05). Does a negative result mean the original result was a false positive? Or that the attempted replication was a false negative? The interpretation of negative statistical results is itself controversial, particularly as they tend to keep to Fisherian tests, and effect sizes are often fuzzy. When RCTs fail to replicate observational studies, the presumption is that, were the effect genuine, the RCTs would have found it. That is why they are taken as an indictment of the observational study. But here, you could argue, the replication of the earlier research is not obviously a study that checks its correctness. Yet that would be to overlook the strengthened features of the replications in the 2011 project: they are preregistered, and are designed to have high power (against observed effect sizes). What is more, they are free of some of the "perverse incentives" of usual research. In particular, the failed replications are guaranteed to be published in a collective report. They will not be thrown in file drawers, even if negative results ensue.

Some ironic consequences immediately enter in thinking about the project. The replication researchers in psychology are the same people who hypothesize that a large part of the blame for lack of replication may be traced to the reward structure: to incentives to publish surprising and sexy studies, coupled with an overly flexible methodology opening the door to promiscuous QRPs. Call this the *flexibility, rewards, and bias* hypothesis. Supposing this hypothesis is correct, as is quite plausible, what happens when non-replication becomes sexy and publishable? Might non-significance become the new significance? *Science* likely wouldn't have published individual failures to replicate, but they welcomed the splashy OSC report of the poor rate of replication they uncovered, as well as back-and-forth updates by critics. Brand new fields of meta-research open up for replication specialists, all ostensibly under the appealing banner of improving psychology. Some ask: should authors be prey to results conducted by a self-selected group – results that could obviously impinge rather unfavorably on them? Many say no and even liken the enterprise to a witch hunt. Kahneman (2014) called for "a new etiquette" requiring original authors to be consulted on protocols:

... tension is inevitable when the replicator does not believe the original findings and intends to show that a reported effect does not exist. The relationship between replicator and author is then, at best, politely adversarial. The relationship is also radically asymmetric: the replicator is in the offense, the author plays defense. The threat is one-sided because of the strong presumption in scientific discourse that more recent news is more believable. (p. 310)

It's not hard to find potential conflicts of interest and biases on both sides. There are the replicators' attitudes – not only toward the claim under study, but toward the very methodology used to underwrite it – usually statistical significance tests. Every failed replication is seen (by some) as one more indictment of the method (never minding its use in showing irreplication). There's the replicator's freedom to stop collecting data once minimal power requirements are met, and the fact that subjects – often students, whose participation is required – are aware of the purpose of the study, revealed at the end. (They are supposed to keep it confidential over the life of the experiment, but is that plausible?) On the other hand, the door may be open too wide for the original author to blame any failed replication on lack of fidelity to nuances of the original study. Lost in the melee is the question of whether any constructive criticism is emerging.

Incidentally, here's a case where it might be argued that loss and cost functions are proper, since the outcome goes beyond statistical inference to reporting a failure to replicate Jane's study, perhaps overturning her life's research.

What Might a Real Replication Revolution in Psychology Require?

Even absent such concerns, the program seems to be missing the real issues that leap out at the average reader of the reports. The replication attempts in psychology stick to what might be called "purely statistical" issues: can we get a low P-value or not? Even in the absence of statistical flaws, research conclusions may be disconnected from the data used in their testing, especially when experimentally manipulated variables serve as proxies for variables of theoretical interest. A serious (and ongoing) dispute arose when a researcher challenged the team who failed to replicate her hypothesis that subjects "primed" with feelings of cleanliness, sometimes through unscrambling soap-related words, were less harsh in judging immoral such bizarre actions as whether it is acceptable to eat your dog after it has been run over. A focus on the P-value computation ignores the larger question of the methodological adequacy of the leap from the statistical to the substantive. Is unscrambling soap-related words an adequate proxy for the intended cleanliness variable? The less said about eating your run-over dog, the better. At this point Bayesians might argue, "We know these theories are implausible, we avoid the inferences by invoking our disbeliefs." That can work in some cases, except that the researchers find them plausible, and, more than that, can point to an entire literature on related studies, say, connecting physical and moral purity or impurity (part of "embodied cognition") e.g., Schnall et al. 2008. The severe tester shifts the unbelievability assignment. What's unbelievable is supposing their experimental method provides evidence for purported effects! Some philosophers look to these experiments on cleanliness and morality, and many others, to appraise their philosophical theories "experimentally."[7] Whether or not this is an advance over philosophical argument, philosophers should be taking the lead in critically evaluating the methodology, in psychology and, now, in philosophy.

Our skepticism is not a denial that we may often use statistical tests to infer a phenomenon quite disparate from the experimental manipulations. Even an artificial lab setting can teach us about a substantive phenomenon "in the wild" so long as there are *testable implications* for the statistically modeled experiment. The famous experiments by Harlow, showing that monkeys prefer a cuddly mom to a wire mesh mom that supplies food (Harlow 1958), are perfectly capable of letting us argue from coincidence to what matters to actual monkeys. Experiments in social psychology are rarely like that.

The "replication revolution in psychology" won't be nearly revolutionary enough until they subject to testing the methods and measurements intended to link statistics with what they really want to know. If you are an ordinary skeptical reader, outside psychology, you're probably flummoxed that researchers blithely assume that role playing by students, unscrambling of words, and those long-standing 5, 7, or 10 point questionnaires are really measuring the intended psychological attributes. Perhaps it's taboo to express this. Imagine that Stapel had not simply fabricated his data, and he'd found that students given a mug of M&M's emblazoned with the word "capitalism" ate statistically significantly more candy than those with a scrambled word on their mug– as one of his make-believe studies proposed (Stapel 2014, pp. 127–8). Would you think you were seeing the effects of greed in action?

Psychometrician Joel Michell castigates psychology for having bought the operationalist Stevens' (1946, p. 667) "famous definition of measurement as

---

[7] The experimental philosophy movement should be distinguished from the New Experimentalism in philosophy.

'the assignment of numerals to objects or events according to rules'", a gambit he considers a deliberate and "pathological" ploy to deflect "attention from the issue of whether psychological attributes are quantitative" to begin with (Michell 2008, p. 9). It's easy enough to have a rule for assigning numbers on a Likert questionnaire, say on degrees of moral opprobrium (never OK, sometimes OK, don't know, always OK) if it's not required to have an independent source of its validity. (Are the distances between units really equal, as statistical analysis requires?) I prefer not to revisit studies against which it's easy to take pot shots. Here's a plausible phenomenon, confined, fortunately, to certain types of people.

Macho Men: Falsifying Inquiries

I have no doubts that certain types of men feel threatened by the success of their female partners, wives, or girlfriends – more so than the other way around. I've even known a few. Some of my female students, over the years, confide that their boyfriends were angered when they got better grades than they did! I advise them to drop the bum immediately if not sooner. The phenomenon is backed up by field statistics (e.g., on divorce and salary differentials where a woman earns more than a male spouse, Thaler 2013[8]). As we used H (the statistical hypothesis), and H^* (a corresponding causal claim), let's write this more general phenomenon as $\mathcal{H}$. Can this be studied in the lab? Ratliff and Oishi (2013) "examined the influence of a romantic partner's success or failure on one's own implicit and explicit self-esteem" (p. 688). Their statistical studies show that

> H: "men's implicit self-esteem is lower when a partner succeeds than when a partner fails." (ibid.)

To take the weakest construal, H is the statistical alternative to a "no effect" null H_0. The "treatment" is to think and write about a time their partner succeeded at something or failed at something. The effect will be a measure of "self-esteem," obtained either explicitly, by asking: "How do you feel about yourself?" or implicitly, based on psychological tests of positive word associations (with "me" versus "other"). Subjects are randomly assigned to five "treatments": think, write about a time your partner (i) succeeded, (ii) failed, (iii) succeeded when you failed, (iv) failed when you succeeded, or (v) a typical day (control) (ibid., p. 695). Here are a few of the several statistical null hypotheses

---

[8] There are some fairly strong statistics, too, of correlations between wives earning more than their husbands and divorce or marital dissatisfaction – although it is likely the disgruntlement comes from both sides.

(as no significant results are found among women, these allude to males thinking about female partners):

(a) The average implicit self-esteem is no different when subjects think about their partner succeeding (or failing) as opposed to an ordinary day.
(b) The average implicit self-esteem is no different when subjects think about their partner succeeding while the subject fails ("she does better than me").
(c) The average implicit self-esteem is no different when subjects think about their partner succeeding as opposed to failing ("she succeeds at something").
(d) The average explicit self-esteem is no different under any of the five conditions.

These statistical null hypotheses are claims about the distributions from which participants are sampled, limited to populations of experimental subjects – generally students who receive course credit. They merely assert the treated/ non-treateds can be seen to come from the same populations as regards the average effect in question.

None of these nulls are able to be statistically rejected except (c)! Each negative result is anomalous for H. Should they take the research hypothesis as disconfirmed? Or as casting doubt on their test? Or should they focus on the null hypotheses that were rejected, in particular null (c). They opt for the third, viewing their results as "demonstrating that men who thought about their romantic partner's success had lower implicit self-esteem than men who thought about their romantic partner's failure (ibid., p. 698). This is a highly careful wording. It refers only to a statistical effect, restricted to the experimental subjects. That's why I write it as H. Of course they really want to infer a causal claim – the self-esteem of males studied is negatively influenced, on average, by female partner success of some sort H^*. More than that, they'd like the results to be evidence that H^* holds in the population of men in general, and speaks to the higher level theory $\mathcal{H}$.

On the face of it, it's a jumble. We do not know if these negative results reflect negatively on a research causal hypothesis – even limited to the experimental population – or whether the implicit self-esteem measure is actually picking up on something else, or whether the artificial writing assignment is insufficiently relevant to the phenomenon of interest. The auxiliaries linking the statistical and the substantive, the audit of the P-values and the statistical assumptions – all are potential sources of blame as we cast about solving the Duhemian challenge. Things aren't clear enough to say researchers *should* have regarded their research hypothesis as disconfirmed much less falsified. This is the nub of the problem.

What Might a Severe Tester Say?

I'll let her speak:

It appears from failing to reject (a) that our "treatment" has no bearing on the phenomenon of interest. It was somewhat of a stretch to suppose that thinking about her "success" (examples given are dancing, cooking, solving an algebra problem) could really be anything like the day Ann got a raise while Mark got fired. Take null hypothesis (b). It was expected that "she beat me in X" would have a greater negative impact on self-esteem than merely, "she succeeded at X." Remember these are completely different groups of men, thinking about whatever it is they chose to. That the macho man's partner bowled well one day should have been less deflating than her getting a superior score. We confess that the non-significant difference in (b) casts a shadow on whether the intended phenomenon is being picked up at all. We could have interpreted it as supporting our research hypothesis. We could view it as lending "some support to the idea that men interpret 'my partner is successful' as 'my partner is more successful than me'" (ibid., p. 698). We could have reasoned, the two conditions show no difference because any success of hers is always construed by macho man as "she showed me up." This skirts too close to viewing the data through the theory, to a *self-sealing fallacy*. Our results lead us to question that this study is latching onto the phenomenon of interest. In fact, insofar as the general phenomenon $\mathcal{H}$ (males taking umbrage at a partner's superior performance) is plausible, it would imply no effect would be found in this artificial experiment. Thus spake the severe tester.

I want to be clear that I'm not criticizing the authors for not proceeding with the severe tester's critique; it would doubtless be considered outlandish and probably would not be accepted for publication. I deliberately looked at one of the better inquiries that also had a plausible research hypothesis. View this as a futuristic self-critical researcher.

While we're at it, are these implicit self-esteem tests off the table? Why? The authors admit that *explicit* self-esteem was unaffected (in men and women). Surely if explicit self-esteem *had* shown a significant difference, they would have reported it as support for their research hypothesis. Many psychology measurements not only lack a firm, independent check on their validity; if they disagree with more direct measurements, it is easily explained away or even taken as a point in their favor. Why do no differences show up on explicit measures of self-esteem? Available reasons: Men do not want to admit their self-esteem goes down when their partner succeeds, or they might be unaware of it. Maybe so, but this assumes what hasn't been vouchsafed. Why not revisit the subjects at a later date to compare their scores on implicit self-esteem? If we find no difference from their score under the experimental manipulation, we'd have some grounds to deny it was validly measuring the effect of the treatment.

Here's an incentive: They're finding that the replication revolution has not made top psychology journals more inclined to publish non-replications – even of effects they have published. The journals want new, sexy effects. Here's sexy: stringently test (and perhaps falsify) some of the seminal measurements or types of inquiry used in psychology. In many cases we may be able to falsify given studies. If that's not exciting enough, imagine showing some of the areas now studied admit of no robust, generalizable effects. You might say it would be ruinous to set out to question basic methodology. Huge literatures on the "well established" Macbeth effect, and many others besides, might come in for question. I said it would be revolutionary for psychology. Psychometricians are quite sophisticated, but their work appears separate from replication research. Who would want to undermine their own field? Already we hear of psychology's new "spirit of self-flaggelation" (Dominus 2017). It might be an apt job for philosophers of science, with suitable expertise, especially now that these studies are being borrowed in philosophy.[9]

A hypothesis to be considered must always be: the results point to the inability of the study to severely probe the phenomenon of interest. The goal would be to build up a body of knowledge on closing existing loopholes when conducting a type of inquiry. How do you give evidence of "sincerely trying (to find flaws)?" Show that you would find the studies poorly run, if the flaws were present. When authors point to other studies as offering replication, they should anticipate potential criticisms rather than showing "once again I can interpret my data through my favored theory." The scientific status of an inquiry is questionable if it cannot or will not distinguish the correctness of inferences from problems stemming from a poorly run study. What must be subjected to grave risk are assumptions that the experiment was well run. This should apply as well to replication projects, now under way. If the producer of the report is not sufficiently self-skeptical, then we the users must be.

Live Exhibit (vii): Macho Men. Entertainment on this excursion is mostly home grown. A reenactment of this experiment will do. Perhaps hand questionnaires to some of the males after they lose to their partners in shuffle board

---

[9] One of the failed replications was the finding that reading a passage against free will contributes to a proclivity for cheating. Both the manipulation and the measured effects are shaky – never mind any statistical issues.

or ping pong. But be sure to include the most interesting information unreported in the study on self-esteem and partner success. Possibly it was never even looked at: What did the subjects write about? What kind of question would Mr. "My-self-esteem-goes-down-when-she-succeeds" elect to think and write about? Consider some questions that would force you to reinterpret even the statistically significant results.

Exhibit (viii): The Multiverse. Gelman and Loken (2014) call attention to the fact that even without explicitly cherry picking, there is often enough leeway in the "forking paths" between data and inference so that a researcher may be led to a desired inference. Growing out of this recognition is the idea of presenting the results of applying the same statistical procedure, but with different choices along the way (Steegen et al. 2016). They call it a *multiverse analysis*. One lists the different choices thought to be plausible at each stage of data processing. The multiverse displays ". . . which constellation of choices corresponds to which statistical result" (p. 707).

They consider an example from 2012 purporting to show that single women prefer Obama to Romney when they are highly fertile; the reverse when they're at low fertility.

In two studies with relatively large and diverse samples of women, we found that ovulation had different effects on women's religious and political orientation depending on whether women were single or in committed relationships. Ovulation led single women to become more socially liberal, less religious, and more likely to vote for Barack Obama. (Durante et al. 2013, p. 1013)

Unlike the Macho Men study, this one's not intuitively plausible. In fact, it was pummeled so vehemently by the public that it had to be pulled off CNN.[10] Should elaborate statistical criticism be applied to such studies? I had considered them only human interest stories. But Gelman rightly finds in them some general lessons.

One of the choice points in the ovulation study would be where to draw the line at "highly fertile" based on days in a woman's cycle. It wasn't based on any hormone check but on an online questionnaire asking when they'd had their last period. There's latitude in using such information to decide whether to place someone in a low or high fertility group (Steegen et al. 2016, p. 705, find five sets of data that could have been used). It turned out that under the other five choice points, many of the results were not statistically significant. Each of the different consistent combinations of choice points could count as a distinct

---

[10] "Last week CNN pulled a story about a study purporting to demonstrate a link between a woman's ovulation and how she votes. . . The story was savaged online as 'silly,' 'stupid,' 'sexist,' and 'offensive.'" (Bartlett, 2012b)

hypothesis, and you can then consider how many of them are statistically insignificant.

If no strong arguments can be made for certain choices, we are left with many branches of the multiverse that have large P values. In these cases, the only reasonable conclusion on the effect of fertility is that there is considerable scientific uncertainty. One should reserve judgment . . . (ibid., p. 708)

Reserve judgment? If we're to apply our severe testing norms on such examples, and not dismiss them as entertainment only, then we'd go further. Here's another reasonable conclusion: The core presumptions are falsified (or would be with little effort). Say each person with high fertility in the first study is tested for candidate preference next month when they are in the low fertility stage. If they have the same voting preferences, the test is falsified. The spirit of their multiverse analysis is a quintessentially error statistical gambit. Anything that increases the flabbiness in uncovering flaws lowers the severity of the test that has passed (we'll visit P-value adjustments later on). But the onus isn't on us to give them a pass. As we turn to impressive statistical meta-critiques, what can be overlooked is whether the entire inquiry makes any sense. Readers will have many other tomatoes to toss at the ovulation research. Unless the overall program is falsified, the literature will only grow. We don't have to destroy statistical significance tests when what we really want is to show that a lot of studies constitute pseudoscience.

Souvenir G: The Current State of Play in Psychology

Failed replications, we hear, are creating a "cold war between those who built up modern psychology and those" tearing it down with failed replications (Letzter 2016). The severe tester is free to throw some fuel on both fires.

The widespread growth of preregistered studies is all to the good; it's too early to see if better science will result. Still, credit is due to those sticking their necks out to upend the status quo. I say it makes no sense to favor preregistration and also deny the relevance to evidence of optional stopping and outcomes other than the one observed. That your appraisal of the evidence is altered when you actually see the history supplied by the registered report is equivalent to worrying about biasing selection effects when they're not written down; your statistical method should pick up on them.

By reviewing the hypotheses and analysis plans in advance, RRs (registered reports) should also help neutralize P-hacking and HARKing (hypothesizing after the results are known) by authors, and CARKing (critiquing after the results are known) by reviewers with their own investments in the research outcomes, although empirical evidence will be required to confirm that this is the case. (Munafò et al. 2017, p. 5)

The papers are provisionally accepted before the results are in. To the severe tester, that requires the author to explain how she will pinpoint blame for negative results. I see nothing in preregistration, in and of itself, to require that. It would be wrong-headed to condemn CARKing: post-data criticism of assumptions and inquiries into hidden biases might be altogether warranted. For instance, one might ask about the attitude toward the finding conveyed by the professor: what did the students know and when did they know it? Of course, they must not be ad hoc saves of the finding.

The field of meta-research is bursting at the seams: distinct research into changing incentives is underway. The severe tester may be jaundiced to raise qualms, but she doesn't automatically assume that research into incentivizing researchers to behave in a fashion correlated with good science – data sharing, preregistration – is itself likely to improve the original field. Not without thinking through what would be needed to link statistics up with the substantive research problem. In some fields, one wonders if they would be better off ignoring statistical experiments and writing about plausible conjectures about human motivations, prejudices, or attitudes, perhaps backed by interesting field studies. It's when researchers try to test them using sciency methods that the project becomes pseudosciency.

2.7 How to Solve the Problem of Induction Now

Viewing inductive inference as severe testing, the problem of induction is transformed into the problem of showing the existence of severe tests and methods for identifying insevere ones. The trick isn't to have a formal, context-free method that you can show is reliable – as with the traditional problem of induction; the trick is to have methods that alert us when an application is shaky. As a relaxing end to a long tour, our evening speaker on ship, a severe tester, will hold forth on statistics and induction.

Guest Speaker: A Severe Tester on Solving Induction Now

Here's his talk:

For a severe tester like me, the current and future problem of induction is to identify fields and inquiries where inference problems are solved efficiently, and ascertain how obstacles are overcome – or not. You've already assembled the ingredients for this final leg of Tour II, including: lift-off, convergent arguments (from coincidence), pinpointing blame (Duhem's problem), and falsification. Essentially, the updated problem is to show that there exist methods for controlling and assessing error probabilities. Does that seem too easy? The problem has always been rather minimalist: to show at least some reliable methods exist; the idea being that they could then be built upon. Just find me one. They thought enumerative induction was the one, but it's not. I will examine four questions: 1. What warrants inferring a hypothesis that stands up to severe tests? 2. What enables induction (as severe testing) to work? 3. What is Neyman's quarrel with Carnap? and 4. What is Neyman's empirical justification for using statistical models?

1. What Warrants Inferring a Hypothesis that Passes Severe Tests? Suppose it is agreed that induction is severe testing. What warrants moving from H passing a severe test to warranting H? Even with a strong argument from coincidence akin to my weight gain showing up on myriad calibrated scales, there is no logical inconsistency with invoking a hypothesis from *conspiracy*: all these instruments conspire to produce results as if H were true but in fact H is false. The ultra-skeptic may invent a *rigged* hypothesis R:

R: Something else other than H actually explains the data

without actually saying what this something else is. That is, we're imagining the extreme position of someone who simply asserts, H is actually false, although everything is as if it's true. Weak severity alone can block inferring a generic rigged hypothesis R as a way to discount a severely tested H. It can't prevent you from stopping there and never allowing a hypothesis is warranted. (Weak severity merely blocks inferring claims when little if anything has been done to probe them.) Nevertheless, if someone is bound to discount a strong argument for H by rigging, then she will be adopting a highly unreliable method. Why? Because a conspiracy hypothesis can always be adduced! Even with claims that are true, or where problems are solved correctly, you would have no chance of finding this out. I began with the stipulation that we wish to learn. Inquiry that blocks learning is pathological. Thus, because I am a severe tester, I hold both strong and weak severity:

> Data from test T are an indication of, or evidence for, H just to the extent that H has severely passed test T.

"Just to the extent that" indicates the "if then" goes in both directions: a claim that passes with low severity is unwarranted; one that passes with high severity is warranted. The phrase "to the extent that" refers to degrees of severity. That said, evidence requires a decent threshold be met, low severity is lousy evidence. It's still useful to point out in our travels when only weak severity suffices.

2. What Enables Induction (as Severe Testing) to Work: Informal and Quasi-formal Severity. You visited briefly the Exhibit (v) on prions and the deadly brain disease kuru. I'm going to use it as an example of a quasi-formal inquiry. Prions were found to contain a single protein dubbed PrP. Much to their surprise, researchers found PrP in normal cells too – it doesn't always cause disease. Our hero, Prusiner, again worries he'd "made a terrible mistake" (and prions had nothing to do with it). There are four strategies:

(a) Can we trick the phenomenon into telling us what it would be like if it really was a mere artifact (H_0)? Transgenic mice with PrP deliberately knocked out. Were H_0 true, they'd be expected to be infected as much as normal mice – the test hypothesis H_0 would not be rejected. H_0 asserts an *implicationary assumption* – one assumed just for testing. Abbreviate it as an *i-assumption*. It turns out that without PrP, none could be infected. Once PrP is replaced, they can again be infected. They argue, there's evidence to reject the artifact error H_0 because a procedure that would have revealed it fails to do so, and instead consistently finds departures from H_0.

(b) Over a period of more than 30 years, Prusiner and other researchers probed a series of local hypotheses. The levels of our hierarchy of models distinguishes various questions – even though I sketch it horizontally to save space (Figure 2.1). Comparativists deny we can proceed with a single hypothesis, but we do. Each question may be regarded as asking: would such and such be an erroneous interpretation of the data? Say the primary question is protein only or not. The alternatives do not include for the moment other "higher level" explanations about the mechanism of prion infectivity or the like. Given this localization, if H has been severely tested – by which I mean it has passed a severe test – then its denial has passed with low severity. That follows by definition of severity.

(c) Another surprise: the disease-causing form, call it pD, has the same exact amino acids as the normal type, call it pN. What's going on? Notice that a method that precluded exceptions to the central dogma (only nucleic acid directs replication of pathogens) would be incapable of identifying the culprit of prion transmission: the misfolding protein. Prusiner's prion hypothesis H^* is that prions target normal PrP, pinning and flattening their spirals to flip from their usual pN shape into pD, akin to a "deadly Virginia reel in the brain," adding newly formed pD's to the ends each time (Prusiner Labs 2004). When the helix is long enough, it ruptures, sending more pD seeds to convert normal prions. Another i-assumption to subject to the test of experiment. Trouble is, the process is so slow it can take years to develop. Not long ago, they found a way to deceive the natural state of affairs, while not altering what they want to learn: artificially rupture (with ultrasound or other means) the pathogenic prion. It's called protein misfolding cyclical amplification, PMCA. They get huge amounts of pD starting with a minute quantity, even a single molecule, so long as there's lots of normal PrP ready to be infected. All normal prions are converted into diseased prions in vitro. They could infer, with severity, that H^* gives a correct understanding of prion propagation, as well as corroborate the new research tool: They corroborated both at once, not instantly of course but over a period of a few years.

(d) Knowing the exponential rates of amplification associated with a method, researchers can infer, statistically, back to the amount of initial infectivity present – something they couldn't discern before, given the very low concentration of pD in accessible bodily fluids. Constantly improved and even automated, pD can now be detected in living animals for the first time.

What are some key elements? Honest self-criticism of how one may be wrong, deliberate deception to get counterfactual knowledge, conjecturing i-assumptions whose rejection leads to finding something out, and so on. Even researchers who hold different theories about the mechanism of transmission do not dispute PMCA – they can't if they want to learn more in the domain. I'm leaving out the political and personality feuds, but there's a good story there (see Prusiner 2014). I also didn't discuss statistically modeled aspects of prion research, but controlling the mean number of days for incubation allowed a stringent causal argument. I want to turn to statistical induction at a more rudimentary entry point.

3. Neyman's Quarrel with Carnap. Statistics is the *sine qua non* for extending our powers to severely probe. Jerzy Neyman, with his penchant for inductive behavior and performance rather than inductive inference, is often seen as a villain in the statistics battles. So take a look at a paper of his with the tantalizing title: "The Problem of Inductive Inference" (Neyman 1955). Neyman takes umbrage with the way confirmation philosophers, in particular Carnap, view frequentist inference:

... when Professor Carnap criticizes some attitudes which he represents as consistent with my ("frequentist") point of view, I readily join him in his criticism without, however, accepting the responsibility for the criticized paragraphs. (p. 13)

In effect, Neyman says I'd never infer from observing that 150 out of 1000 throws with this die landed on six, "nothing else being known," that future throws will result in around 0.15 sixes, as Carnap alleges I would. This is a version of enumerative induction (or Carnap's straight rule). You need a statistical model! Carnap should view "Statistics as the Frequentist Theory of Induction," says Neyman in a section with this title, here the Binomial model. The Binomial distribution builds on n Bernoulli trials, the success–failure trials (visited in Section 1.4). It just adds up all the ways that number of successes could occur:

$$\Pr(k \text{ out of } n \text{ successes}) = \binom{n}{k} \theta^k (1 - \theta)^{n-k}$$

Carnapians could have formed the straight rule for the Binomial experiment, and argued:

> If an experiment can be generated and modeled Binomially, then sample means can be used to reliably estimate population means.
> An experiment can be modeled Binomially.
> Therefore, we can reliably estimate population means in those contexts.

The reliability comes from controlling the method's error probabilities.

4. What Is Neyman's Empirical Justification for Using Statistical Models?
Neyman pays a lot of attention to the empirical justification for using statistical models. Take his introductory text (Neyman 1952). The models are not invented from thin air. In the beginning there are records of different results and stable relative frequencies with which they occurred. These may be called empirical frequency distributions. There are real experiments that "even if carried out repeatedly with the utmost care to keep conditions constant, yield varying results" (ibid., p. 25). These are real, not hypothetical, experiments, he stresses. Examples he gives are roulette wheels (electrically regulated), tossing coins with a special machine (that gives a constant initial velocity to the coin), the number of disintegrations per minute in a quantity of radioactive matter, and the tendency for properties of organisms to vary despite homogeneous breeding. Even though we are unable to predict the outcome of such experiments, a certain stable pattern of regularity emerges rather quickly, even in moderately long series of trials; usually around 30 or 40 trials suffices. The pattern of regularity is in the relative frequency with which specified results occur.

Neyman takes a toy example: toss a die twice and record the frequency of sixes: 0, 1, or 2. Call this a *paired* trial. Now do this 1000 times. You'll have 1000 paired trials. Put these to one side for a moment. Just consider the entire set of

2000 tosses – *first order* trials Neyman calls these. Compute the relative frequency of sixes out of 2000. It may not be 1/6, due to the structure of the die or the throwing. Whatever it is, call it f. Now go back to the paired trials. Record the relative frequency of six found in paired trial 1, maybe it's 0, the relative frequency of six found in paired trial 2, all the way through your 1000 paired trials. We can then ask: what proportion of the 1000 paired trials had no sixes, what proportion had 1 six, what proportion 2 sixes? We find "the proportions of pairs with 0, 1 and 2 sixes will be, approximately,

$$(1 - f)^2, 2f(1 - f), \text{ and } f^2."$$

Instead of pairs of trials, consider n-fold trials: each trial has n throws of the die. Compute f as before: it is the relative frequency of six in the $1000n$ first order trials. Then, turn to the 1000 n-fold trials, and compute the proportion where six occurs k times (for $k < n$). It will be very nearly equal to

$$\binom{n}{k} f^k (1 - f)^{n-k}.$$

"In other words, the relative frequency" of k out of n successes in the n-fold trials "is connected with the relative frequency of the first order experiments in very nearly the same way as the probability" of k out of n successes in a Binomial trial is related to the probability of success at each trial, θ (Neyman 1952, p. 26).

The above fact, which has been found empirically many times … may be called the empirical law of large numbers. I want to emphasize that this law applies not only to the simple case connected with the binomial formula … but also to other cases. In fact, this law seems to be perfectly general … Whenever the law fails, we explain the failure by suspecting a "lack of randomness" in the first order trials. (ibid., p. 27)

Now consider, not just 1000 repetitions of all n-fold trials, but all. Here, f, the relative frequency of heads is θ in the Binomial probability model with n trials. It is this universe of hypothetical repetitions that our one n-fold sample is a random member of. Figure 2.2 shows the frequency distribution if we chose $n = 100$ and $\theta = 1/6$.

The Law of Large Numbers (LLN) shows we can use the probability derived from the probability model of the experiment to approximate the relative frequencies of outcomes in a series of n-fold trials. The LLN is both an empirical law and a mathematical law. The proofs are based on idealized random samples, but there are certain actual experiments that are well

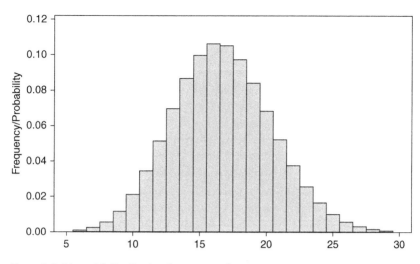

Figure 2.2 Binomial distribution for $n = 100$, $\theta = 1/6$.

approximated by the mathematical law – something we can empirically test (von Mises 1957).

You may bristle at this talk of random experiments, but, as Neyman repeatedly reminds us, these are merely "picturesque" shorthands for results squarely linked up with empirical tests (Neyman 1952, p. 23). We keep to them in order to explicate the issues at the focus of our journey. The justification for applying what is strictly an abstraction is no different from other cases of applied mathematics. We are not barred from fruitfully applying geometry because a geometric point is an abstraction.

"Whenever we succeed in arranging" the data generation such that the relative frequencies adequately approximate the mathematical probabilities in the sense of the LLN, we can say that the probabilistic model "adequately represents the method of carrying out the experiment" (ibid., p. 19). In those cases we are warranted in describing the results of real experiments as random samples from the population given by the probability model. You can reverse direction and ask about f or θ when unknown. Notice that we are modeling something we do, we may do it well or badly. All we need is that mysterious supernatural powers keep their hands off our attempts to carry out inquiry properly, to take one of Peirce's brilliant insights: "the supernal powers withhold their hands and let me alone, and that no mysterious uniformity ... interferes with the action of chance" (2.749) in order to justify induction. *End of talk.*

I wonder if Carnap ever responded to Neyman's grumblings. Why didn't philosophers replace a vague phrase like "if these k out of n successes are all I know about the die" and refer to the Binomial model?, I asked Wesley Salmon in the 1980s. Because, he said, we didn't think the Binomial model could be justified without getting into a circle. But it can be tested empirically. By varying a known Binomial process to violate one of the assumptions deliberately, we develop tests that would very probably detect such violations should they occur. This is the key to justifying induction as severe testing: it corrects its assumptions. Testing the assumption of randomness is independent of estimating θ given that it's random. Salmon and I met weekly to discuss statistical tests of assumptions when I visited the Center for Philosophy of Science at Pittsburgh in 1989. I think I convinced him of this much (or so he said): the confirmation theorists were too hasty in discounting the possibility of warranting statistical model assumptions.

Souvenir H: Solving Induction Is Showing Methods with Error Control

How is the problem of induction transformed if induction is viewed as severe testing? Essentially, it becomes a matter of showing that there exist methods with good error probabilities. The specific task becomes examining the fields or inquiries that are – and are not – capable of assessing and controlling severity. Nowadays many people abjure teaching the different distributions, preferring instead to generate frequency distributions by resampling a given random sample (Section 4.6). It vividly demonstrates what really matters in appealing to probability models for inference, as distinct from modeling phenomena more generally: Frequentist error probabilities are of relevance when frequencies represent the capabilities of inquiries to discern and discriminate various flaws and biases. Where Popper couldn't say that methods probably would have found H false, if it is false, error statistical methods let us go further. (See Mayo 2005a.)

The severity account puts forward a statistical philosophy associated with statistical methods. To see what I mean, recall the Likelihoodist. It's reasonable to suppose that we favor, among pairs of hypotheses, the one that predicts or makes probable the data – proposes the Likelihoodist. The formal Law of Likelihood (LL) is to capture this, and we appraise it according to how well it succeeds, and how well it satisfies the goals of statistical practice. Likewise, the severe tester proposes, there is a pre-statistical plausibility to infer hypotheses to the extent that they have passed stringent tests. The error statistical methodology is the frequentist theory of induction. Here too the statistical philosophy is to be appraised according to how well it captures and supplies rationales for inductive-statistical inference. The rest of our journey will bear this out. Enjoy the concert in the Captain's Central Limit Lounge while the breezes are still gentle, we set out on Excursion 3 in the morn.

Excursion 3 Statistical Tests and Scientific Inference

Itinerary

Tour I Ingenious and Severe Tests

> [T]he impressive thing about [the 1919 tests of Einstein's theory of gravity] is the *risk* involved in a prediction of this kind. If observation shows that the predicted effect is definitely absent, then the theory is simply refuted. The theory is *incompatible with certain possible results of observation* – in fact with results which everybody before Einstein would have expected. This is quite different from the situation I have previously described, [where] . . . it was practically impossible to describe any human behavior that might not be claimed to be a verification of these [psychological] theories. (Popper 1962, p. 36)

The 1919 eclipse experiments opened Popper's eyes to what made Einstein's theory so different from other revolutionary theories of the day: Einstein was prepared to subject his theory to risky tests.[1] Einstein was eager to galvanize scientists to test his theory of gravity, knowing the solar eclipse was coming up on May 29, 1919. Leading the expedition to test GTR was a perfect opportunity for Sir Arthur Eddington, a devout follower of Einstein as well as a devout Quaker and conscientious objector. Fearing "a scandal if one of its young stars went to jail as a conscientious objector," officials at Cambridge argued that Eddington couldn't very well be allowed to go off to war when the country needed him to prepare the journey to test Einstein's predicted light deflection (Kaku 2005, p. 113).

The museum ramps up from Popper through a gallery on "Data Analysis in the 1919 Eclipse" (Section 3.1) which then leads to the main gallery on origins of statistical tests (Section 3.2). Here's our Museum Guide:

> According to Einstein's theory of gravitation, to an observer on earth, light passing near the sun is deflected by an angle, λ, reaching its maximum of $1.75''$ for light just grazing the sun, but the light deflection would be undetectable on earth with the instruments available in 1919. Although the light deflection of stars near the sun (approximately 1 second of arc) *would* be detectable, the sun's glare renders such stars invisible, save during a total eclipse, which "by strange good

---

[1] You will recognize the above as echoing Popperian "theoretical novelty" – Popper developed it to fit the Einstein test.

fortune" would occur on May 29, 1919 (Eddington [1920] 1987, p. 113).

There were three hypotheses for which "it was especially desired to discriminate between" (Dyson et al. 1920 p. 291). Each is a statement about a parameter, the deflection of light at the limb of the sun (in arc seconds): $\lambda = 0''$ (no deflection), $\lambda = 0.87''$ (Newton), $\lambda = 1.75''$ (Einstein). The Newtonian predicted deflection stems from assuming light has mass and follows Newton's Law of Gravity.

The difference in statistical prediction masks the deep theoretical differences in how each explains gravitational phenomena. Newtonian gravitation describes a force of attraction between two bodies; while for Einstein gravitational effects are actually the result of the curvature of spacetime. A gravitating body like the sun distorts its surrounding spacetime, and other bodies are reacting to those distortions.

Where Are Some of the Members of Our Statistical Cast of Characters in 1919? In 1919, Fisher had just accepted a job as a statistician at Rothamsted Experimental Station. He preferred this temporary slot to a more secure offer by Karl Pearson (KP), which had so many strings attached – requiring KP to approve everything Fisher taught or published – that Joan Fisher Box writes: After years during which Fisher "had been rather consistently snubbed" by KP, "It seemed that the lover was at last to be admitted to his lady's court – on conditions that he first submit to castration" (J. Box 1978, p. 61). Fisher had already challenged the old guard. Whereas KP, after working on the problem for over 20 years, had only approximated "the first two moments of the sample correlation coefficient; Fisher derived the relevant distribution, not just the first two moments" in 1915 (Spanos 2013a). Unable to fight in WWI due to poor eyesight, Fisher felt that becoming a subsistence farmer during the war, making food coupons unnecessary, was the best way for him to exercise his patriotic duty.

In 1919, Neyman is living a hardscrabble life in a land alternately part of Russia or Poland, while the civil war between Reds and Whites is raging. "It was in the course of selling matches for food" (C. Reid 1998, p. 31) that Neyman was first imprisoned (for a few days) in 1919. Describing life amongst "roaming bands of anarchists, epidemics" (ibid., p. 32), Neyman tells us, "existence" was the primary concern (ibid., p. 31). With little academic work in statistics, and "since no one in Poland was able to gauge the importance of his statistical work (he was 'sui generis,' as he later described himself)" (Lehmann 1994, p. 398), Polish authorities sent him to University College in

London in 1925/1926 to get the great Karl Pearson's assessment. Neyman and E. Pearson begin work together in 1926.

Egon Pearson, son of Karl, gets his B.A. in 1919, and begins studies at Cambridge the next year, including a course by Eddington on the theory of errors. Egon is shy and intimidated, reticent and diffident, living in the shadow of his eminent father, whom he gradually starts to question after Fisher's criticisms. He describes the psychological crisis he's going through at the time Neyman arrives in London: "I was torn between conflicting emotions: a. finding it difficult to understand R.A.F., b. hating [Fisher] for his attacks on my paternal 'god,' c. realizing that in some things at least he was right" (C. Reid 1998, p. 56). As far as appearances amongst the statistical cast: there are the two Pearsons: tall, Edwardian, genteel; there's hardscrabble Neyman with his strong Polish accent and small, toothbrush mustache; and Fisher: short, bearded, very thick glasses, pipe, and eight children.

Let's go back to 1919, which saw Albert Einstein go from being a little known German scientist to becoming an international celebrity.

3.1   Statistical Inference and Sexy Science: The 1919 Eclipse Test

The famous 1919 eclipse expeditions purported to test Einstein's new account of gravity against the long-reigning Newtonian theory. I get the impression that statisticians consider there to be a world of difference between statistical inference and appraising large-scale theories in "glamorous" or "sexy science." The way it actually unfolds, which may not be what you find in philosophical accounts of theory change, revolves around local data analysis and statistical inference. Even large-scale, sexy theories are made to connect with actual data only by intermediate hypotheses and models. To falsify, or even provide anomalies, for a large-scale theory like Newton's, we saw, is to infer "falsifying hypotheses," which are statistical in nature.

Notably, from a general theory we do not deduce observable data, but at most a general phenomenon such as the Einstein deflection effect due to the sun's gravitational field (Bogen and Woodward 1988). The problem that requires the most ingenuity is finding or inventing a phenomenon, detector, or probe that will serve as a meeting ground between data that can actually be collected and a substantive or theoretical effect of interest. This meeting ground is typically statistical. Our array in Souvenir E provides homes within which relevant stages of inquiry can live. Theories and laws give constraints but the problem at the experimental frontier has much in common with research in fields where there is at most a vague phenomenon and no real theories to speak of.

There are two key stages of inquiry corresponding to two questions within the broad umbrella of *auditing an inquiry*:

(i) is there a deflection effect of the amount predicted by Einstein as against Newton (the "Einstein effect")?
(ii) is it attributable to the sun's gravitational field as described in Einstein's hypothesis?

A distinct third question, "higher" in our hierarchy, in the sense of being more theoretical and more general, is: is GTR an adequate account of gravity as a whole? These three are often run together in discussions, but it is important to keep them apart.

The first is most directly statistical. For one thing, there's the fact that they don't observe stars just grazing the sun but stars whose distance from the sun is at least two times the solar radius, where the predicted deflection is only around 1″ of arc. They infer statistically what the deflection would have been for starlight near the sun. Second, they don't observe a deflection, but (at best) photographs of the positions of certain stars at the time of the eclipse. To "observe" the deflection, if any, requires inferring what the positions of these same stars would have been were the sun's effect absent, a "control" as it were. Eddington remarks:

The bugbear of possible systematic error affects all investigations of this kind. How do you know that there is not something in your apparatus responsible for this apparent deflection? ... To meet this criticism, a different field of stars was photographed ... at the same altitude as the eclipse field. If the deflection were really instrumental, stars on these plates should show relative displacements of a similar kind to those on the eclipse plates. But on measuring these check-plates no appreciable displacements were found. That seems to be satisfactory evidence that the displacement ... is not due to differences in instrumental conditions. ([1920] 1987, p. 116)

If the check plates can serve as this kind of a control, the researchers are able to use a combination of theory, controls, and data to transform the original observations into an approximate linear relationship between two observable variables and use least squares to estimate the deflection. The position of each star photographed at the eclipse (the eclipse plate) is compared to its normal position photographed at night (months before or after the eclipse), when the effect of the sun is absent (the night plate). Placing the eclipse and night plates together allows the tiny distances to be measured in the x and y directions (Figure 3.1). The estimation had to take account of how the two plates are

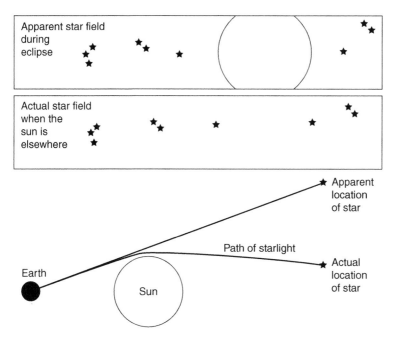

Figure 3.1 Light deflection.

accidentally clamped together, possible changes in the scale – due mainly to differences in the focus between the exposure of the eclipse and the night plates – on a set of other plate parameters, and, finally, on the light deflection.

The general technique was known to astronomers from determining the angle of stellar parallax, "for which much greater accuracy is required" (ibid., pp. 115–16). (The relation between a star position and the sun changes as the earth moves around the sun, and the angle formed is its parallax.) Somewhat like the situation with Big Data, scientists already had a great deal of data on star positions and now there's a highly theoretical question that can be probed with a known method. Still, the eclipse poses unique problems of data analysis, not to mention the precariousness of bringing telescopes on expeditions to Sobral in North Brazil and Principe in the Gulf of Guinea (West Africa).

The problem in (i) is reduced to a statistical one: the observed mean deflections (from sets of photographs) are Normally distributed around the predicted mean deflection μ. The proper way to frame this as a statistical test is to choose one of the values as H_0 and define composite H_1 to include alternative values of interest. For instance, the Newtonian "half deflection" can specify the H_0: $\mu \leq 0.87$, and the H_1: $\mu > 0.87$ includes the Einsteinian value of

1.75. Hypothesis H_0 also includes the third value of potential interest, $\mu = 0$: no deflection.[2] After a good deal of data analysis, the two eclipse results from Sobral and Principe were, with their standard errors,

> Sobral: the eclipse deflection $= 1.98'' \pm 0.18''$.
>
> Principe: the eclipse deflection $= 1.61'' \pm 0.45''$.

The actual report was in probable errors in use at the time, 0.12 and 0.30 respectively, where 1 probable error equals 0.68 standard errors. A sample mean differs from a Normal population mean by one or more probable errors (in either direction) 50% of the time.

It is usual to allow a margin of safety of about twice the probable error on either side of the mean. The evidence of the Principe plates is thus just about sufficient to rule out the possibility of the 'half deflection,' and the Sobral plates exclude it with practical certainty. (Eddington [1920]1987, p. 118)

The idea of reporting the "probable error" is of interest to us. There is no probability assignment to the interval, it's an error probability *of the method*. To infer μ = observed mean $\pm$ 1 probable error is to use a method that 50% of the time correctly covers μ. Two probable errors wouldn't be considered much of a margin of safety these days, being only ~1.4 standard errors. Using the term "probable error" might be thought to encourage misinterpretation – and it does – but it's not so different from the current use of "margin of error."

A text by Ghosh et al. (2010, p. 48) presents the Eddington results as a two-sided Normal test of H_0: $\mu = 1.75$ (the Einstein value) vs. H_1: $\mu \neq 1.75$, with a lump of prior probability given to the point null. If any theoretical prediction were to get a lump at this stage, it is Newton's. The vast majority of Newtonians, understandably, regarded Newton as far more plausible, never mind the small well-known anomalies, such as being slightly off in its prediction of the orbit of the planet Mercury. Few could even understand Einstein's radically different conception of space and time.

Interestingly, the (default) Bayesian statistician Harold Jeffreys was involved in the eclipse experiment in 1919. He lauded the eclipse results as finally putting the Einstein law on firm experimental footing – despite his low Bayesian prior in GTR (Jeffreys 1919). Actually, even the experimental footing did not emerge until the 1960s (Will 1986). The eclipse tests, not just those of 1919, but all eclipse tests of the deflection effect, failed to give very precise results. Nothing like a stringent estimate of the deflection effect

---

[2] "A ray of light nicking the edge of the sun, for example, would bend a minuscule 1.75 arcseconds – the angle made by a right triangle 1 inch high and 1.9 *miles* long" (Buchen 2009).

emerged until the field was rescued by radioastronomical data from quasars (quasi-stellar radio sources). This allowed testing the deflection using radio waves instead of light waves, and without waiting for an eclipse.

Some Popperian Confusions About Falsification and Severe Tests

Popper lauds GTR as sticking its neck out, bravely being ready to admit its falsity were the deflection effect not found (1962, pp. 36–7). Even if no deflection effect had been found in the 1919 experiments, it would have been blamed on the sheer difficulty in discerning so small an effect. This would have been entirely correct. Yet many Popperians, perhaps Popper himself, get this wrong. Listen to Popperian Meehl:

> [T]he stipulation beforehand that one will be pleased about substantive theory T when the numerical results come out as forecast, but will not necessarily abandon it when they do not, seems on the face of it to be about as blatant a violation of the Popperian commandment as you could commit. For the investigator, in a way, is doing . . . what astrologers and Marxists and psychoanalysts allegedly do, playing 'heads I win, tails you lose.' (Meehl 1978, p. 821)

There is a confusion here, and it's rather common. A successful result may rightly be taken as evidence for a real effect H, even though failing to find the effect would not, and should not, be taken to refute the effect, or as evidence against H. This makes perfect sense if one keeps in mind that a test might have had little chance to detect the effect, even if it exists.

One set of eclipse plates from Sobral (the astrographic plates) was sufficiently blurred by a change of focus in the telescope as to preclude any decent estimate of the standard error (more on this case later). Even if all the 1919 eclipse results were blurred, this would at most show no deflection had been found. This is not automatically evidence there's no deflection effect.[3] To suppose it is would violate our minimal principle of evidence: the probability of failing to detect the tiny effect with the crude 1919 instruments is high – even if the deflection effect exists.

Here's how the severity requirement cashes this out: Let H_0 assert the Einstein effect is absent or smaller than the predicted amount, and H_1 that the deflection exists. An observed failure to detect a deflection "accords with" H_0, so the first severity requirement holds. But there's a high probability of this occurring even if H_0 is false and H_1 true (whether as explained in GTR or other theory). The point really reflects the asymmetry of falsification and corroboration (Section 2.1): if the deflection effect passes an audit, then it is a genuine

---

[3] To grasp this, consider that a single black swan proves the hypothesis H: some swans are not white, even though a white swan would not be taken as strong evidence for H's denial. H's denial would be that all swans are white.

anomaly for Newton's half deflection – only one is needed. Yet not finding an anomaly in 1919 isn't grounds for supposing no deflection anomalies exist. Alternatively, you can see this as an unsound but valid deductive argument (*modus tollens*):

> If GTR, then the deflection effect is observed in the 1919 eclipse tests.
> No deflection is observed in the 1919 eclipse tests.
> Therefore ~GTR (or evidence against GTR).

Because the first premise of this valid argument is false, the argument is unsound. By contrast, once instruments were available to powerfully detect any deflection effects, a no-show would have to be taken against its existence, and thus against GTR. In fact, however, a deflection was observed in 1919, although the accuracy was only 30%. Either way, Popperian requirements are upheld, even if some Popperians get this wrong.

George Barnard on the Eclipse Tests

The first time I met George Barnard in 1985, the topics of the 1919 eclipse episode and the N-P vs. Fisher battles were front and center. The focus of his work on the eclipse was twofold: First, "to draw attention to a reasonably accessible instance ... where the inferential processes can be seen at work – and in the mind of someone who, (unlike so many physicists!) had taken the trouble to familiarise himself thoroughly with mathematical statistics" (Barnard 1971, p. 294). He is alluding to Eddington. Of course that was many years ago. Barnard's second reason is to issue a rebuke to Neyman! – or at least to a crude performance construal often associated with Neyman (ibid., p. 300). Barnard's point is that bad luck with the weather resulted in the sample size of usable photographs being very different from what could have been planned. They only used results where enough stars could be measured to apply least squares regression reliably (at least equal to the number of unknown parameters – six). Any suggestion that the standard errors "be reduced because in a repetition of the experiment" more usable images might be expected, "would be greeted with derision" (ibid., p. 295). Did Neyman say otherwise? In practice, Neyman describes cases where he rejects the data as unusable because of failed assumptions (e.g., Neyman 1977, discussing a failed randomization in a cloud seeding experiment).

Clearly, Barnard took Fisher's side in the N-P vs. Fisher disputes; he wanted me to know he was the one responsible for telling Fisher that Neyman had converted "his" significance tests into tools for acceptance sampling, where only long-run performance matters (Pearson 1955 affirms this). Pearson was kept out of it. The set of hypothetical repetitions used in obtaining the relevant error probability, in Barnard's view, should consist of "results of reasonably similar precision" (1971, p. 300). This is a very interesting idea, and it will come up again.

Big Picture Inference: Can Other Hypotheses Explain the Observed Deflection?

Even to the extent that they had found a deflection effect, it would have been fallacious to infer the effect "attributable to the sun's gravitational field." The question (ii) must be tackled: A statistical effect is not a substantive effect. Addressing the causal attribution demands the use of the eclipse data as well as considerable background information. Here we're in the land of "big picture" inference: the inference is "given everything we know". In this sense, the observed effect is used and is "non-novel" (in the use-novel sense). Once the deflection effect was known, imprecise as it was, it had to be used. Deliberately seeking a way to explain the eclipse effect while saving Newton's Law of Gravity from falsification isn't the slightest bit pejorative – so long as each conjecture is subject to severe test. Were *any* other cause to exist that produced a considerable fraction of the deflection effect, that alone would falsify the Einstein hypothesis (which asserts that *all* of the 1.75″ are due to gravity) (Jeffreys 1919, p. 138). That was part of the riskiness of the GTR prediction.

It's Not How Plausible, but How Well Probed

One famous case was that of Sir Oliver Lodge and his proposed "ether effect." Lodge was personally invested in the Newtonian ether, as he believed it was through the ether that he was able to contact departed souls, in particular his son, Raymond. Lodge had "preregistered" in advance that if the eclipse results showed the Einstein deflection he would find a way to give a Newtonian explanation (Lodge 1919). Others, without a paranormal bent, felt a similar allegiance to Newton. "We owe it to that great man to proceed very carefully in modifying or retouching his Law of Gravitation" (Silberstein 1919, p. 397). But respect for Newton was kept out of the data analysis. They were free to try and try again with Newton-saving factors because, unlike in pejorative seeking, it would be extremely difficult for any such factor to pass if false – given the standards available and insisted on by the relevant community of scientists. Each Newton-saving hypothesis collapsed on the basis of a one-two punch: the magnitude of effect that could have been due to the conjectured factor is far too small to account for the eclipse effect; and were it large enough to account for the eclipse effect, it would have blatantly false or contradictory implications elsewhere. Could the refraction of the sun's corona be responsible (as one scientist proposed)? Were it sufficient to explain the deflection, then comets would explode when they pass near the sun, which they do not! Or take another of Lodge's ether modification hypotheses. As scientist Lindemann put it:

Sir Oliver Lodge has suggested that the deflection of light might be explained by assuming a change in the effective dielectric constant near a gravitating body. ... It sounds quite promising at first ... The difficulty is that one has in each case to adopt a different constant in the law, giving the dielectric constant as a function of the gravitational field, unless some other effect intervenes. (1919, p. 114)

This would be a highly insevere way to retain Newton. These criticisms combine quantitative and qualitative severity arguments. We don't need a precise quantitative measure of how frequently we'd be wrong with such ad hoc finagling. The Newton-saving factors might have been plausible but they were unable to pass severe tests. Saving Newton this way would be bad science.

As is required under our demarcation (Section 2.3): the 1919 players were able to embark upon an inquiry to pinpoint the source for the Newton anomaly. By 1921, it was recognized that the deflection effect was real, though inaccurately measured. Further, the effects revealed (corona effect, shadow effect, lens effect) were themselves used to advance the program of experimental testing of GTR. For instance, learning about the effect of the sun's corona (corona effect) not only vouchsafed the eclipse result, but pointed to an effect that could not be ignored in dealing with radioastronomy. Time and space prevents going further, but I highly recommend you return at a later time. For discussion and references, see Mayo (1996, 2010a, e).

The result of all the analysis was merely evidence of a small piece of GTR: an Einstein-like deflection effect. The GTR "passed" the test, but clearly they couldn't infer GTR severely. Even now, only its severely tested parts are accepted, at least to probe relativistic gravity. John Earman, in criticism of me, observes:

[W]hen high-level theoretical hypotheses are at issue, we are rarely in a position to justify a judgment to the effect that $Pr(E|\sim H \& K) \ll 0.5$. If we take H to be Einstein's general theory of relativity and E to be the outcome of the eclipse test, then in 1918 and 1919 physicists were in no position to be confident that the vast and then unexplored space of possible gravitational theories denoted by $\sim$GTR does not contain alternatives to GTR that yield that same prediction for the bending of light as GTR. (Earman 1992, p. 117)

A similar charge is echoed by Laudan (1997), Chalmers (2010), and Musgrave (2010). For the severe tester, being prohibited from regarding GTR as having passed severely – especially in 1918 and 1919 – is just what an account ought to do. (Do you see how this relates to our treatment of irrelevant conjunctions in Section 2.2?)

From the first exciting results to around 1960, GTR lay in the doldrums. This is called the period of *hibernation* or stagnation. Saying it remained uncorroborated or inseverely tested does not mean GTR was deemed scarcely true, improbable, or implausible. It hadn't failed tests, but there were too few link-ups between the highly mathematical GTR and experimental data. Uncorroborated is very different from disconfirmed. We need a standpoint that lets us express being at that stage in a problem, and viewing inference as severe testing gives us one. Soon after, things would change, leading to the Renaissance from 1960 to 1980. We'll pick this up at the end of Sections 3.2 and 3.3. To segue into statistical tests, here's a souvenir.

Souvenir I: So What Is a Statistical Test, Really?

So what's in a statistical test? First there is a question or problem, a piece of which is to be considered statistically, either because of a planned experimental design, or by embedding it in a formal statistical model. There are (A) hypotheses, and a set of possible outcomes or data; (B) a measure of accordance or discordance, fit, or misfit, $d(X)$ between possible answers (hypotheses) and data; and (C) an appraisal of a relevant distribution associated with $d(X)$. Since we want to tell what's true about tests now in existence, we need an apparatus to capture them, while also offering latitude to diverge from their straight and narrow paths.

(A) *Hypotheses.* A statistical hypothesis H_i is generally couched in terms of an unknown parameter θ. It is a claim about some aspect of the process that might have generated the data, $x_0 = (x_1, \ldots, x_n)$, given in a model of that process. Statistical hypotheses assign probabilities to various outcomes x "computed under the supposition that H_i is correct (about the generating mechanism)." That is how to read $f(x; H_i)$, or as I often write it: $\Pr(x; H_i)$. This is just an analytic claim about the assignment of probabilities to x stipulated in H_i.

In the GTR example, we consider n IID Normal random variables: $(X_1, \ldots, X_n)$ that are $N(\mu, \sigma^2)$. Nowadays, the GTR value for $\lambda = \mu$ is set at 1, and the test might be of $H_0: \mu \leq 1$ vs. $H: \mu > 1$. The hypothesis of interest will typically be a claim C posed after the data, identified within the predesignated parameter spaces.

(B) *Distance function and its distribution.* A function of the sample d(X), the *test statistic*, reflects how well or poorly the data ($X = x_0$) accord with the hypothesis H_0, which serves as a reference point. The term "test statistic" is generally reserved for statistics whose distribution can be computed under the main or test hypothesis. If we just want to speak of a statistic measuring distance, we'll call it that.

It is the observed distance d(x_0) that is described as "significantly different" from the null hypothesis H_0. I use x to say something general about the data, whereas x_0 refers to a fixed data set.

(C) *Test rule T.* Some interpretative move or methodological rule is required for an account of inference. One such rule might be to infer that x is evidence of a discrepancy δ from H_0 just when d(x) $\geq c$, for some value of c. Thanks to the requirement in (B), we can calculate the probability that {d(X) $\geq c$} under the assumption that H_0 is true. We want also to compute it under various discrepancies from H_0, whether or not there's an explicit specification of H_1. Therefore, we can calculate the probability of inferring evidence for discrepancies from H_0 when in fact the interpretation would be erroneous. Such an *error probability* is given by the probability distribution of d(X) – its *sampling distribution* – computed under one or another hypothesis.

To develop an account adequate for solving foundational problems, special stipulations and even reinterpretations of standard notions may be required. (D) and (E) reflect some of these.

(D) *A key role of the distribution* of d(X) will be to characterize the probative abilities of the inferential rule for the task of unearthing flaws and misinterpretations of data. In this way, error probabilities can be used to assess the severity associated with various inferences. We are able to consider outputs outside the N-P and Fisherian schools, including "report a Bayes ratio" or "infer a posterior probability" by leaving our measure of agreement or disagreement open. We can then try to compute an associated error probability and severity measure for these other accounts.

(E) *Empirical background assumptions.* Quite a lot of background knowledge goes into implementing these computations and interpretations. They are guided by the goal of assessing severity for the primary inference or problem, housed in the manifold steps from planning the inquiry, to data generation and analyses.

We've arrived at the N-P gallery, where Egon Pearson (actually a hologram) is describing his and Neyman's formulation of tests. Although obviously the museum does not show our new formulation, their apparatus is not so different.

3.2 N-P Tests: An Episode in Anglo-Polish Collaboration

We proceed by setting up a specific hypothesis to test, H_0 in Neyman's and my terminology, the null hypothesis in R. A. Fisher's ... in choosing the test, we take into account alternatives to H_0 which we believe possible or at any rate consider it most important to be on the look out for ...Three steps in constructing the test may be defined:

Step 1. We must first specify the set of results ...

Step 2. We then divide this set by a system of ordered boundaries ...

such that as we pass across one boundary and proceed to the next, we come to a class of results which makes us more and more inclined, on the information available, to reject the hypothesis tested in favour of alternatives which differ from it by increasing amounts.

Step 3. We then, if possible, associate with each contour level the chance that, if H_0 is true, a result will occur in random sampling lying beyond that level ...

In our first papers [in 1928] we suggested that the likelihood ratio criterion, λ, was a very useful one ... Thus Step 2 proceeded Step 3. In later papers [1933–1938] we started with a fixed value for the chance, ε, of Step 3 ... However, although the mathematical procedure may put Step 3 before 2, we cannot put this into operation before we have decided, under Step 2, on the guiding principle to be used in choosing the contour system. That is why I have numbered the steps in this order. (Egon Pearson 1947, p. 173)

In addition to Pearson's 1947 paper, the museum follows his account in "The Neyman–Pearson Story: 1926–34" (Pearson 1970). The subtitle is "Historical Sidelights on an Episode in Anglo-Polish Collaboration"!

We meet Jerzy Neyman at the point he's sent to have his work sized up by Karl Pearson at University College in 1925/26. Neyman wasn't that impressed:

Neyman found ... [K.]Pearson himself surprisingly ignorant of modern mathematics. (The fact that Pearson did not understand the difference between independence and lack of correlation led to a misunderstanding that nearly terminated Neyman's stay ...) (Lehmann 1988, p. 2)

Thus, instead of spending his second fellowship year in London, Neyman goes to Paris where his wife Olga ("Lola") is pursuing a career in art, and where he could attend lectures in mathematics by Lebesque and Borel. "[W]ere it not for Egon Pearson [whom I had briefly met while in London], I would have probably drifted to my earlier passion for [pure mathematics]" (Neyman quoted in Lehmann 1988, p. 3).

What pulled him back to statistics was Egon Pearson's letter in 1926. E. Pearson had been "suddenly smitten" with doubt about the justification of

tests then in use, and he needed someone with a stronger mathematical background to pursue his concerns. Neyman had just returned from his fellowship
years to a hectic and difficult life in Warsaw, working multiple jobs in applied
statistics.

[H]is financial situation was always precarious. The bright spot in this difficult period
was his work with the younger Pearson. Trying to find a unifying, logical basis which
would lead systematically to the various statistical tests that had been proposed by
Student and Fisher was a 'big problem' of the kind for which he had hoped . . . (ibid., p. 3)

N-P Tests: Putting Fisherian Tests on a Logical Footing

For the Fisherian simple or "pure" significance test, alternatives to the null
"lurk in the undergrowth but are not explicitly formulated probabilistically"
(Mayo and Cox 2006, p. 81). Still there are constraints on a Fisherian test
statistic. Criteria for the test statistic $d(X)$ are

   (i) it reduces the data as much as possible;
  (ii) the larger $d(x_0)$ the further the outcome from what's expected under H_0,
       with respect to the particular question;
 (iii) the P-value can be computed $p(x_0)=\Pr(d(X) \geq d(x_0); H_0)$.

Fisher, arch falsificationist, sought test statistics that would be *sensitive* to
discrepancies from the null. Desiderata (i)–(iii) are related, as emerged clearly
from N-P's work.

   Fisher introduced the idea of a parametric statistical model, which may be
written $M_\theta(x)$. Karl Pearson and others had been prone to mixing up a
parameter θ, say the mean of a population, with a sample mean $\bar{x}$. As
a result, concepts that make sense for statistic $\bar{X}$, like having a distribution,
were willy-nilly placed on a fixed parameter θ. Neyman and Pearson [N-P]
gave mathematical rigor to the components of Fisher's tests and estimation.
The model can be represented as a pair (S, Θ) where S denotes the set of all
possible values of the *sample* $X = (X_1, \ldots, X_n)$ – one such value being the data x_0
$= (x_1, \ldots, x_n)$ – and Θ denotes the set of all possible values of the unknown
parameter(s) θ. In hypothesis testing, Θ is used as shorthand for the family of
probability distributions or, in continuous cases, densities *indexed* by θ.
Without the abbreviation, we'd write the full model as

$$M_\theta(x) := \{f(x; \theta), \theta \in \Theta\},$$

where $f(x; \theta)$, for all $x \in S$, is the distribution (or density) of the sample.
We don't test all features of the model at once; it's part of the test specification

to indicate which features (parameters) of the model are under test. The *generic form* of *null* and *alternative* hypotheses is

$$H_0: \theta \in \Theta_0 \text{ vs. } H_1: \theta \in \Theta_1,$$

where (Θ_0, Θ_1) constitute subsets of Θ that partition Θ. Together, Θ_0 and Θ_1 exhaust the parameter space. N-P called H_0 the *test hypothesis*, which is preferable to null hypothesis, since for them it's on par with alternative H_1; but for brevity and familiarity, I mostly call H_0 the null. I follow A. Spanos' treatment.

Lambda Criterion

What were Neyman and Pearson looking for in their joint work from 1928? They sought a criterion for choosing, as well as generating, sensible test statistics. Working purely on intuition, which they later imbued with a justification, N-P employ the likelihood ratio. Pearson found the spark of the idea from correspondence with Gosset, known as Student, but we will see that generating good tests requires much more than considering alternatives.

How can we consider the likelihood ratio of hypotheses when one or both can contain multiple values of the parameter? They consider the maximum values that the likelihood could take over ranges of the parameter space. In particular, they take the maximum likelihood over all possible values of θ in the entire parameter space Θ (not Θ_1), and compare it to the maximum over the restricted range of values in Θ_0, to form the ratio

$$\Lambda(X) = \frac{\max_{\theta \in \Theta} L(X; \theta)}{\max_{\theta \in \Theta_0} L(X; \theta)}.$$

Let's look at this. The numerator is the value of θ that makes the data x most probable over the entire parameter space. It is the *maximum likelihood estimator* for θ. Write it as $\hat{\theta}$. The denominator is the value of θ that maximizes the probability of x restricted just to the members of the null Θ_0. It may be called the *restricted* likelihood. Write it as $\tilde{\theta}$:

$$\Lambda(X) = \frac{L(\hat{\theta}\text{-unrestricted})}{L(\tilde{\theta}\text{-restricted})}.$$

Suppose that looking through the entire parameter space Θ we cannot find a θ value that makes the data more probable than if we restrict ourselves to the parameter values in Θ_0. Then the restricted likelihood in the denominator is large, making the ratio $\Lambda(X)$ small. Thus, a small $\Lambda(X)$ corresponds to H_0 being in accordance with the data (Wilks 1962, p. 404). It's a matter of convenience which way one writes the ratio. In the one we've chosen, following Aris Spanos (1986, 1999), the larger the $\Lambda(X)$, the more discordant the data are from H_0. This suggests the null would be rejected whenever

$$\Lambda(X) \geq k_\alpha$$

for some value of k_α.

So far all of this was to form the distance measure $\Lambda(X)$. It's looking somewhat the same as the Likelihoodist account. Yet we know that the additional step 3 that error statistics demands is to compute the probability of $\Lambda(X)$ under different hypotheses. Merely reporting likelihood ratios does not produce meaningful control of errors; nor do likelihood ratios mean the same thing in different contexts. So N-P consider the probability distribution of $\Lambda(X)$, and they want to ensure the probability of the event $\{\Lambda(X) \geq k_\alpha\}$ is sufficiently small under H_0. They set k_α so that

$$\Pr(\Lambda(X) \geq k_\alpha; H_0) = \alpha$$

for small α. Equivalently, they want to ensure high probability of accordance with H_0 just when it adequately describes the data generation process. Note the complement:

$$\Pr(\Lambda(X) < k_\alpha; H_0) = (1 - \alpha).$$

The event statement to the left of ";" does not reverse positions with H_0 when you form the complement, H_0 stays where it is.

The set of data points leading to $(\Lambda(X) \geq k_\alpha)$ is what N-P call the *critical region* or *rejection region* of the test $\{x\colon \Lambda(X) \geq k_\alpha\}$ – the set of outcomes that will be taken to reject H_0 or, in our terms, to infer a discrepancy from H_0 in the direction of H_1. Specifying the test procedure, in other words, boils down to specifying the rejection (of H_0) region.

Monotonicity. Following Fisher's goal of maximizing sensitivity, N-P seek to maximize the capability of detecting discrepancies from H_0 when they exist. We need the sampling distribution of $\Lambda(X)$, but in practice, $\Lambda(X)$ is rarely in a form that one could easily derive this. $\Lambda(X)$ has to be transformed in clever ways to yield a test statistic $d(X)$, a function of the sample that has a known distribution under H_0. A general trick to finding a suitable test statistic $d(X)$ is to find a function $h(\cdot)$ of $\Lambda(X)$ that is *monotonic* with respect to a statistic $d(X)$. The greater $d(X)$ is, the greater the likelihood ratio; the smaller d(X) is, the smaller the likelihood ratio. Having transformed $\Lambda(X)$ into the test statistic d(X), the rejection region becomes

$$\text{Rejection Region, RR} := \{x: \text{d}(x) \geq c_\alpha\},$$

the set of data points where d(x) $\geq c_\alpha$. All other data points belong to the "non-rejection" or "acceptance" region, NR. At first Neyman and Pearson introduced an "undecided" region, but tests are most commonly given such that the RR and NR regions exhaust the entire sample space S. The term "acceptance," Neyman tells us, was merely shorthand: "The phrase 'do not reject H' is longish and cumbersome ... My own preferred substitute for 'do not reject H' is 'no evidence against H is found'" (Neyman 1976, p. 749). That is the interpretation that should be used.

The use of the $\Lambda(\cdot)$ criterion began as E. Pearson's intuition. Neyman was initially skeptical. Only later did he show it could be the basis for good and even optimal tests.

Having established the usefulness of the Λ-criterion, we realized that it was essential to explore more fully the sense in which it led to tests which were likely to be effective in detecting departures from the null hypothesis. So far we could only say that it seemed to appeal to intuitive requirements for a good test. (E. Pearson 1970 p. 470, I replace λ with Λ)

Many other desiderata for good tests present themselves.

We want a higher and higher value for $\Pr(\text{d}(X) \geq c_\alpha; \theta_1)$ as the discrepancy $(\theta_1 - \theta_0)$ increases. That is, the larger the discrepancy, the easier (more probable) it should be to detect it. This came to be known as the *power function*. Likewise, the power should increase as the sample size increases, and as the variability decreases. The point is that Neyman and Pearson did not start out with a conception of optimality. They groped for criteria that intuitively made sense and that reflected Fisher's tests and theory of estimation. There are some early papers in 1928, but the N-P classic result isn't until the paper in 1933.

Powerful Tests. Pearson describes the days when he and Neyman are struggling to compare various different test statistics – Neyman is in Poland, he is in England. Pearson found himself simulating power for different test statistics and tabling the results. He calls them "empirical power functions." Equivalently, he made tables of the complement to the empirical power function: "what was tabled was the percentage of samples for which a test at 5 percent level failed to establish significance, as the true mean shifted from μ_0 by steps of $\sigma/\sqrt{n}$ (ibid. p. 471). He's construing the test's capabilities in terms of percentage of samples. The formal probability distributions serve as short-cuts to cranking out the percentages. "While the results were crude, they show that our thoughts were turning towards the justification of tests in terms of power"(ibid.).

While Pearson is busy experimenting with simulated power functions, Neyman writes to him in 1931 of difficulties he is having in more complicated cases, saying: I found a test in which, paradoxically, "*the true hypothesis will be rejected more often than some of the false ones*. I told Lola [his wife] that we had invented such a test. She said: 'good boys!'" (ibid. p. 472). A test should have a higher probability of leading to a rejection of H_0 when H_1: $\theta \in \Theta_1$ than when H_0: $\theta \in \Theta_0$. After Lola's crack, pretty clearly, they would insist on *unbiased tests*: the probability of rejecting H_0 when it's true or adequate is always less than that of rejecting it when it's false or inadequate. There are direct parallels with properties of good estimators of θ (although we won't have time to venture into that).

Tests that violate unbiasedness are sometimes called "worse than useless" (Hacking 1965, p. 99), but when you read for example in Gigerenzer and Marewski (2015) that N-P found Fisherian tests "worse than useless" (p. 427), there is a danger of misinterpretation. N-P aren't bad-mouthing Fisher. They know he wouldn't condone this, but want to show that without making restrictions explicit, it's possible to end up with such unpalatable tests. In the case of two-sided tests, the additional criterion of unbiasedness led to uniformly most powerful (UMP) unbiased tests.

Consistent Tests. Unbiasedness by itself isn't a sufficient property for a good test; it needs to be supplemented with the property of *consistency*. This requires that, as the sample size n increases without limit, the probability of detecting any discrepancy from the null hypothesis (the power) should approach 1. Let's consider a test statistic that is unbiased yet inconsistent. Suppose we are testing the mean of a Normal distribution with σ known. The test statistic to which the Λ gives rise is

$$d(X) = \sqrt{n}(\bar{x} - \theta_0)/\sigma.$$

Say that, rather than using the sample mean $\bar{x}$, we use the average of the first and last values. This is to estimate the mean θ as $\hat{\theta} = 0.5(X_1 + X_n)$. The test statistic is then $\sqrt{2}(\hat{\theta} - \theta_0)/\sigma$. This is an unbiased estimator of θ. The distribution of $\hat{\theta}$ is $N(\theta, \sigma^2/2)$. Even though this is unbiased and enables control of the Type I error, it is inconsistent. The result of looking only at two outcomes is that the power does not increase as n increases. The power of this test is much lower than a test using the sample mean for any $n > 2$. If you come across a criticism of tests, make sure *consistency* is not being violated.

Historical Sidelight. Except for short visits and holidays, their work proceeded by mail. When Pearson visited Neyman in 1929, he was shocked at the conditions in which Neyman and other academics lived and worked in Poland. Numerous letters from Neyman describe the precarious position in his statistics lab: "You may have heard that we have in Poland a terrific crisis in everything" [1931] (C. Reid 1998, p. 99). In 1932, "I simply cannot work; the crisis and the struggle for existence takes all my time and energy" (Lehmann 2011, p. 40). Yet he managed to produce quite a lot. While at the start, the initiative for the joint work was from Pearson, it soon turned in the other direction with Neyman leading the way.

By comparison, Egon Pearson's greatest troubles at the time were personal: He had fallen in love "at first sight" with a woman engaged to his cousin George Sharpe, and she with him. She returned the ring the very next day, but Egon still gave his cousin two years to win her back (C. Reid 1998, p. 86). In 1929, buoyed by his work with Neyman, Egon finally declares his love and they are set to be married, but he let himself be intimidated by his father, Karl, deciding "that I could not go against my family's opinion that I had stolen my cousin's fiancée ... at any rate my courage failed" (ibid., p. 94). Whenever Pearson says he was "suddenly smitten" with doubts about the justification of tests while gazing on the fruit station that his cousin directed, I can't help thinking he's also referring to this woman (ibid., p. 60). He was lovelorn for years, but refused to tell Neyman what was bothering him.

N-P Tests in Their Usual Formulation: Type I and II Error Probabilities and Power

Whether we accept or reject or remain in doubt, say N-P (1933, p. 146), it must be recognized that we can be wrong. By choosing a distance measure $d(X)$ wherein the probability of different distances may be computed, if the source of the data is H_0, we can determine the probability of an erroneous rejection of H_0 – a Type I error.

The test specification that dovetailed with the Fisherian tests in use began by ensuring the probability of a Type I error – an erroneous rejection of the null – is fixed at some small number, α, the *significance level* of the test:

Type I error probability = $\Pr(d(X) \geq c_\alpha; H_0) \leq \alpha$.

Compare the Type I error probability and the *P*-value:

P-value: $\Pr(d(X) \geq d(x_0); H_0) = p(x_0)$.

So the N-P test could easily be given in terms of the P-value:

Reject H_0 iff $p(x_0) \leq \alpha$.

Equivalently, the rejection (of H_0) region consists of those outcomes whose P-value is less than or equal to α. Reflecting the tests commonly used, N-P suggest the Type I error be viewed as the "more important" of the two. Let the relevant hypotheses be $H_0: \theta = \theta_0$ vs. $H_1: \theta > \theta_0$.

The Type II error is failing to reject the null when it is false to some degree. The test leads you to declare "no evidence of discrepancy from H_0" when H_0 is false, and a discrepancy exists. The alternative hypothesis H_1 contains more than a single value of the parameter, it is *composite*. So, abbreviate by $ß(\theta_1)$: the Type II error probability assuming $\theta = \theta_1$, for θ_1 values in the alternative region H_1:

Type II error probability (at θ_1) $= \Pr(d(X) < c_\alpha; \theta_1) = ß(\theta_1)$, for $\theta_1 \in \Theta_1$.

In Figure 3.2, this is the area to the left of c_α, the vertical dotted line, under the H_1 curve. The shaded area, the complement of the Type II error probability (at θ_1), is the *power* of the test (at θ_1):

Power of the test (POW) (at θ_1) $= \Pr(d(X) \geq c_\alpha; \theta_1)$.

This is the area to the right of the vertical dotted line, under the H_1 curve, in Figure 3.2. Note $d(x_0)$ and c_α are always approximations expressed as decimals. For continuous cases, Pr is the probability density.

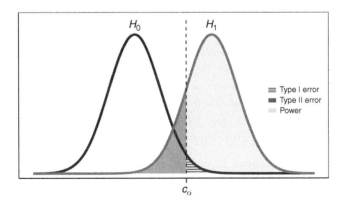

Figure 3.2 Type II error and power.

A *uniformly most powerful* (UMP) N-P test of a hypothesis at level α is one that minimizes $ß(\theta_1)$, or, equivalently, maximizes the power for all $\theta > \theta_0$. One reason alternatives are often not made explicit is the property of being a best test for any alternative. We'll explore power, an often-misunderstood creature, further in Excursion 5.

Although the manipulations needed to derive a test statistic using a monotonic mapping of the likelihood ratio can be messy, it's exhilarating to deduce them. Wilks (1938) derived a general asymptotic result, which does not require such manipulations. He showed that, under certain regularity conditions, as n goes to infinity one can define the asymptotic test, where "~" denotes "is distributed as".

$$2\ln\Lambda(\mathbf{X}) \sim \chi^2(r), \text{ under } H_0, \text{ with rejection region RR} := \{\mathbf{x}: 2\ln\Lambda(\mathbf{x}) \geq c_\alpha\},$$

where $\chi^2(r)$ denotes the chi-square distribution with r degrees of freedom determined by the restrictions imposed by H_0.[4] The monotonicity of the likelihood ratio condition holds for familiar models including one-parameter variants of the Normal, Gamma, Beta, Binomial, Negative Binomial, Poisson (the Exponential family), the Uniform, Logistic, and others (Lehmann 1986). In a wide variety of tests, the Λ principle gives tests with all of the intuitively desirable test properties (see Spanos 1999 and 2019, chapter 13).

Performance versus Severity Construals of Tests

"The work [of N-P] quite literally transformed mathematical statistics" (C. Reid 1998, p. 104). The idea that appraising statistical methods revolves around optimality (of some sort) goes viral. Some compared it "to the effect of the theory of relativity upon physics" (ibid.). Even when the optimal tests were absent, the optimal properties served as benchmarks against which the performance of methods could be gauged. They had established a new pattern for appraising methods, paving the way for Abraham Wald's decision theory, and the seminal texts by Lehmann and others. The rigorous program overshadowed the more informal Fisherian tests. This came to irk Fisher. Famous feuds between Fisher and Neyman erupted as to whose paradigm would reign supreme. Those who sided with Fisher erected examples to show that tests could satisfy predesignated criteria and long-run error control while leading to counterintuitive tests in specific cases. That was Barnard's point on the eclipse

---

[4] The general likelihood ratio $\Lambda(X)$ should be contrasted with the simple likelihood ratio associated with the well-known Neyman–Pearson (N-P) lemma, which assumes that the parameter space Θ includes only two values, i.e., $\Theta := (\theta_0, \theta_1)$. In such a case no estimation is needed because one can take the simple likelihood ratio. Even though the famous lemma for UMP tests uses the highly artificial case of point against point hypotheses (θ_0, θ_1), it is erroneous to suppose the recommended tests are intended for this case. A UMP test, after all, alludes to all the possible parameter values, so just picking two and ignoring the others would not be UMP.

experiments (Section 3.1): no one would consider the class of repetitions as referring to the hoped-for 12 photos, when in fact only some smaller number were usable. We'll meet up with other classic chestnuts as we proceed.

N-P tests began to be couched as formal mapping rules taking data into "reject H_0" or "do not reject H_0" so as to ensure the probabilities of erroneous rejection and erroneous acceptance are controlled at small values, independent of the true hypothesis and regardless of prior probabilities of parameters. Lost in this *behavioristic* formulation was how the test criteria naturally grew out of the requirements of probative tests, rather than good long-run performance. Pearson underscores this in his paper (1947) in the epigraph of Section 3.2: Step 2 comes before Step 3. You must first have a sensible distance measure. Since tests that pass muster on performance grounds can simultaneously serve as probative tests, the severe tester breaks out of the behavioristic prison. Neither Neyman nor Pearson, in their applied work, was wedded to it. Where performance and probativeness conflict, probativeness takes precedent. Two decades after Fisher allegedly threw Neyman's wood models to the floor (Section 5.8), Pearson (1955) tells Fisher: "From the start we shared Professor Fisher's view that in scientific enquiry, a statistical test is 'a means of learning'" (p. 206):

... it was not till after the main lines of this theory had taken shape with its necessary formalization in terms of critical regions, the class of admissible hypotheses, the two sources of error, the power function, etc., that the fact that there was a remarkable parallelism of ideas in the field of acceptance sampling became apparent. Abraham Wald's contributions to decision theory of ten to fifteen years later were perhaps strongly influenced by acceptance sampling problems, but that is another story. (ibid., pp. 204–5)

In fact, the tests as developed by Neyman–Pearson began as an attempt to obtain tests that Fisher deemed intuitively plausible, and this goal is easily interpreted as that of computing and controlling the severity with which claims are inferred.

Not only did Fisher reply encouragingly to Neyman's letters during the development of their results, it was Fisher who first informed Neyman of the split of K. Pearson's duties between himself and Egon, opening up the possibility of Neyman's leaving his difficult life in Poland and gaining a position at University College in London. Guess what else? Fisher was a referee for the all-important N–P 1933 paper, and approved of it.

To Neyman it has always been a source of satisfaction and amusement that his and Egon's fundamental paper was presented to the Royal Society by Karl Pearson, who was hostile and skeptical of its contents, and favorably refereed by the formidable Fisher, who was later to be highly critical of much of the Neyman–Pearson theory. (C. Reid 1998, p. 103)

Souvenir J: UMP Tests

Here are some familiar Uniformly Most Powerful (UMP) unbiased tests that fall out of the Λ criterion (letting μ be the mean):

(1) One-sided Normal test. Each X_i is NIID, $N(\mu, \sigma^2)$, with σ known: $H_0: \mu \leq \mu_0$ against $H_1: \mu > \mu_0$.

$$d(X) = \sqrt{n}(\overline{X} - \mu_0)/\sigma, \ RR(\alpha) = \{x: d(x) \geq c_\alpha\}.$$

Evaluating the Type I error probability requires the distribution of $d(X)$ under $H_0: d(X) \sim N(0,1)$.

Evaluating the Type II error probability (and power) requires the distribution of $d(X)$ under $H_1[\mu = \mu_1]$:

$$d(X) \sim N(\delta_1, 1), \text{ where } \delta_1 = \sqrt{n}(\mu_1 - \mu_0)/\sigma.$$

(2) One-sided Student's t test. Each X_i is NIID, $N(\mu, \sigma^2)$, σ unknown: $H_0: \mu \leq \mu_0$ against $H_1: \mu > \mu_0$:

$$d(X) = \sqrt{n}(\overline{X} - \mu_0)/s, \quad RR(\alpha) = \{x: d(x) \geq c_\alpha\},$$

$$s^2 = \left[\frac{1}{(n-1)}\right]\sum(X_i - \overline{X})^2.$$

Two-sided Normal test of the mean $H_0: \mu = \mu_0$ against $H_1: \mu \neq \mu_0$:

$$d(X) = \sqrt{n}(\overline{X} - \mu_0)/s, \quad RR(\alpha) = \{x: |d(x)| \geq c_\alpha\}.$$

Evaluating the Type I error probability requires the distribution of $d(X)$ under $H_0: d(X) \sim St(n-1)$, the Student's t distribution with $(n-1)$ degrees of freedom (df).

Evaluating the Type II error probability (and power) requires the distribution of $d(X)$ under $H_1[\mu = \mu_1]: d(X) \sim St(\delta_1, n-1)$, where $\delta_1 = \sqrt{n}(\mu_1 - \mu_0)/\sigma$ is the non-centrality parameter.

This is the UMP, unbiased test.

(3) The difference between two means (where it is assumed the variances are equal):

$H_0: \gamma := \mu_1 - \mu_2 = \gamma_0$ against $H_1: \gamma_1 \neq \gamma_0$.
A Uniformly Most Powerful Unbiased (UMPU) test is defined by

$$\tau(\mathbf{Z}) = \frac{\sqrt{n}\left[(\overline{X}_n - \overline{Y}_n) - \gamma_0\right]}{s\sqrt{2}}, \text{RR} = \left\{\mathbf{z}: |\tau(\mathbf{z})| \geq c_\alpha\right\}.$$

Under H_0: $\tau(\mathbf{Z}) = \dfrac{\sqrt{n}\left[(\overline{X}_n - \overline{Y}_n) - \gamma_0\right]}{s\sqrt{2}} \sim \text{St}(2n-2),$

under $H_1[\gamma = \gamma_1]$: $\tau(\mathbf{Z}) \sim \text{St}(\delta_1; 2n-2)$, $\delta_1 = \dfrac{\sqrt{n}\,(\gamma_1 - \gamma_0)}{\sigma\sqrt{2}}$, for $\gamma_1 \neq \gamma_0$.

Many excellent sources of types of tests exist, so I'll stop with these.

Exhibit (i): N-P Methods as Severe Tests: First Look (Water Plant Accident).
There's been an accident at a water plant where our ship is docked, and the cooling system had to be repaired. It is meant to ensure that the mean temperature of discharged water stays below the temperature that threatens the ecosystem, perhaps not much beyond 150 degrees Fahrenheit. There were 100 water measurements taken at randomly selected times and the sample mean $\bar{x}$ computed, each with a known standard deviation $\sigma = 10$. When the cooling system is effective, each measurement is like observing $X \sim N(150, 10^2)$. Because of this variability, we expect different 100-fold water samples to lead to different values of $\overline{X}$, but we can deduce its distribution. If each $X \sim N(\mu = 150, 10^2)$ then $\overline{X}$ is also Normal with $\mu = 150$, but the standard deviation of $\overline{X}$ is only $\sigma/\sqrt{n} = 10/\sqrt{100} = 1$. So $\overline{X} \sim N(\mu = 150, 1)$.
It is the distribution of $\overline{X}$ that is the relevant sampling distribution here. Because it's a large random sample, the sampling distribution of $\overline{X}$ is Normal or approximately so, thanks to the Central Limit Theorem. Note the mean of the sampling distribution of $\overline{X}$ is the same as the underlying mean, both are μ. The frequency link was *created* by randomly selecting the sample, and we assume for the moment it was successful. Suppose they are testing

H_0: $\mu \leq 150$ vs. H_1: $\mu > 150$.

The test rule for $\alpha = 0.025$ is

Reject H_0: iff $\overline{X} \geq 150 + c_\alpha \sigma/\sqrt{100} = 150 + 1.96(1) = 151.96$,

since $c_\alpha = 1.96$.

For simplicity, let's go to the 2-standard error cut-off for rejection:

Reject H_0 (infer there's an indication that $\mu > 150$) iff $\overline{X} \geq 152$.

The test statistic $d(\mathbf{x})$ is a standard Normal variable: $Z = \sqrt{100}(\overline{X} - 150)/10 = \overline{X} - 150$, which, for $\bar{x} = 152$, is 2. The area to the right of 2 under the standard Normal is around 0.025.

Now we begin to move beyond the strict N-P interpretation. Say $\bar{x}$ is just significant at the 0.025 level ($\bar{x} = 152$). What warrants taking the data as indicating $\mu > 150$ is not that they'd rarely be wrong in repeated trials on cooling systems by acting this way – even though that's true. There's a good indication that it's not in compliance right now. Why? *The severity rationale*: Were the mean temperature no higher than 150, then over 97% of the time their method would have resulted in a lower mean temperature than observed. Were it clearly in the safe zone, say $\mu = 149$ degrees, a lower observed mean would be even more probable. Thus, $\bar{x} = 152$ indicates *some* positive discrepancy from H_0 (though we don't consider it rejected by a single outcome). They're going to take another round of measurements before acting. In the context of a policy action, to which this indication might lead, some type of loss function would be introduced. We're just considering the evidence, based on these measurements; all for illustrative purposes.

Severity Function

I will abbreviate "the severity with which claim C passes test T with data x":

SEV(test T, outcome x, claim C).

Reject/Do Not Reject will be interpreted inferentially, in this case as an indication or evidence of the presence or absence of discrepancies of interest.

Let us suppose we are interested in assessing the severity of $C: \mu > 153$. I imagine this would be a full-on emergency for the ecosystem!

Reject H_0. Suppose the observed mean is $\bar{x} = 152$, just at the cut-off for rejecting H_0:

$$d(x_0) = \sqrt{100}(152 - 150)/10 = 2.$$

The data reject H_0 at level 0.025. We want to compute

SEV$(T, \bar{x} = 152, C: \mu > 153)$.

We may say: "the data accord with $C: \mu > 153$," that is, severity condition (S-1) is satisfied; but severity requires there to be at least a reasonable probability of a worse fit with C if C is false (S-2). Here, "worse fit with C" means $\bar{x} \leq 152$ (i.e., $d(x_0) \leq 2$). Given it's continuous, as with all the following examples, $<$ or $\leq$ give the same result. The context indicates which is more useful. This probability must be high for C to pass severely; if it's low, it's BENT.

Table 3.1 Reject in test T+: $H_0: \mu \leq 150$ vs. H_1:
$\mu > 150$ with $\bar{x} = 152$

Claim $\mu > \mu_1$	Severity $\Pr(\bar{X} \leq 152; \mu = \mu_1)$
$\mu > 149$	0.999
$\mu > 150$	0.97
$\mu > 151$	0.84
$\mu > 152$	0.5
$\mu > 153$	0.16

We need $\Pr(\bar{X} \leq 152; \mu > 153$ is false$)$. To say $\mu > 153$ is false is to say $\mu \leq 153$. So we want $\Pr(\bar{X} \leq 152; \mu \leq 153)$. But we need only evaluate severity at the point $\mu = 153$, because this probability is even greater for $\mu < 153$:

$$\Pr(\bar{X} \leq 152; \mu = 153) = \Pr(Z \leq -1) = 0.16.$$

Here, $Z = \sqrt{100}(152 - 153)/10 = -1$. Thus $\text{SEV}(T, \bar{x} = 152, C : \mu > 153)$ $= 0.16$. Very low. Our minimal severity principle blocks $\mu > 153$ because it's fairly probable (84% of the time) that the test would yield an even larger mean temperature than we got, if the water samples came from a body of water whose mean temperature is 153. Table 3.1 gives the severity values associated with different claims, given $\bar{x} = 152$. Call tests of this form T+

In each case, we are making inferences of form $\mu > \mu_1 = 150 + \gamma$, for different values of γ. To merely infer $\mu > 150$, the severity is 0.97 since $\Pr(\bar{X} \leq 152; \mu = 150) = \Pr(Z \leq 2) = 0.97$. While the data give an indication of non-compliance, $\mu > 150$, to infer $C: \mu > 153$ would be making mountains out of molehills. In this case, the observed difference just hit the cut-off for rejection. N-P tests leave things at that coarse level in computing power and the probability of a Type II error, but severity will take into account the actual outcome. Table 3.2 gives the severity values associated with different claims, given $\bar{x} = 153$.

If "the major criticism of the Neyman–Pearson frequentist approach" is that it fails to provide "error probabilities fully varying with the data," as J. Berger alleges (2003, p. 6), then we've answered the major criticism.

Non-rejection. Now suppose $\bar{x} = 151$, so the test does not reject H_0. The standard formulation of N-P (as well as Fisherian) tests stops there. But we want to be alert to a fallacious interpretation of a "negative" result: inferring there's no positive discrepancy from $\mu = 150$. No (statistical) evidence of non-compliance isn't evidence of compliance; here's why. We have (S-1): the data

Table 3.2 Reject in test T+: H_0: $\mu \leq 150$ vs.
H_1: $\mu > 150$ with $\bar{x} = 153$

Claim $\mu > \mu_1$	Severity (with $\bar{X} = 153$) $Pr(\bar{X} \leq 153; \mu = \mu_1)$
$\mu > 149$	~1
$\mu > 150$	0.999
$\mu > 151$	0.97
$\mu > 152$	0.84
$\mu > 153$	0.5

Table 3.3 Non-reject in test T+: H_0: $\mu \leq 150$ vs.
H_1: $\mu > 150$ with $\bar{x} = 151$

Claim $\mu \leq \mu_1$	Severity $Pr(\bar{X} > 151; \mu = \mu_1)$
$\mu \leq 150$	0.16
$\mu \leq 150.5$	0.3
$\mu \leq 151$	0.5
$\mu \leq 152$	0.84
$\mu \leq 153$	0.97

"accord with" H_0, but what if the test had little capacity to have alerted us to discrepancies from 150? The alert comes by way of "a worse fit" with H_0 – namely, a mean $\bar{x} = 151$. Condition (S-2) requires us to consider $Pr(\bar{X} > 151; \mu = 150)$, which is only 0.16. To get this, standardize $\bar{X}$ to obtain a standard Normal variate: $Z = \sqrt{100}(151 - 150)/10 = 1$; and $Pr(\bar{X} > 151$; $\mu = 150) = 0.16$. Thus, SEV(T+, $\bar{x} = 151$, C: $\mu \leq 150$) = low(0.16). Table 3.3 gives the severity values associated with different inferences of form $\mu \leq \mu_1 = 150 + \gamma$, given $\bar{x} = 151$.

Can they at least say that $\bar{x} = 151$ is a good indication that $\mu \leq 150.5$? No, SEV(T+, $\bar{x} = 151$, C: $\mu \leq 150.5$) $\simeq$ 0.3, [$Z = 151 - 150.5 = 0.5$]. But $\bar{x} = 151$ is a good indication that $\mu \leq 152$ and $\mu \leq 153$ (with severity indications of 0.84 and 0.97, respectively).

You might say, assessing severity is no different from what we would do with a judicious use of existing error probabilities. That's what the severe tester says. Formally speaking, it may be seen merely as a good rule of thumb to avoid fallacious interpretations. What's new is the statistical philosophy behind it.

We no longer seek either probabilism or performance, but rather using relevant error probabilities to assess and control severity.[5]

3.3   How to Do All N-P Tests Do (and More) While a Member of the Fisherian Tribe

When Karl Pearson retired in 1933, he refused to let his chair go to Fisher, so they split the department into two: Fisher becomes Galton Chair and Head of Eugenics, while Egon Pearson becomes Head of Applied Statistics. They are one floor removed (Fisher on top)! The Common Room had to be "carefully shared," as Constance Reid puts it: "Pearson's group had afternoon tea at 4; and at 4:30, when they were safely out of the way, Fisher and his group trouped in" (C. Reid 1998, p. 114). Fisher writes to Neyman in summer of 1933 (cited in Lehmann 2011, p. 58):

> You will be interested to hear that the Dept. of Statistics has now been separated officially from the Galton Laboratory. I think Egon Pearson is designated as Reader in Statistics. This arrangement will be much laughed at, but it will be rather a poor joke ... I shall not lecture on statistics, but probably on 'the logic of experimentation'.

Finally E. Pearson was able to offer Neyman a position at University College, and Neyman, greatly relieved to depart Poland, joins E. Pearson's department in 1934.[6]

Neyman doesn't stay long. He leaves London for Berkeley in 1938, and develops the department into a hothouse of statistics until his death in 1981. His first Ph.D. student is Erich Lehmann in 1946. Lehmann's *Testing Statistical Hypotheses*, 1959, the canonical N-P text, developed N-P methods very much in the mode of the N-P-Wald, behavioral-decision language. I find it interesting that even Neyman's arch opponent, subjective Bayesian Bruno de Finetti, recognized that "inductive behavior ... that was for Neyman simply a slogan underlining and explaining the difference between his own, the Bayesian and the Fisherian formulations" became, with Wald's work,

---

[5] Initial developments of the severity idea were Mayo (1983, 1988, 1991, 1996). In Mayo and Spanos (2006, 2011), it was developed much further.

[6] "By the fall of 1932 there appeared to be several reasons why Neyman might never become a professor in Poland. One was his subject matter, which was not generally recognized as an academic specialty. Another was the fact that he was married to a Russian – and an independent, outspoken Russian who lived on occasion apart from her husband, worked and painted in Paris, traveled on a freighter as a nurse for the adventure of it, and sometimes led tourist excursions into the Soviet Union." (C. Reid 1998, p. 105).

"something much more substantial." De Finetti called this "the involuntarily destructive aspect of Wald's work" (1972, p. 176). Cox remarks:

[T]here is a distinction between the Neyman–Pearson formulation of testing regarded as clarifying the meaning of statistical significance via hypothetical repetitions and that same theory regarded as in effect an instruction on how to implement the ideas by choosing a suitable α in advance and reaching different decisions accordingly. The interpretation to be attached to accepting or rejecting a hypothesis is strongly context-dependent . . . (Cox 2006a, p. 36)

If N-P long-run performance concepts serve to clarify the meaning of statistical significance tests, yet are not to be applied literally, but rather in some inferential manner – call this the *meaning* vs. *application distinction* – the question remains – how?

My answer, in terms of severity, may be used whether you prefer the N-P tribe (tests or confidence intervals) or the Fisherian tribe. What would that most eminent Fisherian, Sir David Cox, say? In 2004, in a session we were in on statistical philosophy, at the semi-annual Lehmann conference, we asked: Was it possible to view "Frequentist Statistics as a Theory of Inductive Inference"? If this sounds familiar it's because it echoes a section from Neyman's quarrel with Carnap (Section 2.7), but how does a Fisherian answer it? We began "with the core elements of significance testing in a version very strongly related to but in some respects different from both Fisherian and Neyman–Pearson approaches, at least as usually formulated" (Mayo and Cox 2006, p. 80). First, there is no suggestion that the significance test would typically be the only analysis reported. Further, we agree that "the justification for tests will not be limited to appeals to long-run behavior but will instead identify an inferential or evidential rationale" (ibid., p. 81).

With N-P results available, it became easier to understand why intuitively useful tests worked for Fisher. N-P and Fisherian tests, while starting from different places, "lead to the same destination" (with few exceptions) (Cox 2006a, p. 25). Fisher begins with seeking a test statistic that reduces the data as much as possible, and this leads him to a *sufficient* statistic. Let's take a side tour to sufficiency.

Exhibit (ii): Side Tour of Sufficient Statistic. Consider n independent trials $X := (X_1, X_2, \ldots, X_n)$ each with a binary outcome (0 or 1), where the probability of success is an unknown constant θ associated with Bernoulli trials. The number of successes in n trials, $Y = X_1 + X_2 + \cdots + X_n$ is Binomially distributed with parameters θ and n. The sample mean, which is just $\overline{X} = Y/n$, is a natural estimator of θ with a highly desirable property: it is *sufficient*, i.e., it is a function of the *sufficient* statistic Y. Intuitively, a sufficient statistic reduces the n-dimensional sample X into a statistic of much smaller dimensionality without losing any relevant information for inference purposes. Y reduces the n-fold outcome x to one dimension: the number of successes in n trials. The parameter of the Binomial model θ also has one dimension (the probability of success on each trial).

Formally, a statistic Y is said to be sufficient for θ when the distribution of the sample is no longer a function of θ when conditioned on Y, i.e., $f(x \mid y)$ does not depend on θ,

$$f(x; \theta) = f(y; \theta) \, f(x|y).$$

Knowing the distribution of the sufficient statistic Y suffices to compute the probability of any data set x. The test statistic $d(X)$ in the Binomial case is $\sqrt{n}(\overline{X} - \theta_0)/\sigma$, $\sigma = \sqrt{[\theta(1 - \theta)]}$ and, as required, gets larger as $\overline{X}$ deviates from θ_0. Thanks to $\overline{X}$ being a function of the sufficient statistic Y, it is the basis for a test statistic with maximal sensitivity to inconsistencies with the null hypothesis.

The Binomial experiment is equivalent to having been given the data $x_0 = (x_1, x_2, \ldots, x_n)$ in two stages (Cox and Mayo 2010, p. 285):

First, you're told the value of Y, the number of successes out of n Bernoulli trials. Then an inference can be drawn about θ using the sampling distribution of Y.

Second, you learn the value of the specific data, e.g., the first k trials are successes, the rest failure. The second stage is equivalent to observing a realization of the conditional distribution of X given $Y = y$. If the model is appropriate then "the second phase is equivalent to a random draw from a totally known distribution." All permutations of the sequence of successes and failures are equally probable (ibid., pp 284–5).

"Because this conditional distribution is totally known, it can be used to assess the validity of the assumed model." (ibid.) Notice that for a given x *within* a given Binomial experiment, the ratio of likelihoods at two different values of θ depends on the data only through Y. This is called the *weak likelihood principle* in contrast to the general (or strong) LP in Section 1.5.

Principle of Frequentist Evidence, FEV

Returning to our topic, "Frequentist Statistics as a Theory of Inductive Inference," let me weave together three threads: (1) the Frequentist Principle of Evidence (Mayo and Cox 2006), (2) the divergent interpretations growing out of Cox's taxonomy of test hypotheses, and (3) the links to statistical inference as severe tests. As a starting point, we identified a general principle that we dubbed the Frequentist Principle of Evidence, FEV:

> *FEV(i)*: x is … evidence against H_0 (i.e., evidence of a discrepancy from H_0), if and only if, were H_0 a correct description of the mechanism generating x, then, with high probability, this would have resulted in a less discordant result than is exemplified by x. (Mayo and Cox 2006, p. 82; substituting x for y)

This sounds wordy and complicated. It's much simpler in terms of a quantitative difference as in significance tests. Putting FEV(i) in terms of formal P-values, or test statistic d (abbreviating d(X)):

> *FEV(i)*: x is evidence against H_0 (i.e., evidence of discrepancy from H_0), if and only if the P-value $\Pr(d \geq d_0; H_0)$ is very low (equivalently, $\Pr(d < d_0; H_0) = 1 - P$ is very high).

(We used "strong evidence", although I would call it a mere "indication" until an appropriate audit was passed.) Our minimalist claim about bad evidence, no test (BENT) can be put in terms of a corollary (from contraposing FEV(i)):

> *FEV(ia)*: x is poor evidence against H_0 (poor evidence of discrepancy from H_0), if there's a high probability the test would yield a more discordant result, if H_0 is correct.

Note the one-directional 'if' claim in FEV(ia). We wouldn't want to say this is the only way x can be BENT.

Since we wanted to move away from setting a particular small P-value, we refer to "P-small" (such as 0.05, 0.01) and "P-moderate", or "not small" (such as 0.3 or greater). We need another principle in dealing with non-rejections or insignificant results. They are often imbued with two very different false interpretations: one is that (a) non-significance indicates the truth of the null, the other is that (b) non-significance is entirely uninformative.

The difficulty with (a), regarding a modest P-value as evidence in favor of H_0, is that accordance between H_0 and x may occur even if rivals to H_0 seriously different from H_0 are true. This issue is particularly acute when the capacity to have detected discrepancies is low. However, as against (b), null results have an important role ranging from the scrutiny of substantive theory – setting bounds to parameters to scrutinizing the capability of a method for finding something out. In sync with our "arguing from error" (Excursion 1), we may infer a discrepancy from H_0 is absent if our test very probably would have alerted us to its presence (by means of a more significant P-value).

FEV(ii): A moderate P-value is evidence of the absence of a discrepancy δ from H_0, only if there is a high probability the test would have given a worse fit with H_0 (i.e., smaller P-value) were a discrepancy δ to exist (ibid., pp. 83–4).

This again is an "if-then" or conditional claim. These are canonical pieces of statistical reasoning, in their naked form as it were. To dress them up to connect with actual questions and problems of inquiry requires context-dependent, background information.

FIRST Interpretations: Fairly Intimately Related to the Statistical Test – Cox's Taxonomy

> In the statistical analysis of scientific and technological data, there is virtually always external information that should enter in reaching conclusions about what the data indicate with respect to the primary question of interest. Typically, these background considerations enter not by a probability assignment but by identifying the question to be asked, designing the study, interpreting the statistical results and relating those inferences to primary scientific ones ... (Mayo and Cox 2006, p. 84)

David Cox calls for an interpretive guide between a test's mathematical formulation and substantive applications: "I think that more attention to these rather vague general issues is required if statistical ideas are to be used in the most fruitful and wide-ranging way" (Cox 1977, p. 62). There are aspects of the context that go beyond the mathematics but which are Fairly Intimately Related to the Statistical Test (FIRST) interpretations. I'm distinguishing these FIRST interpretations from wider substantive inferences, not that there's a strict red line difference.

While warning that "it is very bad practice to summarise an important investigation solely by a value of P" (1982, p. 327), Cox gives a rich taxonomy of null hypotheses that recognizes how significance tests can function as part of complex and context-dependent inquiries (1977, pp. 51–2). Pure or simple Fisherian significance tests (with no explicit alternative) are housed within the taxonomy, not separated out as some radically distinct entity. If his taxonomy had been incorporated into the routine exposition of tests, we could have avoided much of the confusion we are still suffering with. The proper way to view significance tests acknowledges a variety of problem situations:

- Are we testing parameter values within some overriding model? (fully embedded)
- Are we merely checking if a simplification will do? (nested alternative)
- Do we merely seek the direction of an effect already presumed to exist? (dividing)
- Would a model pass an audit of its assumptions? (test of assumption)
- Should we worry about data that appear anomalous for a theory that has already passed severe tests? (substantive)

Although Fisher, strictly speaking, had only the null hypothesis, and context directed an appropriate test statistic, the result of such a selection is that the test is sensitive to a type of discrepancy. Even if they only become explicit after identifying a test statistic – which some regard as more basic (e.g., Senn) – we may still regard them as alternatives.

Sensitivity Achieved or Attained

For a Fisherian like Cox, a test's power only has relevance pre-data, in planning tests, but, like Fisher, he can measure "sensitivity":

In the Neyman–Pearson theory of tests, the sensitivity of a test is assessed by the notion of *power*, defined as the probability of reaching a preset level of significance ... for various alternative hypotheses. In the approach adopted here the assessment is via the distribution of the random variable *P*, again considered for various alternatives. (Cox 2006a, p. 25)

This is the key: Cox will measure sensitivity by a function we may abbreviate as $\Pi(\gamma)$. Computing $\Pi(\gamma)$ may be regarded as viewing the *P*-value as a statistic. That is:

$$\Pi(\gamma) = \Pr(P \leq p_{obs}; \mu_0 + \gamma).$$

The alternative is $\mu_1 = \mu_0 + \gamma$. Using the *P*-value distribution has a long history and is part of many approaches. Given the *P*-value inverts the distance, it is clearer and less confusing to formulate $\Pi(\gamma)$ in terms of the test statistic *d*. $\Pi(\gamma)$ is very similar to *power* in relation to alternative μ_1, except that $\Pi(\gamma)$ considers the observed difference rather than the N-P cut-off c_α:

$$\Pi(\gamma) = \Pr(d \geq d_0; \mu_0 + \gamma),$$

$$POW(\gamma) = \Pr(d \geq c_\alpha; \mu_0 + \gamma).$$

Π may be called a "sensitivity function," or we might think of $\Pi(\gamma)$ as the "attained power" to detect discrepancy γ (Section 5.3). The nice thing about power is that it's always in relation to an observed difference from a test or null hypothesis, which gives it a reference. Let's agree that Π will always relate to an observed difference from a designated test hypothesis H_0.

Aspects of Cox's Taxonomy

I won't try to cover Cox's full taxonomy, which has taken different forms. I propose that the main delineating features are, first, whether the null and alternative exhaust the answers or parameter space for the given question, and, second, whether the null hypothesis is considered a viable basis for a substantive research claim, or merely as a reference for exploring the way in which it is false. None of these are hard and fast distinctions, but you'll soon see why they are useful. I will adhere closely to what Cox has said about the taxonomy and the applications of FEV; all I add is a proposed synthesis. I restrict myself now to a single parameter. We assume the P-value has passed an audit (except where noted).

1. **Fully embedded.** Here we have exhaustive parametric hypotheses governed by a parameter θ, such as the mean deflection of light at the 1919 eclipse, or the mean temperature. H_0: $\mu = \mu_0$ vs. H_1: $\mu > \mu_0$ is a typical N-P setting. Strictly speaking, we may have $\theta = (\mu,k)$ with additional parameters k to be estimated. This formulation, Cox notes, "will suggest the most sensitive test statistic, essentially equivalent to the best estimate of μ" (Cox 2006a, p. 37).

A. P-value is modest (not small): Since the data accord with the null hypothesis, FEV directs us to examine the probability of observing a result more discordant from H_0 if $\mu = \mu_0 + \gamma$: $\Pi(\gamma) = \Pr(d \geq d_0; \mu_0 + \gamma)$.

If that probability is very high, following FEV(ii), the data indicate that $\mu < \mu_0 + \gamma$.

Here $\Pi(\gamma)$ gives the severity with which the test has probed the discrepancy γ. So we don't merely report "no evidence against the null," we report a discrepancy that can be ruled out with severity. "This avoids unwarranted interpretations of consistency with H_0 with insensitive tests ... [and] is more relevant to specific data than is the notion of power" (Mayo and Cox 2006, p. 89).

B. P-value is small: From FEV(i), a small P-value indicates evidence of *some* discrepancy $\mu > \mu_0$ since $\Pr(d < d_0; H_0) = 1 - P$ is large. This is the basis for ordinary (statistical) falsification.

However, we add, "if a test is so sensitive that a P-value as small as or even smaller than the one observed is probable even when $\mu \leq \mu_0 + \gamma$, then a small value of P" is poor evidence of a discrepancy from H_0 in excess of γ (ibid.). That is, from FEV(ia), if $\Pi(\gamma) = \Pr(d \geq d_0; \mu_0 + \gamma)$ = moderately high (greater than 0.3, 0.4, 0.5), then there's poor grounds for inferring $\mu > \mu_0 + \gamma$. This is equivalent to saying the SEV($\mu > \mu_0 + \gamma$) is poor.

There's no need to set a sharp line between significance or not in this construal – extreme cases generally suffice. FEV leads to an inference as to both what's indicated, and what's not indicated. Both are required by a severe tester. Go back to our accident at the water plant. The non-significant result, $\bar{x} = 151$ in testing $\mu \leq 150$ vs. $\mu > 150$, only attains a P-value of 0.16. SEV($C: \mu > 150.5$) is 0.7 (Table 3.3). Not terribly high, but if that discrepancy was of interest, it wouldn't be ignorable. A reminder: we are not making inferences about point values, even though we need only compute Π at a point. In this first parametric pigeonhole, confidence intervals can be formed, though we wouldn't limit them to the typical 0.95 or 0.99 levels.[7]

2. **Nested alternative** (non-exhaustive). In a second pigeonhole an alternative statistical hypothesis H_1 is considered not "as a possible base for ultimate interpretation but as a device for determining a suitable test statistic" (Mayo and Cox 2006, p. 85). Erecting H_1 may be only a convenient way to detect small departures from a given statistical model. For instance, one may use a quadratic model H_1 to test the adequacy of a linear relation. Even though polynomial regressions are a poor base for final analysis, they are very convenient and interpretable for detecting small departures from linearity. (ibid.)

Failing to reject the null (moderate P-value) might be taken to indicate the simplifying assumption is adequate; whereas rejecting the null (small P-value) is not evidence for alternative H_1. That's because there are lots of non-linear models not probed by this test. The H_0 and H_1 do not exhaust the space.

A. *P-value is modest (not small):* At best it indicates adequacy of the model in the respects well probed; that is, it indicates the absence of discrepancies that, very probably, would have resulted in a smaller P-value.

B. *P-value small:* This indicates discrepancy from the null in the direction of the alternative, but it is unwarranted to infer the particular H_1 insofar as other non-linear models could be responsible.

We are still employing the FEV principle, even where it is qualitative.

---

[7] "A significance test is defined by a set of [critical] regions $[w_\alpha]$ satisfying the following essential requirements. First,

$$w_{\alpha_1} \subset w_{\alpha_2} \text{ if } \alpha_1 < \alpha_2;$$

this is to avoid such nonsense as saying that data depart significantly from H_0 at the 1% level but not at the 5% level." Next "we require that, for all α, $\Pr(Y \in w_\alpha; H_0) = \alpha$." (Cox and Hinkley 1974, pp. 90–1)

3. **Dividing nulls:** $H_0: \mu = \mu_0$ vs. $H_1: \mu > \mu_0$ and $H_0: \mu = \mu_0$ vs. $H_1: \mu < \mu_0$. In this pigeonhole, we may already know or suspect the null is false and discrepancies exist: but which? Suppose the interest is comparing two or more treatments. For example, compared with a standard, a new drug may increase or may decrease survival rates.

The null hypothesis of zero difference *divides* the possible situations into two qualitatively different regions with respect to the feature tested. To look at both directions, one combines two tests, the first to examine the possibility that $\mu > \mu_0$, say, the second for $\mu < \mu_0$. The overall significance level is twice the smaller P-value, because of a "selection effect." One may be wrong in two ways. It is standard to report the observed direction and double the initial P-value (if it's a two-sided test).

While a small P-value indicates a direction of departure (e.g., which of two treatments is superior), failing to get a small P-value here merely tells us the data do not provide adequate evidence even of the direction of any difference. Formally, the statistical test may look identical to the fully embedded case, but the nature of the problem, and your background knowledge, yields a more relevant construal. These interpretations are still FIRST. You can still report the upper bound ruled out with severity, bringing this case closer to the fully embedded case (Table 3.4).

4. **Null hypotheses of model adequacy.** In "auditing" a P-value, a key question is: how can I check the model assumptions hold adequately for the data in hand? We distinguish two types of tests of assumptions (Mayo and Cox 2006, p. 89): (i) omnibus and (ii) focused.

(i) With a general *omnibus* test, a group of violations is checked all at once. For example: H_0: IID (independent and identical distribution) assumptions hold vs. H_1: IID is violated. The null and its denial exhaust the possibilities, for the question being asked. However, sensitivity can be so low that failure to reject may be uninformative. On the other hand, a small P-value indicates H_1: there's a departure *somewhere*. The very fact of its low sensitivity indicates that when the alarm goes off something's there. But where? Duhemian problems loom. A subsequent task would be to pin this down.

(ii) A *focused* test is sensitive to a specific kind of model inadequacy, such as a lack of independence. This lands us in a situation analogous to the non-exhaustive case in "nested alternatives." Why? Suppose you erect an alternative H_1 describing a particular type of non-independence, e.g., Markov. While a small P-value indicates some departure, you cannot infer H_1 so long as various alternative models, not probed by this test, could account for it.

Table 3.4 FIRST Interpretations

Taxon	Remarks	Small P-value	P-value Not Small
1. Fully embedded exhaustive	H_1 may be the basis for a substantive interpretation	Indicates $\mu > \mu_0 + \gamma$ iff $\Pi(\gamma)$ is low	If $\Pi(\gamma)$ is high, there's poor indication of $\mu > \mu_0 + \gamma$
2. Nested alternatives non-exhaustive	H_1 is set out to search departures from H_0	Indicates discrepancy from H_0 but not grounds to infer H_1	Indicates H_0 is adequate in respect probed
3. Dividing exhaustive	$\mu \leq \mu_0$ vs. $\mu > \mu_0$; a discrepancy is presumed, but in which direction?	Indicates direction of discrepancy If $\Pi(\gamma)$ low, $\mu > \mu_0 + \gamma$ is indicated	Data aren't adequate even to indicate direction of departure
4. Model assumptions (i) omnibus exhaustive	e.g., non-parametric runs test for IID (may have low power)	Indicates departure from assumptions probed, but not specific violation	Indicates the absence of violations the test is capable of detecting
Model assumptions (ii) focused non-exhaustive	e.g., parametric test for specific type of dependence	Indicates departure from assumptions in direction of H_1 but can't infer H_1	Indicates the absence of violations the test is capable of detecting

It may only give suggestions for alternative models to try. The interest may be in the effect of violated assumptions on the primary (statistical) inference if any. We might ask: Are the assumptions sufficiently questionable to preclude using the data for the primary inference? After a lunch break at Einstein's Cafe, we'll return to the museum for an example of that.

Scotching a Famous Controversy

At a session on the replication crisis at a 2015 meeting of the Society for Philosophy and Psychology, philosopher Edouard Machery remarked as to how, even in so heralded a case as the eclipse tests of GTR, one of the results didn't replicate the other two. The third result pointed, not to Einstein's prediction, but as Eddington ([1920]1987) declared, "with all too good agreement to the 'half-deflection,' that is to say, the Newtonian value" (p. 117). He was alluding to a famous controversy that has grown up surrounding the allegation that Eddington selectively ruled out data that supported the Newtonian "half-value" against the Einsteinian one. Earman and Glymour (1980), among others, alleged that Dyson and Eddington threw out the results unwelcome for GTR for political purposes ("... one of the chief benefits to be derived from the eclipse results was a rapprochement between German and British scientists and an end to talk of boycotting German science" (p. 83)).[8] Failed replication may indeed be found across the sciences, but this particular allegation is mistaken. The museum's display on "Data Analysis in the 1919 Eclipse" shows a copy of the actual notes penned on the Sobral expedition *before* any data analysis:

May 30, 3 a.m., four of the astrographic plates were developed ... It was found that there had been a serious change of focus ... This change of focus can only be attributed to the unequal expansion of the mirror through the sun's heat ... It seems doubtful whether much can be got from these plates. (Dyson et al. 1920, p. 309)

Although a fair amount of (unplanned) data analysis was required, it was concluded that there was no computing a usable standard error of the estimate. The hypothesis:

The data x_0 (from Sobral astrographic plates) were due to systematic distortion by the sun's heat, not to the deflection of light,

passes with severity. An even weaker claim is all that's needed: we can't compute a valid estimate of error. And notice how very weak the claim to be corroborated is!

---

[8] Barnard was surprised when I showed their paper to him, claiming it was a good example of why scientists tended not to take philosophers seriously. But in this case even the physicists were sufficiently worried to reanalyze the experiment.

The mirror distortion hypothesis hadn't been predesignated, but it is altogether justified to raise it in auditing the data: It could have been chewing gum or spilled coffee that spoilt the results. Not only that, the same data hinting at the mirror distortion are to be used in testing the mirror distortion hypothesis (though differently modeled)! That sufficed to falsify the requirement that there was no serious change of focus (scale effect) between the eclipse and night plates. Even small systematic errors are crucial because the resulting scale effect from an altered focus quickly becomes as large as the Einstein predicted effect. Besides, the many staunch Newtonian defenders would scarcely have agreed to discount an apparently pro-Newtonian result.

The case was discussed and soon settled in the journals of the time: the brouhaha came later. It turns out that, if these data points are deemed usable, the results actually point to the Einsteinian value, not the Newtonian value. A reanalysis in 1979 supports this reading (Kennefick 2009). Yes, in 1979 the director of the Royal Greenwich Observatory took out the 1919 Sobral plates and used a modern instrument to measure the star positions, analyzing the data by computer.

[T]he reanalysis provides after-the-fact justification for the view that the real problem with the Sobral astrographic data was the difficulty . . . of separating the scale change from the light deflection. (Kennefick 2009, p. 42)

What was the result of this herculean effort to redo the data analysis from 60 years before?

Ironically, however, the 1979 paper had no impact on the emerging story that something was fishy about the 1919 experiment . . . so far as I can tell, the paper has never been cited by anyone except for a brief, vague reference in Stephen Hawking's *A Brief History of Time* [which actually gets it wrong and was corrected]. (ibid.) [9]

The bottom line is, there was no failed replication; there was one set of eclipse data that was unusable.

5. **Substantively based hypotheses.** We know it's fallacious to take a statistically significant result as evidence in affirming a substantive theory, even if that theory predicts the significant result. A qualitative version of FEV, or, equivalently, an appeal to severity, underwrites this. Can failing to reject statistical null hypotheses ever inform about substantive claims? Yes. First consider how, in the midst of auditing, there's a concern to test a claim: is an apparently anomalous result real or spurious?

---

[9] Data from ESA's Gaia mission should enable light deflection to be measured with an accuracy of 2×10^{-6} (Mignard and Klioner 2009, p. 308).

Finding cancer clusters is sometimes compared to our Texas Marksman drawing a bull's-eye around the shots after they were fired into the barn. They often turn out to be spurious. Physical theory, let's suppose, suggests that because the quantum of energy in non-ionizing electromagnetic fields, such as those from high-voltage transmission lines, is much less than is required to break a molecular bond, there should be no carcinogenic effect from exposure to such fields. Yet a cancer association was reported in 1979 (Wertheimer and Leeper 1979). Was it real? In a randomized experiment where two groups of mice are under identical conditions except that one group is exposed to such a field, the null hypothesis that the cancer incidence rates in the two groups are identical may well be true. Testing this null is a way to ask: was the observed cancer cluster really an anomaly for the theory? Were the apparently anomalous results for the theory genuine, it is expected that H_0 would be rejected, so if it's not, it cuts against the reality of the anomaly. Cox gives this as one of the few contexts where a reported small P-value alone might suffice.

This wouldn't entirely settle the issue, and our knowledge of such things is always growing. Nor does it, in and of itself, show the flaw in any studies purporting to find an association. But several of these pieces taken together can discount the apparent effect with severity. It turns out that the initial researchers in the 1979 study did not actually measure magnetic fields from power lines; when they were measured no association was found. Instead they used the wiring code in a home as a proxy. All they really showed, it may be argued, was that people who live in the homes with poor wiring code tend to be poorer than the control (Gurney et al. 1996). The study was biased. Twenty years of study continued to find negative results (Kheifets et al. 1999). The point just now is not when to stop testing – more of a policy decision – or even whether to infer, as they did, that there's no evidence of a risk, and no theoretical explanation of how there could be. It is rather the role played by a negative statistical result, given the background information that, if the effects were real, these tests were highly capable of finding them. It amounts to a failed replication (of the observed cluster), but with a more controlled method. If a well-controlled experiment fails to replicate an apparent anomaly for an independently severely tested theory, it indicates the observed anomaly is spurious. The indicated severity and potential gaps are recorded; the case may well be reopened. Replication researchers might take note.

Another important category of tests that Cox develops, is what he calls testing *discrete families of models*, where there's no nesting. In a nutshell, each model is taken in turn to assess if the data are compatible with one, both, or neither of the possibilities (Cox 1977, p. 59). Each gets its own severity assessment.

Who Says You Can't Falsify Alternatives in a Significance Test?

Does the Cox–Mayo formulation of tests change the logic of significance tests in any way? I don't think so and neither does Cox. But it's different from some of the common readings. Nothing turns on whether you wish to view it as a revised account. SEV goes a bit further than FEV, and I do not saddle Cox with it. The important thing is how you get a nuanced interpretation, and we have barely begun our travels! Note the consequences for a familiar bugaboo about falsifying alternatives to significance tests. Burnham and Anderson (2014) make a nice link with Popper:

While the exact definition of the so-called "scientific method" might be controversial, nearly everyone agrees that the concept of "falsifiability" is a central tenant [sic] of empirical science (Popper 1959). It is critical to understand that historical statistical approaches (i.e., P values) leave no way to "test" the alternative hypothesis. The alternative hypothesis is never tested, hence cannot be rejected or falsified! ... Surely this fact alone makes the use of significance tests and P values bogus. Lacking a valid methodology to reject/falsify the alternative science hypotheses seems almost a scandal. (p. 629)

I think we *should* be scandalized. But not for the reasons alleged. Fisher emphasized that, faced with a non-significant result, a researcher's attitude wouldn't be full acceptance of H_0 but, depending on the context, more like the following:

The possible deviation from truth of my working hypothesis, to examine which test is appropriate, seems not to be of sufficient magnitude to warrant any immediate modification.

   Or, ... the body of data available so far is not by itself sufficient to demonstrate their [the deviations] reality. (Fisher 1955, p. 73)

Our treatment cashes out these claims, by either indicating the magnitudes ruled out statistically, or inferring that the observed difference is sufficiently common, even if spurious.

   If you work through the logic, you'll see that in each case of the taxonomy the alternative may indeed be falsified. Perhaps the most difficult one is ruling out model violations, but this is also the one that requires a less severe test, at least with a reasonably robust method of inference. So what do those who repeat this charge have in mind? Maybe they mean: you cannot falsify an alternative, if you don't specify it. But specifying a directional or type of alternative is an outgrowth of specifying a test statistic. Thus we still have the implicit alternatives in Table 3.4, all of which are open to being falsified with severity. It's a key part of test specification to indicate which claims or features of a model are being tested. The charge might stand if a point null is known to be false, for in those cases we can't say μ is precisely 0, say. In that case you wouldn't want to infer it. One can still set upper bounds for how far off an adequate hypothesis can be. Moreover, there are many cases in science where a point null *is* inferred severely.

Nordtvedt Effect: Do the Earth and Moon Fall With the Same Acceleration?

We left off Section 3.1 with GTR going through a period of "stagnation" or "hibernation" after the early eclipse results. No one knew how to link it up with experiment. Discoveries around 1959–1960 sparked the "golden era" or "renaissance" of GTR, thanks to quantum mechanics, semiconductors, lasers, computers, and pulsars (Will 1986, p. 14). The stage was set for new confrontations between GTR's experiments; from 1960 to 1980, a veritable "zoo" of rivals to GTR was erected, all of which could be constrained to fit the existing test results.

Not only would there have been too many alternatives to report a pairwise comparison of GTR, the testing had to manage without having full-blown alternative theories of gravity. They could still ask, as they did in 1960: How could it be a mistake to regard the existing evidence as good evidence for GTR (or even for the deflection effect)?

They set out a scheme of parameters, the Parameterized Post Newtonian (PPN) framework, that allowed experimental relativists to describe violations to GTR's hypotheses – discrepancies with what it said about specific gravitational phenomena. One parameter is λ – the curvature of spacetime. An explicit goal was to prevent researchers from being biased toward accepting GTR prematurely (Will 1993, p. 10). These alternatives, by the physicist's own admission, were set up largely as straw men to either set firmer constraints on estimates of parameters, or, more interestingly, find violations. They could test 10 or 20 or 50 rivals without having to develop them! The work involved local statistical testing and estimation of parameters describing curved space.

Interestingly, these were non-novel hypotheses set up after the data were known. However rival theories had to be *viable*; they had to (1) account for experimental results already severely passed and (2) be able to show the relevance of the data for gravitational phenomena. They would have to be able to analyze and explore data about as well as GTR. They needed to permit stringent probing to learn more about gravity. (For an explicit list of requirements for a viable theory, see Will 1993, pp. 18–21.[10])

---

[10] While a viable theory can't just postulate the results ad hoc, "this does not preclude 'arbitrary parameters' being required for gravitational theories to accord with experimental results" (Mayo 2010a, p. 48).

All the viable members of the zoo of GTR rivals held the *equivalence principle (EP)*, roughly the claim that bodies of different composition fall with the same accelerations in a gravitational field. This principle was inferred with severity by passing a series of null hypotheses (examples include the Eötvös experiments) that assert a zero difference in the accelerations of two differently composed bodies. Because these null hypotheses passed with high precision, it was warranted to infer that: "gravity is a phenomenon of curved spacetime," that is, it must be described by a "metric theory of gravity" (ibid., p. 10). Those who deny we can falsify non-nulls take note: inferring that an adequate theory must be relativistic (even if not necessarily GTR) was based on inferring a point null with severity! What about the earth and moon, examples of self-gravitating bodies? Do they also fall at the same rate?

While long corroborated for solar system tests, the equivalence principle (later the weak equivalence principle, WEP) was untested for such massive self-gravitating bodies (which requires the *strong equivalence principle*). Kenneth Nordtvedt discovered in the 1960s that in one of the most promising GTR rivals, the Brans–Dicke theory, the moon and earth fell at different rates, whereas for GTR there would be no difference. Clifford Will, the experimental physicist I've been quoting, tells how in 1968 Nordtvedt finds himself on the same plane as Robert Dicke. "Escape for the beleaguered Dicke was unfeasible at this point. Here was a total stranger telling him that his theory violated the principle of equivalence!" (1986 pp. 139–40). To Dicke's credit, he helped Nordtvedt design the experiment. A new parameter to describe the Nordtvedt effect was added to the PPN framework, i.e., η. For GTR, $\eta = 0$, so the statistical or substantive null hypothesis tested is that $\eta = 0$ as against $\eta \neq 0$ for rivals.

How can they determine the rates at which the earth and moon are falling? Thank the space program. It turns out that measurements of the round trip travel times between the earth and moon (between 1969 and 1975) enable the existence of such an anomaly for GTR to be probed severely (and the measurements continue today). Because the tests were sufficiently sensitive, these measurements provided good evidence that the Nordtvedt effect is absent, set upper bounds to the possible violations, and provided evidence for the correctness of what GTR says with respect to this effect.

So the old saw that we cannot falsify $\eta \neq 0$ is just that, an old saw. Critics take Fisher's correct claim, that failure to reject a null isn't automatically evidence for its correctness, as claiming we never have such evidence. Even he says it lends some weight to the null (Fisher 1955). With the N-P test, the null and alternative needn't be treated asymmetrically. In testing $H_0: \mu \geq \mu_0$ vs. $H_1: \mu < \mu_0$, a rejection falsifies a claimed increase.[11] Nordtvedt's null result added weight to GTR, not in rendering it more probable, but in extending the domain for which GTR gives a satisfactory explanation. It's still provisional in the sense that gravitational phenomena in unexplored domains could introduce certain couplings that, strictly speaking, violate the strong equivalence principle. The error statistical standpoint describes the state of information at any one time, with indications of where theoretical uncertainties remain.

You might discover that critics of a significance test's falsifying ability are themselves in favor of methods that preclude falsification altogether! Burnham and Anderson raised the scandal, yet their own account provides only a comparative appraisal of fit in model selection. No falsification there.

Souvenir K: Probativism

[A] fundamental tenet of the conception of inductive learning most at home with the frequentist philosophy is that inductive inference requires building up incisive arguments and inferences by putting together several different piece-meal results ... the payoff is an account that approaches the kind of full-bodied arguments that scientists build up in order to obtain reliable knowledge and understanding of a field. (Mayo and Cox 2006, p. 82)

The error statistician begins with a substantive problem or question. She jumps in and out of piecemeal statistical tests both formal and quasi-formal. The pieces are integrated in building up arguments from coincidence, informing background theory, self-correcting via blatant deceptions, in an iterative movement. The inference is qualified by using error probabilities to determine not "how probable," but rather, "how well-probed" claims are, and what has been poorly probed. What's wanted are ways to measure how far off what a given theory says about a phenomenon can be from what a "correct" theory would need to say about it by setting bounds on the possible violations.

An account of testing or confirmation might entitle you to confirm, support, or rationally accept a large-scale theory such as GTR. One is free to reconstruct episodes this way – after the fact – but as a forward-looking account, they fall far short. Even if somehow magically it was known in 1960 that GTR was true, it wouldn't snap experimental relativists out of their doldrums because they still couldn't be said to have understood gravity, how it behaves, or how to use one severely affirmed piece to opportunistically probe entirely distinct areas.

---

[11] Some recommend "equivalence testing" where $H_0: \mu \geq \mu_0$ or $\mu \leq -\mu_0$ and rejecting both sets bounds on μ. One might worry about low-powered tests, but it isn't essentially different from setting upper bounds for a more usual null. (For discussion see Lakens 2017, Senn 2001a, 2014, R. Berger and Hsu 1996, R. Berger 2014, Wellek 2010).

Learning from evidence turns not on appraising or probabilifying large-scale theories but on piecemeal tasks of data analysis: estimating backgrounds, modeling data, and discriminating signals from noise. Statistical inference is not radically different from, but is illuminated by, sexy science, which increasingly depends on it. Fisherian and N-P tests become parts of a cluster of error statistical methods that arise in full-bodied science. In Tour II, I'll take you to see the (unwarranted) carnage that results from supposing they belong to radically different philosophies.

Tour II It's The Methods, Stupid

> There is perhaps in current literature a tendency to speak of the
> Neyman–Pearson contributions as some static system, rather than as part
> of the historical process of development of thought on statistical theory which
> is and will always go on. (Pearson 1962, p. 276)

This goes for Fisherian contributions as well. Unlike museums, we won't
remain static.

The lesson from Tour I of this Excursion is that Fisherian and
Neyman–Pearsonian tests may be seen as offering clusters of methods appro-
priate for different contexts within the large taxonomy of statistical inquiries.
There is an overarching pattern:

> Just as with the use of measuring instruments, applied to the specific case, we employ
> the performance features to make inferences about aspects of the particular thing that is
> measured, aspects that the measuring tool is appropriately capable of revealing. (Mayo
> and Cox 2006, p. 84)

This information is used to ascertain what claims have, and have not, passed
severely, post-data. Any such proposed inferential use of error probabilities
gives considerable fodder for criticism from various tribes of Fisherians,
Neyman–Pearsonians, and Bayesians. We can hear them now:

- N-P theorists can only report the preset error probabilities, and can't use
  P-values post-data.
- A Fisherian wouldn't dream of using something that skirts so close to power
  as does the "sensitivity function" $\Pi(\gamma)$.
- Your account cannot be evidential because it doesn't supply posterior
  probabilities to hypotheses.
- N-P and Fisherian methods preclude any kind of inference since they use
  "the sample space" (violating the LP).

How can we reply? To begin, we need to uncover how the charges originate in
traditional philosophies long associated with error statistical tools. That's the
focus of Tour II.

Only then do we have a shot at decoupling traditional philosophies from
those tools in order to use them appropriately today. This is especially so when

the traditional foundations stand on such wobbly grounds, grounds largely rejected by founders of the tools. There is a philosophical disagreement between Fisher and Neyman, but it differs importantly from the ones that you're presented with and which are widely accepted and repeated in scholarly and popular treatises on significance tests. Neo-Fisherians and N-P theorists, keeping to their tribes, forfeit notions that would improve their methods (e.g., for Fisherians: explicit alternatives, with corresponding notions of sensitivity, and distinguishing statistical and substantive hypotheses; for N-P theorists, making error probabilities relevant for inference in the case at hand).

The spadework on this tour will be almost entirely conceptual: we won't be arguing for or against any one view. We begin in Section 3.4 by unearthing the basis for some classic counterintuitive inferences thought to be licensed by either Fisherian or N-P tests. That many are humorous doesn't mean disentangling their puzzles is straightforward; a medium to heavy shovel is recommended. We can switch to a light to medium shovel in Section 3.5: excavations of the evidential versus behavioral divide between Fisher and N-P turn out to be mostly built on sand. As David Cox observes, Fisher is often more performance-oriented in practice, but not in theory, while the reverse is true for Neyman and Pearson. At times, Neyman exaggerates the behavioristic conception just to accentuate how much Fisher's tests need reining in. Likewise, Fisher can be spotted running away from his earlier behavioristic positions just to derogate the new N-P movement, whose popularity threatened to eclipse the statistics program that was, after all, his baby. Taking the polemics of Fisher and Neyman at face value, many are unaware how much they are based on personality and professional disputes. Hearing the actual voices of Fisher, Neyman, and Pearson (F and N-P), you don't have to accept the gospel of "what the founders really thought." Still, there's an entrenched history and philosophy of F and N-P: A thick-skinned jacket is recommended. On our third stop (Section 3.6) we witness a bit of magic. The very concept of an error probability gets redefined and, hey presto!, a reconciliation between Jeffreys, Fisher, and Neyman is forged. Wear easily removed shoes and take a stiff walking stick. The Unificationist tribes tend to live near underground springs and lakeshore bounds; in the heady magic, visitors have been known to accidentally fall into a pool of quicksand.

3.4 Some Howlers and Chestnuts of Statistical Tests

> The well-known definition of a statistician as someone whose aim in life is to be wrong in exactly 5 per cent of everything they do misses its target. (Sir David Cox 2006a, p. 197)

Showing that a method's stipulations could countenance absurd or counter-intuitive results is a perfectly legitimate mode of criticism. I reserve the term "howler" for common criticisms based on logical fallacies or conceptual mis-understandings. Other cases are better seen as chestnuts – puzzles that the founders of statistical tests never cleared up explicitly. Whether you choose to see my "howler" as a "chestnut" is up to you. Under each exhibit is the purported basis for the joke.

Exhibit (iii): Armchair Science. *Did you hear the one about the statistical hypothesis tester* . . . who claimed that observing "heads" on a biased coin that lands heads with probability 0.05 is evidence of a statistically significant improvement over the standard treatment of diabetes, on the grounds that such an event occurs with low probability (0.05)?

The "armchair" enters because diabetes research is being conducted solely by flipping a coin. The joke is a spin-off from Kadane (2011):

Flip a biased coin that comes up heads with probability 0.95, and tails with probability 0.05. If the coin comes up tails reject the null hypothesis. Since the probability of rejecting the null hypothesis if it is true is 0.05, this is a valid 5 percent level test. It is also very robust against data errors; indeed it does not depend on the data at all. It is also nonsense, of course, but nonsense allowed by the rules of significance testing. (p. 439)

Basis for the joke: Fisherian test requirements are (allegedly) satisfied by any method that rarely rejects the null hypothesis.

But are they satisfied? I say no. The null hypothesis in Kadane's example can be in any field, diabetes, or the mean deflection of light. (Yes, Kadane affirms this.) He knows the test entirely ignores the data, but avers that "it has the property that Fisher proposes" (Kadane 2016, p. 1). Here's my take: in significance tests and in scientific hypotheses testing more generally, data can disagree with H only by being counter to what would be expected under the assumption that H is correct. An improbable series of coin tosses or plane crashes does not count as a disagreement from hypotheses about diabetes or light deflection. In Kadane's example, there is accordance so long as a head occurs – but this is a nonsensical distance measure. Were someone to tell you that any old improbable event (three plane crashes in one week) tests a hypothesis about light deflection, you would say that person didn't under-stand the meaning of testing in science or in ordinary life. You'd be right (for some great examples, see David Hand 2014).

Kadane knows it's nonsense, but thinks the only complaint a significance tester can have is its low power. What's the power of this "test" against any alternative? It's just the same as the probability it rejects, period, namely, 0.05. So an N-P tester could at least complain. Now I agree that bad tests may still be tests; but I'm saying Kadane's is no test at all. If you want to insist Fisher permits this test, fine, but I don't think that's a very generous interpretation. As egregious as is this howler, it is instructive because it shows like nothing else the absurdity of a crass performance view that claims: reject the null and infer evidence of a genuine effect, so long as it is done rarely. Crass performance is bad enough, but this howler commits a further misdemeanor: It overlooks the fact that a test statistic $d(x)$ must track discrepancies from H_0, becoming bigger (or smaller) as discrepancies increase (I list it as (ii) in Section 3.2). With any sensible distance measure, a misfit with H_0 must be *because* of the falsity of H_0. The probability of "heads" under a hypothesis about light deflection isn't even defined, because deflection hypotheses do not assign probabilities to coin-tossing trials. Fisher wanted test statistics to reduce the data from the generating mechanism, and here it's not even from the mechanism.

Kadane regards this example as "perhaps the most damaging critique" of significance tests (2016, p. 1). Well, Fisher can get around this easily enough.

Exhibit (iv): Limb-sawing Logic. *Did you hear the one about significance testers sawing off their own limbs?*

As soon as they reject the null hypothesis H_0 based on a small P-value, they no longer can justify the rejection because the P-value was computed under the assumption that H_0 holds, and now it doesn't.

Basis for the joke: If a test assumes H, then as soon as H is rejected, the grounds for its rejection disappear!

This joke, and I swear it is widely repeated but I won't name names, reflects a serious misunderstanding about ordinary conditional claims. The assumption we use in testing a hypothesis H, statistical or other, is an *implicationary* or *i-assumption*. We have a conditional, say: If H then expect x, with H the antecedent. The entailment from H to x, whether it is statistical or deductive, does not get sawed off after the hypothesis or model H is rejected when the prediction is not borne out. A related criticism is that statistical tests assume the truth of their test or null hypotheses. No, once again, they may serve only as i-assumptions for drawing out implications. The howler occurs when a test hypothesis that serves merely as an i-assumption is purported to be an actual assumption, needed for the inference to go through. A little logic goes a long way toward exposing many of these howlers. As the point is general, we use H.

This next challenge is by Harold Jeffreys. I won't call it a howler because it hasn't, to my knowledge, been excised by testers: it's an old chestnut, and a very revealing one.

Exhibit (v): Jeffreys' Tail Area Criticism. *Did you hear the one about statistical hypothesis testers rejecting H_0 because of outcomes it failed to predict?*
What's unusual about that?
What's unusual is that they do so even when these unpredicted outcomes haven't occurred!

Actually, one can't improve upon the clever statement given by Jeffreys himself. Using *P*-values, he avers, implies that *"a hypothesis that may be true may be rejected because it has not predicted observable results that have not occurred"* (1939/1961 p. 385).

Basis for the joke: The *P*-value, $\Pr(d \geq d_0; H_0)$, uses the "tail area" of the curve under H_0. d_0 is the observed difference, but $\{d \geq d_0\}$ includes differences even further from H_0 than d_0.

This has become the number one joke in comical statistical repertoires. Before debunking it, let me say that Jeffreys shows a lot of admiration for Fisher: "I have in fact been struck repeatedly in my own work . . . to find that Fisher had already grasped the essentials by some brilliant piece of common sense, and that his results would either be identical with mine or would differ only in cases where we should both be very doubtful" (ibid., p. 393). The famous quip is funny because it seems true, yet paradoxical. Why consider more extreme outcomes that didn't occur? The non-occurrence of more deviant results, Jeffreys goes on to say, "might more reasonably be taken as evidence for the law [in this case, H_0], not against it" (ibid., p. 385). The implication is that considering outcomes beyond d_0 is to unfairly discredit H_0, in the sense of finding more evidence against it than if only the actual outcome d_0 is considered.

The opposite is true.

Considering the tail area makes it harder, not easier, to find an outcome statistically significant (although this isn't the only function of the tail area). Why? Because it requires not merely that $\Pr(d = d_0; H_0)$ be small, but that $\Pr(d \geq d_0; H_0)$ be small. This alone squashes the only sense in which this could be taken as a serious criticism of tests. Still, there's a legitimate question about why the tail area probability is relevant. Jeffreys himself goes on to give it a rationale: "If mere improbability of the observations, given the hypothesis, was the criterion, any hypothesis whatever would be rejected. Everybody rejects the conclusion" (ibid., p. 385), so some other criterion is needed. Looking at the tail area supplies one, another would be a prior, which is Jeffreys' preference.

It's worth reiterating Jeffreys' correctly pointing out that "everybody rejects" the idea that the improbability of data under H suffices for evidence against H.

Shall we choose priors or tail areas? Jeffreys chooses default priors. Interestingly, as Jeffreys recognizes, for Normal distributions "the tail area represents the probability, given the data" that the actual discrepancy is in the direction opposite to that observed – d_0 is the wrong "sign" (ibid., p. 387). (This relies on a uniform prior probability for the parameter.) This connection between P-values and posterior probabilities is often taken as a way to "reconcile" them, at least for one-sided tests (Sections 4.4, 4.5). This was not one of Fisher's given rationales.

Note that the joke talks about outcomes the null does not predict – just what we wouldn't know without an assumed test statistic or alternative. One reason to evoke the tail area in Fisherian tests is to determine what H_0 "has not predicted," that is, to identify a sensible test statistic $d(x)$. Fisher, strictly speaking, has only the null distribution, with an implicit interest in tests with sensitivity of a given type. Fisher discusses this point in relation to the lady tasting tea (1935a, pp. 14–15). Suppose I take an observed difference d_0 as grounds to reject H_0 on account of it's being improbable under H_0, when in fact larger differences (larger d values) are even more probable under H_0. Then, as Fisher rightly notes, the improbability of the observed difference would be a poor indication of underlying discrepancy. (In N-P terms, it would be a biased test.) Looking at the tail area would reveal this fallacy; whereas it would be readily committed, Fisher notes, in accounts that only look at the improbability of the observed outcome d_0 under H_0.

When E. Pearson (1970) takes up Jeffreys' question: "Why did we use tail-area probabilities . . .?", his reply is that "this interpretation was not part of our approach" (p. 464). Tail areas simply fall out of the N-P desiderata of good tests. Given the lambda criterion one needed to decide at what point H_0

should be regarded as no longer tenable, that is where should one choose to bound the rejection region? To help in reaching this decision it appeared that the probability of falling into the region chosen, if H_0 were true, was one necessary piece of information. (ibid.)

So looking at the tail area could be seen as the result of formulating a sensible distance measure (for Fisher), or erecting a good critical region (for Neyman and Pearson).

Pearson's reply doesn't go far enough; it does not by itself explain why reporting the probability of falling into the rejection region is relevant for *inference*. It points to a purely performance-oriented justification that I know Pearson shied away from: It ensures data fall in a critical region rarely under H_0 and sufficiently often under alternatives in H_1 – but this tends to be left as a pre-data, performance goal (recall Birnbaum's Conf, Souvenir D). It is often alleged the N-P tester only reports whether or not x falls in the rejection region. Why are N-P collapsing all outcomes in this region?

In my reading, the error statistician does not collapse the result beyond what the minimal sufficient statistic requires for the question at hand. From our Translation Guide, Souvenir C, considering $(d(X) \geq d(x_0))$ signals that we're interested in the method, and we insert "the test procedure would have yielded" before $d(X)$. We report what was observed x_0 and the corresponding $d(x_0)$ – or d_0 – but we require the methodological probability, via the sampling distribution of $d(X)$ – abbreviated as d. This could mean looking at other stopping points, other endpoints, and other variables. We require that with high probability our test would have warned us if the result could easily have come about in a universe where the test hypothesis is true, that is $\Pr(d(X) < d(x_0); H_0)$ is high. Besides, we couldn't throw away the detailed data, since they're needed to audit model assumptions.

To conclude this exhibit, considering the tail area does not make it easier to reject H_0 but harder. Harder because it's not enough that the outcome be improbable under the null, outcomes even greater must be improbable under the null. $\Pr(d(X) = d(x_0); H_0)$ could be small while $\Pr(d(X) \geq d(x_0); H_0)$ not small. This leads to blocking a rejection when it should be because it means the test could readily produce even larger differences under H_0. Considering other possible outcomes that could have arisen is essential for assessing the test's capabilities. To understand the properties of our inferential tool is to understand what it would do under different outcomes, under different conjectures about what's producing the data. (Yes, the sample space matters post-data.) I admit that neither Fisher nor N-P adequately pinned down an inferential justification for tail areas, but now we have.

A bit of foreshadowing of a later shore excursion: some argue that looking at $d(X) \geq d(x_0)$ actually *does* make it easier to find evidence against H_0. How can that be? Treating $(1 - \beta)/\alpha$ as a kind of likelihood ratio in favor of an alternative over the null, then fed into a Likelihoodist or Bayesian algorithm, it can appear that way. Stay tuned.

Exhibit (vi): Two Measuring Instruments of Different Precisions. *Did you hear about the frequentist who, knowing she used a scale that's right only half the time, claimed her method of weighing is right 75% of the time?*

> She says, "I flipped a coin to decide whether to use a scale that's right 100% of the time, or one that's right only half the time, so, overall, I'm right 75% of the time." (She wants credit because she could have used a better scale, even knowing she used a lousy one.)

Basis for the joke: An N-P test bases error probabilities on all possible outcomes or measurements that could have occurred in repetitions, but did not.

As with many infamous pathological examples, often presented as knock-down criticisms of all of frequentist statistics, this was invented by a frequentist, Cox (1958). It was a way to highlight what could go wrong in the case at hand, if one embraced an unthinking behavioral-performance view. Yes, error probabilities are taken over hypothetical repetitions of a process, but not just any repetitions will do. Here's the statistical formulation.

We flip a fair coin to decide which of two instruments, E_1 or E_2, to use in observing a Normally distributed random sample Z to make inferences about mean θ. E_1 has variance of 1, while that of E_2 is 10^6. Any randomizing device used to choose which instrument to use will do, so long as it is irrelevant to θ. This is called a *mixture* experiment. The full data would report both the result of the coin flip and the measurement made with that instrument. We can write the report as having two parts: First, which experiment was run and second the measurement: (E_i, z), $i = 1$ or 2.

In testing a null hypothesis such as $\theta = 0$, the same z measurement would correspond to a much smaller P-value were it to have come from E_1 rather than from E_2: denote them as $p_1(z)$ and $p_2(z)$, respectively. The overall significance level of the mixture: $[p_1(z) + p_2(z)]/2$, would give a misleading report of the precision of the actual experimental measurement. The claim is that N-P statistics would report the average P-value rather than the one corresponding to the scale you actually used! These are often called the unconditional and the conditional test, respectively. The claim is that the frequentist statistician must use the unconditional test.

Suppose that we know we have observed a measurement from E_2 with its much larger variance:

The unconditional test says that we can assign this a higher level of significance than we ordinarily do, because if we were to repeat the experiment, we might sample some quite different distribution. But this fact seems irrelevant to the interpretation of an observation which we know came from a distribution [with the larger variance]. (Cox 1958, p. 361)

Once it is known which E_i has produced z, the P-value or other inferential assessment should be made with reference to the experiment actually run. As we say in Cox and Mayo (2010):

The point essentially is that the marginal distribution of a P-value averaged over the two possible configurations is misleading for a particular set of data. It would mean that an individual fortunate in obtaining the use of a precise instrument in effect sacrifices some of that information in order to rescue an investigator who has been unfortunate enough to have the randomizer choose a far less precise tool. From the perspective of interpreting the specific data that are actually available, this makes no sense. (p. 296)

To scotch his famous example, Cox (1958) introduces a principle: weak conditionality.

Weak Conditionality Principle (WCP): If a mixture experiment (of the aforementioned type) is performed, then, if it is known which experiment produced the data, inferences about θ are *appropriately drawn in terms of the sampling behavior* in the experiment known to have been performed (Cox and Mayo 2010, p. 296).

It is called weak conditionality because there are more general principles of conditioning that go beyond the special case of mixtures of measuring instruments.

While conditioning on the instrument actually used seems obviously correct, nothing precludes the N-P theory from choosing the procedure "which is best on the average over both experiments" (Lehmann and Romano 2005, p. 394), and it's even possible that the average or unconditional power is better than the conditional. In the case of such a conflict, Lehmann says relevant conditioning takes precedence over average power (1993b). He allows that in some cases of acceptance sampling, the average behavior may be relevant, but in scientific contexts the conditional result would be the appropriate one (see Lehmann 1993b, p. 1246). Context matters. Did Neyman and Pearson ever weigh in on this? Not to my knowledge, but I'm sure they'd concur with N-P tribe leader Lehmann. Admittedly, if your goal in life is to attain a precise α level, then when discrete distributions preclude this, a solution would be to flip a coin to decide the border-line cases! (See also Example 4.6, Cox and Hinkley 1974, pp. 95–6; Birnbaum 1962 p. 491.)

Is There a Catch?

The "two measuring instruments" example occupies a famous spot in the pantheon of statistical foundations, regarded by some as causing "a subtle earthquake" in statistical foundations. Analogous examples are made out in terms of confidence interval estimation methods (Tour III, Exhibit (viii)). It is a warning to the most behavioristic accounts of testing from which we have already distinguished the present approach. Yet justification for the condition-ing (WCP) is fully within the frequentist error statistical philosophy, for contexts of scientific inference. There is no suggestion, for example, that only the particular data set be considered. That would entail abandoning the sampling distribution as the basis for inference, and with it the severity goal. Yet we are told that "there is a catch" and that WCP leads to the Likelihood Principle (LP)!

It is not uncommon to see statistics texts argue that in frequentist theory one is faced with the following dilemma: either to deny the appropriateness of conditioning on the precision of the tool chosen by the toss of a coin, or else to embrace the strong likelihood principle, which entails that frequentist sampling distributions are irrelevant to inference once the data are obtained. This is a false dilemma. Conditioning is warranted to achieve objective frequentist goals, and the [weak] conditionality principle coupled with sufficiency does not entail the strong likelihood principle. The 'dilemma' argument is therefore an illusion. (Cox and Mayo 2010, p. 298)

There is a large literature surrounding the argument for the Likelihood Principle, made famous by Birnbaum (1962). Birnbaum hankered for something in between radical behaviorism and throwing error probabilities out the window. Yet he himself had apparently proved there is no middle ground (if you accept WCP)! Even people who thought there was something fishy about Birnbaum's "proof" were discomfited by the lack of resolution to the paradox. It is time for post-LP philosophies of inference. So long as the Birnbaum argument, which Savage and many others deemed important enough to dub a "breakthrough in statistics," went unanswered, the frequentist was thought to be boxed into the pathological examples. She is not.

In fact, I show there is a flaw in his venerable argument (Mayo 2010b, 2013a, 2014b). That's a relief. Now some of you will howl, "Mayo, not everyone agrees with your disproof! Some say the issue is not settled." Fine, please explain where my refutation breaks down. It's an ideal brainbuster to work on along the promenade after a long day's tour. Don't be dismayed by the fact that it has been accepted for so long. But I won't revisit it here.

3.5 P-values Aren't Error Probabilities Because Fisher Rejected Neyman's Performance Philosophy

> Both Neyman–Pearson and Fisher would give at most lukewarm support to standard significance levels such as 5% or 1%. Fisher, although originally recommending the use of such levels, later strongly attacked any standard choice. (Lehmann 1993b, p. 1248)

> Thus, Fisher rather incongruously appears to be attacking his own past position rather than that of Neyman and Pearson. (Lehmann 2011, p. 55)

By and large, when critics allege that Fisherian P-values are not error probabilities, what they mean is that Fisher wanted to interpret them in an evidential manner, not along the lines of Neyman's long-run behavior. I'm not denying there is an important difference between using error probabilities inferentially and behavioristically. The truth is that N-P and Fisher used

P-values and other error probabilities in both ways.[1] What they didn't give us is a clear account of the former. A big problem with figuring out the "he said/they said" between Fisher and Neyman–Pearson is that "after 1935 so much of it was polemics" (Kempthorne 1976) reflecting a blow-up which had to do with professional rivalry rather than underlying philosophy. Juicy details later on.

We need to be clear on the meaning of an error probability. A method of statistical inference moves from data to some inference about the source of the data as modeled. Associated error probabilities refer to the probability the method outputs an erroneous interpretation of the data. Choice of test rule pins down the particular error; for example, it licenses inferring there's a genuine discrepancy when there isn't (perhaps of a given magnitude). The test method is given in terms of a test statistic $d(X)$, so the error probabilities refer to the probability distribution of $d(X)$, the sampling distribution, computed under an appropriate hypothesis. Since we need to highlight subtle changes in meaning, call these ordinary "frequentist" error probabilities. (I can't very well call them error statistical error probabilities, but that's what I mean.)[2] We'll shortly require subscripts, so let this be error probability$_1$. Formal error probabilities have almost universally been associated with N-P statistics, and those with long-run performance goals. I have been disabusing you of such a straightjacketed view; they are vital in assessing how well probed the claim in front of me is. Yet my reinterpretation of error probabilities does not change their mathematical nature.

We can attach a frequentist performance assessment to any inference method. Post-data, these same error probabilities can, though they need not, serve to quantify the severity associated with an inference. Looking at the mathematics, it's easy to see the *P*-value as an error probability. Take Cox and Hinkley (1974):

For given observations **y** we calculate $t = t_{obs} = t(\mathbf{y})$, say, and the *level of significance* p_{obs} by $p_{obs} = \Pr(T \geq t_{obs}; H_0)$.

... Hence p_{obs} is the probability that we would mistakenly declare there to be evidence against H_0, were we to regard the data under analysis as just decisive against H_0. (p. 66)

Thus p_{obs} would be the Type I error probability associated with the test procedure consisting of finding evidence against H_0 when reaching p_{obs}.[3]

---

[1] Neyman (1976) said he was "not aware of a conceptual difference between a 'test of a statistical hypothesis' and a 'test of significance' and uses these terms interchangeably" (p. 737). We will too, with qualifications as needed.

[2] Thanks to the interpretation being fairly intimately related to the test, we get the error probabilities (formal or informal) attached to the interpretation.

[3] Note that p_{obs} and t_{obs} are the same as our p_0 and d_0. (or $d(x_0)$)

Thus the P-value equals the corresponding Type I error probability. [I've been using upper case P, but it's impossible to unify the literature.] Listen to Lehmann, speaking for the N-P camp:

[I]t is good practice to determine not only whether the hypothesis is accepted or rejected at the given significance level, but also to determine the smallest significance level ... at which the hypothesis would be rejected for the given observation. This number, the so-called *P-value* gives an idea of how strongly the data contradict the hypothesis. It also enables others to reach a verdict based on the significance level of their choice. (Lehmann and Romano 2005, pp. 63–4)

N-P theorists have no compunctions in talking about N-P tests using attained significance levels or P-values. Bayesians Gibbons and Pratt (1975) echo this view:

The P-value can then be interpreted as the smallest level of significance, that is, the 'borderline level', since the outcome observed would ... not [be] significant at any smaller levels. Thus it is sometimes called the 'level attained' by the sample ... Reporting a P-value ... permits each individual to choose his own ... maximum tolerable probability for a Type I error. (p. 21)

Is all this just a sign of texts embodying an inconsistent hybrid? I say no, and you should too.

A certain tribe of statisticians professes to be horrified by the remarks of Cox and Hinkley, Lehmann and Romano, Gibbons and Pratt and many others. That these remarks come from leading statisticians, members of this tribe aver, just shows the depth of a dangerous "confusion over the evidential content (and mixing) of p's and α's" (Hubbard and Bayarri 2003, p. 175). On their view, we mustn't mix what they call "evidence and error": F and N-P are incompatible. For the rest of this tour, we'll alternate between the museum and engaging the Incompatibilist tribes themselves. When viewed through the tunnel of the Incompatibilist statistical philosophy, these statistical founders appear confused.

The distinction between evidence (p's) and error (α's) is not trivial ... it reflects the fundamental differences between Fisher's ideas on significance testing and inductive inference, and [N-P's] views on hypothesis testing and inductive behavior. (Hubbard and Bayarri 2003, p. 171)

What's fascinating is that the Incompatibilists admit it's the philosophical difference they're on about, not a mathematical one. The paper that has become *the* centerpiece for the position in this subsection is Berger and Sellke (1987). They ask:

Can P values be justified on the basis of how they perform in repeated use? We doubt it. For one thing, how would one measure the performance of P values? With significance tests and confidence intervals, they are either right or wrong, so it is possible to talk about error rates. If one introduces a decision rule into the situation by saying that H_0 is rejected when the P value < 0.05, then of course the classical error rate is 0.05. (p. 136)

Good. Then we can agree a P-value is, mathematically, an error probability. Berger and Sellke are merely opining that Fisher wouldn't have *justified* their use on grounds of error rate performance. That's different. Besides, are we so sure Fisher wouldn't sully himself with crass error probabilities, and dichotomous tests? Early on at least, Fisher appears as a behaviorist par excellence. That he is later found "attacking his own position," as Lehmann puts it, is something else.

Mirror Mirror on the Wall, Who's the More Behavioral of Them All?

N-P were striving to emulate the dichotomous interpretation they found in Fisher:

It is open to the experimenter to be more or less exacting in respect of the smallness of the probability he would require before he would be willing to admit that his observations have demonstrated a positive result. It is obvious that an experiment would be useless of which no possible result would satisfy him. ... It is usual and convenient for the experimenters to take 5 per cent as a standard level of significance, in the sense that they are prepared to ignore all results which fail to reach this standard, and, by this means, to eliminate from further discussion the greater part of the fluctuations which chance causes have introduced into their experimental results. (Fisher 1935a, pp. 13–14)

Fisher's remark can be taken to justify the tendency to ignore negative results or stuff them in file drawers, somewhat at odds with his next lines, the ones that I specifically championed in Excursion 1: "we may say that a phenomenon is experimentally demonstrable when we know how to conduct an experiment which will rarely fail to give us a statistically significant result..." (1935a, p. 14).[4] This would require us to keep the negative results around for a while. How else could we see if we are rarely failing, or often succeeding?

What I mainly want to call your attention to now are the key phrases "willing to admit," "satisfy him," "deciding to ignore." What are these, Neyman asks, but actions or behaviors? He'd learned from R. A. Fisher! So, while many take

---

[4] Fisher, in a 1926 paper, gives another nice rendering: "A scientific fact should be regarded as experimentally established only if a properly designed experiment *rarely fails* to give this level of significance. The very high odds sometimes claimed for experimental results should usually be discounted, for inaccurate methods of estimating error have far more influence than has the particular standard of significance chosen" (pp. 504–5).

the dichotomous "up-down" spirit of tests as foreign to Fisher, it is not foreign at all. Again from Fisher (1935a):

Our examination of the possible results of the experiment has therefore led us to a statistical test of significance, by which these results are divided into two classes with opposed interpretations ... those which show a significant discrepancy from a certain hypothesis; ... and on the other hand, results which show no significant discrepancy from this hypothesis. (pp. 15–16)

No wonder Neyman could counter Fisher's accusations that he'd turned his tests into tools for inductive behavior by saying, in effect, look in the mirror (for instance, in the acrimonious exchange of 1955–6, 20 years after the blow-up): Pearson and I were only systematizing your practices for how to interpret data, taking explicit care to prevent untoward results that you only managed to avoid on intuitive grounds!

Fixing Significance Levels. What about the claim that N-P tests fix the Type I error probability in advance, whereas P-values are post-data? Doesn't *that* prevent a P-value from being an error probability? First, we must distinguish between fixing the significance level for a test prior to data collection, and fixing a threshold to be used across one's testing career. Fixing α and power is part of specifying a test with reasonable capabilities of answering the question of interest. Having done so, there's nothing illicit about reporting the *achieved* or *attained* significance level, and it is even recommended by Lehmann. As for setting a threshold for habitual practice, that's actually more Fisher than N-P.

Lehmann is flummoxed by the association of fixed levels of significance with N-P since "[U]nlike Fisher, Neyman and Pearson (1933, p. 296) did not recommend a standard level but suggested that 'how the balance [between the two kinds of error] should be struck must be left to the investigator'" (Lehmann 1993b, p. 1244). From their earliest papers, Neyman and Pearson stressed that the tests were to be "used with discretion and understanding" depending on the context (Neyman and Pearson 1928, p. 58). In a famous passage, Fisher (1956) raises the criticism – but without naming names:

A man who 'rejects' a hypothesis provisionally, as a matter of habitual practice, when the significance is at the 1% level or higher, will certainly be mistaken in not more than 1% of such decisions. For when the hypothesis is correct he will be mistaken in just 1% of these cases, and when it is incorrect he will never be mistaken in rejection ... However, the calculation is absurdly academic, for in fact no scientific worker has a fixed level of significance at which from year to year, and in all circumstances, he rejects hypotheses; he rather gives his mind to each particular case in the light of his evidence and his ideas. (pp. 44–5)

It is assumed Fisher is speaking of N-P, or at least Neyman. But N-P do not recommend such habitual practice.

Long Runs Are Hypothetical. What about the allegation that N-P error probabilities allude to actual long-run repetitions, while the *P*-value is a *hypothetical* distribution? N-P error probabilities are also about hypothetical would-be's. Each sample of size n gives a single value of the test statistic $d(X)$. Our inference is based on this one sample. The third requirement (Pearson's "Step 3") for tests is that we be able to compute the distribution of $d(X)$, under the assumption that the world is approximately like H_0, and under discrepancies from H_0. Different outcomes would yield different $d(X)$ values, and we consider the frequency distribution of $d(X)$ over hypothetical repetitions.

At the risk of overkill, the sampling distribution is all about hypotheticals: the relative frequency of outcomes under one or another hypothesis. These also equal the relative frequencies assuming you really did keep taking samples in a long run, tiring yourself out in the process. It doesn't follow that the value of the hypothetical frequencies depends on referring to, much less actually carrying out, that long run. A statistical hypothesis has implications for some hypothetical long run in terms of how frequently this or that would occur. A statistical test uses the data to check how well the predictions are met. The sampling distribution is the testable meeting-ground between the two.

The same pattern of reasoning is behind resampling from the one and only sample in order to generate a sampling distribution. (We meet with resampling in Section 4.10.) The only gap is to say why such a hypothetical (or counterfactual) is relevant for inference in the case at hand. Merely proposing that error probabilities give a vague "strength of evidence" to an inference won't do. Our answer is that they capture the capacities of tests, which in turn tell us how severely tested various claims may be said to be.

It's Time to Get Beyond the "Inconsistent Hybrid" Charge

Gerd Gigerenzer is a wonderful source of how Fisherian and N-P methods led to a statistical revolution in psychology. He is famous for, among much else, arguing that the neat and tidy accounts of statistical testing in social science texts are really an inconsistent hybrid of elements from N-P's behavioristic philosophy and Fisher's more evidential approach (Gigerenzer 2002, p. 279). His tribe is an offshoot of the Incompatibilists, but with a Freudian analogy to illuminate the resulting tension and anxiety that a researcher is seen to face.

N-P testing, he says, "functions as the Superego of the hybrid logic" (ibid., p. 280). It requires alternatives, significance levels, and power to be prespecified, while strictly outlawing evidential or inferential interpretations about the truth of a particular hypothesis. The Fisherian "Ego gets things done ... and gets papers published" (ibid.). Power is ignored, and the level of significance is found after the experiment, cleverly hidden by rounding up to the nearest standard level. "The Ego avoids ... exact predictions of the alternative hypothesis, but claims support for it by rejecting a null hypothesis" and in the end is "left with feelings of guilt and shame for having violated the rules" (ibid.). Somewhere in the background lurks his Bayesian Id, driven by wishful thinking into misinterpreting error probabilities as degrees of belief.

As with most good caricatures, there is a large grain of truth in Gigerenzer's Freudian metaphor – at least as the received view of these methods. I say it's time to retire the "inconsistent hybrid" allegation. Reporting the attained significance level is entirely legitimate and is recommended in N-P tests, so long as one is not guilty of other post-data selections causing *actual P*-values to differ from *reported* or nominal ones. By failing to explore the inferential basis for the stipulations, there's enormous unclarity as to what's being disallowed and why, and what's mere ritual or compulsive hand washing (as he might put it (ibid., p. 283)). Gigerenzer's Ego might well *deserve* to feel guilty if he has chosen the hypothesis, or characteristic to be tested, based on the data, or if he claims support for a research hypothesis by merely rejecting a null hypothesis – the illicit NHST animal. A post-data choice of test statistic may be problematic, but not an attained significance level.

Gigerenzer recommends that statistics texts teach the conflict and stop trying "to solve the conflict between its parents by denying its parents" (2002, p. 281). I, on the other hand, think we should take responsibility for interpreting the tools according to their capabilities. Polemics between Neyman and Fisher, however lively, taken at face value, are a highly unreliable source; we should avoid chiseling into even deeper stone the hackneyed assignments of statistical philosophy – "he's inferential, he's an acceptance sampler." The consequences of the "inconsistent hybrid" allegation are dire: both schools are caricatures, robbed of features that belong in an adequate account.

Hubbard and Bayarri (2003) are a good example of this; they proclaim an N-P tester is forbidden – forbidden! – from reporting the observed P-value. They eventually concede that an N-P test "could be defined equivalently in terms of the p value ... the null hypothesis should be rejected if the observed $p < \alpha$, and accepted otherwise" (p. 175). But they aver "no matter how small the p value is, the appropriate report is that the procedure guarantees a $100\alpha\%$ false rejection of the null on repeated use" (ibid.). An N-P tester must robotically obey the reading that has grown out of the Incompatibilist tribe to which they belong. A user must round up to the predesignated α. This type of prohibition gives a valid guilt trip to Gigerenzer's Ego; yet the hang-up stems from the Freudian metaphor, not from Neyman and Pearson, who say:

it is doubtful whether the knowledge that P_z [the P-value associated with test statistic z] was really 0.03 (or 0.06) rather than 0.05, ... would in fact ever modify our judgment ... regarding the origin of a single sample. (Neyman and Pearson 1928, p. 27)

But isn't it true that rejection frequencies needn't be indicative of the evidence against a null? Yes. Kadane's example, if allowed, shows how to get a small rejection frequency with no evidence. But this was to be a problem for Fisher, solved by N-P (even if Kadane is not fond of them either). Granted, even in tests not so easily dismissed, crude rejection frequencies differ from an evidential assessment, especially when some of the outcomes leading to rejection vary considerably in their evidential force. This is the lesson of Cox's famous "two machines with different precisions." Some put this in terms of selecting the relevant reference set which "need not correspond to all possible repetitions of the experiment" (Kalbfleisch and Sprott 1976, p. 272). We've already seen that relevant conditioning is open to a N-P tester. Others prefer to see it as a matter of adequate model specification. So once again it's not a matter of Fisher vs. N-P.

I'm prepared to admit Neyman's behavioristic talk. Mayo (1996, Chapter 11) discusses: "Why Pearson rejected the (behavioristic) N-P theory" (p. 361). Pearson does famously declare that "the behavioristic conception is Neyman's not mine" (1955, p. 207). Furthermore, Pearson explicitly addresses "the situation where statistical tools are applied to an isolated investigation of considerable importance ..." (1947, p. 170).

In other and, no doubt, more numerous cases there is no repetition of the same type of trial or experiment, but all the same we can and many of us do use the same test rules ... Why do we do this? ... Is it because the formulation of the case in terms of hypothetical repetition helps to that clarity of view needed for sound judgment?

Or is it because we are content that the application of a rule, now in this investigation, now in that, should result in a long-run frequency of errors in judgment which we control at a low figure? (ibid., p. 172)

While tantalizingly leaving the answer dangling, it's clear that for Pearson: "the formulation of the case in terms of hypothetical repetition helps to that clarity of view needed for sound judgment" (ibid.) in learning about the particular case at hand. He gives an example from his statistical work in World War II:

Two types of heavy armour-piercing naval shell of the same caliber are under consideration; they may be of different design or made by different firms ... Twelve shells of one kind and eight of the other have been fired; two of the former and five of the latter failed to perforate the plate ... (Pearson 1947, 171)

Starting from the basis that individual shells will never be identical in armour-piercing qualities, ... he has to consider how much of the difference between (i) two failures out of twelve and (ii) five failures out of eight is likely to be due to this inevitable variability. (ibid.)

He considers what other outcomes could have occurred, and how readily, in order to learn what variability alone is capable of producing.[5] Pearson opened the door to the evidential interpretation, as I note in 1996, and now I go further.

Having looked more carefully at the history before the famous diatribes, and especially at Neyman's applied work, I now hold that Neyman largely rejected it as well! Most of the time, anyhow. But that's not the main thing. Even if we couldn't point to quotes and applications that break out of the strict "evidential versus behavioral" split: *we* should be the ones to interpret the methods for inference, and supply the statistical philosophy that directs their right use.

Souvenir L: Beyond Incompatibilist Tunnels

What people take away from the historical debates is Fisher (1955) accusing N-P, or mostly Neyman, of converting his tests into acceptance sampling rules more appropriate for five-year plans in Russia, or making money in the USA, than for science. Still, it couldn't have been too obvious that N-P distorted his tests, since Fisher tells us only in 1955 that it was Barnard who explained that, despite agreeing mathematically in very large part, there is this distinct philosophical position. Neyman suggests that his terminology was to distinguish what he (and Fisher!) were doing from the attempts to define a unified rational measure of belief on hypotheses. N-P both denied there was such a thing. Given Fisher's vehement disavowal of subjective Bayesian probability, N-P thought nothing of crediting Fisherian tests as a step in the development of "inductive behavior" (in their 1933 paper).

The myth of the radical difference in either methods or philosophy is a myth. Yet, as we'll see, the hold it has over people continues to influence the use and discussion of tests. It's based almost entirely on sniping between Fisher and Neyman from 1935 until Neyman leaves for the USA in 1938. Fisher didn't engage much with statistical developments during World War II. Barnard describes Fisher as cut off "by some mysterious personal or political agency. Fisher's isolation occurred, I think, at a particularly critical

---

[5] Pearson said that a statistician has an α and a β side, the former alludes to what they say in theory, the latter to what they do in practice. In practice, even Neyman, so often portrayed as performance-oriented, was as inferential as Pearson.

time, when opportunities existed for a fruitful fusion of ideas stemming from Neyman and Pearson and from Fisher" (Barnard 1985, p. 2). Lehmann observes that Fisher kept to his resolve not to engage in controversy with Neyman until the highly polemical exchange of 1955 at age 65. Fisher alters some of the lines of earlier editions of his books. For instance, Fisher's disinterest in the attained P-value was made clear in *Statistical Methods for Research Workers* (SMRW) (1934a, p. 80):

... in practice we do not want to know the exact value of P for any observed value of [the test statistic], but, in the first place, whether or not the observed value is open to suspicion.

If P is between .1 and .9 there is certainly no reason to suspect the hypothesis tested. If it is below .02 it is strongly indicated that the hypothesis fails to account for the whole of the facts. We shall not often be astray if we draw a conventional line at .05.

Lehmann explains that it was only "fairly late in life, Fisher's attitude had changed" (Lehmann 2011, p. 52). In the 13th edition of SMRW, Fisher changed his last sentence to:

The actual value of P obtainable ... indicates the strength of the evidence against the hypothesis. [Such a value] is seldom to be disregarded. (p. 80)

Even so, this at most suggests how the methodological (error) probability is thought to provide a measure of evidential strength – it doesn't abandon error probabilities. There's a deeper reason for this backtracking by Fisher; I'll save it for Excursion 5. One other thing to note: F and N-P were creatures of their time. Their verbiage reflects the concern with "operationalism" and "behaviorism," growing out of positivistic and verificationist philosophy. I don't deny the value of tracing out the thrust and parry between Fisher and Neyman in these excursions. None of the founders solved the problem of an inferential interpretation of error probabilities – though they each offered tidbits. Their name-calling: "you're too mechanical," "no *you* are," at most shows, as Gigerenzer and Marewski observe, that they all rejected mechanical statistics (2015, p. 422).

The danger is when one group's interpretation is the basis for a historically and philosophically "sanctioned" reinterpretation of one or another method. Suddenly, rigid rules that the founders never endorsed are imposed. Through the Incompatibilist philosophical tunnel, as we are about to see, these reconstruals may serve as an effective way to dismiss the entire methodology – both F and N-P. After completing this journey, you shouldn't have to retrace this "he said/they said" dispute again. It's the methods, stupid.

3.6   Hocus-Pocus: *P*-values Are Not Error Probabilities, Are Not Even Frequentist!

> Fisher saw the *p* value as a measure of evidence, not as a frequentist evaluation. Unfortunately, as a measure of evidence it is very misleading. (Hubbard and Bayarri 2003, p. 181)

This entire tour, as you know, is to disentangle a jungle of conceptual issues, not to defend or criticize any given statistical school. In sailing forward to scrutinize Incompatibilist tribes who protest against mixing *p*'s and *α*'s, we need to navigate around a pool of quicksand. They begin by saying *P*-values are for evidence and inference, unlike error probabilities. N-P error probabilities are too performance oriented to be measures of evidence. In the next breath we're told *P*-values aren't good measures of evidence either. A good measure of evidence, it's assumed, should be probabilist, in some way, and *P*-values disagree with probabilist measures, be they likelihood ratios, Bayes factors, or posteriors. If you reinterpret error probabilities, they promise, you can make peace with all tribes. Whether we get on firmer ground or sink in a marshy swamp will have to be explored.

Berger's Unification of Jeffreys, Neyman, and Fisher

With "reconciliation" and "unification" in the air, Jim Berger, a statistician deeply influential in statistical foundations, sets out to see if he can get Fisher, Neyman, and (non-subjective) Bayesian Jeffreys to agree on testing (2003). A compromise awaits, if we nip and tuck the meaning of "error probability" (Section 3.5). If you're an N-P theorist and like your error probability$_1$, you can keep it he promises, but he thinks you will want to reinterpret it. It then becomes possible to say that a *P*-value is not an error probability (full stop), meaning it's not the newly defined error probability$_2$. What's error probability$_2$? It's a type of posterior probability in a null hypothesis, conditional on the outcome, given a prior. It may still be frequentist in some sense. On this reinterpretation, *P*-values are not error probabilities. Neither are N-P Type I and II, α and β. Following the philosopher's clarifying move via subscripts, there is error probability$_1$ – the usual frequentist notion – and error probability$_2$ – notions from probabilism that had never been called error probabilities before.

In commenting on Berger (2003), I noted my surprise at his redefinition (Mayo 2003b). His reply: "Why should the frequentist school have exclusive right to the term 'error probability?' It is not difficult to simply add the designation 'frequentist' (or Type I or Type II) or 'Bayesian' to the term to differentiate between the schools" (Berger 2003, p. 30). That would work splendidly. So let error probability$_2$ = Bayesian error probability. Frankly, I didn't think Bayeslans would want the term. In a minute, however, Berger will claim they alone are the true frequentist error probabilities! If you feel yourself sinking in a swamp of sliding meanings, remove your shoes, flip onto your back atop your walking stick and you'll stop sinking. Then, you need only to pull yourself to firm land. (See Souvenir M.)

The Bayes Factor. In 1987, Berger and Sellke said that in order to consider P-values as error probabilities we need to introduce a decision or test rule. Berger (2003) proposes such a rule and error probability$_2$ is born. In trying to merge different methodologies, there's always a danger of being biased in favor of one, begging the question against the others. From the severe tester's perspective, this is what happens here, but so deftly that you might miss it if you blink.[6]

His example involves $X_1, \ldots, X_n$ IID data from $N(\theta, \sigma^2)$, with σ^2 known, and the test is of two simple hypotheses $H_0: \theta = \theta_0$ and $H_1: \theta = \theta_1$. Consider now their two P-values: "for $i = 0, 1$, let p_i be the p-value in testing H_i against the other hypothesis" (ibid., p. 6). Then reject H_0 when $p_0 \leq p_1$, and accept H_0 otherwise. If you reject H_0 you next compute the posterior probability of H_0 using one of Jeffreys' default priors giving 0.5 to each hypothesis. The computation rests on the *Bayes factor* or likelihood ratio $B(x) = \Pr(x|H_0)/\Pr(x|H_1)$:

$$\Pr(H_0|x) = B(x)/[1 + B(x)].$$

The priors drop out, being 0.5. As before, x refers to a generic value for X.

This was supposed to be something Fisher would like, so what happened to P-values? They have a slight walk-on part: the rejected hypothesis is the one that has the lower P-value. Its value is irrelevant, but it directs you to which posterior to compute. We might understand his Bayesian error probabilities this way: If I've rejected H_0, I'd be wrong if H_0 were true, so $\Pr(H_0|x)$ is a probability of being wrong about H_0. It's the *Bayesian Type I error probability$_2$*. If instead you reject H_1, then you'd be wrong if H_1 were true. So in that case you report the Bayesian Type II error probability$_2$, which would be $\Pr(H_1|x) = 1/[1 + B(x)]$. Whatever you think of these, they're quite different from error probability$_1$, which does not use priors in H_i.

Sleight of Hand? Surprisingly, Berger claims to give a "dramatic illustration of the nonfrequentist nature of P-values" (ibid., p. 3). Wait a second, how did they become *non-frequentist*? What he means is that the P-value can be shown to disagree with the special posterior probability for H_0, defined as error

---

[6] We are forced to spend more time on P-values than one would wish simply because so many of the criticisms and proposed reforms are in terms of them.

probability$_2$. They're not called Bayesian error probabilities any more but frequentist conditional error probabilities (CEPs). Presto! A brilliant sleight of hand.

This 0.5 prior is not supposed to represent degree of belief, but it is Berger's "objective" default Bayesian prior. Why does he call it frequentist? He directs us to an applet showing if we imagine randomly selecting our test hypothesis from a population of null hypotheses, 50% of which are true, the rest false, and then compute the relative frequency of true nulls conditional on its having been rejected at significance level p, we get a number that is larger than p. This violates what he calls the frequentist principle (not to be confused with FEV):

> *Berger's frequentist principle*: $\Pr(H_0$ true $\mid H_0$ rejected at level $p)$ should equal p.

This is very different from what a P-value gives us, namely, $\Pr(P \le p; H_0) = p$ (or $\Pr(d(X) \ge d(x_0); H_0) = p)$.

He actually states the frequentist principle more vaguely; namely, that the reported error probability should equal the actual one, but the computation is to error probability$_2$. If I'm not being as clear as possible, it's because Berger isn't, and I don't want to prematurely saddle him with one of at least two interpretations he moves between. For instance, Berger says the urn of nulls applet is just a heuristic, showing how it could happen. So suppose the null was randomly selected from an urn of nulls 50% of which are true. Wouldn't 0.5 be its frequentist prior? One has to be careful. First consider a legitimate frequentist prior. Suppose I selected the hypothesis H_0: that the mean temperature in the water, θ, is 150 degrees (Section 3.2). I can see this value resulting from various features of the lake and cooling apparatus, and identify the relative frequency that θ takes different values. $\{\Theta = \theta\}$ is an event associated with random variable Θ. Call this an *empirical* or *frequentist* prior just to fix the notion. What's imagined in Berger's applet is very different. Here the analogy is with diagnostic screening for disease, so I will call it that (Section 5.6). We select one null from an urn of nulls, which might include all hypotheses from a given journal, a given year, or lots of other things.[7] If 50% of the nulls in this urn are true, the experiment of

---

[7] It is ironic that it's in the midst of countering a common charge that he requires repeated sampling from the same population that Neyman (1977) talks about a series of distinct scientific inquiries (presumably independent) with Type I and Type II error probabilities (for specified alternatives) $\alpha_1, \alpha_2, \ldots, \alpha_n, \ldots$ and $\beta_1, \beta_2, \ldots, \beta_n, \ldots$

I frequently hear a particular regrettable remark ... that the frequency interpretation of either the level of significance α or of power $(1 - \beta)$ is only possible when one deals many times WITH THE SAME HYPOTHESIS H, TESTED AGAINST THE SAME ALTERNATIVE. (Neyman 1977, 109, his use of capitals)

randomly selecting a null from the urn could be seen as a Bernoulli trial with two outcomes: a null that is true or false. The probability of selecting a null that has the property "true" is 0.5. Suppose I happen to select H_0: $\theta = 150$, the hypothesis from the accident at the water plant. It would be incorrect to say 0.5 was the relative frequency that $\theta = 150$ would emerge with the empirical prior. So there's a frequentist computation, but it differs from what Neyman's empirical Bayesian would assign it. I'll come back to this later (Excursion 6).

Suppose instead we keep to the default Bayesian construal that Berger favors. The priors come from one or another conventional assignment. On this reading, his frequentist principle is: the P-value should equal the default posterior on H_0. That is, a reported P-value should equal error probability$_2$. By dropping the designation "Bayesian" that he himself recommended "to differentiate between the schools" (p. 30), it's easy to see how confusion ensues.

Berger emphasizes that the confusion he is on about "is different from the confusion between a P-value and the posterior probability of the null hypothesis" (p. 4). What confusion? That of thinking P-values are frequentist error probabilities$_2$ – but he has just introduced the shift of meaning! But the only way error probability$_2$ inherits a frequentist meaning is by reference to the heuristic (where the prior is the proportion of true nulls in a hypothetical urn of nulls), giving a diagnostic screening posterior probability. The subscripts are a lifesaver for telling what's true when definitions shift about throughout an argument. The frequentist had only ever wanted error probabilities$_1$ – the ones based solely on the sampling distribution of d(X). Yet now he declares that error probability$_2$ – Bayesian error probability – is the only real or relevant frequentist error probability! If this is the requirement, preset α, β aren't error probabilities either.

It might be retorted, however, that this was to be a compromise position. We can't dismiss it out of hand because it requires Neyman and Fisher to become default Bayesians. To smoke the peace pipe, everyone has to give a little. According to Berger, "Neyman criticized p-values for violating the frequentist principle." (p. 3) With Berger's construal, it is not violated. So it appears Neyman gets something. Does he? We know N-P used P-values, and never saw them as non-frequentist; and surely Neyman wouldn't be criticizing a P-value for not being equal to a default (or other) posterior probability. Hence Nancy Reid's quip: "the Fisher/Jeffreys agreement is essentially to have Fisher"

From the Central Limit Theorem, Neyman remarks:

The relative frequency of the first kind of errors will be close to the arithmetic mean of numbers $\alpha_1, \alpha_2, \ldots, \alpha_m \ldots$ Also the relative frequency of detecting the falsehood of the hypotheses tested, when false $\ldots$ will differ but little from the average of [the corresponding powers, for specified alternatives].

kowtow to Jeffreys (N. Reid 2003). The surest sign that we've swapped out meanings are the selling points.

Consider the Selling Points

"Teaching statistics suddenly becomes easier . . . it is considerably less important to disabuse students of the notion that a frequentist error probability is the probability that the hypothesis is true, given the data" (Berger 2003, p. 8), since his error probability$_2$ actually has that interpretation. We are also free of having to take into account the stopping rule used in sequential tests (ibid.). As Berger dangles his tests in front of you with the labels "frequentist," "error probabilities," and "objectivity," there's one thing you know: if the methods enjoy the simplicity and freedom of paying no price for optional stopping, you'll want to ask if they're also controlling error probabilities$_1$. When that handwringing disappears, unfortunately, so does our assurance that we block inferences that have passed with poor severity.

Whatever you think of default Bayesian tests, Berger's error probability$_2$ differs from N-P's error probability$_1$. N-P requires controlling the Type I and II error probabilities at low values regardless of prior probability assignments. The scrutiny here is not of Berger's recommended tests – that comes later. The scrutiny here is merely to shine a light on the type of shifting meanings that our journey calls for. Always carry your walking stick – it serves as a metaphorical subscript to keep you afloat.

Souvenir M: Quicksand Takeaway

The howlers and chestnuts of Section 3.4 call attention to: the need for an adequate test statistic, the difference between an i-assumption and an actual assumption, and that tail areas serve to raise, and not lower, the bar for rejecting a null hypothesis. The stop in Section 3.5 pulls back the curtain on one front of typical depictions of the N-P vs. Fisher battle, and Section 3.6 disinters equivocal terms in a popular peace treaty between the N-P, Fisher, and Jeffreys tribes. Of these three stops, I admit that the last may still be murky. One strategy we used to clarify are subscripts to distinguish slippery terms. Probabilities of Type I and Type II errors, as well as P-values, are defined exclusively in terms of the sampling distribution of $d(X)$, under a statistical hypothesis of interest. That's error probability$_1$. Error probability$_2$, in addition to requiring priors, involves conditioning on the particular outcome, with the hypothesis varying. There's no consideration of the sampling distribution of $d(X)$, if you've conditioned on the actual outcome. A second strategy is to consider the selling points of the new "compromise" construal, to gauge what it's asking you to buy.

Here's from our guidebook:

> You're going to need to be patient. Depending on how much quick-sand is around you, it could take several minutes or even hours to slowly, methodically get yourself out ...
>
> *Relax.* Quicksand usually isn't more than a couple feet deep ... If you panic you can sink further, but if you relax, your body's buoyancy will cause you to float.
>
> Breathe deeply ... It is impossible to "go under" if your lungs are full of air (WikiHow 2017).

In later excursions, I promise, you'll get close enough to the edge of the quicksand to roll easily to hard ground. More specifically, all of the terms and arguments of Section 3.6 will be excavated.

Tour III Capability and Severity: Deeper Concepts

From the itinerary: A long-standing family feud among frequentists is between hypotheses tests and confidence intervals (CIs), but in fact there's a clear duality between the two. The dual mission of the first stop (Section 3.7) of this tour is to illuminate both CIs and severity by means of this duality. A key idea is arguing from the capabilities of methods to what may be inferred. The severity analysis seamlessly blends testing and estimation. A typical inquiry first tests for the existence of a genuine effect and then estimates magnitudes of discrepancies, or inquires if theoretical parameter values are contained within a confidence interval. At the second stop (Section 3.8) we reopen a highly controversial matter of interpretation that is often taken as settled. It relates to statistics and the discovery of the Higgs particle – displayed in a recently opened gallery on the "Statistical Inference in Theory Testing" level of today's museum.

3.7 Severity, Capability, and Confidence Intervals (CIs)

It was shortly before Egon offered him a faculty position at University College starting 1934 that Neyman gave a paper at the Royal Statistical Society (RSS) which included a portion on confidence intervals, intending to generalize Fisher's fiducial intervals. With K. Pearson retired (he's still editing *Biometrika* but across campus with his assistant Florence David), the tension is between E. Pearson, along with remnants of K.P.'s assistants, and Fisher on the second and third floors, respectively. Egon hoped Neyman's coming on board would melt some of the ice.

Neyman's opinion was that "Fisher's work was not really understood by many statisticians ... mainly due to Fisher's very condensed form of explaining his ideas" (C. Reid 1998, p. 115). Neyman sees himself as championing Fisher's goals by means of an approach that gets around these expository obstacles. So Neyman presents his first paper to the Royal Statistical Society (June, 1934), which includes a discussion of confidence intervals, and, as usual, comments (later published) follow. Arthur Bowley (1934), a curmudgeon on the K.P. side of the aisle, rose to thank the speaker. Rubbing his hands together in gleeful anticipation of a blow against Neyman by Fisher, he declares: "I am very glad Professor Fisher is present, as it is his work that Dr Neyman has accepted and incorporated. . . . I am not at all sure that the 'confidence' is not a confidence trick" (p.132). Bowley was to be disappointed. When it was Fisher's turn, he was full of praise. "Dr Neyman . . . claimed to have generalized the argument of fiducial probability, and he had every reason to be proud of the line of argument he had developed for its perfect clarity" (Fisher 1934c, p.138). Caveats were to come later (Section 5.7). For now, Egon was relieved:

Fisher had on the whole approved of what Neyman had said. If the impetuous Pole had not been able to make peace between the second and third floors of University College, he had managed at least to maintain a friendly foot on each! (C. Reid 1998, p. 119)

CIs, Tests, and Severity. I'm always mystified when people say they find P-values utterly perplexing while they regularly consume polling results in terms of confidence limits. You could substitute one for the other.

Suppose that 60% of 100 voters randomly selected from a population U claim to favor candidate Fisher. An estimate of the proportion of the population who favor Fisher, θ, at least at this point in time, is typically given by means of confidence limits. A 95% confidence interval for θ is $\bar{x} \pm 1.96\sigma_{\bar{x}}$ where $\bar{x}$ is the observed proportion and we estimate $\sigma_{\bar{x}}$ by plugging $\bar{x}$ in for θ to get $\sigma_{\bar{x}} = \sqrt{[0.60\,(0.40)/100]} = 0.048$. The 95% CI limits for $\theta = 0.6 \pm 0.09$ using the Normal approximation. The lower limit is 0.51 and the upper limit is 0.69. Often, 0.09 is reported as the *margin of error*. We could just as well have asked, having observed $\bar{x} = 0.6$,

> what value of θ would 0.6 be statistically significantly greater than at the 0.025 level, and what value of θ would 0.6 be statistically significantly less than at the 0.025 level?

The two answers would yield 0.51 and 0.69, respectively. So infer $\theta > 0.51$ and infer $\theta < 0.69$ (against their denials), each at level 0.025, for a combined error probability of 0.05.

Not only is there a duality between confidence interval estimation and tests, they were developed by Jerzy Neyman at the same time he was developing tests! The 1934 paper in the opening to this tour builds on Fisher's fiducial intervals dated in 1930, but he'd been lecturing on it in Warsaw for a few years already. Providing upper and lower confidence limits shows the range of plausible values for the parameter and avoids an "up/down" dichotomous tendency of some users of tests. Yet, for some reason, CIs are still often used in a dichotomous manner: rejecting μ values excluded from the interval, accepting (as plausible or the like) those included. There's the tendency, as well, to fix the confidence level at a single $1 - \alpha$, usually 0.9, 0.95, or 0.99. Finally, there's the adherence to a performance rationale: the estimation method will cover the true θ 95% of the time in a series of uses. We will want a much more nuanced, inferential construal of CIs. We take some first steps toward remedying these shortcomings by relating confidence limits to tests and to severity.

To simply make these connections, return to our test T+, an IID sample from a Normal distribution, H_0: $\mu \leq \mu_0$ against H_1: $\mu > \mu_0$. In a CI estimation procedure, an observed statistic is used to set an upper or lower (one-sided) bound, or both upper and lower (two-sided) bounds for parameter μ. Good and best properties of tests go over into good or best properties of corresponding confidence intervals. In particular, the uniformly most powerful (UMP) test T+ corresponds to a uniformly most accurate lower confidence bound (see Lehmann and Romano 2005, p. 72). The $(1 - \alpha)$ uniformly most accurate (UMA) lower confidence bound for μ, which I write as $\hat{\mu}_{1-\alpha}(\overline{X})$, corresponding to test T+ is

$$\mu > \overline{X} - c_\alpha(\sigma/\sqrt{n}),$$

where $\overline{X}$ is the sample mean, and the area to the right of c_α under the standard Normal distribution is α. That is $\Pr(Z \geq c_\alpha) = \alpha$ where Z is the standard Normal statistic. Here are some useful approximate values for c_α:

α	0.5	0.16	0.05	0.025	0.02	0.005	0.001
c_α	0	1	1.65	1.96	2	2.5	3

The Duality

"Infer: $\mu > \overline{X} - 2.5(\sigma/\sqrt{n})$" alludes to the rule for inferring; it is the CI *estimator*. Substituting $\overline{x}$ for $\overline{X}$ yields an *estimate*. Here are some abbreviations, alluding throughout to our example of a UMA estimator:

A *generic* $1 - \alpha$ lower confidence interval estimator is $\mu > \hat{\mu}_{1-\alpha}(\overline{X}) = \mu > \overline{X} - c_\alpha(\sigma/\sqrt{n})$.

A *specific* $1 - \alpha$ lower confidence interval estimate is $\mu > \hat{\mu}_{1-\alpha}(\overline{x}) = \mu > \overline{x} - c_\alpha(\sigma/\sqrt{n})$.

The corresponding value for α is close enough to 0.005 to allow $c_{0.005} = 2.5$ (it's actually closer to 0.006). The impressive thing is that, regardless of the true value of μ, these rules have high coverage probability. If, for any observed $\overline{x}$, in our example, you shout out

$$\mu > \overline{X} - 2.5(\sigma/\sqrt{n}),$$

your assertions will be correct 99.5% of the time. The specific inference results from plugging in $\bar{x}$ for $\bar{X}$. The specific 0.995 lower limit = $\hat{\mu}_{0.995}(\bar{x}) = \bar{x} - 2.5(\sigma/\sqrt{n})$, and the specific 0.995 estimate is $\mu > \hat{\mu}_{0.995}(\bar{x})$. This inference is qualified by the error probability of the method, namely the confidence level 0.995. But the upshot of this qualification is often misunderstood. Let's have a new example to show the duality between the lower confidence interval estimator $\mu > \hat{\mu}_{1-\alpha}(\bar{X})$ and the *generic* (α level) test T+ of form: $H_0: \mu \leq \mu_0$ against $H_1: \mu > \mu_0$. The "accident at the water plant" has a nice standard error of 1, but that can mislead about the role of sample size n. Let $\sigma = 1$, $n = 25$, $\sigma_{\bar{X}} = (\sigma/\sqrt{n}) = 0.2$. (Even though we'd actually have to estimate σ, the logic is the same and it's clearer.) I use $\sigma/\sqrt{n}$ rather than $\sigma_{\bar{X}}$ when a reminder of sample size seems needed.

Work backwards. Suppose we've collected the 25 samples and observed sample mean $\bar{x} = 0.6$. (The 0.6 has nothing to do with the polling example at the outset.) For what value of μ_0 would $\bar{x} = 0.6$ exceed μ_0 by $2.5\sigma_{\bar{X}}$? Since $2.5\sigma_{\bar{X}} = 0.5$, the answer is $\mu = 0.1$. If we were testing $H_0: \mu \leq 0.1$ vs. $H_1: \mu > 0.1$ at level 0.005, we'd reject with this outcome. The corresponding 0.995 lower estimate would be

$$\mu > 0.1.$$

(see Note 1).

Now for the duality. $\bar{X}$ is not statistically significantly greater than any μ value larger than 0.1 (e.g., 0.15, 0.2, etc.) at the 0.005 level. A test of form T+ would fail to reject each of the values in the CI interval at the 0.005 level, with $\bar{x} = 0.6$. Since this is continuous, it does not matter if the cut-off is at 0.1 or greater than or equal to 0.1.[1] By contrast, if we were testing μ_0 values 0.1 or less (T+: $H_0: \mu \leq 0.1$ against $H_1: \mu > 0.1$), these nulls *would* be rejected by $\bar{x} = 0.6$ at the 0.005 level (or even lower for values less than 0.1). That is, under the supposition that the data were generated from a world where $H_0: \mu \leq 0.1$, at least 99.5% of the time a *smaller* $\bar{X}$ than what was observed (0.6) would occur:

$$\Pr(\bar{X} < 0.6; \mu = 0.1) = 0.995.$$

The probability of observing $\bar{X} \geq 0.6$ would be low, 0.005.

Severity Fact (for test T+): Taking an outcome $\bar{x}$ that just reaches the α level of significance ($\bar{x}_\alpha$)as warranting $H_1: \mu > \mu_0$ with severity $(1 - \alpha)$

---

[1] To avoid confusion, note the duality is altered accordingly. If we set out the test rule for T + $H_0: \mu \leq \mu_0$ vs $H_1: \mu > \mu_0$ as reject H_0: iff $\bar{X} \geq \mu_0 + c_\alpha(\sigma/\sqrt{n})$, then we do not reject H_0 iff $\bar{X} < \mu_0 + c_\alpha(\sigma/\sqrt{n})$. This is the same as $\mu_0 > \bar{X} - c_\alpha(\sigma/\sqrt{n})$, the corresponding lower CI bound. If the test rule is $\bar{X} > \mu_0 + c_\alpha(\sigma/\sqrt{n})$, the corresponding lower bound is $\mu_0 \geq \bar{X} - c_\alpha(\sigma/\sqrt{n})$.

is mathematically the same as inferring $\mu > \bar{x} - c_\alpha(\sigma/\sqrt{n})$ at level $(1 - \alpha)$.

Hence, there's an intimate mathematical relationship between severity and confidence limits. However, severity will break out of the fixed $(1 - \alpha)$ level, and will supply a non-behavioristic rationale that is now absent from confidence intervals.[2]

Severity and Capabilities of Methods

Begin with an instance of our "Fact": To take an outcome that just reaches the 0.005 significance level as warranting H_1 with severity 0.995, is the same as taking the observed $\bar{x}$ and inferring μ just exceeds the 99.5 and lower confidence bound: $\mu > 0.1$. My justification for inferring $\mu > 0.1$ (with $\bar{x} = 0.6$) is this. Suppose my inference is false. Take the smallest value that renders it false, namely $\mu = 0.1$. Were $\mu = 0.1$, then the test very probably would have resulted in a smaller observed $\bar{X}$ than I got (0.6). That is, 99.5% of the time it would have produced a result *less discordant* with claim $\mu > 0.1$ than what I observed. (For μ values less than 0.1 this probability is increased.) Given that the method was highly *in*capable of having produced a value of $\bar{X}$ as large as 0.6, if $\mu \leq 0.1$, we argue that there is an indication at least (if not full blown evidence) that $\mu > 0.1$. The severity with which $\mu > 0.1$ "passes" (or is indicated by) this test is approximately 0.995.

Some caveats: First, throughout this exercise, we are assuming these values are "audited," and the assumptions of the model permit the computations to be licit. Second, we recognize full well that we merely have a single case, and inferring a genuine experimental effect requires being able to produce such impressive results somewhat regularly. That's why I'm using the word "indication" rather than evidence. Interestingly though, you don't see the same admonition against "isolated" CIs as with tests. (Rather than repeating these auditing qualifications, I will assume the context directs the interpretation.)

Severity versus Performance. The severity interpretation differs from both the construals that are now standard in confidence interval theory: The first is the *coverage probability* construal, and the second I'm calling *rubbing-off*. The coverage probability rationale is straightforwardly performance oriented. The rationale for the rule: infer

---

[2] For the computations, in test T+: H_0: $\mu \leq \mu_0$ against H_1: $\mu > \mu_0$. Suppose the observed $\bar{x}$ just reaches the c_α cut-off: $\bar{x} = \mu_0 + c_\alpha \sigma_{\bar{X}}$. The $(1 - \alpha)$ CI lower bound, CI_L, is $\mu > \bar{X} - c_\alpha \sigma_{\bar{X}}$. So Pr(test T+ does not reject H_0; $\mu = \text{CI}_L$) = Pr($\bar{X} < \mu_0 + c_\alpha \sigma_{\bar{X}}$; $\mu = \mu_0$). Standardize $\bar{X}$ to get Z: $Z = [(\mu_0 + c_\alpha \sigma_{\bar{X}}) - \mu_0](1/\sigma_{\bar{X}}) = c_\alpha$. So the severity for $\mu > \mu_0$ = Pr(test T+ does not reject H_0; $\mu = \text{CI}_L$) = Pr($Z < c_\alpha$) = $(1 - \alpha)$.

$$\mu > \overline{X} - 2.5\sigma/\sqrt{n},$$

is simply that you will correctly cover the true value at least 99.5% of the time in repeated use (we can allow the repetitions to be actual or hypothetical):

$$\Pr(\mu > (\overline{X} - 2.5\sigma/\sqrt{n}); \mu) = 0.995.$$

Aside: The equation above is not treating μ as a random variable, although it might look that way. $\overline{X}$ is the random variable. It's the same as asserting $\Pr(\overline{X} \geq \mu + 2.5(\sigma/\sqrt{n}); \mu) = 0.005$. Is this performance-oriented interpretation really all you can say? The severe tester says no. Here's where different interpretive philosophies enter.

Cox and Hinkley (1974) do not adhere to a single choice of $1 - \alpha$. Rather, to assert a 0.995 CI estimate, they say, is to follow:

> ... a procedure that would be wrong only in a proportion α of cases, in hypothetical repeated applications, whatever may be the true value μ. Note that this is a hypothetical statement that gives an empirical meaning, which in principle can be checked by experiment, rather than a prescription for using confidence limits. In particular, we do not recommend or intend that a fixed value α_0 should be chosen in advance and the information in the data summarized in the single assertion $[\mu > \hat{\mu}_{1-\alpha}]$. (p. 209, μ is substituted for their θ)

We have the *meaning versus application* gap again, which severity strives to close. "[W]e define procedures for assessing evidence that are calibrated by how they would perform were they used repeatedly. In that sense they do not differ from other measuring instruments" (Cox 2006a, p. 8). Yet this performance is not the immediate justification for the measurement in the case at hand. What I mean is, it's not merely that if you often use a telescope with good precision, your measurements will have a good track record – no more than with my scales (in Section 1.1). Rather, the thinking is, knowing how they would perform lets us infer how they're performing now. Good long-run properties "rub-off" in some sense on the case at hand (provided at least they are the relevant ones).

It's not so clear what's being rubbed off. You can't say the probability it's correct *in this case* is 0.995, since either it's right or not. That's why "confidence" is introduced. Some people say from the fact that the procedure is rarely wrong we may assign a low probability to its being wrong in the case at hand. First, this is dangerously equivocal, since the probability properly attaches to the method of inferring. Some espouse it as an informal use of "probability" outside of statistics, for instance, that confidence is "the degree of belief of a rational person that the confidence interval covers the parameter" (Schweder and Hjort 2016, p. 11). They call this "epistemic probability." My main gripe is that neither epistemic probability, whatever it is, nor performance gives a report of well-testedness associated with the claim at hand.

By providing several limits at different values, we get a more informative assessment, sometimes called a confidence distribution (CD). An early reference is Cox (1958). "The set of all confidence intervals at different levels of probability... [yields a] confidence distribution" (Cox 1958, p. 363). We'll visit others later. The severe tester still wants to nudge the CD idea; whether it's a large or small nudge is unclear because members of CD tribes are unclear. By and large, they're either a tad bit too performance oriented or too close to a form of probabilism for a severe tester. Recall I've said I don't see the severity construal out there, so I don't wish to saddle anyone with it. If that is what some CD tribes intend, great.

The severity logic is the counterfactual reasoning: Were μ less than the 0.995 lower limit, then it is very probable (> 0.995) that our procedure would yield a smaller sample mean than 0.6. This probability gives the severity. To echo Popper, $\mu > \hat{\mu}_{1-\alpha}$ is corroborated (at level 0.995) because it may be presented as a *failed attempt to falsify* it statistically. The severe testing philosophy hypothesizes that this is how humans reason. It underwrites formal error statistics as well as day-to-day reasoning.

Exhibit (vii): Capability. Let's see how severity is computed for the CI claim $(\mu > \hat{\mu}_{0.995})$ with $\bar{x} = 0.6$:

1. The particular assertion h is $\mu > 0.1 (\hat{\mu}_{0.995} = 0.1)$.
2. $\bar{x} = 0.6$ accords with h, an assertion about a positive discrepancy from 0.1.
3. Values of $\overline{X}$ less than 0.6 accord less well with h. So we want to compute the probability $(\overline{X} < 0.6)$ just at the point that makes h false: $\mu = 0.1$.
   Pr(method would yield $\overline{X} < 0.6; 0.1) = 0.995$.
4. From (3), SEV($\mu > 0.1$) = 0.995 (or we could write $\geq$, but our convention will be to write =).

Although we are moving between values of the parameter and values of $\overline{X}$, so long as we are careful, there is no illegitimacy. We can see that CI limits follow severity reasoning. For general lower $1 - \alpha$ limits, with small level α:

The inference of interest is h: $\mu > \hat{\mu}_{1-\alpha}$.
Since Pr(method would yield $\overline{X} < \bar{x}; \mu = \hat{\mu}_{1-\alpha}) = (1 - \alpha)$,
   it follows that SEV(h) = $(1 - \alpha)$.

(Lower case h emphasizes these are typically members of the full alternative in a test.) Table 3.5 gives several examples.

Perhaps "capability or incapability" of the method can serve to get at what's rubbing off. The specific moral I've been leading up to can be read right off the Table, as we vary the value for α (from 0.001 to 0.84) and form the corresponding lower confidence bound from $\hat{\mu}_{0.999}$ to $\hat{\mu}_{0.16}$.

> The higher the test's capability to produce such large (or even larger) differences as we observe, under the assumption $\mu = \hat{\mu}$, the *less* severely tested is assertion $\mu > \hat{\mu}$. (See Figure 3.3.)

The third column of Table 3.5 gives the complement to the severity assessment: the capability of a more extreme result, which in this case is α: $\Pr(\overline{X} > \overline{x}; \mu = \hat{\mu}_{1-\alpha}) = \alpha$. This is the Π function – the attained sensitivity in relation to μ: $\Pi(\gamma)$ (section 3.3) – but there may be too many moving parts to see this simply right away. You can return to it later.

We do not report a single, but rather several confidence limits, and the corresponding inferences of form h_1. Take the third row. The 0.975 lower limit that would be formed from $\overline{x} = 0.6$, $\hat{\mu}_{0.975}$, is $\mu = 0.2$. The estimate takes the form $\mu > 0.2$. Moreover, the observed mean, 0.6, is statistically significantly greater than 0.2 at level 0.025. Since $\mu = 0.2$ would very probably produce $\overline{X} < 0.6$, the severe tester takes the outcome as a good indication of $\mu \geq 0.2$. I want to draw your attention to the fact that the probability of producing an $\overline{X} \geq 0.6$ ranges from 0.005 to 0.5 for values of μ between 0.1 and the observed $\overline{x} = 0.6$. It never exceeds 0.5. To see this compute $\Pr(\overline{X} \geq 0.6; \mu = \mu')$ letting μ' range from 0.1 to 0.6. We standardize $\overline{X}$ to get $Z = (\overline{X} - \mu')/(\sigma/\sqrt{n})$ which is N(0,1). To find $\Pr(\overline{X} \geq 0.6; \mu = \mu')$, compute $Z = (0.6 - \mu')/0.2$ and use the areas under the standard Normal curve to get $\Pr(Z \geq z_0)$, μ' ranging from 0.1 to 0.6.

Do you notice it is only for negative z values that the area to the right of z exceeds 0.5? The test only begins to have more than 50% capability of generating observed means as large as 0.6, when μ is larger than 0.6. An important benchmark enters. The lower 0.5 bound $\hat{\mu}_{0.5}$ is 0.6. Since a result even larger than observed is brought about 50% of the time when $\mu = 0.6$, we rightly *block* the inference to $\mu > 0.6$.

Go to the next to last row: using a lower confidence limit at level 0.31! Now nobody goes around forming confidence bounds at level 0.5, let alone 0.31, but they might not always realize that's what they're doing! We could give a performance-oriented justification: the inference to $\mu > 0.7$ from $\overline{x} = 0.6$ is an instance of a rule that errs 31% of the time. Or we could use counterfactual,

Table 3.5 Lower confidence limit with $\bar{x} = 0.6$, α ranging from 0.001 to 0.84 in T+: $\sigma = 1$, $n = 25$, $\sigma_{\bar{x}} = (\sigma/\sqrt{n}) = 0.2$, $\bar{x} = 0.6$

α	c_α	$\hat{\mu}_{1-\alpha}$	$h_1{:}\mu{>}\hat{\mu}_{1-\alpha}$	$\Pr(\bar{X} \geq 0.6; \mu{=}\hat{\mu}_{1-\alpha}){=}\alpha$	SEV(h_1)
0.001	3	0	$(\mu > 0)$	0.001	0.999
0.005	2.5	0.1	$(\mu > 0.1)$	0.005	0.995
0.025	2	0.2	$(\mu > 0.2)$	0.025	0.975
0.07	1.5	0.3	$(\mu > 0.3)$	0.07	0.93
0.16	1	0.4	$(\mu > 0.4)$	0.16	0.84
0.3	0.5	0.5	$(\mu > 0.5)$	0.3	0.7
0.5	0	0.6	$(\mu > 0.6)$	0.5	0.5
0.69	−0.5	0.7	$(\mu > 0.7)$	0.69	0.31
0.84	−1	0.8	$(\mu > 0.8)$	0.84	0.16

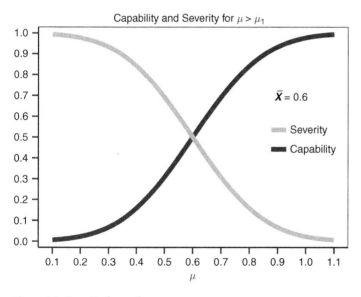

Figure 3.3 Severity for $\mu{>}\hat{\mu}$.

severity reasoning: even if μ were only 0.7, we'd get a larger $\bar{X}$ than we observed a whopping 69% of the time. Our observed $\bar{X}$ is terrible grounds to suppose μ must exceed 0.7. If anything, we're starting to get an indication that $\mu < 0.7$! Observe that, with larger α, the argument is more forceful by emphasizing >, rather than ≥, but it's entirely correct either way, as it is continuous.

In grasping the duality between tests and confidence limits we consider the *general form* of the test in question, here we considered T+. Given the general form, we imagine the test hypotheses varying, with a fixed outcome $\bar{x}$. Considering other instances of the general test T+ is a heuristic aid in interpreting confidence limits using the idea of statistical inference as severe testing. We will often allude to confidence limits to this end. However, the way the severe tester will actually use the duality is best seen as a post-data way to ask about various discrepancies indicated. For instance, in testing H_0: $\mu \leq 0$ vs. H_1: $\mu > 0$, we may wish to ask, post-data, about a discrepancy such as $\mu > 0.2$. That is, we ask, for each of the inferences, how severely passed it is.

Granted this interval estimator has a nice pivot. If I thought the nice cases weren't the subject of so much misinterpretation, I would not start there. But there's no chance of seeing one's way into more complex cases if we are still hamstrung by the simple ones. In fact, the vast majority of criticism and proposed reforms revolve around our test T+ and two-sided variants. If you grasp the small cluster of the cases that show up in the debates, you'll be able to extend the results. The severity interpretation enables confidence intervals to get around some of their current problems. Let's visit a few of them now. (See also Excursion 4 Tour II, Excursion 5 Tour II.)

Exhibit (viii): Vacuous and Empty Confidence Intervals: Howlers and Chestnuts. *Did you hear the one about the frequentist who reports a confidence level of 0.95 despite knowing the interval must contain the true parameter value?*
Basis for the joke: it's possible that CIs wind up being vacuously true: including all possible parameter values. "Why call it a 95 percent CI if it's known to be true?" the critics ask. The obvious, performance-based, answer is that the confidence level refers to the probability the method outputs true intervals; it's not an assignment of probability to the specific interval. It's thought to be problematic only by insisting on a probabilist interpretation of the confidence level. Jose Bernardo thinks that the CI user "should be subject to some re-education using well-known, standard counterexamples. . . . conventional 0.95-confidence regions may actually consist of the whole real line" (Bernardo 2008, p. 453). Not so.

Cox and Hinkley (1974, p.226) proposed interpreting confidence intervals, or their corresponding confidence limits (lower or upper), as the set of parameter values consistent at the confidence level.

This interpretation of confidence intervals also scotches criticisms of examples where, due to given restrictions, it can happen that a $(1 - \alpha)$ estimate contains all possible parameter values. Although such an inference is 'trivially true,' it is scarcely vacuous in our construal. That all parameter values are consistent with the data is an informative statement about the limitations of the data to detect discrepancies at the particular level. (Cox and Mayo 2010, p. 291)

Likewise it can happen that all possible parameter points are inconsistent with the data at the $(1 - \alpha)$ level. Criticisms of "vacuous" and empty confidence intervals stem from a probabilist construal of $(1 - \alpha)$ as the degree of support, belief, or probability attached to the particular interval; but this construal isn't endorsed by CI interval methodology. There is another qualification to add: the error probability computed must be relevant. It must result from the relevant sampling distribution.

Pathological Confidence Set. Here's a famous chestnut that is redolent of Exhibit (vi) in Section 3.4 (Cox's 1958 two measuring instruments with different precisions). It is usually put in terms of a "confidence set" with $n = 2$. It could also be put in the form of a test. Either way, it is taken to question the relevance of error statistical assessments in the case at hand (e.g., Berger and Wolpert 1988, Berger 2003, p. 6). Two independent and identically distributed observations are to be made represented by random variables X_1, X_2. Each X can take either value $\psi - 1$ or $\psi + 1$ with probability of 0.5, where ψ is the unknown parameter to be estimated using the data. The data can result in both outcomes being the same, or both different.

Consider the second case: With both different, we know they will differ by 2. A possibility might be $\langle 9, 11 \rangle$. Right away, we know ψ must be 10. What luck! We know we're right to infer ψ is 10. To depict this case more generally, the two outcomes are $x_1 = x' - 1$ and $x_2 = x' + 1$, for some value x'.

Consider now that the first case obtains. We are not so lucky. The two outcomes are the same: $x_1 = x_2$ (maybe they're both 9 or whatever). What should we infer about the value of parameter ψ? We know ψ is either $x_1 - 1$ or $x_1 + 1$ (e.g., 8 or 10); each accords equally well with the data. Say we infer ψ is $x_1 - 1$. The method is correct with probability 0.5. Averaging over the two possibilities, the probability of an erroneous inference is 0.25. Now suppose I was lucky and observed two different outcomes. Then I know the value of ψ so it makes no sense to infer "ψ is $(x_1 + x_2)/2$" while attaching a confidence coefficient of 0.75.

You see the pattern. The example is designed so that some outcomes yield much more information than others. As with Cox's "two measuring instruments," the data have two parts: First, an indication of whether the two outcomes are the same or different; second, the observed result. Let A be an indicator of the first part: $A = 1$ if both are the same (unlucky); $A = 2$ if the sample values differ by 2 (lucky!). The full data may be represented as (A, x). The distribution of A is fixed independently of the parameter of interest: $Pr(A = 1) = Pr(A = 2) = 0.5$. It is an example of an *ancillary* statistic. However, learning whether $A = 1$ or $A = 2$ is very informative as to the precision achieved by the inference. Thus the relevant properties associated with the particular inference would be conditional on the value of A.

The tip-off that we're dealing with a problem case is this: The sufficient statistic S has two parts (A, X), that is it has *dimension* 2. But there's only one parameter ψ. Without getting into the underlying theory, this alone indicates that S has a property known as being *incomplete*, opening the door to different P-values or confidence levels when calculated conditionally on the value of A. In particular, the marginal distribution of a P-value averaged over the two possibilities $(0.5(0) + 0.5(0.5) = 0.25)$ would be misleading for any particular set of data. Instead we condition on the value of A obtained. David Cox calls this process "*technical conditioning to induce relevance* of the frequentist probability to the inference at hand" (Cox and Mayo 2010, pp. 297–8).

Such examples have other noteworthy features: the ancillary part A gives a sneaky way of assigning a probability to "being correct" in the subset of cases given by the value of A. It's an example of what Fisher called "recognizable subsets." By careful artifice, the event that "a random variable A takes a given value a" is equivalent to "the data were generated by a hypothesized parameter value." So the probability of $A = a$ gives the probability a hypothesis is true. Aris Spanos considers these examples "rigged" for this reason, and he discusses these and several other famous pathological examples (Spanos 2012).

Even putting pathologies aside, is there any reason the frequentist wouldn't do the sensible thing and report on how well probed the inference is once A is known? No. Certainly a severe testing theorist would.

Live Exhibit (ix). What Should We Say When Severity Is Not Calculable?
In developing a system like severity, at times a conventional decision must be made. However, the reader can choose a different path and still work within this system.

What if the test or interval estimation procedure does not pass the audit? Consider for the moment that there has been optional stopping, or cherry picking, or multiple testing. Where these selection effects are well understood, we may adjust the error probabilities so that they do pass the audit. But what if the moves are so tortuous that we can't reliably make the adjustment? Or perhaps we don't feel secure enough in the assumptions? Should the severity for $\mu > \mu_0$ be low or undefined?

You are free to choose either. The severe tester says SEV$(\mu > \mu_0)$ is low. As she sees it, having evidence requires a minimum threshold for severity, even without setting a precise number. If it's close to 0.5, it's quite awful. But if it cannot be computed, it's also awful, since the onus on the researcher is to satisfy the minimal requirement for evidence. I'll follow her: If we cannot compute the severity even approximately (which is all we care about), I'll say it's low, along with an explanation as to why: It's low because we don't have a clue how to compute it!

A probabilist, working with a single "probability pie" as it were, would take a low probability for H as giving a high probability to $\sim H$. By contrast we wish to clearly distinguish between having poor evidence for H and having good evidence for $\sim H$. Our way of dealing with bad evidence, no test (BENT) allows us to do that. Both SEV(H) and SEV$(\sim H)$ can be low enough to be considered lousy, even when both are computable.

Souvenir N: Rule of Thumb for SEV

> Can we assume that if SEV$(\mu > \mu_0)$ is a high value, $1 - \alpha$, then SEV$(\mu \leq \mu_0)$ is α?

Because the claims $\mu > \mu_0$ and $\mu \leq \mu_0$ form a partition of the parameter space, and because we are assuming our test has passed (or would pass) an audit, else these computations go out the window, the answer is yes.

> If SEV$(\mu > \mu_0)$ is high, then SEV$(\mu \leq \mu_0)$ is low.

The converse need not hold – given the convention we just saw in Exhibit (ix). At the very least, "low" would not exceed 0.5.

A rule of thumb (for test T+ or its dual CI):

- If we are pondering a claim that an observed difference from the null seems *large* enough to indicate $\mu > \mu'$, we want to be sure the test was highly capable of producing *less* impressive results, were $\mu = \mu'$.
- If, by contrast, the test was highly capable of producing *more* impressive results than we observed, even in a world where $\mu = \mu'$, then we block an inference to $\mu > \mu'$ (following weak severity).

This rule will be at odds with some common interpretations of tests. Bear with me. I maintain those interpretations are viewing tests through "probabilist-colored" glasses, while the correct error-statistical view is this one.

3.8 The Probability Our Results Are Statistical Fluctuations: Higgs' Discovery

One of the biggest science events of 2012–13 was the announcement on July 4, 2012 of evidence for the discovery of a Higgs-like particle based on a "5-sigma observed effect." With the March 2013 data analysis, the 5-sigma difference grew to 7 sigmas, and some of the apparent anomalies evaporated. In October 2013, the Nobel Prize in Physics was awarded jointly to François Englert and Peter W. Higgs for the "theoretical discovery of a mechanism" behind the particle experimentally discovered by the collaboration of thousands of scientists (on the ATLAS and CMS teams) at CERN's Large Hadron Collider in Switzerland. Yet before the dust had settled, the very nature and rationale of the 5-sigma discovery criterion began to be challenged among scientists and in the popular press. Because the 5-sigma standard refers to a benchmark from frequentist significance testing, the discovery was immediately imbued with controversies that, at bottom, concern statistical philosophy.

Why a 5-sigma standard? Do significance tests in high-energy particle (HEP) physics escape the misuses of P-values found in social and other sciences? Of course the main concern wasn't all about philosophy: they were concerned that their people were being left out of an exciting, lucrative, many-years project. But unpacking these issues is philosophical, and that is the purpose of this last stop of Excursion 3. I'm an outsider to HEP physics, but that, apart from being fascinated by it, is precisely why I have chosen to discuss it. Anyone who has come on our journey should be able to decipher the more public controversies about using P-values.

I'm also an outsider to the International Society of Bayesian Analysis (ISBA), but a letter was leaked to me a few days after the July 4, 2012 announcement, prompted by some grumblings raised by a leading subjective Bayesian, Dennis Lindley. The letter itself was sent around to the ISBA list by statistician Tony O'Hagan. "Dear Bayesians," the letter began. "We've heard a lot about the Higgs boson."

Why such an extreme evidence requirement? We know from a Bayesian perspective that this only makes sense if (a) the existence of the Higgs boson . . . has extremely small prior probability and/or (b) the consequences of erroneously announcing its discovery are dire in the extreme. (O'Hagan 2012)

Neither of these seemed to be the case in his opinion: "[Is] the particle physics community completely wedded to frequentist analysis? If so, has anyone tried to explain what bad science that is?" (ibid.).

Bad science? Isn't that a little hasty? HEP physicists are sophisticated with their statistical methodology: they'd seen too many bumps disappear. They want to ensure that before announcing a new particle has been discovered that, at the very least, the results being spurious is given a run for its money. Significance tests, followed by confidence intervals, are methods of choice here for good reason. You already know that I favor moving away from traditional interpretations of statistical tests and confidence limits. But some of the criticisms, and the corresponding "reforms," reflect misunderstandings, and the knottiest of them all concerns the very meaning of the phrase (in the title of Section 3.8): "the probability our results are merely statistical fluctuations." Failing to clarify it may well impinge on the nature of future big science inquiry based on statistical models. The problem is a bit delicate, and my solution is likely to be provocative. You may reject my construal, but you'll see what it's like to switch from wearing probabilist, to severe testing, glasses.

The Higgs Results

Here's a quick sketch of the Higgs statistics. (I follow the exposition by physicist Robert Cousins (2017). See also Staley (2017). There is a general model of the detector within which researchers define a "global signal strength" parameter μ "such that H_0: $\mu = 0$ corresponds to the background-only hypothesis and $\mu = 1$ corresponds to the [Standard Model] SM Higgs boson signal in addition to the background" (ATLAS collaboration 2012c). The statistical test may be framed as a one-sided test:

$$H_0: \mu = 0 \text{ vs. } H_1: \mu > 0.$$

The test statistic $d(X)$ data records how many *excess events* of a given type are "observed" (from trillions of collisions) in comparison to what would be expected from background alone, given in standard deviation or sigma units. Such excess events give a "signal-like" result in the form of bumps off a smooth curve representing the "background" alone.

The improbability of the different $d(X)$ values, its sampling distribution, is based on simulating what it would be like under H_0 fortified with much cross-checking of results. These are converted to corresponding probabilities under a standard Normal distribution. The probability of observing results as extreme as or more extreme than 5 sigmas, under H_0, is approximately 1 in 3,500,000! Alternatively, it is said that the probability that the results were just a statistical fluke (or fluctuation) is 1 in 3,500,000.

Why such an extreme evidence requirement, Lindley asked. Given how often bumps disappear, the rule for interpretation, which physicists never intended to be rigid, is something like: if $d(X) \geq 5$ sigma, infer discovery, if $d(X) \geq 2$ sigma, get more data.

Now "deciding to announce" the results to the world, or "get more data" are actions all right, but each corresponds to an evidential standpoint or inference: infer there's evidence of a genuine particle, and infer that spurious bumps had not been ruled out with high severity, respectively.

What "the Results" Really Are

You know from the Translation Guide (Souvenir C) that $Pr(d(X) \geq 5; H_0)$ is to be read Pr (the test procedure would yield $d(X) \geq 5; H_0$). Where do we record Fisher's warning that we can only use P-values to legitimately indicate a genuine effect by demonstrating an *experimental phenomenon*. In good sciences and strong uses of statistics, "the results" may include demonstrating the "know-how" to generate results that rarely fail to be significant. Also important is showing the test passes an audit (it isn't guilty of selection biases, or violations of statistical model assumptions). "The results of test T" incorporates the entire display of know-how and soundness. That's what the severe tester means by Pr(test T would produce $d(X) \geq d(x_0); H_0$). So we get:

> Fisher's Testing Principle: To the extent that you know how to bring about results that rarely fail to be statistically significant, there's evidence of a genuine experimental effect.

There are essentially two stages of analysis. The first stage is to test for a genuine Higgs-like particle, the second, to determine its properties (production mechanism, decay mechanisms, angular distributions, etc.). Even though the SM Higgs sets the signal parameter to 1, the test is going to be used to learn about the value of any discrepancy from 0. Once the null is rejected at the first stage, the second stage essentially shifts to learning the particle's properties, and using them to seek discrepancies from a new null hypothesis: the SM Higgs.

The *P*-Value Police

The July 2012 announcement gave rise to a flood of buoyant, if simplified, reports heralding the good news. This gave ample grist for the mills of P-value critics. Statistician Larry Wasserman playfully calls them the "P-Value Police"(2012a) such as Sir David Spiegelhalter (2012), a professor of the Public's Understanding of Risk at the University of Cambridge. Their job was to examine if reports by journalists and scientists could be seen to be misinterpreting the sigma levels as posterior probability assignments to the various models and claims. Thumbs up or thumbs down! Thumbs up went to the ATLAS group report:

A statistical combination of these channels and others puts the significance of the signal at 5 sigma, meaning that *only one experiment in 3 million would see an apparent signal this strong in a universe without a Higgs*. (2012a, emphasis added)

Now HEP physicists have a term for an apparent signal that is actually produced due to chance variability alone: a *statistical fluctuation* or *fluke*. Only one experiment in 3 million would produce so strong a background fluctuation. ATLAS (2012b) calls it the "background fluctuation probability." By contrast, Spiegelhalter gave a thumbs down to:

> There is less than a one in 3 million chance that their results are a statistical fluctuation.

If they had written "would be" instead of "is" it would get thumbs up. Spiegelhalter's ratings are generally echoed by other Bayesian statisticians. According to them, the thumbs down reports are guilty of misinterpreting the *P*-value as a posterior probability on H_0.

A careful look shows this is not so. H_0 does not say the observed results are due to background alone; H_0 does not say the result is a fluke. It is just $H_0: \mu = 0$. Although if H_0 were true it *follows* that various results would occur with specified probabilities. In particular, it entails (along with the rest of the background) that large bumps are improbable.

It may in fact be seen as an ordinary error probability:

(1) Pr(test T would produce d(X) $\geq$ 5; H_0) $\leq$ 0.0000003.

The portion within the parentheses is how HEP physicists understand "a 5-sigma fluctuation." Note (1) is not a conditional probability, which involves a prior probability assignment to the null. It is not

Pr(test T would produce d(X) $\geq$ 5 and H_0)/Pr(H_0).

Only random variables or their values are conditioned upon. This may seem to be nit-picking, and one needn't take a hard line on the use of "conditional." I mention it because it may explain part of the confusion here. The relationship between the null hypothesis and the test results is intimate: the assignment of probabilities to test outcomes or values of d(X) "under the null" may be seen as a tautologous statement.

Since it's not just a single result, but also a dynamic test display, we might even want to emphasize a fortified version:

(1)* Pr(test T would display $d(X) \geq 5; H_0) \leq 0.0000003$.

Critics may still object that (1), even fortified as (1)*, only entitles saying:

> There is less than a one in 3 million chance of a fluctuation (at least as strong as in their results).

It does not entitle one to say:

> There is less than a one in 3 million chance that *their results* are a statistical fluctuation.

Let's compare three "ups" and three "downs" to get a sense of the distinction that leads to the brouhaha:

Ups

U-1. The probability of the background alone fluctuating up by this amount or more is about one in 3 million. (CMS 2012)
U-2. Only one experiment in 3 million would see an apparent signal this strong in a universe described in H_0.
U-3. The probability that their signal would result by a chance fluctuation was less than one chance in 3 million.

Downs

D-1. The probability their results were due to the background fluctuating up by this amount or more is about one in 3 million.
D-2. One in 3 million is the probability the signal is a false positive – a fluke produced by random statistical fluctuation.
D-3. The probability that their signal was the result of a statistical fluctuation was less than one chance in 3 million.

The difference is that the thumbs down allude to "this" signal or "these" data are due to chance or is a fluctuation. Critics might say the objection to "this" is that the P-value refers to a difference as great or greater – a tail area. But if the probability of $\{d(X) \geq d(x)\}$ is low under H_0, then $Pr(d(X) = d(x); H_0)$ is even lower. We've dealt with this back with Jeffreys' quip (Section 3.4). No statistical account recommends going from improbability of a point result on a continuum under H to rejecting H. The Bayesian looks to the prior probability in H and its alternatives. The error statistician looks to the general procedure. The notation $\{d(X) \geq d(x)\}$ is used to signal the latter.

But if we're talking about the procedure, the critic rightly points out, we are not assigning probability to these particular data or signal. True, but that's the way frequentists always give probabilities to general events, whether they have occurred, or we are contemplating a hypothetical excess of 5 sigma that might occur. It's always treated as a generic type of event. We are never considering the probability "the background fluctuates up this much on Wednesday July 4, 2012," except as that is construed as a type of collision result at a type of detector, and so on. It's illuminating to note, at this point:

[t]he key distinction between Bayesian and sampling theory statistics is the issue of what is to be regarded as random and what is to be regarded as fixed. To a Bayesian, parameters are random and data, once observed, are fixed. . . (Kadane 2011, p. 437)

Kadane's point is that "[t]o a sampling theorist, data are random even after being observed, but parameters are fixed" (ibid.). When an error statistician speaks of the probability that the results standing before us are a mere statistical fluctuation, she is referring to a methodological probability: the probability the method used would produce data displays (e.g., bumps) as impressive as these, under the assumption of H_0. If you're a Bayesian probabilist D-1 through D-3 appear to be assigning a probability to a hypothesis (about the parameter) because, since the data are known, only the parameter remains unknown. But they're to be scrutinizing a non-Bayesian procedure here. Whichever approach you favor, my point is that they're talking past each other. To get beyond this particular battle, this has to be recognized.

The Real Problem with D-1 through D-3. The error probabilities in U-1 through U-3 are straightforward. In the Higgs experiment, the needed computations are based on simulating relative frequencies of events where $H_0: \mu = 0$ (given a detector model). In terms of the corresponding P-value:

(1) Pr(test T would produce a P-value ≤ 0.0000003; H_0) ≤ 0.0000003.

D-1, 2, 3 are just slightly imprecise ways of expressing U-1, 2, 3. So what's the objection to D-1, 2, 3? It's the danger some find in moving from such claims to their complements. If I say there's a 0.0000003 probability their results are due to chance, some infer there's a 0.999999 (or whatever) probability their results are not due to chance – are not a false positive, are not a fluctuation. And those claims are wrong. If $Pr(A; H_0) = p$, for some assertion A, the probability of the complement is $Pr(\text{not-A}; H_0) = 1 - p$. In particular:

(1) Pr(test T would *not* display a *P*-value ≤ 0.0000003; H_0) > 0.9999993.

There's no transposing! That is, the hypothesis after the ";" does not switch places with the event to the left of ";"! But despite how the error statistician hears D-1 through D-3, I'm prepared to grant the corresponding U claims are safer. I assure you that my destination is not merely refining statistical language, but when critics convince practitioners that they've been speaking Bayesian prose without knowing it (as in Molière), the consequences are non-trivial. I'm about to get to them.

Detaching Inferences Uses an Implicit Severity Principle

Phrases such as "the probability our results are a statistical fluctuation (or fluke) is very low" are common enough in HEP – although physicists tell me it's the science writers who reword their correct U-claims as slippery D-claims. Maybe so. But if you follow the physicist's claims through the process of experimenting and modeling, you find they are alluding to proper error probabilities. You may think they really mean an illicit posterior probability assignment to "real effect" or H_1 if you think that statistical inference takes the form of probabilism. In fact, if you're a Bayesian probabilist, and assume the statistical inference must have a posterior probability, or a ratio of posterior probabilities, you will regard U-1 through U-3 as legitimate but irrelevant to inference; and D-1 through D-3 as relevant only by misinterpreting *P*-values as giving a probability to the null hypothesis H_0.

If you are an error statistician (whether you favor a behavioral performance or a severe probing interpretation), even the correct claims U-1 through U-3 are not statistical inferences! They are the (statistical) justifications associated with implicit statistical inferences, and even though HEP practitioners are well aware of them, they should be made explicit. Such inferences can take many forms, such as those I place in brackets:

U-1. The probability of the background alone fluctuating up by this amount or more is about one in 3 million.
[Thus, our results are not due to background fluctuations.]

U-2. Only one experiment in 3 million would see an apparent signal this strong in a universe [where H_0 is adequate].
[Thus H_0 is not adequate.]

U-3. The probability that their signal would result by a chance fluctuation was less than one in 3.5 million.

[Thus the signal was not due to chance.]

The formal statistics moves from

(1) Pr(test T produces d(X) ≥ 5; H_0) < 0.0000003

to

(2) there is strong evidence for
(first) (2a) a genuine (non-fluke) discrepancy from H_0;
(later) (2b) H^*: a Higgs (or a Higgs-like) particle.

They move in stages from indications, to evidence, to discovery. Admittedly, moving from (1) to inferring (2) relies on the implicit assumption of error statistical testing, the severity principle. I deliberately phrase it in many ways. Here's yet another, in a Popperian spirit:

> *Severity Principle* (from low P-value) Data provide evidence for a genuine discrepancy from H_0 (just) to the extent that H_0 would (very probably) have survived, were H_0 a reasonably adequate description of the process generating the data.

What *is* the probability that H_0 would have "survived" (and not been falsified) at the 5-sigma level? It is the probability of the complement of the event {d(X) ≥ 5}, namely, {d(X) < 5} under H_0. Its probability is correspondingly 1 − 0.0000003. So the overall argument starting from a fortified premise goes like this:

> (1)* With probability 0.9999997, the bumps would be smaller, would behave like statistical fluctuations: disappear with more data, wouldn't be produced at both CMS and ATLAS, in a world adequately modeled by H_0.

They did not disappear, they grew (from 5 to 7 sigma). So,

> (2a) infer there's evidence of H_1: non-fluke, or (2b) infer H^*: a Higgs (or a Higgs-like) particle.

There's always the error statistical qualification of the inference in (2), given by the relevant methodological probability. Here it is a report of the stringency or severity of the test that the claim has passed, as given in (1)*: 0.9999997. We might even dub it the severity coefficient. Without making the underlying principle of testing explicit, some critics assume the argument is all about the reported P-value. It's a mere stepping stone to an inductive inference that is detached.

Members of a strict (N-P) behavioristic tribe might reason as follows: If you follow the rule of behavior: Interpret 5-sigma bumps as a real effect (a discrepancy from 0), you'd erroneously interpret data with probability less than 0.0000003 – a very low *error probability*. Doubtless, HEP physicists are keen to avoid repeating such mistakes as apparently finding particles that move faster than light, only to discover some problem with the electrical wiring (Reich 2012). I claim the specific evidential warrant for the 5-sigma Higgs inferences aren't low long-run errors, but being able to detach an inference based on a stringent test or a *strong* argument from coincidence.[3]

Learning How Fluctuations Behave: The Game of Bump-Hunting

Dennis Overbye (2013) wrote an article in the *New York Times*: "Chasing the Higgs," based on his interviews with spokespeople Fabiola Gianotti (ATLAS) and Guido Tonelli (CMS). It's altogether common, Tonelli explains, that the bumps they find are "random flukes" – spuriously significant results – "So 'we crosscheck everything' and 'try to kill' any anomaly that might be merely random."

One bump on physicists' charts . . . was disappearing. But another was blooming like the shy girl at a dance. . . . nobody could remember exactly when she had come in. But she was the one who would marry the prince . . . It continued to grow over the fall until it had reached the 3-sigma level – the chances of being a fluke [spurious significance] were less than 1 in 740, enough for physicists to admit it to the realm of "evidence" of something, but not yet a discovery. (Overbye 2013)

What's one difference between HEP physics and fields where most results are claimed to be false? HEP physicists don't publish on the basis of a single, isolated (nominal) *P*-value. That doesn't mean promising effects don't disappear. "'We've made many discoveries,' Dr. Tonelli said, 'most of them false'" (ibid.).

Look Elsewhere Effect (LEE). The null hypothesis is formulated to correspond to regions where an excess or bump is found. Not knowing the mass region in advance means "the local *p*-value did not include the fact that 'pure chance' had lots of opportunities . . . to provide an unlikely occurrence" (Cousins 2017, p. 424). So here a nominal (they call it local) *P*-value is assessed at a particular, data-determined, mass. But the probability of so impressive a difference anywhere in a mass range – the global

---

[3] The inference to (2) is a bit stronger than merely falsifying the null because certain properties of the particle must be shown at the second stage.

P-value – would be greater than the local one. "The original concept of '5σ' in HEP was therefore mainly motivated as a (fairly crude) way to account for a multiple trials factor ... known as the 'Look Elsewhere Effect'" (ibid. p. 425). HEP physicists often report both local and global *P*-values.

Background information enters, not via prior probabilities of the particles' existence, but as to how researchers might be led astray. "If they were flukes, more data would make them fade into the statistical background ... If not, the bumps would grow in slow motion into a bona fide discovery" (Overbye 2013). So, they give the bump a hard time, they stress test, look at multiple decay channels, and they hide the details of the area they found it from the other team. When two independent experiments find the same particle signal at the same mass, it helps to overcome the worry of multiple testing, strengthening an argument from coincidence.

Once the null is rejected, the job shifts to testing if various parameters agree with the SM predictions.

This null hypothesis of no Higgs (or Higgs-like) boson was definitively rejected upon the announcement of the observation of a new boson by both ATLAS and CMS on July 4, 2012. The confidence intervals for signal strength θ ... were in reasonable agreement with the predictions for the SM Higgs boson. Subsequently, much of the focus shifted to measurements of ... production and decay mechanisms. For measurements of continuous parameters, ... the tests ... use the frequentist duality ... between interval estimation and hypothesis testing. One constructs (approximate) confidence intervals and regions for parameters ... and checks whether the predicted values for the SM Higgs boson are within the confidence regions. (Cousins 2017, p. 414)

Now the corresponding null hypothesis, call it H_0^2, is the SM Higgs boson

$$H_0^2: \text{SM Higgs boson: } \mu = 1$$

and discrepancies from it are probed and estimated with confidence intervals. The most important role for statistical significance tests is actually when results are insignificant, or the *P*-values are not small: *negative* results. They afford a standard for blocking inferences that would be made too readily. In this episode, they arose to

(a) block precipitously declaring evidence of a new particle;
(b) rule out values of various parameters, e.g., spin values that would preclude its being "Higgs-like," and various mass ranges of the particle.

While the popular press highlighted the great success for the SM, the HEP physicists, at both stages, were vigorously, desperately seeking to uncover BSM (Beyond the Standard Model) physics.

Once again, the background knowledge of fluke behavior was central to curbing their enthusiasm about bumps that hinted at discrepancies with the new null: H_0^2: $\mu = 1$. Even though July 2012 data gave evidence of the existence of a Higgs-like particle – where calling it "Higgs-like" still kept the door open for an anomaly with the "plain vanilla" particle of the SM – they also showed some hints of such an anomaly.

Matt Strassler, who, like many, is longing to find evidence for BSM physics, was forced to concede: "The excess (in favor of BSM properties) has become a bit smaller each time … That's an unfortunate sign, if one is hoping the excess isn't just a statistical fluke" (2013a). Or they'd see the bump at ATLAS … and not CMS. *"Taking all of the LHC's data, and not cherry picking … there's nothing here that you can call 'evidence'"* for the much sought BSM (Strassler 2013b). They do not say the cherry-picked results 'give evidence, but disbelief in BSM physics lead us to discount it,' as Royall's Likelihoodist may opt to. They say: "There's nothing here that you can call evidence."

Considering the frequent statistical fluctuations, and the hot competition between the ATLAS and CMS to be first, a tool for when to "curb their enthusiasm" is exactly what was wanted. So, this negative role of significance tests is crucial for denying BSM anomalies are real, and setting upper bounds for these discrepancies with the SM Higgs. Since each test has its own test statistic, I'll use g(x) rather than d(x).

> *Severity Principle (for non-significance):* Data provide evidence to rule out a discrepancy δ^* to the extent that a larger g(x_0) would very probably have resulted if δ were as great as δ^*.

This can equivalently be seen as inferring confidence bounds or applying FEV. The particular value of δ^* isn't so important at this stage. What happens with negative results here is that the indicated discrepancies get smaller and smaller as do the bumps, and just vanish. These were not genuine effects, even though there's no falsification of BSM.

Negative results in HEP physics are scarcely the stuff of file drawers, a serious worry leading to publication bias in many fields. Cousins tells of the wealth of papers that begin "Search for …" (2017, p. 412). They are regarded as important and informative – if only in ruling out avenues for theory development. There's another idea for domains confronted with biases against publishing negative results.

Back to O'Hagan and a 2015/2016 Update

O'Hagan published a digest of responses a few days later. When it was clear his letter had not met with altogether enthusiastic responses, he backed off, admitting that he had only been being provocative with the earlier letter. Still, he declares, the Higgs researchers would have been better off avoiding the "ad hoc" 5 sigma by doing a proper (subjective) Bayesian analysis. "They would surely be willing to [announce SM Higgs discovery] if they were, for instance, 99.99 percent certain" [SM Higgs] existed. Wouldn't it be better to report

$$Pr(SM\ Higgs|data) = 0.9999?$$

Actually, no. Not if it's taken as a formal probability rather than a chosen way to abbreviate: the reality of the SM Higgs has passed a severe test. Physicists believed in a Higgs particle before building the big billion-dollar collider. Given the perfect predictive success of the SM, and its simplicity, such beliefs would meet the familiar standards for plausibility. But that's very different from having evidence for a discovery, or information about the characteristics of the particle. Many aver they didn't expect it to have so small a mass, 125 GeV. In fact, given the unhappy consequences some find with this low mass, some researchers may well have gone back and changed their prior probabilities to arrive at something more sensible (more "natural" in the parlance of HEP). Yet, their strong argument from coincidence via significance tests prevented the effect from going away.

O'Hagan/Lindley admit that a subjective Bayesian model for the Higgs would require prior probabilities to scads of high dimensional "nuisance" parameters of the background and the signal; it would demand multivariate priors, correlations between parameters, joint priors, and the ever worrisome Bayesian catchall factor: $Pr(data|not\text{-}H^*)$. Lindley's idea of subjectively eliciting beliefs from HEP physicists is rather unrealistic here.

Now for the update. When the collider restarted in 2015, it had far greater collider energies than before. On December 15, 2015 something exciting happened: "ATLAS and CMS both reported a small 'bump' in their data" at a much higher energy level than the Higgs: 750 GeV (compared to 125 GeV) (Cartlidge 2016). "As this unexpected bump could be the first hint of a new massive particle that is not predicted by the Standard Model of particle physics, the data generated hundreds of theory papers that attempt to explain the signal" (ibid.). I believe it was 500.

The significance reported by CMS is still far below physicists' threshold for a discovery: 5 sigma, or a chance of around 3 in 10 million that the signal is a statistical fluke. (Castelvecchi and Gibney 2016)

We might replace "the signal" with "a signal like this" to avoid criticism. While more stringent than the usual requirement, the "we're not that impressed" stance kicks in. It's not so very rare for even more impressive results to occur by background alone. As the data come in, the significance levels will either grow or wane with the bumps:

Physicists say that by June, or August [2016] at the latest, CMS and ATLAS should have enough data to either make a statistical fluctuation go away – if that's what the excess is – or confirm a discovery. (Castelvecchi and Gibney 2016)

Could the Bayesian model wind up in the same place? Not if Lindley/ O'Hagan's subjective model merely keeps updating beliefs in the already expected parameters. According to Savage, "The probability of 'something else' . . . is definitely very small" (Savage 1962, p. 80). It would seem to require a long string of anomalies before the catchall is made sufficiently probable to start seeking new physics. Would they come up with a particle like the one they were now in a frenzy to explain? Maybe, but it would be a far less efficient way for discovery than the simple significance tests.

I would have liked to report a more exciting ending for our tour. The promising bump or "resonance" disappeared as more data became available, drowning out the significant indications seen in April. Its reality was falsified.

Souvenir O: Interpreting Probable Flukes

There are three ways to construe a claim of the form: A small P-value indicates it's improbable that the results are statistical flukes.

(1) The person is using an informal notion of probability, common in English. They mean a small P-value gives grounds (or is evidence) of a genuine discrepancy from the null. Under this reading there is no fallacy. Having inferred H^*: Higgs particle, one may say informally, "so probably we have experimentally demonstrated the Higgs," or "probably, the Higgs exists."

"So probably" H_1 is merely qualifying the grounds upon which we assert evidence for H_1.

(2) An ordinary error probability is meant. When particle physicists associate a 5-sigma result with claims like "it's highly improbable our results are a statistical fluke," the reference for "our results" includes: the overall display of bumps, with significance growing with more and better data, along with satisfactory crosschecks. Under this reading, again, there is no fallacy.

To turn the tables on the Bayesians a bit, maybe they're illicitly sliding from what may be inferred from an entirely legitimate high probability. The reasoning is this: With probability 0.9999997, our methods would show that the bumps disappear, under the assumption the data are due to background H_0. The bumps don't disappear but grow. Thus, infer H^*: real particle with thus and so properties. Granted, unless you're careful about forming probabilistic complements, it's safer to adhere to the claims along the lines of U-1 through U-3. But why not be careful in negating D claims? An interesting phrase ATLAS sometimes uses is in terms of "the background fluctuation probability": "This observation, which has a significance of 5.9 standard deviations, corresponding to a background fluctuation probability of 1.7×10^{-9}, is compatible with ... the Standard Model Higgs boson" (2012b, p.1).

(3) The person is interpreting the P-value as a posterior probability of null hypothesis H_0 based on a prior probability distribution: $p = \Pr(H_0|x)$. Under this reading there is a fallacy. Unless the P-value tester has explicitly introduced a prior, it would be "ungenerous" to twist probabilistic assertions into posterior probabilities. It would be a kind of "confirmation bias" whereby one insists on finding a sentence among many that could be misinterpreted Bayesianly.

ASA 2016 Guide: Principle 2 reminds practitioners that P-values aren't Bayesian posterior probabilities, but it slides into questioning an interpretation sometimes used by practitioners – including Higgs researchers:

P-values do not measure (a) the probability that the studied hypothesis is true, or (b) the probability that the data were produced by random chance alone. (Wasserstein and Lazar 2016, p. 131)[4]

---

[4] The ASA 2016 Guide's Six Principles:

1. P-values can indicate how incompatible the data are with a specified statistical model.
2. P-values do not measure the probability that the studied hypothesis is true, or the probability that the data were produced by random chance alone.

I insert the (a), (b), absent from the original principle 2, because, while (a) is true, phrases along the lines of (b) should not be equated to (a).

Some might allege that I'm encouraging a construal of P-values that physicists have bent over backwards to avoid! I admitted at the outset that "the problem is a bit delicate, and my solution is likely to be provocative." My question is whether it is legitimate to criticize frequentist measures from a perspective that assumes a very different role for probability. Let's continue with the ASA statement under principle 2:

Researchers often wish to turn a *p*-value into a statement about the truth of a null hypothesis, or about the probability that random chance produced the observed data. The *p*-value is neither. It is a statement about data in relation to a specified hypothetical explanation, and is not a statement about the explanation itself. (Wasserstein and Lazar 2016, p. 131)

Start from the very last point: what does it mean, that it's not "about the explanation"? I think they mean it's not a posterior probability on a hypothesis, and that's correct. The P-value is a methodological probability that can be used to quantify "how well probed" rather than "how probable." Significance tests can be the basis for, among other things, falsifying a proposed explanation of results, such as that they're "merely a statistical fluctuation." So the statistical inference that emerges is surely a statement about the explanation. Even proclamations issued by high priests – especially where there are different axes to grind – should be taken with severe grains of salt.

As for my provocative interpretation of "probable fluctuations," physicists might aver, as does Cousins, that it's the science writers who take liberties with the physicists' careful U-type statements, turning them into D-type statements. There's evidence for that, but I think physicists may be reacting to criticisms based on how things look from Bayesian probabilists' eyes. For a Bayesian, once the data are known, they are fixed; what's

---

3. Scientific conclusions and business or policy decisions should not be based only on whether a *p*-value passes a specific threshold.
4. Proper inference requires full reporting and transparency.
5. A *p*-value, or statistical significance, does not measure the size of an effect or the importance of a result.
6. By itself, a *p*-value does not provide a good measure of evidence regarding a model or hypothesis.

These principles are of minimal help when it comes to understanding and using P-values. The first thing that jumps out is the absence of any mention of P-values as error probabilities. (Fisher-N-P Incompatibilist tribes might say "they're not!" In tension with this is the true claim (under #4) that cherry picking results in spurious P-values; p. 132.) The ASA effort has merit, and should be extended and deepened.

random is an agent's beliefs or uncertainties on what's unknown – namely the hypothesis. For the severe tester, considering the probability of $\{d(X) \geq d(x_0)\}$ is scarcely irrelevant once $d(x_0)$ is known. It's the way to determine, following the severe testing principles, whether the null hypothesis can be falsified. ATLAS reports, on the basis of the P-value display, that "these results provide conclusive evidence for the discovery of a new particle with mass [approximately 125 GeV]" (ATLAS collaboration 2012b, p. 15).

Rather than seek a high probability that a suggested new particle is real; the scientist wants to find out if it disappears in a few months. As with GTR (Section 3.1), at no point does it seem we want to give a high formal posterior probability to a model or theory. We'd rather vouchsafe some portion, say the SM model with the Higgs particle, and let new data reveal, perhaps entirely unexpected, ways to extend the model further. The open-endedness of science must be captured in an adequate statistical account. Most importantly, the 5-sigma report, or corresponding P-value, strictly speaking, *is not the statistical inference*. Severe testing premises – or something like them – are needed to move from statistical data plus background (theoretical and empirical) to detach inferences with lift-off.

Excursion 4 Objectivity and Auditing

Itinerary

Tour I The Myth of "The Myth of Objectivity"

> Objectivity in statistics, as in science more generally, is a matter of both aims and methods. Objective science, in our view, aims to find out what is the case as regards aspects of the world [that hold] independently of our beliefs, biases and interests; thus objective methods aim for the critical control of inferences and hypotheses, constraining them by evidence and checks of error. (Cox and Mayo 2010, p. 276)

Whenever you come up against blanket slogans such as "no methods are objective" or "all methods are equally objective and subjective" it is a good guess that the problem is being trivialized into oblivion. Yes, there are judgments, disagreements, and values in any human activity, which alone makes it too trivial an observation to distinguish among very different ways that threats of bias and unwarranted inferences may be controlled. Is the objectivity–subjectivity distinction really toothless, as many will have you believe? I say no. I know it's a meme promulgated by statistical high priests, but you agreed, did you not, to use a bit of *chutzpah* on this excursion? Besides, cavalier attitudes toward objectivity are at odds with even more widely endorsed grass roots movements to promote replication, reproducibility, and to come clean on a number of sources behind illicit results: multiple testing, cherry picking, failed assumptions, researcher latitude, publication bias and so on. The moves to take back science are rooted in the supposition that we can more objectively scrutinize results – even if it's only to point out those that are BENT. The fact that these terms are used equivocally should not be taken as grounds to oust them but rather to engage in the difficult work of identifying what there is in "objectivity" that we won't give up, and shouldn't.

The Key Is Getting Pushback! While knowledge gaps leave plenty of room for biases, arbitrariness, and wishful thinking, we regularly come up against data that thwart our expectations and disagree with the predictions we try to foist upon the world. We get pushback! This supplies objective constraints on which our critical capacity is built. Our ability to recognize when data fail to match anticipations affords the opportunity to systematically improve our orientation. Explicit attention needs to be paid to communicating results to set the stage for others to check, debate, and extend the inferences reached.

Which conclusions are likely to stand up? Where do the weakest parts remain? Don't let anyone say you can't hold them to an objective account.

Excursion 2, Tour II led us from a Popperian tribe to a workable demarcation for scientific inquiry. That will serve as our guide now for scrutinizing the myth of the myth of objectivity. First, good sciences put claims to the test of refutation, and must be able to embark on an inquiry to pin down the sources of any apparent effects. Second, refuted claims aren't held on to in the face of anomalies and failed replications; they are treated as refuted in further work (at least provisionally); well-corroborated claims are used to build on theory or method: science is not just stamp collecting. The good scientist deliberately arranges inquiries so as to capitalize on pushback, on effects that will not go away, on strategies to get errors to ramify quickly and force us to pay attention to them. The ability to register how hunting, optional stopping, and cherry picking alter their error-probing capacities is a crucial part of a method's objectivity. In statistical design, day-to-day tricks of the trade to combat bias are consciously amplified and made systematic. It is not because of a "disinterested stance" that we invent such methods; it is that we, quite competitively and self-interestedly, want our theories to succeed in the market place of ideas.

Admittedly, that desire won't suffice to incentivize objective scrutiny if you can do just as well producing junk. Successful scrutiny is very different from success at grants, getting publications and honors. That is why the reward structure of science is so often blamed nowadays. New incentives, gold stars and badges for sharing data and for resisting the urge to cut corners are being adopted in some fields. Fortunately, for me, our travels will bypass lands of policy recommendations, where I have no special expertise. I will stop at the perimeters of scrutiny of methods which at least provide us citizen scientists armor against being misled. Still, if the allure of carrots has grown stronger than the sticks, we need stronger sticks.

Problems of objectivity in statistical inference are deeply intertwined with a jungle of philosophical problems, in particular with questions about what objectivity demands, and disagreements about "objective versus subjective" probability. On to the jungle!

4.1 Dirty Hands: Statistical Inference Is Sullied with Discretionary Choices

> If all flesh is grass, kings and cardinals are surely grass, but so is everyone else and we have not learned much about kings as opposed to peasants. (Hacking 1965, p. 211)

Trivial platitudes can appear as convincingly strong arguments that everything is subjective. Take this one: No human learning is pure so anyone who demands objective scrutiny is being unrealistic and demanding immaculate inference. This is an instance of Hacking's "all flesh is grass." In fact, Hacking is alluding to the subjective Bayesian de Finetti (who "denies the very existence of the physical property [of] chance" (ibid.)). My one-time colleague, I. J. Good, used to poke fun at the frequentist as "denying he uses any judgments!" Let's admit right up front that every sentence can be prefaced with "agent x judges that," and not sweep it under the carpet (SUTC) as Good (1976) alleges. Since that can be done for any statement, it cannot be relevant for making the distinctions in which we are interested, and we know can be made, between warranted or well-tested claims and those so poorly probed as to be BENT. You'd be surprised how far into the thicket you can cut your way by brandishing this blade alone.

It is often urged that, however much we may aim at objective constraints, we can never have clean hands, free of the influence of beliefs and interests. We invariably sully methods of inquiry by the entry of background beliefs and personal judgments in their specification and interpretation. The real issue is not that a human is doing the measuring; the issue is whether that which is being measured is something we can reliably use to solve some problem of inquiry. An inference done by machine, untouched by human hands, wouldn't make it objective in any interesting sense. There are three distinct requirements for an objective procedure of inquiry:

1. *Relevance:* It should be relevant to learning about what is being measured; having an uncontroversial way to measure something is not enough to make it relevant to solving a knowledge-based problem of inquiry.
2. *Reliably capable:* It should not routinely declare the problem solved when it is not (or solved incorrectly); it should be capable of controlling reports of erroneous solutions to problems with reliability.
3. *Able to learn from error:* If the problem is not solved (or poorly solved) at a given point, the method should set the stage for pinpointing why.

Yes, there are numerous choices in collecting, analyzing, modeling, and drawing inferences from data, and there is often disagreement about how they should be made, and about their relevance for scientific claims. Why suppose that this introduces subjectivity into an account, or worse, means that all accounts are in the same boat as regards subjective factors? It need not, and they are not. An account of inference shows itself to be objective precisely in how it steps up to the plate in handling potential threats to objectivity.

Dirty Hands Argument. To give these arguments a run for their money, we should try to see why they look so plausible. One route is to view the reasoning as follows:

1. A variety of human judgments go into specifying experiments, tests, and models.
2. Because there is latitude and discretion in these specifications, which may reflect a host of background beliefs and aims, they are "subjective."
3. Whether data are taken as evidence for a statistical hypothesis or model depends on these subjective methodological choices.
4. Therefore, statistical methods and inferences are invariably subjective, if only in part.

The mistake is to suppose we are incapable of critically scrutinizing how discretionary choices influence conclusions. It is true, for example, that choosing a very insensitive test for detecting a risk δ' will give the test low probability of detecting such discrepancies even if they exist. Yet I'm not precluded from objectively determining this. Setting up a test with low power against δ' might be a product of your desire not to find an effect for economic reasons, of insufficient funds to collect a larger sample, or of the inadvertent choice of a bureaucrat. Or ethical concerns may have entered. But our critical evaluation of what the resulting data do and do not indicate need not itself be a matter of economics, ethics, or what have you.

Idols of Objectivity. I sympathize with disgruntlement about phony objectivity and false trappings of objectivity. They grow out of one or another philosophical conception about what objectivity requires – even though you will almost surely not see them described that way. It's the curse of logical positivism, but also its offshoots in post-positivisms. If it's thought objectivity is limited to direct observations (whatever they are) plus mathematics and logic, as the typical positivist, then it's no surprise to wind up worshiping what Gigerenzer and Marewski (2015) call "the idol of a universal method." Such a method is to supply a formal, ideally mechanical, rule to process statements of observations and hypotheses – translated into a neutral observation language. Can we translate Newtonian forces and Einsteinian curved spacetime into a shared observation language? The post-positivists, rightly, said no. Yet giving up on logics of induction and theory-neutral languages, they did not deny these were demanded by objectivity. They only decided that they were unobtainable. Genuine objectivity goes by the board, replaced by various stripes of relativism and constructivism, as well as more extreme forms of anarchism and postmodernism.[1]

From the perspective of one who has bought the view that objectivity is limited to math, logic, and fairly direct observations (the dial now points to 7), methods that go beyond these appear "just as" subjective as another. They may augment their rather thin gruel with an objectivity arising from social or political negotiation, or a type of consensus, but that's to give away the goat far too soon. The result is to relax the core stipulations of scientific objectivity. To be clear: There are authentic problems that threaten objectivity. Let's not allow outdated philosophical accounts to induce us into giving it up.

What about the fact that different methods yield different inferences, for example that Richard Royall won't infer the composite $\mu > 0.2$ while N-P testers will? I have no trouble understanding why, if you define inference as comparative likelihoods, the results disagree with error statistical tests. Running different analyses on the same data can be the best way to unearth flaws. However, objectivity is an important criterion in appraising such rival statistical accounts.

Objectivity and Observation. In facing objectivity skeptics, you might remind them of parallels between learning from statistical experiments and learning from observations in general. The problem in objectively interpreting observations is that observations are always relative to the particular instrument or observation scheme employed. But we are often aware not only of the fact that observation schemes influence what we observe but also of how: How much noise are they likely to introduce? How might we subtract it out?

The result of a statistical method need only (and should only) be partly determined by the specifications of a given method (e.g., the cut-off for statistical significance); it is also determined by the underlying scientific phenomenon, as modeled. What enables objective learning is the possibility of taking into account *how* test specifications color results as we intervene in phenomena of interest. Don't buy the supposition that the word "arbitrary" always belongs in front of "convention." That my weight shows up as k pounds is a convention in the USA. Still, *given the convention*, the readout of k pounds is a matter of how much I weigh. I cannot simply ignore the additional weight as due to an arbitrary convention, even if I wanted to.

---

[1] See Larry Laudan's (1996) ingenious analysis of how much the positivists and post-positivists share in "The Sins of the Fathers."

How Well Have You Probed *H* versus How Strongly Do (or Should) You Believe It?

When Albert Einstein was asked "What if Eddington had not found evidence of the deflection effect?", Einstein famously replied, "Then I would feel sorry for the dear Lord; the theory is correct." Some might represent this using subjective Bayesian resources. Einstein had a strong prior conviction in GTR – a negative result might have moved his belief down a bit but it would still be plenty high. Such a reconstruction may be found useful. If we try to cash it out as a formal probability, it isn't so easy. Did he assign high prior to the deflection effect being 1.75″, or also to the underlying theoretical picture of curved spacetime (which is really the basis of his belief)? A formal probability assignment works better for individual events than for assessing full-blown theories, but let us assume that it could be done. What matters is that Einstein would also have known the deflection hypothesis had not been well probed, that is, it had not yet passed a severe test in 1919. An objective account of statistics needs to distinguish how probable (believable, plausible, supported) a claim is from how well it has been probed. This remains true whether the focus is on a given set of data, several sets, or, given everything I know, what I called "big picture" inference.

Having distinguished our aim – appraising how stringently and responsibly probed a claim *H* is by the results of a given inquiry – from that of determining *H*'s plausibility or belief-worthiness, it's easy to allow that different methodologies and criteria are called for in pursuing these two goals. Recall the root of *probare* is to demonstrate or show.

Some argue that "discretionary choices" in tests, which Neyman himself tended to call "subjective," lead us to subjective probabilities. A weak version goes: since you can't avoid discretionary choices in getting the data and model, how can you complain about subjective degrees of belief in the resulting inference? This is weaker than arguing you must use subjective probabilities; it argues merely that doing so *is no worse than* discretion. It still misses the point.

First, as we saw in exposing the "dirty hands" argument, even if discretionary judgments can introduce subjectivity, they need not. Second, not all discretionary judgments are in the same boat when it comes to being open to severe testing of their own. E. Pearson imagines he

might quote at intervals widely different Bayesian probabilities for the same set of states, simply because I should be attempting what would be for me impossible and resorting to guesswork. It is difficult to see how the matter could be put to experimental test. (Pearson 1962, pp. 278–9)

A stronger version of the argument goes on a slippery slope from the premise of discretion in data generation and modeling to the conclusion: statistical inference is a matter of subjective belief. How does that work? One variant involves a subtle slide from "our models are merely objects of belief," to "statistical inference is a matter of degrees of belief." From there it's a short step to "statistical inference is a matter of subjective probability." It is one thing to allow talk of our models as objects of belief and quite another to maintain that our task is to model beliefs.

This is one of those philosophical puzzles of language that might set some people's eyes rolling. If I believe in the deflection effect then that effect is the object of my belief, but only in the sense that my belief is about said effect. Yet if I'm inquiring into the deflection effect, I'm not inquiring into beliefs about the effect. The philosopher of science Clark Glymour (2010, p. 335) calls this a shift from phenomena (content) to *epiphenomena* (degrees of belief). Popper argues that the key confusion all along was sliding from the degree of the rationality (or warrantedness) of a belief, to the degree of rational belief (1959, p. 407).

Or take subjectivist Frank Lad. To him,

... so-called 'statistical models' are not real entities that merit being estimated. To the extent that models mean anything, they are models of someone's (some group's) considered uncertain opinion about observable quantities. (Lad 2006, p. 443)

Notice the slide from uncertainty or partial knowledge of quantities in models, to models being *models of opinions*. I'm not saying Lad is making a linguistic error. He appears instead to embrace a positivist philosophy of someone like Bruno de Finetti. De Finetti denies we can put probabilities on general claims because we couldn't settle bets on them. If it's also maintained that scientific inference takes the form of a subjective degree of belief, then we cannot infer general hypotheses – such as statistical hypotheses. Are we to exclude them from science as so much meaningless metaphysics?

When current-day probabilists echo such stances, it's a good bet they would react with horror at the underlying logical positivist philosophy. So how do you cut to the chase without sinking into a philosophical swamp? You might ask: Are you saying statistical models are just models of beliefs and opinions? They are bound to say no. So press on and ask: Are you saying they are mere approximations, and we hold fallible beliefs and opinions about them? They're likely to agree. But the error statistician holds this as well!

What's Being Measured versus My Ability to Test It. You will sometimes hear it claimed that anyone who says their probability assignments to hypotheses are subjective must also call the use of any model subjective because it too is based on my choice of specifications. It's important not to confuse two notions of subjective. The first concerns what's being measured, and for the Bayesian, at least the subjective Bayesian, probability represents a subject's strength of belief. The second sense of subjective concerns whether the measurement is checkable or testable. Nor does latitude for disagreement entail untestability. An intriguing analysis of objectivity and subjectivity in statistics is Gelman and Hennig (2017).

4.2 Embrace Your Subjectivity

The classical position of the subjective Bayesian aims at inner coherence or consistency rather than truth or correctness. Take Dennis Lindley:

> I am often asked if the method gives the *right* answer: or, more particularly, how do you know if you have got the *right* prior. My reply is that I don't know what is meant by 'right' in this context. The Bayesian theory is about *coherence*, not about right or wrong. (Lindley 1976, p. 359)

There's no reason to suppose there is a correct degree of belief to hold. For Lindley, $Pr(H|x)$ "is your belief about $[H]$ when you know $[x]$" (Lindley 2000, p. 302, substituting Pr for P; H for A and x for B). My opinions are my opinions and your opinions are yours. How do I criticize your prior degrees of belief? As Savage said, "[T]he Bayesian outlook reinstates opinion in statistics – in the guise of the personal probabilities of events ..." (Savage 1961, p. 577). Or again, "The concept of personal probability ... seems to those of us who have worked with it an excellent model for the concept of opinion" (ibid., pp. 581–2).[2] That might be so, but what if we are not trying to model opinions, but instead insist on meeting requirements for objective scrutiny? For these goals, inner coherence or consistency among your beliefs is not enough. One can be consistently wrong, as everyone knows (or should know).

If you're facing a radical skeptic of all knowledge, a radical relativist, postmodernist, social-constructivist, or anarchist, there may be limited room to maneuver. The position may be the result of a desire to shock or be camp (as Feyerabend or Foucault) or give voice to political interests. The position may be mixed with, or at least dressed in the clothes of, philosophy: We are locked in a world of appearances seeing mere shadows of an "observer independent reality." Our bold activist learner, who imaginatively creates abstract models that give him pushback, terrifies them. Calling it unholy metaphysics may actually reflect their inability to do the math.

---

[2] I will not distinguish personalists and subjectivists, even though I realize there is a history of distinct terms.

Progress of Philosophy

To the error statistician, radical skepticism is a distraction from the pragmatic goal of understanding how we do manage to learn, and finding out how we can do it better. Philosophy does make progress. Logical positivism was developed and embraced when Einstein's theory rocked the Newtonian worldview. Down with metaphysics! All must be verifiable by observation. But there are no pure observations, no theory-neutral observational languages, no purely formal rules of confirmation holding between any statements of observation and hypotheses. Popper sees probabilism as a holdover from a watered down verificationism, ". . . under the influence of the mistaken view that science, unable to attain certainty, must aim at a kind of '*Ersatz*' – at the highest attainable probability" (Popper 1959, p. 398 (Appendix IX)). Even in the face of the "statistical crisis in science," by and large, scientists aren't running around terrified that our cherished theories of physics will prove wrong: they expect even the best ones are incomplete, and several rival metaphysics thrive simultaneously. In genetics, we have learned to cut, delete, and replace genes in human cells with the new CRISPR technique discovered by Jennifer Doudna and Emmanuelle Charpentier (2014). The picture of the knower limited by naked observations no longer has any purchase, if it ever did.

Some view the Big Data revolution, with its focus on correlations rather than causes, as a kind of return to theory-free neopositivism. Theory freedom and black-box modeling might work well for predicting what color website button is most likely to get me to click. AI has had great successes. We've also been learning how theory-free prediction techniques come up short when it comes to scientific understanding.

Loss and Cost Functions

The fact that we have interests, and that costs and values may color our interpretation of data, does not mean they should be part of the scientific interpretation of data. Frequent critics of statistical significance tests, economists Stephen Ziliak and Deirdre McCloskey, declare, in relation to me, that "a notion of a severe test without a notion of a loss function is a diversion from the main job of science" (2008a, p. 147). It's unclear if this is meant as a vote for a N-P type of balancing of error probabilities, or for a full-blown decision theoretic account. If it is the latter, with subjective prior probabilities in hypotheses, we should ask: Whose losses? Whose priors? The drug company? The patient? They may lie hidden in impressive Big Data algorithms as Cathy O'Neil (2016) argues.

Remember that a severity appraisal is always in relation to a question or problem, and that problem could be a decision, within a scientific inquiry or wholly personal. In the land of science, we'd worry that to incorporate into an inference on genomic signatures, say, your expected windfall from patenting it would let it pass without a severe probe. So if that is what they mean, I disagree, and so should anyone interested in blocking flagrant biases. Science is already politicized enough. Besides, in order for my assessment of costs to be adequate, I've got to get the science approximately correct first – wishing and hoping don't suffice (as Potti and Nevins discovered in Excursion 1).

We can agree with Ziliak and McCloskey if all they mean is that in deciding if a treatment, say hormone replacement therapy (HRT), is right for me, then a report on how it improves skin elasticity ignoring, say, the increase in cardiac risk, is likely irrelevant for my decision.

Some might eschew all this as naïve: scientists cannot help but hold on to beliefs based on private commitments, costs, and other motivations. We may readily agree. Oliver Lodge, our clairvoyant, had a keen interest in retaining a Newtonian ether to justify conversations with his son, Raymond, on "the other side" (Section 3.1). Doubtless Lodge, while accepting the interpretation of the deflection experiments, could never really bring himself to disbelieve in the ether. It might not even have been healthy, psychologically, for him to renounce his belief. Yet, the critical assessment of each of his purported ether explanations had nothing to do with this. Perhaps one could represent his personal assessment using a high prior in the ether, or a high cost to relinquishing belief in it. Yet everyone understands the difference between *adjudicating* disagreements on evidential grounds and producing a psychological conversion, or making it worthwhile financially, as when a politician's position "evolves" if the constituency demands it.

"Objective" (Default, Non-subjective) Bayesians

The desire for a Bayesian omelet while only breaking "objective" eggs gives rise to default Bayesianism or, if that sounds too stilted, default/non-subjective.[3] Jim Berger is one of the leaders:

I feel that there are a host of practical and sociological reasons to use the label 'objective' for priors of model parameters that appropriately reflect a lack of subjective information ... [None of the other names] carries the simplicity or will carry the same weight outside of statistics as 'objective.'... we should start systematically accepting the 'objective Bayes' name before it is co-opted by others. (Berger 2006, p. 387)

---

[3] Aside: should an author stop to explain every joke, as some reviewers seem to think? I don't think so, but you can look up "omelet Savage 1961."

The holy grail of truly "uninformative" priors has been abandoned – what is uninformative under one parameterization can be informative for another. (For example, "if θ is uniform e^{θ} has an improper exponential distribution, which is far from flat"; Cox 2006a, p. 73.) Moreover, there are competing systems for ensuring the data are weighed more heavily than the priors. As we will see, so-called "equipoise" assignments may be highly biased. For the error statistician, as long as an account is restricted to priors and likelihoods, it still leaves out the essential ingredient for objectivity: the sampling distribution, the basis for error probabilities and severity assessments. Classical Bayesians, both subjective and default, reject this appeal to "frequentist objectivity" as solely rooted in claims about long-run performance. Failure to craft a justification in terms of probativeness means that there's uncharted territory, waiting to be developed. Fortunately I happen to have my own maps, rudimentary perhaps, but enough for our excavations.

Beyond Persuasion and Coercion

The true blue subjectivists regard the call to free Bayesianism from beliefs as a cop-out. As they see it, statisticians ought to take responsibility for their personal assessments.

To admit that my model is personal means that I must persuade you of the reasonableness of my assumptions in order to convince you ... To claim objectivity is to try to coerce you into consenting, without requiring me to justify the basis for the assumptions. (Kadane 2006, p. 434)

The choice should not be persuasion or coercion. Perhaps the persuasion ideal served at a time when a small group of knowledgeable Bayesians could be counted on to rigorously critique each other's outputs. Now we have massive data sets and powerful data-dredging tools. What about the allegation of coercion? I guess being told it's an Objective prior (with a capital O) can sound coercive. Yet anyone who has met Jim Berger will be inclined to agree that the line between persuasion and coercion is quite thin. His assurances that we're controlling error probabilities (even if he's slipped into error probability$_2$) can feel more seductive than coercive (Excursion 3).

Wash-out Theorems

If your prior beliefs are not too extreme, and if model assumptions hold, then if you continually observe data on H and update by Bayes' Theorem, in some long run the posteriors will converge – assuming your beliefs about the likelihoods providing a random sample are correct. It isn't just that these wash-out theorems have limited guarantees or that they depend on agents assigning non-zero priors to the same set of hypotheses, or that even with non-extreme prior probabilities, and any body of evidence, two scientists can have posteriors that differ by arbitrary amounts (Kyburg 1992, p. 146); it's that appeals to consilience of beliefs in an asymptotic long run have little relation to the critical appraisal that we demand regarding the case at hand. The error statistician, and the rest of us, can and will raise criticisms of bad evidence, no test (BENT) regarding today's study. Ironically, the Bayesians appeal to a long run of repeated applications of Bayes' Theorem to argue that their priors would wash out eventually. Look who is appealing to long runs! Fisher's response to the possibility of priors washing out is that far from showing their innocuousness, "we may well ask what the expression is doing in our reasoning at all, and whether, if it were altogether omitted, we could not without its aid draw whatever inferences may, with validity, be inferred from the data" (Fisher 1934b, p. 287).

Take Account of "Subjectivities"

This is radically ambiguous! A well-regarded position about objectivity in science is that it is best promoted by excluding personal opinions, biases, preferences, and interests; if you can't exclude these, you ought at least to *take account of them*. How should you do this? It seems obvious it should be done in a way that *excludes biasing influences from claims as to what the data have shown*. Or if not, their influence should be made explicit in a report of findings. There's a very different way of "taking account of" them: To wit: view them as beliefs in the claim of inquiry, quantify them probabilistically, and blend them into the data. If they are to be excluded, they can't at the same time be blended; one can't have it both ways. Consider some exhibits regarding taking account of biases.

Exhibit (i): **Prior Probabilities Let Us Be Explicit about Bias.** There's a constellation of positions along these lines, but let's consider Nate Silver, the well-known pollster and data analyst. I was sitting in the audience when he gave the invited president's address for the American Statistical Association in 2013. He told us the reason he favored the Bayesian philosophy of statistics is that people – journalists in particular – should be explicit about the biases, prior conceptions, wishes, and goals that invariably enter into the collection and interpretation of data.

How would this work, say in *FiveThirtyEight*, the online statistically based news source of which Silver is editor-in-chief? Perhaps it would go like this: if a journalist is writing on, say, GM foods, she should declare at the outset she believes their risks are exaggerated (or the other way around). Then the reader can understand that her selection and interpretation of facts was through the lens of the "GM is safe" theory. Isn't this tantamount to saying she's unable to evaluate the data impartially – belying the goal of news based on "hard numbers"? Perhaps to some degree this is true. However, if people are inclined to see the world using tunnel vision, what's the evidence they'd be able or willing to be explicit about their biases? Imagine for the moment they would. Suppose further that prior probabilities are to be understood as expressing these biases – say the journalist's prior probability in GM risks is low.

Now if the prior was kept separate, readers could see if the data alone point to increased GM risks. If so, they reveal how the journalist's priors biased the results. But if only the posterior probability was reported, they cannot. Even reporting the priors may not help if it's complicated, which, to an untutored reader, they always are. Further, how is the reader to even trust the likelihoods? Even if they could be, why would the reader want the journalist to blend her priors – described by Silver as capturing biases – into the data? It would seem to be just the opposite. Someone might say they checked the insensitivity of an inference over a range of priors. That can work in some cases, but remember they selected the priors to look at. To you and me, these points seem to go without saying, but in today's environment, it's worth saying them.[4]

Exhibit (ii): Prior Probabilities Allow Combining Background Information with Data. In a published, informal spoken exchange between Cox and me, the question of background information arose.

COX: Fisher's resolution of this issue in the context of the design of experiments was essentially that in designing an experiment you do have all sorts of prior information, and you use that to set up a good experimental design. Then when you come to analyze it, you do not use the prior information. In fact you have very clever ways of making sure that your analysis is valid even if the prior information is totally wrong. (Cox and Mayo 2011, p. 104–5)

MAYO: But they should use existing knowledge.

COX: Knowledge yes … It's not evidence that should be used if let's say a group of surgeons claim we are very, very strongly convinced, maybe to probability 0.99, that this surgical procedure works and is good for patients, without inquiring where the 0.99 came from. It's a very dangerous line of argument. But not unknown. (ibid., p. 107)

Elsewhere, Cox remarks (2006a, p. 200):

Expert opinion that is not reasonably firmly evidence-based may be forcibly expressed but is in fact fragile. The frequentist approach does not ignore such evidence but separates it from the immediate analysis of the specific data under consideration.

---

[4] Silver recognizes that groupthink created an echo chamber during the 2016 election in the USA.

Admittedly, frequentists haven't been clear enough as to the informal uses of background knowledge, especially at the stage of "auditing." They leave themselves open to the kind of challenge Andrew Gelman (2012) puts to Cox, in reference to Cox and Mayo (2011).

Surely, Gelman argues, there are cases where the background knowledge is so strong that it should be used in the given inference.

Where did Fisher's principle go wrong here? The answer is simple – and I think Cox would agree with me here. We're in a setting where the prior information is much stronger than the data. ... it is essential to use prior information (even if not in any formal Bayesian way) to interpret the data and generalize from sample to population. (Gelman 2012, p. 53)

Now, in the same short paper, Gelman, who identifies as Bayesian, declares: "Bayesians Want Everybody Else to be Non-Bayesian."

Bayesian inference proceeds by taking the likelihoods from different data sources and then combining them with a prior (or, more generally, a hierarchical model). The likelihood is key. ... No funny stuff, no posterior distributions, just the likelihood... I don't want everybody coming to me with their posterior distribution – I'd just have to divide away their prior distributions before getting to my own analysis. (ibid., p. 54)

No funny stuff, no posterior distributions, says Gelman. Thus, he too is recommending the priors and likelihoods be kept separate, at least for this purpose (scrutinizing an inquiry using background).

So is he agreeing or disagreeing with Cox? Perhaps Gelman is saying: don't combine the prior with the likelihood, but allow well-corroborated background to be used as grounds for scrutinizing, or, in my terms, conducting an "audit" of, the statistical inference. A statistical inference fails an audit if either the statistical assumptions aren't adequately met, or the error probabilities are invalidated by biasing selection effects. In that case there's no real disagreement with Cox's use of background. Still, there is something behind Gelman's lament that deserves to be made explicit. There's no reason for the frequentist to restrict background knowledge to pre-data experimental planning and test specification. We showed how the background gives the context for a FIRST interpretation in Section 3.3. Audits also employ background, and may likely be performed by a different party than those who designed and conducted the study. This would not be a Bayesian updating to a posterior probability, but would use any well-corroborated background knowledge in auditing. A background repertoire of the slings and arrows known to threaten the type of inquiry may show a statistical inference fails an audit, or ground suspicion that it would fail an audit.

Exhibit (iii): Use Knowledge of a Repertoire of Mistakes. The situation is analogous, though not identical, when background knowledge shows a hypothesized effect to have been falsified: since the effect doesn't exist, any claim to have found it is due to some flaw; unless there was a special interest in pinpointing it, that would suffice. This is simple deductive reasoning. It's fair to say that experimental ESP was falsified some time in the 1980s, even though one can't point to a single bright line event. You might instead call it a "degenerating program" (to use Lakatos' term): anomalies regularly must be explained away by ad hoc means. In each case, Perci Diaconis (1978), statistician and magician, explains that "controls often are so loose that no valid statistical analysis is possible. Some common problems are multiple end points, subject cheating, and unconscious sensory cueing" (p. 131). There may be a real effect, but it's not ESP. It may be that Geller bent the spoon when you weren't looking, or that flaws entered in collecting, selecting, and reporting data. A severe tester would infer that experimental ESP doesn't exist, that the purported reality of the effect had been falsified on these grounds.

Strictly speaking, even falsifications may be regarded as provisional, and the case reopened. Human abilities could evolve. However, anyone taking up an effect that has been manifested only with highly questionable research practices or insevere tests, must, at the very least, show they have avoided the well-known tricks in the suitcase of mistakes that a researcher in the field should be carrying. If they do not, or worse, openly flout requirements to avoid biasing selection effects, then they haven't given a little bit of evidence – as combining prior and likelihood could allow – but rather an inference that's BENT. A final exhibit:

Exhibit (iv): Objectivity in Epistemology. Kent Staley is a philosopher of science who has developed the severity account based on error statistics (he calls it the ES account), linking it to more traditional distinctions in epistemology, notably between "internalist" and "externalist" accounts. In a paper with Aaron Cobb:

> ... there seems to be a resemblance between *ES* and a paradigmatically externalist account of justification in epistemology. Just as Alvin Goldman's reliabilist theory makes justification rest on the tendency of a belief-forming process to produce true rather than false beliefs (Goldman 1986, 1999), *ES* links the justification of an inference to its having resulted from a testing procedure with low error probabilities (Woodward 2000). (Staley and Cobb 2011, p. 482)

The last sentence would need to read "low error probabilities relevant for satisfying severity," since low error probabilities won't suffice for a good test. My problem with the general epistemological project of giving necessary and sufficient conditions for knowledge or justified belief or the like is that it does not cash out terms such as "reliability" by alluding to actual methods. The project is one of definition. That doesn't mean it's not of interest to try and link to the more traditional epistemological project to see where it leads. In so doing, Staley and Cobb are right to note that the error-statistician will not hold a strictly externalist view of justification. The trouble with "externalism" is that it makes it appear that a claim (or "belief" as many prefer), is justified so long as a severity relationship SEV holds between data, hypotheses, and a test. It needn't be able to be shown or known. The internalist view, like the appeal to inner coherence in subjective Bayesianism, has a problem in showing how internally justified claims link up to truth. The analytical internal/external distinction isn't especially clear, but from the perspective of that project, Staley and Cobb are right to view ES as a "hybrid" view. In the ES view, the reliability of a method is independent of what anybody knows, but the knower or group of knowers must be able to respond to skeptical challenges such as: you're overlooking flaws, you haven't taken precautions to block errors and so on. They must display the ability to put to rest reasonable skeptical challenges. (Not just any skeptical doubts count, as discussed in solving induction in Section 2.7.) This is an integral part of being an adequate scientific researcher in a domain. (We can sidestep the worry epistemologists might voice that this precludes toddlers from having knowledge; even toddlers can non-verbally display their know-how.) Without showing a claim has been well probed, it has not been well corroborated. Warranting purported severity claims is the task of auditing.

There are interesting attempts to locate objectivity in science in terms of the diversity and clout of the members of the social groups doing the assessing (Longino 2002). Having the stipulated characteristics might even correlate with producing good assessments, but it seems to get the order wrong (Miller 2008). It's necessary to first identify the appropriate requirements for objective criticism. What matters are methods whose statistical properties may be shown in relation to probes on real experiments and data.

Souvenir P: Transparency and Informativeness

There are those who would replace objectivity with the fashionable term "transparency." Being transparent about what was done and how one got from the raw data to the statistical inferences certainly promotes objectivity, provided I can use that information to critically appraise the inference. For example, being told about stopping rules, cherry picking, altered endpoints, and changed variables is useful in auditing your error probabilities. Simmons, Nelson, and Simonsohn (2012) beg researchers to "just say it," if you didn't p-hack or commit other QRPs. They offer a "21 word solution" that researchers can add to a Methods section: "We report how we determined our sample size, all data exclusions (if any), all manipulations, and all measures in the study (p. 4)." If your account doesn't use error probabilities, however, it's unclear how to use reports of what would alter error probabilities.

You can't make your inference objective merely announcing your choices and your reasons; there needs to be a system in place to critically evaluate that information. It should not be assumed the scientist is automatically to be trusted. Leading experts might arrive at rival statistical inferences, each being transparent as to their choices of a host of priors and models. What then? It's likely to descend into a battle of the experts. Salesmanship, popularity, and persuasiveness are already too much a part of what passes for knowledge. On the other hand, if well-understood techniques are provided for critical appraisal of the elements of the statistical inference, then transparency could have real force.

One last thing. Viewing statistical inference as severe testing doesn't mean our sole goal is severity. "Shun error" is not a terribly interesting rule to follow. To merely state tautologies is to state objectively true claims, but they are vacuous. We are after the dual aims of severity and informativeness. Recalling Popper, we're interested in "improbable" claims – claims with high information content that can be subjected to more stringent tests, rather than low content claims. Fisher had said that in testing causal claims you should "make [your] theories elaborate by which he meant . . . [draw out] implications" for many different phenomena, increasing the chance of locating any flaw (Mayo and Cox 2006, p. 264). As I see it, the goals of stringent testing and informative theories are mutually reinforcing. Let me explain.

To attain stringent tests, we seek strong arguments from coincidence, and "corroborative tendrils" in order to triangulate results. In so doing, we seek to make our theories more general as Fisher said. A more general claim not only has more content, opening itself up to more chances of failing, it enables cross-checks to ensure that a mistake not caught in one place is likely to ramify somewhere else. A hypothesis H^* with greater depth or scope than another H may be said to be at a "higher level" than H in my horizontal "hierarchy" (Figure 2.1). For instance, the full GTR is at a higher level than the individual hypothesis about light deflection; and current theories about prion diseases are at a higher level than Prusiner's initial hypotheses limited to kuru. If a higher level theory H^* is subjected to tests with good capacity (high probability) of finding errors, it would be necessary to check and rule out more diverse phenomena than the more limited lower level hypothesis H. Were H^* to nevertheless pass tests, then it does so with higher severity than does H.

Tour II Rejection Fallacies: Who's Exaggerating What?

Comedian Jackie Mason will be doing his shtick this evening in the ship's theater: a one-man show consisting of a repertoire of his "Greatest Hits" without a new or updated joke in the mix. A sample:

> If you want to eat nothing, eat nouvelle cuisine. Do you know what it means? No food. The smaller the portion the more impressed people are, so long as the food's got a fancy French name, haute cuisine. An empty plate with sauce!

You'll get the humor only once you see and hear him (Mayo 2012b). As one critic (Logan 2012) wrote, Mason's jokes "offer a window to a different era," one whose caricatures and biases one can only hope we've moved beyond. It's one thing for Jackie Mason to reprise his greatest hits, another to reprise statistical foibles and howlers which could leave us with radical changes to science. Among the tribes we'll be engaging: Large n, Jeffreys–Lindley, and Spike and Smear.

How Could a Group of Psychologists Be so Wrong? I'll carry a single tome in our tour: Morrison and Henkel's 1970 classic, *The Significance Test Controversy*. Some abuses of the proper interpretation of significance tests were deemed so surprising even back then that researchers in psychology conducted studies to try to understand how this could be. Notably, Rosenthal and Gaito (1963) discovered that statistical significance at a given level was often fallaciously taken as evidence of a greater discrepancy from the null hypothesis the larger the sample size n. In fact, it is indicative of *less* of a discrepancy from the null than if it resulted from a smaller sample size.

What is shocking is that these psychologists indicated substantially greater confidence or belief in results associated with the larger sample size for the same p values. According to the theory, especially as this has been amplified by Neyman and Pearson (1933), the probability of rejecting the null hypothesis for any given deviation from null and p values increases as a function of the number of observations. The rejection of the null hypothesis when the number of cases is small speaks for a more dramatic effect in the population ... The question is, how could a group of psychologists be so wrong? (Bakan 1970, p. 241)

(Our convention is for "discrepancy" to refer to the parametric, not the observed, difference. Their use of "deviation" from the null alludes to our "discrepancy.")

As statistician John Pratt notes, "the more powerful the test, the more a just significant result favors the null hypothesis" (1961, p. 166). Yet we still often hear: "The thesis implicit in the [N-P] approach, [is] that a hypothesis may be rejected with increasing confidence or reasonableness as the power of the test increases" (Howson and Urbach 1993, p. 209). In fact, the thesis implicit in the N-P approach, as Bakan remarks, is the opposite! The fallacy is akin to making mountains out of molehills according to severity (Section 3.2):

> *Mountains out of Molehills* (MM) *Fallacy* (large *n* problem): The fallacy of taking a rejection of H_0, just at level P, with larger sample size (*higher power*) as indicative of a greater discrepancy from H_0 than with a smaller sample size.

Consider an analogy with two fire alarms: The first goes off with a sensor liable to pick up on burnt toast; the second is so insensitive it doesn't kick in until your house is fully ablaze. You're in another state, but you get a signal when the alarm goes off. Which fire alarm indicates the greater extent of fire? Answer: the second, less sensitive one. When the sample size increases it alters what counts as a *single sample*. It is like increasing the sensitivity of your fire alarm. It is true that a large enough sample size triggers the alarm with an observed mean that is quite "close" to the null hypothesis. But, if the test rings the alarm (i.e., rejects H_0) even for tiny discrepancies from the null value, then the alarm is poor grounds for inferring larger discrepancies. Now this is an analogy, you may poke holes in it. For instance, a test must have a large enough sample to satisfy model assumptions. True, but our interpretive question can't even get started without taking the P-values as legitimate and not spurious.

4.3 Significant Results with Overly Sensitive Tests: Large *n* Problem

> "[W]ith a large sample size virtually every null hypothesis is rejected, while with a small sample size, virtually no null hypothesis is rejected. And we generally have very accurate estimates of the sample size available without having to use significance testing at all!" (Kadane 2011, p. 438).

P-values are sensitive to sample size, but to see this as a problem is to forget what significance tests are for. We want consistent tests, so that as *n* increases the probability of discerning any discrepancy from the null (i.e., the power) increases. The fact that the test would eventually uncover any discrepancy there may be, regardless of how small, doesn't mean there always is such a discrepancy, by the way. (Another little confusion repeated in the form of "all null hypotheses are false.") Let's focus on the example of Normal testing, T+ with H_0: $\mu \leq 0$ vs. H_1: $\mu > 0$ letting $\sigma = 1$. It's precisely to bring out the effect of sample size that many prefer to write the statistic as

$$d(X) = \sqrt{n}(\overline{X} - 0)/\sigma$$

rather than

$$d(X) = (\overline{X} - 0)/\sigma_{\overline{X}},$$

where $\sigma_{\overline{X}}$ abbreviates $(\sigma/\sqrt{n})$.

T+ rejects H_0 (at the 0.025 level) iff the sample mean $\overline{X} \geq 0 + 1.96(\sigma/\sqrt{n})$. As n increases, a single $(\sigma/\sqrt{n})$ unit decreases. Thus the value of $\overline{X}$ required to reach significance decreases as n increases.

The test's goal is to distinguish observed effects due to ordinary expected variability under H_0 with those that cannot be readily explained by mere noise. If the inter-ocular test will do, you don't need statistics. As the sample size increases, the ordinary expected variability decreases. The severe tester takes account of the sample size in interpreting the discrepancy indicated. The test is like a thermostat, a fire alarm, or the mesh size in a fishing net. You choose the sensitivity, and it does what you told it to do.

Keep in mind that the hypotheses entertained are not point values, but discrepancies. Informally, for a severe tester, each corresponds to an assertion of form: there's evidence of a discrepancy at least this large, but there's poor evidence it's as large as thus and so. Let's compare statistically significant results at the same level but with different sample sizes.

Consider the 2-standard deviation cut-off for $n = 25, 100, 400$ in test T+, $\sigma = 1$ (Figure 4.1).

Let $\overline{x}_{0.025}$ abbreviate the sample mean that is just statistically significant at the 0.025 level in each test. With $n = 25$, $\overline{x}_{.025} = 2(1/5)$; with $n = 100$, $\overline{x}_{0.025} = 2(1/10)$; with $n = 400$, $\overline{x}_{0.025} = 2(1/20)$. So the cut-offs for rejection are 0.4, 0.2, and 0.1, respectively.

Again, alterations of the sample size change what counts as one unit. If you treat identical values of $(\overline{X} - \mu_0)/\sigma$ the same, ignoring $\sqrt{n}$, you will misinterpret your results. With large enough n, the cut-off for rejection can be so close to the null value as to lead some accounts to regard it as evidence *for* the null. This is the Jeffreys–Lindley paradox that we'll be visiting this afternoon (Section 4.4).

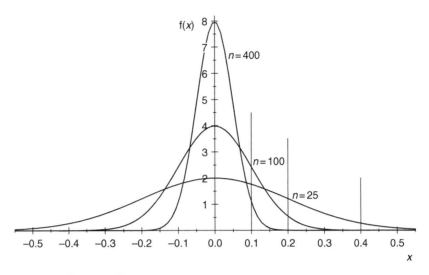

Figure 4.1 $\bar{X} \sim N\,(\mu, \sigma^2/n)$ for $n = 25, 100, 400$.

Exhibit (v): Responding to a Familiar Chestnut

Did you hear the one about the significance tester who rejected H_0 in favor of H_1 even though the result makes H_0 more likely than H_1?

I follow the treatment in Elliott Sober (2008, p. 56), who is echoing Howson and Urbach (1993, pp. 208–9), who are echoing Lindley (1957). The only difference is that I will allude to a variant on test T+: H_0: $\mu = 0$ vs. H_1: $\mu > 0$ with $\sigma = 1$. "Another odd property of significance tests," says Sober, "concerns the way in which they are sensitive to sample size." Suppose you are applying test T+ with null H_0: $\mu = 0$. If your sample size is $n = 25$, and you choose $\alpha = 0.025$, you will reject H_0 whenever $\bar{x} \geq 0.4$. If you examine $n = 100$, and choose the same value for α, you will reject H_0 whenever $\bar{x} \geq 0.2$. And if you examine $n = 400$, again with $\alpha = 0.025$, you will reject H_0 whenever $\bar{x} \geq 0.1$. "As sample size increases" the sample mean $\bar{x}$ must be closer and closer to 0 for you *not* to reject H_0. "This may not seem strange until you add the following detail. Suppose the alternative to H_0 is the hypothesis" H_1: $\mu = 0.24$. "The Law of Likelihood now entails that observing" $\bar{x} < 0.12$ favors H_0 over H_1, so in particular $\bar{x} = 0.1$ favors H_0 over H_1 (Section 1.4).

Your reply: Hold it right at "add the following detail." You're observing that the significance test disagrees with a Law of Likelihood appraisal to a point vs. point test: H_0: $\mu = 0$ vs. H_1: $\mu = 0.24$. We require the null and alternative hypotheses to exhaust the space of parameters, and these don't. Nor are our inferences to points, but rather to inequalities about discrepancies. That said, we're prepared to consider your example, and make short work of it. We're testing H_0: $\mu \leq 0$ vs. H_1: $\mu > 0$ (one could equally view it as testing H_0: $\mu = 0$ vs. H_1: $\mu > 0$). The outcome $\bar{x} = 0.1$, while indicating *some* positive discrepancy from 0, offers bad evidence and an insevere test for inferring μ as great as 0.24. Since $\bar{x} = 0.1$ rejects H_0, we say the result *accords* with H_1. The severity associated with inference $\mu \geq 0.24$ asks: what's the probability of observing $\overline{X} \leq 0.1$ – i.e., a result *more discordant* with H_1 assuming $\mu = 0.24$.

SEV($\mu \geq 0.24$) with $\bar{x} = 0.1$ and $n = 400$ is computed as $\Pr(\overline{X} < 0.1; \mu = 0.24)$. Standardizing $\overline{X}$ yields $Z = \sqrt{400}\,(0.1 - 0.24)/1 = 20(-0.14) = -2.8$. So SEV($\mu \geq 0.24$) = 0.003! Were μ as large as 0.24, we'd have observed a larger observed mean than we did with 0.997 probability! It's terrible evidence for H_1: $\mu = 0.24$.

This is redolent of the Binomial example in discussing Royall (Section 1.4). To underscore the difference between the Likelihoodist's comparative appraisal and the significance tester, you might go further. Consider an alternative that the Likelihoodist takes as favored over H_0: $\mu = 0$ with $\bar{x} = 0.1$, namely, the maximum likely alternative H_1: $\mu = 0.1$. This is one of our key benchmarks for a discrepancy that's poorly indicated. To the Likelihoodist, inferring that H_1: $\mu = 0.1$ "is favored" over H_0: $\mu = 0$ makes sense, whereas to infer a discrepancy of 0.1 from H_0 is highly *un*warranted for a significance tester.[1] Our aims are very different.

We can grant this: starting with any value for $\bar{x}$, however close to 0, there's an n such that $\bar{x}$ is statistically significantly greater than 0 at a chosen level. If one understands the test's intended task, this is precisely what is wanted. How large would n need to be so that 0.02 is statistically significant at the 0.025 level (still retaining $\sigma = 1$)?

Answer: Setting $0.02 = 2(1/\sqrt{n})$ and solving for n yields $n = 10{,}000$.[2]

Statistics won't tell you what magnitudes are of relevance to you. No matter, we can critique results and purported inferences.

---

[1] SEV($\mu \geq 0.1$) with $\bar{x} = 0.1$ and $n = 400$ is computed by considering $\Pr(\overline{X} < 0.1; \mu = 0.1)$. Standardizing $\overline{X}$ yields $z = \sqrt{400}\,(0.1 - 0.1)/1 = 0$. So SEV($\mu \geq 0.1$) = 0.5!

[2] Let's use this to illustrate the MM fallacy: Compare (i) $n = 100$ and (ii) $n = 10{,}000$ in the same test T+. With $n = 100$, 1SE = 0.1, with $n = 10{,}000$, 1SE = 0.01. The just 0.025 significant outcomes in the two tests are (i) $\bar{x} = 0.2$ and (ii) $\bar{x} = 0.02$. Consider the 0.93 lower confidence bound for each. Subtracting 1.5 SE from the outcome yields $\mu > 0.5(1/\sqrt{n})$: (i) for $n = 100$, the inferred 0.93 lower estimate is. $\mu > 0.5(1/5) = 0.05$, (ii) for $n = 10{,}000$, the inferred 0.93 lower estimate is $\mu > 0.5(1/100) = 0.005$. So a difference that is just statistically significant at the same level, 0.025, permits inferring $\mu > 0.05$ when $n = 25$, but only $\mu > 0.005$ when $n = 10{,}000$ Section 3.7.

Exhibit (vi): Reforming the Reformers on Confidence Intervals. You will be right to wonder why some of the same tribes who raise a ruckus over P-values – to the extent, in some cases, of calling for a "test ban" – are cheerleading for confidence intervals (CIs), given there is a clear duality between the two (Section 3.7). What they're really objecting to is a dichotomous use of significance tests where the report is "significant" or not at a predesignated significance level. I completely agree with this objection, and reject the dichotomous use of tests (which isn't to say there are no contexts where an "up/down" indication is apt). We should reject Unitarianism where a single method with a single interpretation must be chosen. Ironically, some of the most outspoken CI leaders use them in the dichotomous fashion (rightly) deplored when it comes to testing.

Geoffrey Cumming, an acknowledged tribal leader on CIs, tells us that "One-sided CIs are analogous to one-tailed tests but, as usual, the estimation approach is better" (2012, p. 109). Well, it might be better, but like hypothesis testing, it calls for supplements and reinterpretations as begun in Section 3.7.

Our one-sided test T+ (H_0: $\mu \leq 0$ vs. H_1: $\mu > 0$, and $\sigma = 1$) at $\alpha = 0.025$ has as its dual the one-sided (lower) 97.5% general confidence interval: $\mu > \overline{X} - 2(1/\sqrt{n})$ – rounding to 2 from 1.96. So you won't have to flip back pages, here's a quick review of the notation we developed to avoid the common slipperiness with confidence intervals. We abbreviate the generic lower limit of a $(1 - \alpha)$ confidence interval as $\hat{\mu}_{1-\alpha}(\overline{X})$ and the particular limit as $\hat{\mu}_{1-\alpha}(\overline{x})$. The general estimating procedure is: Infer $\mu > \hat{\mu}_{1-\alpha}(\overline{X})$. The particular estimate is $\mu > \hat{\mu}_{1-\alpha}(\overline{x})$. Letting $\alpha = 0.025$ we have: $\mu > \overline{x} - 2(1/\sqrt{n})$. With $\alpha = 0.05$, we have $\mu > \overline{x} - 1.65(1/\sqrt{n})$.

Cumming's interpretation of CIs and confidence levels points to their performance-oriented construal: "In the long run 95% of one-sided CIs will include the population mean ... We can say we're 95% confident our one-sided interval includes the true value ... meaning that for 5% of replications the [lower limit] will exceed the true value" (Cumming 2012, p. 112). What does it mean to be 95% confident in the particular interval estimate for Cumming? "It means that the values in the interval are plausible as true values for μ, and that values outside the interval are relatively implausible – though not impossible" (ibid., p. 79). The performance properties of the method rub off in a plausibility assessment of some sort.

The test that's dual to the CI would "accept" those parameter values within the corresponding interval, and reject those outside, all at a single predesignated confidence level $1 - \alpha$. Our main objection to this is it gives the misleading idea that there's evidence for each value in the interval, whereas, in fact, the interval simply consists of values that aren't rejectable, were one testing at the α level. Not being a rejectable value isn't the same as having evidence for that value. Some values are close to being rejectable, and we should convey this. Standard CIs do not.

To focus on how CIs deal with distinguishing sample sizes, consider again the three instances of test T+ with (i) $n = 25$, (ii) $n = 100$, and (iii) $n = 400$. Imagine the observed mean $\bar{x}$ from each test just hits the significance level 0.025. That is, (i) $\bar{x} = 0.4$, (ii) $\bar{x} = 0.2$, and (iii) $\bar{x} = 0.1$. Form 0.975 confidence interval estimates for each:

(i) for $n = 25$, the inferred estimate is $\mu > \hat{\mu}_{0.975}$, that is, $\mu > \bar{x} - 2(1/5)$;
(ii) for $n = 100$, the inferred estimate is $\mu > \hat{\mu}_{0.975}$, that is, $\mu > \bar{x} - 2(1/10)$;
(iii) for $n = 400$, the inferred estimate is $\mu > \hat{\mu}_{0.975}$, that is, $\mu > \bar{x} - 2(1/20)$.

Substituting $\bar{x}$ in all cases, we get the same one-sided confidence interval:

$$\mu > 0.$$

Cumming writes them as [0, infinity). How are the CIs distinguishing them?

They are not. The construal is dichotomous: in or out, plausible or not. Would we really want to say "the values in the interval are plausible as true values for μ"? Clearly not, since that includes values to infinity. I don't want to step too hard on the CI champion's toes, since CIs are in the frequentist, error statistical tribe. Yet, to avoid fallacies, this standard use of CIs won't suffice. Severity directs you to avoid taking your result as indicating a discrepancy beyond what's warranted. For an example, we can show the same inference is poorly indicated with $n = 400$, while fairly well indicated when $n = 100$. For a poorly indicated claim, take our benchmark for severity of 0.5; for fairly well, 0.84:

For $n = 400$, $\bar{x}_{0.025} = 0.1$, so $\mu > 0.1$ is poorly indicated;

For $n = 100$, $\bar{x}_{0.025} = 0.2$, and $\mu > 0.1$ is fairly well indicated.

The reasoning based on severity is counterfactual: were μ less than or equal to 0.1, it is fairly probable, 0.84, that a smaller $\overline{X}$ would have occurred. This is not part of the standard CI account, but enables the distinction we want. Another move would be for a CI advocate to require we always compute a two-sided interval. The upper 0.975 bound would reflect the greater sensitivity with increasing sample sizes:

(i) $n = 25$: (0, 0.8],   (ii) $n = 100$: (0, 0.4],   (iii) $n = 400$: (0, 0.2].

But we cannot just deny one-sided tests, nor does Cumming. In fact, he encourages their use: "it's unfortunate they are usually ignored" (2012, p. 113). (He also says he is happy for people to decide afterwards whether to report it as a one- or two-sided interval (ibid., p. 112), only doubling α, which I do not mind.) Still needed is a justification for bringing in the upper limit when applying a one-sided estimator, and severity supplies it. You should always be interested in at least *two benchmarks*: discrepancies well warranted and those terribly warranted. In test T+, our handy benchmark for the terrible is to set the lower limit to $\bar{x}$. The severity for $(\mu > \bar{x})$ is 0.5. Two side notes:

First I grant it would be wrong to charge Cumming with treating all parameter values within the confidence interval *on par*, because he does suggest distinguishing them by their likelihoods (by how probable each renders the outcome). Take just the single 0.975 lower CI bound with $n = 100$ and $\bar{x} = 0.2$. A μ value closer to the observed 0.2 has higher likelihood (in the technical sense) than ones close to the 0.975 lower limit 0. For example, $\mu = 0.15$ is more likely than $\mu = 0.05$. However, this moves away from CI reasoning (toward likelihood comparisons). The claim $\mu > 0.05$ has a *higher* confidence level (0.93) than does $\mu > 0.15$ (0.7)[3] even though the point hypothesis $\mu = 0.05$ is less likely than $\mu = 0.15$ (the latter is closer to $\bar{x} = 0.2$ than is the former). Each point in the lower CI corresponds to a different lower bound, each associated with a different confidence level, and corresponding severity assessment. That's how to distinguish them.

Second there's an equivocation, or at least a potential equivocation, in Cumming's assertion "that for [2.5%] of replications the [lower limit] will exceed the true value" (Cumming 2012, p. 112 replacing 5% with 2.5%). This is not a true claim if "lower limit" is replaced by a *particular* lower limit: $\hat{\mu}_{0.025}(\bar{x})$, it holds only for the *generic* lower limit $\hat{\mu}_{0.025}(\bar{X})$. That is, we can't say μ exceeds zero 2.5% of the time, which would be to assign a probability of 0.975 to $\mu > 0$. Yet this misinterpretation of CIs is legion, as we'll see in a historical battle about fiducial intervals (Section 5.8).

4.4   Do *P*-Values Exaggerate the Evidence?

"Significance levels overstate the evidence against the null hypothesis," is a line you may often hear. Your first question is:

> What do you mean by overstating the evidence against a hypothesis?

Several (honest) answers are possible. Here is one possibility:

---

[3] Subtract 1.5 SE and 0.5 SE from $\bar{x} = 0.2$, respectively.

What I mean is that when I put a lump of prior weight π_0 of $1/2$ on a point null H_0 (or a very small interval around it), the P-value is smaller than my Bayesian posterior probability on H_0.

More generally, the "*P-values exaggerate*" criticism typically boils down to showing that if inference is appraised via one of the probabilisms – Bayesian posteriors, Bayes factors, or likelihood ratios – the evidence against the null (or against the null and in favor of some alternative) isn't as big as $1 - P$.

You might react by observing that: (a) P-values are not intended as posteriors in H_0 (or Bayes ratios, likelihood ratios) but rather are used to determine if there's an indication of discrepancy from, or inconsistency with, H_0. This might only mean it's worth getting more data to probe for a real effect. It's not a degree of belief or comparative strength of support to walk away with. (b) Thus there's no reason to suppose a P-value should match numbers computed in very different accounts, that differ among themselves, and are measuring entirely different things. Stephen Senn gives an analogy with "height and stones":

> ... [S]ome Bayesians in criticizing P-values seem to think that it is appropriate to use a threshold for significance of 0.95 of the probability of the alternative hypothesis being true. This makes no more sense than, in moving from a minimum height standard (say) for recruiting police officers to a minimum weight standard, declaring that since it was previously 6 foot it must now be 6 stone. (Senn 2001b, p. 202)

To top off your rejoinder, you might ask: (c) Why assume that "the" or even "a" correct measure of evidence (relevant for scrutinizing the P-value) is one of the probabilist ones?

All such retorts are valid, and we'll want to explore how they play out here. Yet, I want to push beyond them. Let's be open to the possibility that evidential measures from very different accounts can be used to scrutinize each other.

Getting Beyond "I'm Rubber and You're Glue". The danger in critiquing statistical method X from the standpoint of the goals and measures of a distinct school Y, is that of falling into begging the question. If the P-value is exaggerating evidence against a null, meaning it seems too small from the perspective of school Y, then Y's numbers are too big, or just irrelevant, from the perspective of school X. Whatever you say about me bounces off and sticks to you. This is a genuine worry, but it's not fatal. The goal of this journey is to identify minimal theses about "bad evidence, no test (BENT)" that enable some degree of scrutiny of any statistical inference account – at least on the meta-level. Why assume all schools of statistical inference embrace the minimum severity principle? I don't, and they don't. But by identifying when methods violate severity, we can pull back the veil on at least one source of disagreement behind the battles.

Thus, in tackling this latest canard, let's resist depicting the critics as committing a gross blunder of confusing a P-value with a posterior probability in a null. We resist, as well, merely denying we care about their measure of support. I say we should look at exactly what the critics are on about. When we do, we will have gleaned some short-cuts for grasping a plethora of critical debates. We may even wind up with new respect for what a P-value, the least popular girl in the class, really *does*.

To visit the core arguments, we travel to 1987 to papers by J. Berger and Sellke, and Casella and R. Berger. These, in turn, are based on a handful of older ones (Cox 1977, E, L, & S 1963, Pratt 1965), and current discussions invariably revert back to them. Our struggles through quicksand of Excursion 3, Tour II, are about to pay large dividends.

J. Berger and Sellke, and Casella and R. Berger. Berger and Sellke (1987a) make out the conflict between P-values and Bayesian posteriors by considering the two-sided test of the Normal mean, H_0: $\mu = 0$ vs. H_1: $\mu \neq 0$. "Suppose that $X = (X_1, \ldots, X_n)$, where the X_i are IID $N(\mu, \sigma^2)$, σ^2 known" (p. 112). Then the test statistic $d(X) = \sqrt{n}|\overline{X} - \mu_0|/\sigma$, and the P-value will be twice the P-value of the corresponding one-sided test.

Starting with a lump of prior, generally 0.5, on the point hypothesis H_0, they find the posterior probability in H_0 is larger than the P-value for a variety of different priors on the alternative. However, the result depends entirely on how the remaining 0.5 is allocated or smeared over the alternative (a move dubbed spike and smear). Using what they call a Jeffreys-type prior, the 0.5 is spread out over the alternative parameter values as if the parameter is itself distributed $N(\mu_0, \sigma)$. Now Harold Jeffreys recommends the lump prior only to capture cases where a special value of a parameter is deemed plausible, for instance, the GTR deflection effect $\lambda = 1.75''$, after about 1960. The rationale is to avoid a 0 prior on H_0 and enable it to receive a reasonable posterior probability .

By subtitling their paper "The irreconcilability of P-values and evidence," Berger and Sellke imply that if P-values disagree with posterior assessments, they can't be measures of evidence at all. Casella and R. Berger (1987) retort that "reconciling" is at hand, if you move away from the lump prior. So let's see how this unfolds. I assume throughout, as do the critics, that the P-values are "audited," so that neither selection effects nor violated model assumptions are in question at this stage. I see no other way to engage their arguments.

Table 4.1 Pr($H_0|x$) for Jeffreys-type prior

P one-sided	z_α	10	20	50	100	1000
		\multicolumn{5}{}{n (sample size)}				
0.05	1.645	0.47	0.56	0.65	0.72	0.89
0.025	1.960	0.37	0.42	0.52	0.60	0.82
0.005	2.576	0.14	0.16	0.22	0.27	0.53
0.0005	3.291	0.024	0.026	0.034	0.045	0.124

(From Table 1, J. Berger and T. Sellke (1987) p. 113 using the one-sided P-value)

Table 4.1 gives the values of Pr($H_0|x$). We see that we would declare no evidence against the null, and even evidence for it (to the degree indicated by the posterior) whenever d(x) fails to reach a 2.5 or 3 standard error difference. With $n = 50$, "one can classically 'reject H_0 at significance level $p = 0.05$,' although Pr($H_0|x$) = 0.52 (which would actually indicate that the evidence *favors* H_0)" (J. Berger and Sellke 1987, p. 113).

If $n = 1000$, a result statistically significant at the 0.05 level results in the posterior probability to $\mu = 0$ going up from 0.5 (the lump prior) to 0.82! From their Bayesian perspective, this suggests P-values are exaggerating evidence against H_0. Error statistical testers, on the other hand, balk at the fact that using the recommended priors allows statistically significant results to be interpreted as no evidence against H_0 – or even evidence for it. They point out that 0 is excluded from the two-sided confidence interval at level 0.95. Although a posterior probability doesn't have an error probability attached, a tester can evaluate the error probability credentials of these inferences. Here we'd be concerned with a Type II error: failing to find evidence against the null, and providing a fairly high posterior for it, when it's false (Souvenir I).

Let's use a less extreme example where we have some numbers handy: our water-plant accident. We had $\sigma = 10$, $n = 100$ leading to the nice ($\sigma/\sqrt{n}$) value of 1. Here it would be two-sided, to match their example: H_0: $\mu = 150$ vs. H_1: $\mu \neq 150$. Look at the second entry of the 100 column, the posterior when $z_\alpha = 1.96$. With the Jeffreys prior, perhaps championed by the water coolant company, J. Berger and Sellke assign a posterior of 0.6 to H_0: $\mu = 150$ degrees when a mean temperature of 152 (151.96) degrees is observed – reporting decent evidence the cooling mechanism is working just fine. How often would this occur even if the actual underlying mean temperature is, say, 151 degrees? With a two-sided test, cutting off 2 standard errors on either side, we'd reject whenever either $\overline{X} \geq 152$ or $\overline{X} \leq 148$. The probability of the second is negligible under $\mu = 151$, so the probability we want is

$\Pr(\overline{X} < 152; \mu = 151) = 0.84\ (Z = (152 - 151) = 1)$. The probability of declaring evidence for 150 degrees (with posterior of 0.6 to H_0) even if the true increase is actually 151 degrees is around 0.84; 84% of the time they erroneously fail to ring the alarm, and would boost their probability of $\mu = 150$ from 0.5 to 0.6. Thus, from our minimal severity principle, the statistically significant result can't even be taken as evidence for compliance with 151 degrees, let alone as evidence *for* the null of 150 (Table 3.1).

Is this a problem for them? It depends what you think of that prior. The N-P test, of course, does not use a prior, although, as noted earlier, one needn't rule out a frequentist prior on mean water temperature after an accident (Section 3.2). For now our goal is making out the criticism.

Jeffreys–Lindley "Paradox" or Bayes/Fisher Disagreement

But how, intuitively, does it happen that a statistically significant result corresponds to a Bayes boost for H_0? Go back to J. Berger and Sellke's example of Normal testing of H_0: $\mu = 0$ vs. H_1: $\mu \neq 0$. Some sample mean $\overline{x}$ will be close enough to 0 to increase the posterior for H_0. By choosing a sufficiently large n, even a statistically significant result can correspond to large posteriors on H_0. This is the Jeffreys–Lindley "paradox," which some more aptly call the Bayes/Fisher disagreement. Lindley's famous result dealt with just this example, two-sided Normal testing with known variance. With a lump given to the point null, and the rest appropriately spread over the alternative, an n can be found such that an α significant result corresponds to $\Pr(H_0|x) = (1 - \alpha)$! We can see by extending Table 4.1 to arbitrarily large n, we can get a posterior for the null of 0.95, when the (two-sided) P-value is 0.05. Many say you should decrease the required P-value for significance as n increases; and Cox and Hinkley (1974, p. 397) provide formulas to achieve this and avoid the mismatch. There's nothing in N-P or Fisherian theory to oppose this. I won't do that here, as I want to make out the criticism. We need only ensure that the interpretation takes account of the (obvious) fact that, with a fixed P-value and increasing n, the test is more and more sensitive to smaller and smaller discrepancies. Using a smaller plate at the French restaurant may make the portion appear bigger, but, Jackie Mason notwithstanding, knowing the size of the plate, I can see there's not much there.

Why assign the lump of ½ as prior to the point null? "The choice of $\pi_0 = 1/2$ has obvious intuitive appeal in scientific investigations as being 'objective'" say J. Berger and Sellke (1987, p. 115). But is it? One starts by making H_0 and H_1 equally probable, then the 0.5 accorded to H_1 is spread out over all the values in H_1: "The net result is that all values of $[\mu]$ are far from being equally likely"

(Senn 2015a). Any small group of μ values in H_1 gets a tiny prior. David Cox describes how it happens:

... if a sample say at about the 5% level of significance is achieved, then either H_0 is true or some alternative in a band of order $1/\sqrt{n}$; the latter possibility has, as $n \to \infty$, prior probability of order $1/\sqrt{n}$ and hence at a fixed level of significance the posterior probabilities shift in favour of H_0 as n increases. (Cox 1977, p. 59)

What justifies the lump prior of 0.5?

A Dialogue at the Water Plant Accident

EPA Rep: The mean temperature of the water was found statistically significantly higher than 150 degrees at the 0.025 level.

Spiked Prior Rep: This even strengthens my belief the water temperature's no different from 150. If I update the prior of 0.5 that I give to the null hypothesis, my posterior for H_0 is still 0.6; it's not 0.025 or 0.05, that's for sure.

EPA Rep: Why do you assign such a high prior probability to H_0?

Spiked Prior Rep: If I gave H_0 a value lower than 0.5, then, if there's evidence to reject H_0, at most I would be claiming an improbable hypothesis has become more improbable.

[W]ho, after all, would be convinced by the statement 'I conducted a Bayesian test of H_0, assigning prior probability 0.1 to H_0, and my conclusion is that H_0 has posterior probability 0.05 and should be rejected?' (J. Berger and Sellke 1987, p. 115).

This quote from J. Berger and Sellke is peculiar. They go on to add: "We emphasize this obvious point because some react to the Bayesian–classical conflict by attempting to argue that [prior] π_0 should be made small in the Bayesian analysis so as to force agreement" (ibid.). We should not force agreement. But it's scarcely an obvious justification for a lump of prior on the null H_0 – one which results in a low capability to detect discrepancies – that it ensures, if they *do* reject H_0, there will be a meaningful drop in its probability. Let's listen to the pushback from Casella and R. Berger (1987a), the Berger being Roger now (I use initials to distinguish them).

   The Cult of the Holy Spike and Slab. Casella and R. Berger (1987a) charge that the problem is not P-values but the high prior, and that "concentrating mass on the point null hypothesis is biasing the prior in favor of H_0 as much as possible" (p. 111) whether in one- or two-sided tests. According to them:

The testing of a point null hypothesis is one of the most misused statistical procedures. In particular, in the location parameter problem, the point null hypothesis is more the mathematical convenience than the statistical method of choice. (ibid., p. 106)

Most of the time "there is a direction of interest in many experiments, and saddling an experimenter with a two-sided test would not be appropriate" (ibid.). The "cult of the holy spike" is an expression I owe to Sander Greenland (personal communication).

By contrast, we can reconcile P-values and posteriors in one-sided tests if we use more diffuse priors. (e.g., Cox and Hinkley 1974, Jeffreys 1939/1961, Pratt 1965). In fact, Casella and Berger show that for sensible priors in that case, the P-value is at least as big as the minimum value of the posterior probability on the null, again contradicting claims that P-values exaggerate the evidence.[4]

J. Berger and Sellke (1987) adhere to the spikey priors, but following E, L, & S (1963), they're keen to show that P-values exaggerate evidence even in cases less extreme than the Jeffreys posteriors in Table 4.1. Consider the likelihood ratio of the null hypothesis over the hypothesis most generous to the alternative, they say. This is the point alternative with maximum likelihood, H_{max} – arrived at by setting $\mu = \bar{x}$. Through their tunnel, it's disturbing that even using this likelihood ratio, the posterior for H_0 is still larger than 0.05 – when they give a 0.5 spike to both H_0 and H_{max}. Some recent authors see this as the key to explain today's lack of replication of significant results. Through the testing tunnel, things look different (Section 4.5).

Why Blame Us Because You Can't Agree on Your Posterior? Stephen Senn argues that the reason for the wide range of variation of the posterior is the fact that it depends radically on the choice of alternative to the null and its prior.[5] According to Senn, "... the reason that Bayesians can regard P-values as overstating the evidence against the null is simply a reflection of the fact that Bayesians can disagree *sharply* with each other" (Senn 2002, p. 2442). Senn

---

[4] Casella and R. Berger (1987b) argue, "We would be surprised if most researchers would place even a 10% prior probability of H_0. We hope that the casual reader of Berger and Delampady realizes that the big discrepancies between P-values $P(H_0|x)$ ... are due to a large extent to the large value of [the prior of 0.5 to H_0] that was used." The most common uses of a point null, asserting the difference between means is 0, or the coefficient of a regression coefficient is 0, merely describe a potentially interesting feature of the population, with no special prior believability. "Berger and Delampady admit ..., P-values are reasonable measures of evidence when there is no a priori concentration of belief about H_0" (ibid., p. 345). Thus, "the very argument that Berger and Delampady use to dismiss P-values can be turned around to argue *for* P-values" (ibid., p. 346).

[5] In defending spiked priors, Berger and Sellke move away from the importance of effect size. "Precise hypotheses ... ideally relate to, say, some precise theory being tested. Of primary interest is whether the theory is right or wrong; the amount by which it is wrong may be of interest in developing alternative theories, but the initial question of interest is that modeled by the precise hypothesis test" (1987, p. 136).

illustrates how "two Bayesians having the same prior probability that a hypothesis is true and having seen the same data can come to radically different conclusions because they differ regarding the alternative hypothesis" (Senn 2001b, p. 195). One of them views the problem as a one-sided test and gets a posterior on the null that matches the P-value; a second chooses a Jeffreys-type prior in a two-sided test, and winds up with a posterior to the null of $1 - p$!

Here's a recap of Senn's example (ibid., p. 200): Two scientists A and B are testing a new drug to establish its treatment effect, δ, where positive values of δ are good. Scientist A has a vague prior whereas B, while sharing the same distribution about the probability of positive values of δ, is less pessimistic than A regarding the effect of the drug. If it's not useful, B believes it will have no effect. They "share the same belief that the drug has a positive effect. Given that it has a positive effect, they share the same belief regarding its effect. . . . They differ only in belief as to how harmful it might be." A clinical trial yields a difference of 1.65 standard units, a one-sided P-value of 0.05. The result is that A gives 1/20 posterior probability to H_0: the drug does *not* have a positive effect, while B gives a probability of 19/20 to H_0. B is using the two-sided test with a lump of prior on the null (H_0: $\mu = 0$ vs. H_1: $\mu \neq 0$), while A is using a one-sided test T+ (H_0: $\mu \leq 0$ vs. H_1: $\mu > 0$). The contrast, Senn observes, is that of Cox's distinction between "precise and dividing hypothesis" (Section 3.3). "[F]rom a common belief in the drug's efficacy they have moved in opposite directions" (ibid., pp. 200–201). Senn riffs on Jeffreys' well-known joke that we heard in Section 3.4:

> It would require that a procedure is dismissed [by significance testers] because, when combined with information which it doesn't require and which may not exist, it disagrees with a [Bayesian] procedure that disagrees with itself. (ibid., p. 195)

In other words, if Bayesians disagree with each other even when they're measuring the same thing – posterior probabilities – why be surprised that disagreement is found between posteriors and P-values? The most common argument behind the "P-values exaggerate evidence" appears not to hold water. Yet it won't be zapped quite so easily, and will reappear in different forms.

Exhibit (vii): Contrasting Bayes Factors and Jeffreys–Lindley Paradox. We've uncovered some interesting bones in our dig. Some lead to seductive arguments purporting to absolve the latitude in assigning priors in Bayesian tests. Take Wagenmakers and Grünwald (2006, p. 642): "Bayesian hypothesis tests are often criticized because of their dependence on prior distributions

... [yet] no matter what prior is used, the Bayesian test provides sub-stantially less evidence against H_0 than" P-values, in the examples we've considered. Be careful in translating this. We've seen that what counts as "less" evidence runs from seriously underestimating to overestimating the discrepancy we are entitled to infer with severity. Begin with three types of priors appealed to in some prominent criticisms revolving around the Fisher – Jeffreys disagreement.

1. *Jeffreys-type prior with the "spike and slab" in a two-sided test.* Here, with large enough n, a statistically significant result becomes evidence *for* the null; the posterior to H_0 exceeds the lump prior.
2. *Likelihood ratio most generous to the alternative.* Here, there's a spike to a point null, H_0: $\theta = \theta_0$ to be compared to the point alternative that's maximally likely θ_{max}. Often, both H_0 and H_{max} are given 0.5 priors.
3. *Matching.* Instead of a spike prior on the null, it uses a smooth diffuse prior, as in the "dividing" case. Here, the P-value "is an approximation to the posterior probability that $\theta < 0$" (Pratt 1965, p. 182).

In sync with our attention to high-energy particle physics (HEP) in Section 3.6, consider an example that Aris Spanos (2013b) explores in relation to the Jeffreys–Lindley paradox. The example is briefly noted in Stone (1997).

A large number ($n = 527,135$) of independent collisions that can be of either type A or type B are used to test if the proportion of type A collisions is exactly 0.2, as opposed to any other value. It's modeled as n Bernoulli trials testing H_0: $\theta = 0.2$ vs. H_1: $\theta \neq 0.2$. The observed proportion of type A collisions is scarcely greater than the point null of 0.2:

$$\bar{x} = k/n = 0.20165233, \text{ where } n = 527,135, \ k = 106,298.$$

The significance level against H_0 is small *(so there's evidence against H_0)*
The *test statistic* $d(X) = [\sqrt{n}(\bar{X} - 0.2)/\sigma] = 3$, $\sigma = \sqrt{[\theta(1 - \theta)]}$, which under the null is $\sqrt{[0.2(0.8)]} = 0.4$. The significance level associated with $d(x_0)$ in this two-sided test is

$$\Pr(|d(X)| > |d(x_0)|; H_0) = 0.0027.$$

So the result $\bar{x}$ is highly significant, even though it's scarcely different from the point null.

The Bayes factor in favor of H_0 is high

H_0 is given the spiked prior of 0.5, and the remaining 0.5 is spread equally among the values in H_1. I follow Spanos' computations:[6]

$$\Pr(k|H_0) = \binom{n}{k} 0.2^k (0.8)^{n-k},$$

$$\Pr(k|H_1) = \int_1^0 \binom{n}{k} \theta^k (1-\theta)^{n-k} d\theta = 1/(n+1),$$

where $n = 527{,}135$ and $k = 106{,}298$.

The Bayes factor $B_{01} = \Pr(k|H_0)/\Pr(k|H_1) = 0.000015394/0.000001897$
$$= 8.115.$$

While the likelihood of H_0 in the numerator is tiny, the likelihood of H_1 is even tinier. Since B_{01} in favor of H_0 is 8, which is greater than 1, the posterior for H_0 goes up, even though the outcome is statistically significantly greater than the null.

There's no surprise once you consider the Bayesian question here: compare the likelihood of a result scarcely different from 0.2 being produced by a universe where $\theta = 0.2$ – where this has been given a spiked prior of 0.5 under H_0 – with the likelihood of that result being produced by any θ in a small band of θ values, which have been given a very low prior under H_1. Clearly, $\theta = 0.2$ is more likely, and we have an example of the Jeffreys–Fisher disagreement.

Who should be afraid of this disagreement (to echo the title of Spanos' paper)? Many tribes, including some Bayesians, think it only goes to cast doubt on this particular Bayes factor. Compare it with proposal 2 in Exhibit (vii): the *Likelihood ratio most generous to the alternative*: $\mathrm{Lik}(0.2)/\mathrm{Lik}(\theta_{\max})$. We know the maximally likely value for θ, $\theta_{\max} = \bar{x}$:

$$\bar{x} = k/n = 0.20165233 = \theta_{\max},$$

$$\Pr(k|H_{\max}) = \binom{n}{k} 0.20165233^k (1-0.20165233)^{n-k} = 0.0013694656,$$

$$\mathrm{Lik}(0.2) = 0.000015394, \text{ and } \mathrm{Lik}(\theta_{\max}) = 0.0013694656.$$

Now B_{01} is 0.01 and B_{10}, $\mathrm{Lik}(\theta_{\max})/\mathrm{Lik}(0.2) = 89$.

Why should a result 89 times more likely under alternative $\theta_{\max}$ than under $\theta = 0.2$ be taken as strong evidence *for* $\theta = 0.2$? It shouldn't, according to some, including Lindley's own student, default Bayesian José Bernardo (2010).

---

[6] The spiked prior drops out, so the result is the same as a uniform prior on the null and alternative.

Presumably, the Likelihoodist concurs. There are family feuds within and between the diverse tribes of probabilisms.[7]

Greenland and Poole Given how often spiked priors arise in foundational arguments, it's worth noting that even Bayesians Edwards, Lindman, and Savage (1963, p. 235), despite raising the "*P*-values exaggerate" argument, aver that for Bayesian statisticians, "no procedure for testing a sharp null hypothesis is likely to be appropriate unless the null hypothesis deserves special initial credence." Epidemiologists Sander Greenland and Charles Poole, who claim not to identify with any one statistical tribe, but who often lead critics of significance tests, say:

> Our stand against spikes directly contradicts a good portion of the Bayesian literature, where null spikes are used too freely to represent the belief that a parameter 'differs negligibly' from the null. In many settings ... even a tightly concentrated probability near the null has no basis in genuine evidence. Many scientists and statisticians exhibit quite a bit of irrational prejudice in favor of the null ... (2013, p. 77).

They angle to reconcile *P*-values and posteriors, and to this end they invoke the matching result in # 3, Exhibit (vii). An uninformative prior, assigning equal probability to all values of the parameter, allows the *P*-value to approximate the posterior probability that $\theta < 0$ in one-sided testing ($\theta \leq 0$ vs. $\theta > 0$). In two-sided testing, the posterior probability that θ is on the opposite side of 0 than the observed is *P*/2. They proffer this as a way "to live with" *P*-values. Commenting on them, Andrew Gelman (2013, p. 72) raises this objection:

> [C]onsider what would happen if we routinely interpreted one-sided *P* values as posterior probabilities. In that case, an experimental result that is 1 standard error from zero – that is, exactly what one might expect from chance alone – would imply an 83% posterior probability that the true effect in the population has the same direction as the observed pattern in the data at hand. It does not make sense to me to claim 83% certainty – 5 to 1 odds [to H_1] ...

(The *P*-value is 0.16.) Rather than relying on non-informative priors, Gelman prefers to use prior information that leans towards the null. This avoids as high a posterior to H_1 as when using the matching result.

Greenland and Poole respond that Gelman is overlooking the hazard of "strong priors that are not well founded. ... Whatever our prior opinion and

---

[7] Bernardo shocked his mentor in announcing that the Lindley paradox is really an indictment of the Bayesian computations: "Whether you call this a paradox or a disagreement, the fact that the Bayes factor for the null may be arbitrarily large for sufficiently large *n*, *however relatively unlikely the data may be under* H_0 is, ... deeply disturbing" (Bernardo 2010, p. 59).

its foundation, we still need reference analyses with weakly informative priors to alert us to how much our prior probabilities are driving our posterior probabilities" (2013, p. 76). They rightly point out that, in some circles, giving weight to the null can be the outgrowth of some ill-grounded metaphysics about "simplicity." Or it may be seen as an assumption akin to a presumption of innocence in law. So the question turns on the appropriate prior on the null.

Look what has happened! The problem was simply to express "I'm not impressed" with a result reaching a P-value of 0.16: Differences even larger than 1 standard error are not so very infrequent – they occur 16% of the time – even if there's zero effect. So I'm not convinced of the reality of the effect, based on this result. P-values did their job, reflecting as they do the severity requirement. H_1 has passed a lousy test. That's that. No prior probability assignment to H_0 is needed. Problem solved.

But there's a predilection for changing the problem (if you're a probabilist). Greenland and Poole feel they're helping us to live with P-values without misinterpretation. By choosing the prior so that the P-value matches the posterior on H_0, they supply us "with correct interpretations" (ibid., p. 77) where "correct interpretations" are those where the misinterpretation (of a P-value as a posterior in the null) is not a misinterpretation. To a severe tester, this results in completely changing the problem from an assessment of how well tested the reality of the effect is, with the given data, to what odds I would give in betting, or the like. We land in the same uncharted waters as with other attempts to fix P-values, when we could have stayed on the cruise ship, interpreting P-values as intended.

Souvenir Q: Have We Drifted From Testing Country? (Notes From an Intermission)

Before continuing, let's pull back for a moment, and take a coffee break at a place called Spike and Smear. Souvenir Q records our notes. We've been exploring the research program that appears to show, quite convincingly, that significance levels exaggerate the evidence against a null hypothesis, based on evidential assessments endorsed by various Bayesian and Likelihoodist accounts. We suspended the impulse to deny it can make sense to use a rival inference school to critique significance tests. We sought to explore if there's something to the cases they bring as ammunition to this conflict. The Bayesians say the disagreement between their numbers and P-values is relevant for impugning P-values, so we try to go along with them.

Reflect just on the first argument, pertaining to the case of two-sided Normal testing $H_0: \mu = 0$ vs. $H_0: \mu \neq 0$, which was the most impressive, particularly with $n \geq 50$. It showed that a statistically significant difference from a test hypothesis at familiar levels, 0.05 or 0.025, can correspond to a result that a Bayesian takes as evidence *for* H_0. The prior for this case is the spike and smear, where the smear will be of the sort leading to J. Berger and Sellke's results, or similar. The test procedure is to move from a statistically significant result at the 0.025 level, say, and infer the posterior for H_0.

Now our minimal requirement for data *x* to provide evidence for a claim *H* is that

(S-1) *H* accords with (agrees with) *x*, and

(S-2) there's a reasonable, preferably a high, probability that the procedure would have produced disagreement with *H*, if in fact *H* were false.

So let's apply these severity requirements to the data taken as evidence for H_0 here.

Consider (S-1). Is a result that is 1.96 or 2 standard errors away from 0 in good accord with 0? Well, 0 is excluded from the corresponding 95% confidence interval. That does not seem to be in accord with 0 at all. Still, they have provided measures whereby *x* does accord with H_0, the likelihood ratio or posterior probability on H_0. So, in keeping with the most useful and most generous way to use severity, let's grant (S-1) holds.

What about (S-2)? Has anything been done to probe the falsity of H_0? Let's allow that H_0 is not a precise point, but some very small set of values around 0. This is their example, and we're trying to give it as much credibility as possible. Did the falsity of H_0 have a good chance of showing itself? The falsity of H_0 here is H_1: $\mu \neq 0$. What's troubling is that we found the probability of failing to pick up on population discrepancies as much as 1 standard error in excess of 0 is rather high (0.84) with $n = 100$. Larger sample sizes yield even less capability. Nor are they merely announcing "no discrepancy from 0" in this case. They're finding evidence for 0!

So how did the Bayesian get the bump in posterior probability on the null? It was based on a spiked prior of 0.5 to H_0. All the other points get minuscule priors having to share the remaining 0.5 probability. What was the warrant for the 0.5 prior to H_0? J. Berger and Sellke are quite upfront about it: if they allowed the prior spike to be low, then a rejection of the null would merely be showing an improbable hypothesis got more improbable. "[W]ho, after all, would be convinced," recall their asking: if "my conclusion is that H_0 has posterior probability 0.05 and should be rejected" since it previously had probability, say 0.1 (1987, p. 115). A slight lowering of probability won't cut it. Moving from a low prior to a slightly higher one also lacks punch.

This explains their high prior (at least 0.5) on H_0, but is it evidence for it? Clearly not, nor does it purport to be. We needn't deny there are cases where a theoretical parameter value has passed severely (we saw this in the case of GTR in Excursion 3). But that's not what's happening here. Here they intend for the 0.5 prior to show, *in general*, that statistically significant results problematically exaggerate evidence.[8]

A tester would be worried when the rationale for a spike is to avoid looking foolish when rejecting with a small drop; she'd be worried too by a report: "I don't take observing a mean temperature of 152 in your 100 water samples as indicating it's hotter than 150, because I give a whopping spike to our coolants being in compliance." That is why Casella and R. Berger describe J. Berger and Sellke's spike and smear as maximally biased toward the null (1987a, p. 111). Don't forget the powerful role played by the choice of how to smear the 0.5 over the alternative! Bayesians might reassure us that the high Bayes factor for a point null doesn't depend on the priors given to H_0 and H_1, when what they mean is that it depends only on the priors given to discrepancies under H_1. It was the diffuse prior to the effect size that gave rise to the Jeffreys–Lindley Paradox. It affords huge latitude in what gets supported.

We thought we were traveling in testing territory; now it seems we've drifted off to a different place. It shouldn't be easy to take data as evidence for a claim when that claim is false; but here it is easy (the claim here being H_0). How can this be one of a handful of main ways to criticize significance tests as exaggerating evidence? Bring in a navigator from a Popperian testing tribe before we all feel ourselves at sea:

> Mere supporting instances are as a rule too cheap to be worth having ... any support capable of carrying weight can only rest upon ingenious tests, undertaken with the aim of refuting our hypothesis, if it can be refuted. (Popper 1983, p. 130)

The high spike and smear tactic can't be take as a basis from which to launch a critique of significance tests because it fails rather glaringly a minimum requirement for evidence, let alone a test. We met Bayesians who don't approve of these tests either, and I've heard it said that Bayesian testing is still a work in progress (Bernardo). Yet a related strategy is at the heart of some recommended statistical reforms.

---

[8] In the special case, where there's appreciable evidence for a special parameter, Senn argues that Jeffreys only required H_1's posterior probability to be greater than 0.5. One has, so to speak, used up the prior belief by using the spiked prior (Senn 2015a).

4.5 Who's Exaggerating? How to Evaluate Reforms Based on Bayes Factor Standards

Edwards, Lindman, and Savage (E, L, & S) – who were perhaps first to raise this criticism – say this:

Imagine all the density under the alternative hypothesis concentrated at x, the place most favored by the data. . . .
Even the utmost generosity to the alternative hypothesis cannot make the evidence in favor of it as strong as classical significance levels might suggest. (1963, p. 228)

The example is the Normal testing case of J. Berger and Sellke, but they compare it to a one-tailed test of H_0: $\mu = 0$ vs. H_1: $\mu = \mu_1 = \mu_{max}$ (entirely sensibly in my view). We abbreviate H_1 by H_{max}. Here the likelihood ratio $\text{Lik}(\mu_{max})/\text{Lik}(\mu_0) = \exp[z^2/2]$); the inverse is $\text{Lik}(\mu_0)/\text{Lik}(\mu_{max}) = \exp[-z^2/2]$. I think the former makes their case stronger, yet you will usually see the latter. (I record their values in a Note[9]). What is μ_{max}? It's the observed mean $\bar{x}$, the place most "favored by the data." In each case we consider $\bar{x}$ as the result that is just statistically significant at the indicated P-value, or its standardized z form.

With a P-value of 0.025, H_{max} is "only" 6.84 times as likely as the null. I put quotes around "only" not because I think 6.84 is big; I'm never clear what's to

Table 4.2 Upper Bounds on the Comparative Likelihood

P-value: one-sided	z_α	$\text{Lik}(\mu_{max})/\text{Lik}(\mu_0)$
0.05	1.65	3.87
0.025	1.96	6.84
0.01	2.33	15
0.005	2.58	28
0.0005	3.29	227

[9] The entries for the inverse are useful. This is adapted from Berger and Sellke (1987) Table 3.

P-value: one-sided	z_α	$\text{Lik}(\mu_0)/\text{Lik}(\mu_{max})$
0.05	1.65	0.258
0.025	1.96	0.146
0.01	2.33	0.067
0.005	2.58	0.036
0.0005	3.29	0.0044

count as big until I have information about the associated error probabilities. If you seek to ensure H_{max}: $\mu = \mu_{max}$ is 28 times as likely as is H_0: $\mu = \mu_0$, you need to use a P-value ~0.005, with z value of 2.58, call it 2.6. Compare the corresponding error probabilities. Were there 0 discrepancy from the null, a difference smaller than 1.96 would occur 97.5% of the time; one smaller than 2.6 would occur 99.5% of the time. In both cases, the 95% two-sided and 97.5% confidence intervals entirely exclude 0. The two one-sided lower intervals are $\mu > 0$ and $\mu > $ ~0.64. Both outcomes are good indications of $\mu > 0$: the difference between the likelihood ratios 6.8 and 28 doesn't register as very much when it comes to indicating a positive discrepancy. Surely E, L, & S couldn't expect Bayes factors to match error probabilities when they are the ones who showed how optional stopping can alter the latter and not the former (Section 1.5).

Valen Johnson (2013a,b) offers a way to bring the likelihood ratio more into line with what counts as strong evidence, according to a Bayes factor. He begins with a review of "Bayesian hypotheses tests." "The posterior odds between two hypotheses H_1 and H_0 can be expressed as"

$$\frac{\Pr(H_1|x)}{\Pr(H_0|x)} = BF_{10}(x) \times \frac{\Pr(H_1)}{\Pr(H_0)}.$$

Like classical statistical hypothesis tests, the tangible consequence of a Bayesian hypothesis test is often the rejection of one hypothesis, say H_0, in favor of the second, say H_1. In a Bayesian test, the null hypothesis is rejected if the posterior probability of H_1 exceeds a certain threshold. (Johnson 2013b, pp. 1720-1)

According to Johnson, Bayesians reject hypotheses based on a sufficiently high posterior and "the alternative hypothesis is accepted if $BF_{10} > k$" (ibid., p. 1726, k for his γ). A weaker stance might stop with the comparative report $Lik(\mu_{max})$/ $Lik(\mu_0)$. It's good that he supplies a falsification rule.

Johnson views his method as showing how to specify an alternative hypothesis – he calls it the "implicitly tested" alternative (ibid., p. 1739) – when H_0 is rejected. H_0 and H_1 are each given a 0.5 prior. Unlike N-P, the test does not exhaust the parameter space, it's just two points.

[D]efining a Bayes factor requires the specification of both a null hypothesis and an alternative hypothesis, and in many circumstances there is no objective mechanism for defining an alternative hypothesis. The definition of the alternative hypothesis therefore involves an element of subjectivity and it is for this reason that scientists generally eschew the Bayesian approach toward hypothesis testing. (Johnson 2013a, p. 19313)

He's right that comparative inference, as with Bayes factors, leaves open a wide latitude of appraisals by dint of the alternative chosen, and any associated priors.

Table 4.3 V. Johnson's implicit alternative analysis for T+: H_0: $\mu \leq 0$ vs. H_1: $\mu > 0$

P-value one-sided	z_α	Lik(μ_{max})/Lik(μ_0)	μ_{max}	$\Pr(H_0 \vert x)$	$\Pr(H_{max} \vert x)$
0.05	1.65	3.87	$1.65\sigma/\sqrt{n}$	0.2	0.8
0.025	1.96	6.84	$1.96\sigma/\sqrt{n}$	0.128	0.87
0.01	2.33	15	$2.33\sigma/\sqrt{n}$	0.06	0.94
0.005	2.58	28	$2.58\sigma/\sqrt{n}$	0.03	0.97
0.0005	3.29	227	$3.3\sigma/\sqrt{n}$	0.004	0.996
	$\sqrt{(2 \log k)}$	$exp\left(\frac{z_\alpha^2}{2}\right)$	$z_\alpha\, \sigma/\sqrt{n}$	$1/(1+k)$	$k/(1+k)$

In his attempt to rein in that choice, Johnson offers an illuminating way to relate the Bayes factor and the standard cut-offs for rejection, at least in UMP tests such as this. (He even calls it a uniformly most powerful Bayesian test!) We focus on the cases where we just reach statistical significance at various levels. Setting k as the Bayes factor you want, you can obtain the corresponding cut-off for rejection by computing $\sqrt{(2 \log k)}$: this matches the z_α corresponding to a N-P, UMP one-sided test. The UMP test T+ is of the form: Reject H_0 iff $\overline{X} \geq \overline{x}_\alpha$, where $\overline{x}_\alpha = \mu_0 + z_\alpha\, \sigma/\sqrt{n}$, which is $z_\alpha\sigma/\sqrt{n}$ for the case $\mu_0 = 0$. Thus he gets (2013b, p. 1730)

$$H_1: \mu_1 = \sigma\sqrt{\frac{2 \log k}{n}}.$$

Since this is the alternative under which the observed data, which we are taking to be $\overline{x}_\alpha$, have maximal probability, write it as H_{max} and μ_1 as μ_{max}. The computations are rather neat, see Note 10. (The last row of Table 4.3 gives an equivalent form.) The reason the LR in favor of the (maximal) alternative gets bigger and bigger is that $\Pr(x; H_0)$ is getting smaller and smaller with increasingly large x values.

Johnson's approach is intended to "provide a new form of default, non subjective Bayesian tests" (2013b, p. 1719), and he extends it to a number of other cases as well. Given it has the same rejection region as a UMP error statistical test, he suggests it "can be used to translate the results of classical significance tests into Bayes factors and posterior model probabilities" (ibid.). To bring them into line with the BF, however, you'll need a smaller α level. Johnson recommends levels more like 0.01 or 0.005. *Is there anything lost in translation?* There's no doubt that if you reach a smaller significance level in the same test, the discrepancy you are entitled to infer is larger. You've made the hurdle for rejection higher: any observed mean that makes it over must be larger. It also means that more will fail to make it over the hurdle: the Type II error probability increases. Using the 1.96 cut-off, a discrepancy of 2.46, call it 2.5, will be detected 70% of the time – add 0.5 SE to the cut-off – (the Type II error is 0.3) whereas using a 2.6 cut-off has less than 50% (0.46) chance of detecting a 2.5 discrepancy (Type II error of 0.54!). Which is a better cut-off for rejection? The severe tester eschews rigid cut-offs. In setting up a test, she looks at the worst cases she can live with; post-data she reports the discrepancies well or poorly warranted at the attained levels. (Recall, discrepancy always refers to parameter values.) Johnson proposes to make up for the loss of power by increasing the sample size, but it's not that simple. We know that as sample size increases, the discrepancy indicated by results that reach a given level of significance decreases. Still, you get a Bayes factor and a default posterior probability that you didn't have with ordinary significance tests. What's not to like?

We perform our two-part criticism, based on the minimal severity requirement. The procedure under the looking glass is: having obtained a statistically significant result, say at the 0.005 level, reject H_0 in favor of H_{max}: $\mu = \mu_{max}$. Giving priors of 0.5 to both H_0 and H_{max} you can report the posteriors. Clearly, (S-1) holds: H_{max} accords with $\bar{x}$ – it's equal to it. Our worry is with (S-2). H_0 is being rejected in favor of H_{max}, but should we infer it? The severity associated with inferring μ is as large as μ_{max} is

$$\Pr(Z < z_\alpha; \mu = \mu_{max}) = 0.5.$$

This is our benchmark for poor evidence. So (S-2) doesn't check out. You don't have to use severity, just ask: what confidence level would permit the inference $\mu \geq \mu_{max}$ (answer 0.5). Yet Johnson assigns $\Pr(H_{max}|x) = 0.97$. H_{max} is comparatively more likely than H_0 as $\bar{x}$ moves further from 0 – but that doesn't mean we'd want to infer there's evidence for H_{max}. If we add a column to Table 4.1 for SEV($\mu \geq \mu_{max}$) it would be 0.5 all the way down!

To have some numbers, in our example (H_0: $\mu \leq 0$ vs. H_1: $\mu > 0$), $\sigma = 1$, $n = 25$, and the 0.005 cut-off is $2.58\sigma/\sqrt{n} = 0.51$, round to 0.5. When a significance tester says the difference $\bar{x} = 0.5$ is statistically significantly greater than 0 at the 0.005 level, she isn't saying anything as strong as "there is fairly good evidence that $\mu = 0.5$." Here it gets a posterior of 0.97. While the goal of the reform was to tamp down on "overstated evidence," it appears to do just the opposite from a severe tester's perspective.

How can I say it's lousy if it's the maximally likely estimate? Because there is the variability of the estimator, and statistical inference must take this into account. It's true that the error statistician's inference isn't the point alternative these critics want us to consider (H_{max}), but they're the ones raising the criticism of ostensive relevance to us, and we're struggling in good faith to see what there might be in it. Surely to infer $\mu = 0.5$ is to infer $\mu > 0.4$. Our outcome of 0.5 is 0.5 standard error in excess of 0.4, resulting in SEV($\mu > 0.4$) = 0.7. Still rather poor. Equivalently, it is to form the 0.7 lower confidence limit ($\mu > 0.4$).

Johnson (2013a, p. 19314) calls the 0.5 spikes equipoise, but what happened to the parameter values in between H_0 and H_{max}? Do they get a prior of 0? To be clear, it may be desirable or at least innocuous for a significance tester to require smaller P-values. What is not desirable or innocuous is basing the altered specification on a BF appraisal, if in fact it is an error statistical justification you're after. Defenders of the argument may say, they're just showing the upper bound of evidence that can accrue, even if we imagine being as biased as possible against the null and for the alternative. But are they? A fair assessment, say Casella and R. Berger, wouldn't have the spike prior on the null – yes, it's still there. If you really want a match, why not use the frequentist matching priors for this case? (Prior 3 in Exhibit vii) The spiked prior still has a mismatch between BF and P-value.[10] This is the topic of megateam battles. (Benjamin et al. 2017 and Lakens et al. 2018).

Exhibit (viii): Whether P-values Exaggerate Depends on Philosophy. When a group of authors holding rather different perspectives get together to examine a position, the upshot can take them out of their usual comfort zones. We need more of that. (See also the survey in Hurlbert and Lombardi 2009, and Haig 2016.) Here's an exhibit from Greenland et al. (2016). They greet each member of a list of incorrect interpretations of P-values with "No!", but then make this exciting remark:

---

10 Computations

1. Suppose the outcome is just significant at the α level: $\bar{x} = \mu_0 + z_\alpha \sigma / \sqrt{n}$.
2. So the most likely alternative is H_{max}: $\mu_1 = \bar{x} = \mu_0 + z_\alpha \sigma / \sqrt{n}$.
3. The ratio of the maximum likely alternative H_{max} to the likelihood of H_0 is:

$$\frac{\text{Lik}(x|H_{max})}{\text{Lik}(x|H_0)} = \frac{1}{\exp[-z^2/2]} = \exp[z^2/2].$$

This gives the Bayes factor: BF$_{10}$. (BF$_{01}$ would be $\exp[-z^2/2]$.)

4. Set Lik(x| H_{max})/Lik(x|H_0) = k.
5. So $\exp[z^2/2] = k$.
   Since the natural log (ln) and exp are inverses:

   $\log k = \log(\exp[z^2/2]) = [z^2/2]$;
   $2 \log k = z^2$, so $\sqrt{(2 \log k)} = z$.

There are other interpretations of P values that are controversial, in that whether a categorical "No!" is warranted depends on one's philosophy of statistics and the precise meaning given to the terms involved. The disputed claims deserve recognition if one wishes to avoid such controversy. . . .

For example, it has been argued that P values overstate evidence against test hypotheses, based on directly comparing P values against certain quantities (likelihood ratios and Bayes factors) that play a central role as evidence measures in Bayesian analysis . . . Nonetheless, many other statisticians do not accept these quantities as gold standards, and instead point out that P values summarize crucial evidence needed to gauge the error rates of decisions based on statistical tests (even though they are far from sufficient for making those decisions). Thus, from this frequentist perspective, P values do not overstate evidence and may even be considered as measuring one aspect of evidence . . . with $1 - P$ measuring evidence against the model used to compute the P value. (p. 342)

It's erroneous to fault one statistical philosophy from the perspective of a philosophy with a different and incompatible conception of evidence or inference. The severity principle always evaluates a claim as against its denial within the framework set. In N-P tests, the frame is within a model, and the hypotheses exhaust the parameter space. Part of the problem may stem from supposing N-P tests infer a point alternative, and then seeking that point. Whether you agree with the error statistical form of inference, you can use the severity principle to get beyond this particular statistics battle.

Souvenir R: The Severity Interpretation of Rejection (SIR)

In Tour II you have visited the tribes who lament that P-values are sensitive to sample size (Section 4.3), and they exaggerate the evidence against a null hypothesis (Sections 4.4, 4.5). We've seen that significance tests take into account sample size in order to critique the discrepancies indicated objectively. A researcher may choose to decrease the P-value as n increases, but there's no problem in understanding that the same P-value reached with a larger sample size indicates fishing with a finer mesh. Surely we should not commit the fallacy exposed over 50 years ago.

Here's a summary of the severe tester's interpretation (of a rejection) putting it in terms that seem most clear:

> *SIR: The Severity Interpretation of a Rejection in test T+:* (small *P-value*)
>
> (i): [*Some* discrepancy is indicated]: $d(x_0)$ is a good indication of $\mu > \mu_1 = \mu_0 + \gamma$ if there is a high probability of observing a *less* statistically significant difference than $d(x_0)$ if $\mu = \mu_0 + \gamma$.

N-P and Fisher tests officially give the case with $\gamma = 0$. In that case, what does a small P-value mean? It means the test very probably $(1 - P)$ would have produced a result more in accord with H_0, were H_0 an adequate description of the data-generating process. So it indicates a discrepancy from H_0, especially if I can bring it about fairly reliably. To avoid making mountains out of molehills, it's good to give a second claim about the discrepancies that are *not* indicated:

(ii): [I'm not *that* impressed]: $d(x_0)$ is a poor indication of $\mu > \mu_1 = \mu_0 + \gamma$ if there is a high probability of an even more statistically significant difference than $d(x_0)$ even if $\mu = \mu_0 + \gamma$.

As for the exaggeration allegation, merely finding a single statistically significant difference, even if audited, is indeed weak: it's an indication of *some* discrepancy from a null, a first step in a task of identifying a genuine effect. But, a legitimate significance tester would never condone rejecting H_0 in favor of alternatives that correspond to a low severity or confidence level such as 0.5. Stephen Senn sums it up: "Certainly there is much more to statistical analysis than P-values but they should be left alone rather than being deformed . . . to become second class Bayesian posterior probabilities" (Senn 2015a). Reformers should not be deformers.

There is an urgency here. Not only do some reforms run afoul of the minimal severity requirement, to suppose things are fixed by lowering P-values ignores or downplays the main causes of non-replicability. According to Johnson:

[I]t is important to note that this high rate of nonreproducibility is not the result of scientific misconduct, publication bias, file drawer biases, or flawed statistical designs; it is simply the consequence of using evidence thresholds that do not represent sufficiently strong evidence in favor of hypothesized effects. (2013a, p. 19316)

This sanguine perspective sidesteps the worry about the key sources of spurious statistical inferences: biasing selection effects and violated assumptions, at all levels. (Fortunately, recent reforms admit this; Benjamin et al. 2017.) Catching such misdemeanors requires *auditing*, the topic of Tours III and IV of this Excursion.

Tour III Auditing: Biasing Selection Effects and Randomization

> This account of the rationale of induction is distinguished from others in that it has as its consequences two rules of inductive inference which are very frequently violated . . .The first of these is that the sample must be a random one . . .The other rule is that the character [to be studied] must not be determined by the character of the particular sample taken. (Peirce 1.95)

The biggest source of handwringing about statistical inference these days boils down to the fact it has become very easy to infer claims that have been subjected to insevere tests. High-powered computerized searches and data trolling permit sifting through reams of data, often collected by others, where in fact no responsible statistical assessments are warranted. "We're more fooled by noise than ever before, and it's because of a nasty phenomenon called 'big data'. With big data, researchers have brought cherry picking to an industrial level" (Taleb 2013). Selection effects alter a method's error probabilities and yet a fundamental battle in the statistics wars revolves around their relevance (Section 2.4). We begin with selection effects, and our first stop is to listen in on a court case taking place.

Dr. Paul Hack, CEO of Best Drug Co., is accused of issuing a report on the benefits of drug X that exploits a smattering of questionable research practices (QRPs): It ignores multiple testing, uses data-dependent hypotheses, and is oblivious to a variety of selection effects. What happens shines a bright spotlight on a mix of statistical philosophy and evidence. The case is fictional; any resemblance to an actual case is coincidental.

Round 1 The prosecution marshals their case, calling on a leader of an error statistical tribe who is also a scientist at Best: *Confronted with a lack of statistically significant results on any of 10 different prespecified endpoints (in randomized trials on Drug X), Dr. Hack proceeded to engage in a post-data dredging expedition until he unearthed a subgroup wherein a nominally statistically significant benefit [B] was found. That alone was the basis for a report to share-holders and doctors that Drug X shows impressive benefit on factor B.* Colleagues called up by the prosecution revealed further details: *Dr. Hack had ordered his chief data analyst to "shred and dice the data into tiny julienne slices until he got some positive results" sounding like the adman for*

Chop-o-Matic. The P-value computed from such a post-hoc search cannot be regarded as an actual P-value, yet Dr. Hack performed no adjustment of P-values, nor did he disclose the searching expedition had been conducted. Moreover, we learn, *the primary endpoint describing the basic hypothesis about the mechanism by which Drug X might offer benefits attained a non-significant P-value of 0.52. Despite knowing the FDA would almost certainly not approve drug X based on post-hoc searching, Dr. Hack optimistically reported on profitability for Best, thanks to the "positive" trials on drug X.*

Anyone who trades Biotech stocks knows that when a company reports: 'We failed to meet primary and perhaps secondary endpoints,' the stock is negatively affected, at times greatly. When one company decides to selectively report or be overly optimistic, it's unfair to patients and stockholders.

Round 2 Next to be heard from are defenders of Dr. Hack: *There's no need to adjust for post-hoc data dredging, the fact that significance tests require such adjustments is actually one of their big problems. What difference does it make if Dr. Hack intended to keep trying and trying again until he found something? Intentions are irrelevant to the import of data.* Others insist that: *the position on cherry picking is open to debate, depending on one's philosophy of evidence. For the courts to take sides would set an ominous precedent.* They cite upstanding statisticians who can attest to the irrelevance of such considerations.

Round 3 A second wave of Hack's defenders (which could be the same as in Round 2) pile on, with a list of reasons for P-phobia: *Significance levels exaggerate evidence, force us into dichotomous thinking, are sensitive to sample size, aren't measures of evidence because they aren't comparative reports, and violate the likelihood principle.* Even expert prosecutors, they claim, construe a P-value as the probability the results are due to chance, which is to commit the prosecutor's fallacy (misinterpreting P-values as posterior probabilities), so they are themselves confused.

Dr. Hack's lawyer jumps at the opening before him: *You see there is disagreement among scientists, at a basic philosophical level. To hold my client accountable would be no different than banning free and open discussion of rival interpretations of data amongst scientists.*

Dr. Hack may not get off the hook in the end – at least in fields where best practice manuals encode taking account of selection effects – but that could change if enough people adopt the stance of friends of Hack. In any event, it is not too much of a caricature of actual debates taking place. You, the citizen scientist, have the tough job of sifting through the cacophony. Severity principle in hand, you can at least decipher where the controversy is coming from. To limber up you might disinter the sources of claims of Round 3.

4.6   Error Control is Necessary for Severity Control

> To base the choice of the test of a statistical hypothesis upon an inspection of the observations is a dangerous practice; a study of the configuration of a sample is almost certain to reveal some feature, or features, which are exceptional if the [chance] hypothesis is true. (Pearson and Chandra Sekar 1936, p. 127)

> The likelihood principle implies...the irrelevance of predesignation, of whether an hypothesis was thought of beforehand or was introduced to explain the known effects. (Rosenkrantz 1977, p. 122)

Here we encounter the same source of tribal rivalry first spotted in Excursion I with optional stopping (Tour II). Yet we also allowed that data dependencies, double counting and non-novel results are not always problematic. The advantage of the current philosophy of statistics is that it makes it clear that the problem – *when it is a problem* – is that these gambits alter how well or severely probed claims are. We defined problematic cases as those where data or hypotheses are selected or generated, or a test criterion is specified, in such a way that the minimal severity requirement is violated, altered (without the alteration being mentioned), or unable to be assessed (Section 2.4).

Because they alter the severity, they must be taken account of in auditing a result, which includes checking for (i) selection effects, (ii) violations of model assumptions, and (iii) obstacles to any move from statistical to substantive causal or other theoretical claims.

There is no point in raising thresholds for significance if your methodology does not pick up on biasing selection effects. Yet, surprisingly, that is the case for many of the methods advocated by critics of significance tests, and related error statistical methods.

> Two problems that plague frequentist inference: multiple comparisons and multiple looks, or, as they are more commonly called, *data dredging* and peeking at the data. The frequentist solution to both problems involves adjusting the P value ... But adjusting the measure of evidence because of considerations that have nothing to do with the data defies scientific sense, belies the claim of 'objectivity' that is often made for the P value. (Goodman 1999, p. 1010)

This is epidemiologist Steven Goodman.[1] To his credit, he recognizes the philosophical origins of his position.

Older arguments from Edwards, Lindman, and Savage (1963) (E, L, & S) live on in contemporary forms. The Bayesian psychologist Eric-Jan Wagenmakers tells us:

---

[1] Currently a co-director of the Meta-Research Innovation Center at Stanford (METRICS).

[I]f the sampling plan is ignored, the researcher is able to always reject the null hypothesis, even if it is true. This example is sometimes used to argue that any statistical framework should somehow take the sampling plan into account. Some people feel that 'optional stopping' amounts to cheating . . . This feeling is, however, contradicted by a mathematical analysis. (2007, p. 785)

Being contradicted by mathematics is a heavy burden to overcome. Look closely and you'll see we are referred to E, L, & S. But the "proof" assumes the Likelihood Principle (LP) by which error probabilities drop out (Section 1.5). Error probabilities and severity are altered, but if your account has no antennae to pick up on them, then, to you, there's no effect.

Holders of the LP point fingers at error statisticians for worrying about "the sample space" and "intentions." To leaders of movements keen to rein in researcher flexibility, by contrast, a freewheeling attitude toward data-dependent hypotheses and stopping rules is pegged as a major source of spurious significance levels. Simmons, Nelson, and Simonsohn (2011) list as their first requirement for authors: "Authors must decide the rule for terminating data collection before data collection begins and report this rule in the article" (p. 1362). I'd relax it a little, requiring they report how their stopping plan alters the relevant error probability. So let me raise an "either or" question: Either your methodology picks up on influences on error probing capacities of methods or it does not. If it does, then you are in sync with the minimal severity requirement. We may compare our different ways of satisfying it. If it does not, then we've hit a crucial nerve. If you care, but your method fails to reflect that concern, then a supplement is in order. Opposition in methodology of statistics is fighting over trifles if it papers over this crucial point. If there is to be a meaningful "reconciliation," it will have to be here.

I'm raising this in a deliberate stark, provocative fashion to call attention to an elephant in the room (or on our ship). The widespread concern of a "crisis of replication" has nearly everyone rooting for predesignation of aspects of the data analysis, but if they're not also rooting for error control, what are they cheering for? Let's allow there are other grounds to champion predesignation, so don't accuse me of being unfair . . . *yet*. The tester sticks her neck out: she requires it to have a direct effect on inferential measures.

Paradox of Replication

> We often hear it's too easy to obtain small p-values, yet replication attempts find it difficult to get small p-values with preregistered results. This shows the problem isn't p-values but failing to adjust them for cherry picking, multiple testing, post-data subgroups and other *biasing selection effects*. (Mayo 2016, p. 1)

Criticism assumes a standard. If you criticize hunting and snooping because of the lack of control of false positives (Type I errors), then the assumption is that those matter. Suppose someone claims it's too easy to satisfy standard significance thresholds, while chiming in with those bemoaning the lack of replication of statistically significant results.

CRITIC OF TESTS: It's too easy to get low significance levels.
YOU: Why is it so hard to get small P-values in replication research?

Wait for their answer.

CRITIC: Aside from expected variability in results, there was likely some P-hacking, cherry picking, and other QRPs in the initial studies.
YOU: So, I take it you want methods that pick up on these biasing effects, and you favor techniques to check or avoid them.
CRITIC: Actually I think we should move to methods where selection effects make no difference (Bayes factors, Bayesian posteriors).

Now what? One possibility is a belief in magical thinking, that ignoring biasing effects makes them disappear. That doesn't feel very generous. Or maybe the unwarranted effects really do disappear. Not to the tester. Consider the proposals from Tour II: Bayes ratios, with or without their priors. Imagine my low P-value emerged from the kind of data dredging that threatens the P-value's validity. I go all the way to getting a z-value of 2.5, apparently satisfying Johnson's approach. I erect the maximally likely alternative, it's around 20 times more likely than the null, and am entitled, on this system, to infer a posterior of 0.95 on H_{max}. Error statistical testers would complain that the probability of finding a spurious P-value this way is high; if they are right, as I think they are, then the probability of finding a spurious posterior of 0.95 *is just as high*. That is why Bayes factors are at odds with the severity requirement. I'm not saying in principle they couldn't be supplemented in order to control error probabilities – nor that if you tell Bayesians what you want, they can't arrange it (with priors, shrinkage, or what have you). I'm just saying that the data-dredged hypothesis that finds its way into a significance test can also find its way into a Bayes factor. There's one big difference. I have error statistical grounds to criticize the former. If I switch tribes to one where error probabilities are irrelevant, my grounds for criticism disappear.

The valid criticism of our imaginary Dr. Hack, in Round 1, is this: he purports to have found an effect that would be difficult to generate if there were no genuine discrepancy from the null, when in fact it is easy to generate it. It is frequently brought about in a world where the null hypothesis is true.

The American Statistical Association's statement on P-values (2016, p. 131) correctly warns, "[c]onducting multiple analyses of the data and reporting only those with certain p-values" leads to spurious statistical levels. Their validity is lost, and the alarm goes off when we audit. When Hack's defenders maintain that scientists should not be convicted for engaging in the all-important task of exploration, you can readily agree. But you can still insist the results of explorations be independently tested or separately defended. If you're not controlling error probabilities, however, there's no alarm bell. This leads to the next group, Round 2, declaring that it makes no sense to adjust measures of evidence "because of considerations that have nothing to do with the data," thereby denying the initial charge against poor Dr. Hack, or should I say, lucky Dr. Hack, because the whole matter has now come down to something "quasi-philosophical," a murky business at best. Round 3 piles on with 'P-values are invariably misinterpreted', and no one really likes them much anyway. The P-value is the most unpopular girl in the class and I wouldn't even take you to visit P-value tribes – or "cults" as some call them[2] – I prefer speaking of observed significance levels, if it weren't that they suddenly have occupied so much importance in the statistics wars.

You might say that even if some people deny that selection effects actually alter the "evidence," the question of whether they can be ignored in interpreting data in legal or policy settings is not open for debate. After all, statutes in law and medicine require taking them into account. For example, the *Reference Manual on Scientific Evidence* for lawyers makes it clear that finding post-data subgroups that show impressive effects – when primary and secondary endpoints are insignificant – calls for adjustments. In a chapter by David Kaye and David Freedman, they emphasize the importance of asking:

> *How many tests have been performed?* Repeated testing complicates the interpretation of significance levels. If enough comparisons are made, random error almost guarantees that some will yield 'significant' findings, even when there is no real effect . . .
>
> If a fair coin is tossed a few thousand times, it is likely that at least one string of ten consecutive heads will appear. . . . Ten heads in the first ten tosses means one thing [evidence the coin is biased]; a run of ten heads somewhere along the way to a few thousand tosses of a coin means quite another. (Kaye and Freedman 2011, pp. 127–8)

Nevertheless, statutes can be changed if their rationale is overturned. The issue is not settled. There are regularly cases where defenders emphasize the lack of consensus and argue precisely as defenders in Round 2.[3]

---

[2] For example, Ziliak and McCloskey's *The Cult of Significance* (2008a).
[3] A case that went to the Supreme Court of the United States (SCOTUS): "There is no consensus whether, when, or how to adjust p-values or Type I error rates for multiple testing . . . the issue is not settled among scientists." Rothman, Lash, and Schachtman 2013, pp. 21–2.

Error Control is Only for Long Runs. But wait a minute. I am overlooking the reasons given for ignoring error control, some even in direct reply to Mayo (2016):

Bayesian analysis does not base decisions on error control. Indeed, Bayesian analysis does not use sampling distributions ... As Bayesian analysis ignores counterfactual error rates, it cannot control them. (Kruschke and Liddell 2017, pp. 13, 15)

This recognition may be a constructive step, admitting lack of error control. But on their view, caring about error control could only matter if you were in the business of performance in the long run. Is this true? By the way, the error control is actual; counterfactuals are used in determining what they are. Kruschke and Liddell continue:

[I]f the goal is specifically to control the rate of false alarms when the decision rule is applied repeatedly to imaginary data from the null hypothesis, then, by definition, the analyst must compute a p value and corresponding CIs. ... the analyst must take into account the exact stopping and testing intentions. ... any such procedure does not yield the credibility of parameter values ... but instead yields the probability of imaginary data .... (ibid., p. 15)

What's this about imaginary data? Suppose a researcher reports data x claiming to show impressive improvement in an allergic reaction among patients given a drug. They don't report that they dredged different measures of improvement or searched through post-hoc subgroups among patients. The actual data x only showed positive results on allergic reaction, let's say. Why report imaginary data that could have resulted but didn't? Looking at imaginary data sounds really damning, but when we see what they mean, we're inclined to regard it a scandal to *hide* such information. The problem is not about long runs – lose the alleged time aspect. I can simulate the sampling distribution to see the relative frequencies any day of the week. I can show you if a method had very little capability of reporting bupkis when it should have reported bupkis (nothing). It didn't do its job today. If Kruschke and Liddell's measure of credibility ignores what's regarded as QRPs in common statutes of evidence, you may opt for error control. True, error control is a necessary, not a sufficient, condition for a severity assessment. It follows from the necessity that an account with poor error control can neither falsify nor corroborate with severity.

Direct Protection Requires Skin off Your Nose. Testers are prepared to admit a difference in goals. Rival tribes may say lack of error control is no skin off their noses, since what they want are posterior probabilities and Bayes factors. Severe testers, on the other hand, appeal to statistics for protection against being misled by a cluster of biases and mistakes. A method *directly protects* against a QRP only insofar as its commission *is* skin off its nose. It must show up in the statistical analysis. More than that, there must be a general rationale for the concern. That a Bayesian probabilist avoids selection effects that hurt severity doesn't automatically mean they do so directly because of the severity violation. Moreover, it shouldn't be an option whether to take account of them or not, they must be. The tester holds this even granting there are cases where auditing shows no damage to severity.

Capitalizing on Chance

Gather round to listen to Hanan Selvin, a sociologist writing in 1957, reprinted in Morrison and Henkel (1970):

[W]hen the hypotheses are tested on the same data that suggested them and when tests of significance are based on such data, then a spurious impression of validity may result. The computed level of significance may have almost no relation to the true level . . . Suppose that twenty sets of differences have been examined, that one difference seems large enough to test and that this difference turns out to be 'significant at the 5 percent level.' Does this mean that differences as large as the one tested would occur by chance only 5 percent of the time when the true difference is zero? The answer is *no*, because the difference tested has been *selected* from the twenty differences that were examined. The actual level of significance is not 5 percent, but 64 percent! (Selvin 1970, p. 104)

Selvin would give Dr. Hack a hard time: to ignore a variety of selection effects results in a fallacious computation of the *actual* significance level associated with a given inference. Each of the 20 hypotheses is on different but closely related effects.

Suppose the single property found to have an impressive departure is hypothesis 13 of the 20. The possible results now are the possible factors that might be found to show a 2 standard deviation departure (for a two-sided 0.05 test) from the null. Thus the Type I error probability is the probability of finding at least one such significant difference out of 20, even though all 20 nulls are true. The probability that this procedure yields erroneous rejections differs from, and will be much greater than, 0.05. There are different and many more ways that one can err in this procedure than in testing a single prespecified null hypothesis, and this influences the actual P-value. The *nominal* (or computed) level for H_{13} would ignore the selection and report the P-value associated with a 2 standard deviation difference, 0.05.

Assuming 20 independent samples are drawn from populations having true discrepancies of zero, the probability of attaining at least one nominally statistically significant outcome (in either direction) with N independent tests at the α significance level is $1 - (1 - \alpha)^N$. Test results are treated like Bernoulli trials with probability of "success" (a 0.05 rejection) equal to 0.05 on each trial. The probability of getting no successes in 20 independent trials is $(0.95)^{20}$.

Pr(Test rejects at least one H_i at level 0.05; all H_i true) = $1 - (1 - 0.05)^{20}$
= 0.64.

This would give the *actual* significance level, if we could assume independence; in practice this wouldn't hold, so it would be a conservative value. This is the experiment-wide significance level or *family-wise error rate* (FWER). In terms of P-values, you'd need to compute the probability that the smallest of 20 is as small as yours. The *Bonferroni correction* can ensure this is at most α^* by setting the α-level for each test at α^*/N. Alternatively the correction may be attained by multiplying the P-value by N. You could hunt through N hypotheses and be an "honest hunter," and report the adjusted P-value. Some find the Bonferroni adjustment too strict, not to mention its assumption of independence is unlikely to hold. There's a large literature, and myriad ways to adjust, appropriate for different problem situations. Juliet Shaffer has done considerable work on this (Shaffer 1995).

The need for an adjustment recognizes how various tactics lead reported error probabilities to mischaracterize how readily the test would alert us to blatant deceptions. Skeptical of the inferred inference, we ask: how frequently would this method have alerted me to erroneous claims of this form? If it would almost never alert me to such mistakes, I deny it is good evidence for this particular claim. This is to use error statistical probabilities to assess severity. Even where we would not place much stock on the precise corrected P-value, it's important to have an alert that the unaudited P-value is invalid or questionable.

This illustrates how error statistical methods directly protect against such biasing selection effects: revealing how we could be fooled, as well as self-correct. For the severe tester, outputting H_{13}, ignoring the non-significant others, renders H_{13} poorly tested. You might say, but look there's the hypothesis H_{13}, and data x – shouldn't it speak for itself? No. That's just the evidential-relationship or logicist in you coming out. As with all problematic cases, it's not that the method of hypothesis testing has been refuted or even found flawed. It's the opposite. It's doing its work. Because the method's requirements are distorted so that their logic breaks down, it outputs "spurious," just as it should. The types of cases calling for adjustment tend to be cases where there is a related group of hypotheses – such as effects of a drug. The concern is treating the hypotheses disinterred in explorations in just the *same way* as if they were predesignated.

Adjustment for Selection Goes against Scientific Norms. Epidemiologist Kenneth Rothman denies an adjustment for selection is appropriate because it depends on the assumption that the variable of interest (e.g., treatment with drug X) is unrelated to all 20 (or however many) of the effect variables searched. He calls it the "universal null," which he views as untenable. Why? "[N]o empiricist could comfortably presume that randomness underlies the variability of all observations. Scientists presume instead that the universe is governed by natural laws"(1990, p. 45).

This is interesting; let's examine it. First, the universal null is an i-assumption (argumentative) only. Second, the universal null does not say that observed outcomes are not governed by laws. Suppose the one nominally significant factor is improved mortality (while 19 others are non-significant). Each death has a reason or cause. It's because death can come about for so many different reasons that there's variability, and we try to root out those due to (or at least systematically associated with) the treatment variable. What's alleged to be due to chance is that the group assigned drug X happens to do nominally better on one factor. Reflect on some of the cases we've visited in our two first excursions: the Texas Sharpshooter, the Pickrite Stock method, Dr. Playfair, Lady Tasting Tea, and the angst of Diederik Stapel's review committee. They're all a bit different but share canonical features.

The Texas Sharpshooter vividly shows how selection effects can change the process responsible for your observations. A silly version described by Goldacre is this: "Imagine I am standing near a large wooden barn with an enormous machine gun. I place a blindfold over my eyes and laughing maniacally I fire off many thousands and thousands of bullets into the side of the barn." Circling a cluster of closely placed bullet holes, he declares it evidence of his marksmanship (2008, p. 258). The skill that he's allegedly testing and making inferences about is his ability to shoot when the target is given and fixed, but that's not the skill actually responsible for *the resulting high score*. That would be so even without the blindfold. It's the high score that's due to chance. Analogously, in searching the data, if you draw a line around those treated who happen to show the beneficial effect, you're influencing what's producing your overall score. Given the variability of the outcomes, it's fairly probable that at least one of 20 factors shows a nominally significant association – say decreased mortality – even if drug X does not systematically improve *any* of the 20 outcomes searched in the populations of interest. You may question the plausibility of this universal null, but that doesn't block its argumentative role. It serves as a canonical representation of deception, or a blatant error scenario for which statistical models are so apt. For a severe tester, the mere fact that your method doesn't distinguish a case where the high score is due to the efficacy of the drug and one where it's due to post hoc hunting, suffices to discredit the resulting inference. You haven't sincerely tried to avoid deception. Other data might later turn up to warrant the claimed benefit; this does not stop us from needing to say something about this one data analysis: It's BENT.

False Discovery Rates. The Bonferroni, or a number of related adjustments, is relevant when the nominally significant difference is the basis for a specific inference. What if you're just trying to get some factors for subsequent severe scrutiny, and the concern is with overlooking genuine effects (Type II errors)? For example, the analysis of microarray data involves analyzing thousands of genes to see which ones are "on" or "expressed" in healthy versus diseased tissue. These are called genome-wide association studies (GWAS). A GWAS might test tens of thousands of null hypotheses that disease status and a given genomic expression are statistically independent. The required P-value using the conservative Bonferroni adjustment would be so tiny that you'd miss out on genes worth following up. Background knowledge might suggest a small proportion of the null hypotheses are false, and a culling procedure is needed. A novel approach by Yosef Benjamini and Yoav Hochberg (1995) uses what's called the *false-discovery rate* (FDR): the expected proportion of the N hypotheses tested that are falsely rejected.[4]

Let R be the number of rejected null hypotheses and V the number of erroneously rejected hypotheses. Then the proportion of falsely rejected null hypotheses is V/R. R is observable, V is not, but we can compute the expected value of V/R:

FDR: the expected proportion of the hypotheses tested that are falsely rejected, $E(V/R)$.

Benjamini and Hochberg cite a study from the literature where a new treatment is compared to an existing one in a randomized trial of patients with heart disease. (I'm omitting details.) Although 15 different tests are run, the study does not take account of multiple testing. To apply the FDR method, you rank the 15 P-values from lowest to highest (p. 295):

0.0001, 0.0004, 0.0019, 0.0095, 0.0201, 0.0278, 0.0298, 0.0344, 0.0459, 0.3240, 0.4262, 0.5719, 0.6528, 0.7590, 1.000.

---

[4] This should not be confused with use of this term as a posterior probability, e.g., David Colquhoun (2014). There's no prior used in Benjamini and Hochberg.

Each gets an index number i from 1 to 15: $p_1 = 0.0001$, $p_2 = 0.0004$, ..., $p_{15} =$ 1.000, etc. The Bonferroni correction is $0.05/15 = 0.0033$. So it only allows rejecting the hypotheses corresponding to those with the three lowest P-values. These happen to be reduced allergic reaction and two aspects of bleeding. The fourth corresponds to mortality, so one can't infer a reduction in mortality at the Bonferroni adjusted level.

Let's compare the Bonferroni to applying the FDR method. Choose Q, the desired FDR – before collecting the data – and compute for each p_i its Benjamini and Hochberg critical value: $Q_i/15$. They illustrate with the FDR value of 0.05. Starting with p_{15}, look for the first P-value to satisfy $p_i < 0.05i/15$. In this case it's p_4:

$$p_4 = 0.0095 \leq 4(0.05)/15 = 0.013.$$

Thus, the null hypothesis corresponding to p_4 – no improved mortality – may be rejected using the FDR. Generally, Q is chosen to be much higher than 0.05, say 0.25. In that case you'd look for the first P-value, starting at p_{15}, to satisfy $p_i < 0.25i/15$. Here it's p_9: $p_9 = 0.0459 \leq 9(0.25)/15 = 0.15$. So all null hypotheses corresponding to ranks 1–9 would be rejected using the FDR with $Q = 0.25$.

In screening genes, the immediate task is largely performance: controlling the noise in the network, balancing false positives with low power to detect genes worth following up, rather than inferring, for any specific gene, that there is evidence of its association to a disease.[5] Here the FDR isn't serving as an inferential measure. Another context Benjamini and Hochberg give, where the FDR seems more apt than the FWER, is where an overall decision to recommend a new treatment is correct, so long as it gives improvement on any one of N features:

> We wish therefore to make as many discoveries as possible (which will enhance a decision in favour of the new treatment), subject to control of the FDR. ... a small proportion of errors will not change the overall validity of the conclusion. (ibid., p. 292)

Here the "mistake" would be failing to recommend the new treatment even though it's better on any of these N factors (and presumably no worse). Here the FDR gives the relevant assessment for severity, given what counts as a mistaken inference.

In general, it makes most sense to report the nominal P-value associated with each hypothesis and then indicate if it's rejectable according to the chosen

---

[5] The FDR reflects "the concern to assure reproducible results in the face of selection effects", in contrast to making personal decisions as to what leads to follow (Benjamini 2008, p. 23).

adjustment method and level. Despite the importance I place on avoiding or correcting for selection effects, I do not think much weight should be attached to any particular selected P-value adjustment. It may suffice to register the spuriousness of a nominal P-value. Even searching can yield enough significant results so that the error probability is low. Go back to Selvin who is alluding to FWER (under the assumption of all true nulls):

We have seen that the probability of *at least one* difference 'significant' at the 5 percent level is 0.64. By similar calculation it can be shown that the probability of at least *two* 'significant' differences is 0.26 ... at least three is 0.07, ... at least four is 0.01. (Selvin 1970, pp. 104–5)

The set of differences combined could be seen to be significant at the 1 percent level. Whether this translates into the relevant assessment for severity depends on the context. One may be able to argue, in considering a group of related hypotheses, that if enough non-nulls are found nominally significant, the overall error probability can be low. There is still the problem of actually computing the overall P-values, taking into account correlated biases and dependencies of the factors hunted.[6]

The current state of play in dealing with multiple testing is fascinating and growing daily, but would take us too far afield to discuss. Most embody the performance oriented goal of error control. Westfall and Young (1993), in one of the earliest treatments on P-value adjustments, develop resampling methods to adjust for both multiple testing and multiple modeling (e.g., selecting variables to be included in a regression model). Others devise model-building techniques that control error probabilities as they build (David Hendry 2011). In some cases, it's deemed more appropriate to combine tests through meta-analysis, rather than assess individual inferences, especially where the tests can be considered to be testing the same or appropriately related hypotheses.

There is always room for an appeal procedure. Appealing to everything we know (big picture inference), even the hunter for significance may succeed in arguing that the mistaken interpretation has been avoided *in the case at hand*. Just because the formal statistics has run its course, there are other informal arguments we can turn to. The severe tester builds formal and informal repertoires to classify how the capacities of tests are affected by one or another selection effect. Leading the way is what counts as erroneously solving the

---

[6] Efron (2013, p. 142) describes a study on 102 men, 52 with prostate cancer and 50 controls, in which over 6000 genes are studied. This is akin to 6000 significance tests. His alternative formulation is empirical Bayesian, where the relative frequency of genes of a certain type is known.

problem at hand. We then consider if the capacity of the test to avoid such mistakes is compromised. The fact that our analysis depends on context should not make us feel that we do not know what we are talking about. It's the onus of the researcher to demonstrate the validity of her inference. Inability to compute error probabilities, even approximately, entails low severity in our system.

Exhibit (ix): Infant Training. In the late 1940s, William Sewell sought to investigate a variety of infant training experiences regarding nursing, weaning, and toilet training that, according to widely accepted Freudian psychological theories of the day, were thought advantageous for a child's personality adjustment. Leslie Kish's (1959) discussion is another classic we find in Morrison and Henkel (1970, pp. 127–41).

The researchers in the infant training study conducted 460 statistical significance tests! Out of these 18 were found statistically significant at the 0.05 level or beyond, 11 in the direction expected by the popular psychological account. Sewell denies that we should be just as impressed with the 11 statistically significant results as we would be if they were the only 11 hypotheses to be tested. As Kish points out: "By chance alone one would expect 23 'significant differences' at the 5 percent level. A 'hunter' would report either the 11 or the 18 and not the hundreds of 'misses'" (ibid., p. 138, note 12). A hunter is one who denies the need to take account of the searching.

What's intriguing about this study is that the hunting expedition led to negative results, adjusted to take account of the searching. Here's Sewell (1952, p. 158): "On the basis of the results of this study, the general null hypothesis that the personality adjustments and traits of children who have undergone varying training experiences do not differ significantly cannot be rejected." There were six main hypotheses. For example:

- *"The personality adjustments of the children who were breast fed do not differ* [statistically] *significantly from those of the children who were bottle fed* cannot be rejected." None of the 46 tests were significant (ibid., p. 156).
- *The personality adjustments and traits of the children who were weaned gradually do not differ* [statistically] *significantly from those of the children weaned abruptly* cannot be rejected on the basis of the statistical evidence." Only two of 46 tests were significant (ibid.).

And so it went for children with late versus early induction to bowel training, for those not punished versus those punished for toilet training accidents and others besides (ibid.).

While recognizing the need for a more controlled study in order to falsify the claims, Sewell concludes:

[Psychologists and counselors] have strongly advocated systems of infant care which they believe follow logically from the Freudian position. . . . breast feeding, a prolonged period of nursing, gradual weaning, a self-demand schedule, easy and late bowel and bladder training, . . . freedom from punishment . . . They have assumed that these practices will promote the growth of secure and unneurotic personalities. (ibid., p. 151)

Certainly, the results of this study cast serious doubts on the validity of the psychoanalytic claims regarding the importance of the infant [training] . . . (ibid., pp. 158–9).

In addition to their astuteness in taking account of searching, notice they're not reticent in reporting negative results. Since few if any were statistically significant, Fisher's requirement for demonstrating an effect fails. This casts serious doubt on what was a widely accepted basis for recommending Freudian-type infant training.[7] The presumption of a genuine effect is statistically falsified, or very close to it.

When Searching Doesn't Damage Severity: Explaining a Known Effect

What is sometimes overlooked is that the problem is not that a method is guaranteed to output some claim or other that fit the data, the problem is doing so unreliably. Some criticisms of adjusting for selection are based on examples where severity is *improved* by searching. We should not be tempted to run together examples of pejorative data-dependent hunting with cases that are superficially similar, but unproblematic. For example, searching for a DNA match with a criminal's DNA is somewhat akin to finding a statistically significant departure from a null hypothesis: "one searches through data and concentrates on the one case where a 'match' with the criminal's DNA is found, ignoring the non-matches." (Mayo and Cox 2006; p. 94) Isn't an error statistician forced to adjust for hunting here as well? No.

In illicit hunting and cherry picking, the concern is that of inferring a genuine effect, when none exists; whereas "here there is a known effect or specific event, the criminal's DNA, and reliable procedures are used to track down the specific cause or source" – or so we assume with background knowledge of a low "erroneous match" rate. "The probability is high that we would not obtain a match with person i, if i were not the criminal"; so, by the severity criterion (or FEV), finding the match is good evidence that i is the

---

[7] Some recommend that researchers simply state "with maximum clarity and precision, which hypotheses were developed entirely independently of the data and those which were not, so readers will know how to interpret the results" (Bailar 1991). I concur with Westfall and Young (1993, p. 20) that it is doubtful that a reader will know how to interpret a report: the 11 favorable results have been selected from the 200 tests. Isn't that the purpose of a statistical analysis?

criminal. "Moreover, each non-match found, by the stipulations of the example, virtually excludes that person." Thus, the more such negative results, the stronger is the evidence against i when a match is finally found. Negative results fortify the inferred match. Since "at most one null hypothesis of innocence is false, evidence of innocence on one individual increases, even if only slightly, the chance of guilt of another" (ibid).

Philip Dawid (2000, p. 325) invites his readers to assess whether they are "intuitive Bayesians or intuitive frequentists" by "the extreme case that the data base contains records on everyone in the population." One could imagine this put in terms of 'did you hear about the frequentist who thought finding a non-match with everyone but Sam is poor evidence that Sam is guilty?' The criticism is a consequence of blurring pejorative and non-pejorative cases. It would be absurd to consider the stringency of the probe as diminishing as more non-matches are found. Thus we can remain intuitive frequentist testers. A cartoon shows a man finding his key after searching with the caption "always the last place you look." Searching for your lost key is like the DNA search for a criminal. (Note the echoes with the philosophical dispute about the relevance of novel predictions; Section 2.4.)

If an effect is known to be genuine, then a sought-for and found explanation needn't receive a low severity. Don't misunderstand: not just any explanation of a known effect passes severely, but one mistake – spurious effect – is already taken care of. Once the deflection effect was found to be genuine, it had to be a constraint on theorizing about its cause. In other cases, the trick is to hunt for a way to make the effect manifest in an experiment. When teratogenicity was found in babies whose mothers had been given thalidomide, it took them quite some time to find an animal in which the effect showed up: finally it was replicated in New Zealand rabbits! It is one thing if you are really going where the data *take you*, as opposed to subliminally taking the data where you want them to go. Severity makes the needed distinctions.

Renouncing Error Probabilities Leaves Trenchant Criticisms on the Table

Some of the harshest criticisms of frequentist error-statistical methods these days rest on principles that the critics themselves purport to reject. An example is for a Bayesian to criticize a reported P-value on the grounds that it failed to adjust for searching, while denying searching matters to evidence. If it is what I call a "for thee and not for me" argument, the critic is not being inconsistent. She accepts the "I don't care, but you do" horn. When a Bayesian, like I. J. Good, says a Fisherian but not a Bayesian can cheat with optional stopping, he means that error probabilities aren't the Bayesian's concern (Section 1.5). (Error probabilities without a subscript always refer to error probability$_1$ (Section 3.6) from frequentist methods.) It's perfectly fair to take the "we don't care" horn of my dilemma, and that's a great help in getting beyond the statistics wars. What's unfair is dismissing those who care as fetishizing imaginary data. Critics of error statistics should admit to consequences sufficiently concerning to be ensconced in statutes of best practices. At least if the statistics wars are to become less shrill and more honest.

What should we say about a distinct standpoint you will come across? First a critic berates a researcher for reporting an unadjusted P-value despite hunting and multiple testing. Next, because the critic's own methodology eschews the error statistical rationale on which those concerns rest, she is forced to switch to different grounds for complaining – generally by reporting disbelief in the effect that was hunted. Recall Royall blocking "the deck is made up of aces of diamonds," despite its being most likely (given the one ace of diamonds drawn), by switching to the belief category (Section 1.4). That might work in such trivial cases. In others, it weakens the intended criticism to the point of having it obliterated by those who deserve to be charged with severe testing crimes.

Exhibit (x): Relinquishing Their Strongest Criticism: Bem. There was an ESP study that got attention a few years back (Bem 2011). Anyone choosing to take up an effect that has been demonstrated only with questionable research practices or has been falsified must show they have avoided the well-known tricks. But Bem openly admits he went on a fishing expedition to find results that appear to show an impressive non-chance effect, which he credits to ESP (subjects did better than chance at predicting which erotic picture they'll be shown in the future). The great irony is that Wagenmakers et al. (2011), keen as they are to show "Psychologists Must Change the Way They Analyze Their Data" and trade significance tests for Bayes factors, relinquish their strongest grounds for criticism. While they mention Bem's P-hacking (fishing for a type of picture subjects get right most often), this isn't their basis for discrediting Bem's results. After all, Wagenmakers looks askance at adjusting for selection effects:

P values can only be computed once the sampling plan is fully known and specified in advance. In scientific practice, few people are keenly aware of their intentions, particularly with respect to what to do when the data turn out not to be significant after the first inspection. Still fewer people would adjust their p values on the basis of their intended sampling plan. (Wagenmakers 2007, p. 784)

Rather than insist they ought to adjust, Wagenmakers dismisses a concern with "hypothetical actions for imaginary data" (ibid.). To criticize Bem, Wagenmakers et al. (2011) resort to a default Bayesian prior that makes the null hypothesis comparatively more probable than a chosen alternative (along the lines of Excursion 4, Tour II). Not only does this forfeit their strongest criticism, they give Bem et al. (2011) a cudgel to thwack back at them:

> Whenever the null hypothesis is sharply defined but the prior distribution on the alternative hypothesis is diffused over a wide range of values, as it is in . . . Wagenmakers et al. (2011), it boosts the probability that *any* observed data will be higher under the null hypothesis than under the alternative. This is known as the Lindley-Jeffreys paradox: A frequentist analysis that yields strong evidence in support of the experimental hypothesis can be contradicted by a misguided Bayesian analysis that concludes that the same data are more likely under the null. (p. 717)

Instead of getting flogged, Bem is positioned to point to the flexibility of getting a Bayes factor in favor of the null hypothesis. Rather than showing psychologists should switch, the exchange is a strong argument for why they should stick to error statistical requirements.

P-Values Can't Be Trusted Except When Used to Argue That *P*-values Can't Be Trusted!

There is more than a whiff of inconsistency in proclaiming *P*-values cannot be trusted while in the same breath extolling the uses of statistical significance tests and *P*-values in mounting criticisms of significance tests and P-values. Isn't a critic who denies the entire error statistical methodology, significance test, N-P tests, and confidence intervals, also required to forfeit the results those methods give when they just happen to criticize a given of the tests? How much more so when those criticisms are the basis for charging someone with fraud. Yet that is not what we see.

Uri Simonsohn became a renowned fraud-buster by inventing statistical tests to rigorously make out his suspicions of the work of social psychologists Dirk Smeesters, Lawrence Sanna, and others – "based on statistics alone"– as one of his titles reads. He shows the researcher couldn't have gotten so little variability, or the results are too good to be true – along with a fastidious analysis of numerous papers to rule out, statistically, any benign explanations (Simonsohn 2013). Statistician Richard Gill, often asked for advice on such cases, notes: "The methodology here is not new. It goes back to Fisher (founder of modern statistics) in the 30's. . . The tests of goodness of fit were, again and again, too good" (2014). I expected that tribes who deny the evidential weight of significance tests would come to the defense of the accused, but (to my knowledge) none did.

Note, too, that the argument of the fraud-busters underscores the severity rationale for the case at hand. Critics called in to adjudicate high-profile cases of suspected fraud are not merely trying to ensure they will rarely erroneously pinpoint frauds in the long run. They are making proclamations on the specific case at hand – and in some cases, a person's job depends on it. They will use a cluster of examples to mount a strong argument from coincidence that the data in front of us could not have occurred without finagling. Other tools are used to survey a group of significance tests in a whole field, or by a given researcher. For instance, statistical properties of P-values are employed to ascertain if too many P-values at a given level are attained. These are called P-curves (Simonsohn et al. 2014). Such fraud detection machines at most give an indication about a field or group of studies. Of course, once known, they might themselves be gamed. But it's an intriguing new research field; and it is an interesting fact that when scientists need to warrant serious accusations of bad statistics, if not fraud, they turn to the error statistical reasoning and to statistical tests. If you got rid of them, they'd only have to be reinvented by those who insist on holding others accountable for their statistical inferences.

Exhibit (xi): How Data-dependent Selections Invalidate Error Probability Guarantees. It can be shown that a statistical method directly protects against data-dependent selections by demonstrating how they can cause a breakdown in methods. Philosopher of science Ronald Giere considers Neyman–Pearson interval estimation for a Binomial proportion. If assumptions are met, the sample mean will differ by 2 standard deviations from the true value of θ less than 5 percent of the time, approximately. Giere shows how to make the probability of successful estimates not 0.95 but 0! "This will be sufficient to prove the point [the inadmissibility of this method] because Neyman's theory asserts that the average ratio of success is independent of the constitutions of the populations examined" (Giere 1969, p. 375). Take a population of A's and to each set of n members assign a shared property. The full population has U members where $U > 2n$. Then arbitrarily assign this same property to $U/2 - n$ additional members.

Given a sufficient store of logically independent properties, this can be done for all possible combinations of n A's. The result is a population so constructed that while every possible n-membered sample contains at least one apparent regularity [a property shared by all members of the sample] every independent property has an actual ratio of exactly one-half in the total population. (ibid., p. 376)[8]

The bottom line is, showing how you can distort error probabilities through the efforts of finagling shows the *value* of these methods. It's hard to see how accounts that claim error probabilities are irrelevant can supply such direct protection, although they may *indirectly* block the same fallacies. This remains to be shown.

Souvenir S: Preregistration and Error Probabilities

"One of the best-publicized approaches to boosting reproducibility is preregistration … to prevent cherry picking statistically significant results" (Baker 2016, p. 454). It shouldn't be described as too onerous to carry out. Selection effects alter the outcomes in the sample space, showing up in altered error probabilities. If the sample space (and so error probabilities) is deemed irrelevant post-data, the direct rationale for preregistration goes missing. Worse, in the interest of promoting a methodology that downplays error probabilities, researchers who most deserve lambasting are thrown a handy line of defense. Granted it is often presupposed that error probabilities are relevant only for long-run performance goals. I've been disabusing you of that notion. Perhaps some of the "never error probabilities" tribe will shift their stance now: 'But Mayo, using error probabilities for severity, differs from the official line, which is all about performance.' One didn't feel too guilty denying a concern with error probabilities before. If viewing statistical inference as severe tests yields such a concession, I will consider my project a success. Actually, my immediate goal is less ambitious: to show that looking through the severity tunnel lets you unearth the crux of major statistical battles. In the meantime, no fair critic of error statistics should proclaim error control is all about hidden intentions that a researcher can't be held responsible for. They should be.

4.7   Randomization

> The purpose of randomisation … is to guarantee the validity of the test of significance, this test being based on an estimate of error made possible by replication. (Fisher [1935b]1951, p. 26)

> The problem of analysing the idea of randomization is more acute, and at present more baffling, for subjectivists than for objectivists, more baffling because an ideal subjectivist would not need randomization at all. He would

---

[8] For a miniature example, if $U = 6$ (there are 6 A's in the population) and $n = 2$, there are 15 possible pairs. Each pair is given a property and so is one additional member.

simply choose the specific layout that promised to tell him the most. (Savage 1962, p. 34)

Randomization is a puzzle for Bayesians. The intuitive need for randomization is clear, but there is a standard result that Bayesians need not randomize. (Berry and Kadane 1997, p. 813)

Many Bayesians (though there are some very prominent exceptions) regard it as irrelevant and most frequentists (again there are some exceptions) consider it important. (Senn 2007, p. 34)

There's a nagging voice rarely heard from in today's statistical debates: if an account has no niche for error statistical reasoning, what happens to design principles whose primary purpose is to afford it? Randomization clearly exploits counterfactual considerations of outcomes that could have occurred, so dear to the hearts of error statisticians.

Some of the greatest contributions of statistics to science involve adding additional randomness and leveraging that randomness. Examples are randomized experiments, permutation tests, cross-validation and data-splitting. These are unabashedly frequentist ideas and, while one can strain to fit them into a Bayesian framework, they don't really have a place in Bayesian inference. (Wasserman 2008, p. 465)

One answer is to recognize that, apart from underwriting significance tests and the estimation of the standard error, randomization also has a role in preventing types of biased selections, especially where the context requires convincing others. Although these and other justifications are interesting and important, their defenders tend to regard them as subsidiary and largely optional uses for randomization.

Randomization country is huge; one scarcely does it justice in a single afternoon's tour. Moreover, a deeper look would require I call in a more expert field guide. A glimpse will shed light on core differences that interest us. Let's focus on the random allocation of a treatment or intervention, in particular in comparative treatment-control studies.

The problem with attributing Mary's lack of dementia (by age 80) to her having been taking HRT since menopause is that we don't know what her condition would have been like if she had not been so treated. Moreover, she's just one case and we're interested in treatment effects that are statistical. A factor is sometimes said to statistically contribute to some response where the response on average in the experimental population of treateds would be higher (or lower) than it would have been had they not been treated – in effect comparing two counterfactual populations. Randomized control experiments let us peer into these counterfactual populations by finding out about the difference between the average response in the treated group μ_T and the average response among a control group μ_C. With randomized control trials (RCTs), there is a deliberate introduction of a probabilistic assignment of a treatment of interest, using known chance mechanisms, such as a random number generator. Letting $\Delta = \mu_T - \mu_C$, one may consider what Cox calls a strong ("no effect") null: that the average response is no different or no greater among the treated than among the control group $H_0: \Delta = 0$ (vs. $H_1: \Delta \neq 0$), or a one-sided null: $H_0: \Delta \leq 0$ vs. $H_1: \Delta > 0$. We observe $\bar{x}_T - \bar{x}_C$, where $\bar{x}_T$ and $\bar{x}_C$ are the observed sample means in the treated and control groups, forming the standard test statistic $d^* = (d - \Delta)/\text{SE}$, where SE is the standard error ($d = \bar{x}_T - \bar{x}_C$). Thanks to randomized assignment, we can estimate the standard error and the sampling distribution of d^*.

Under the (strong) null hypothesis, the two groups, treated and control, may be regarded as coming from the same population with respect to mean effect, such as age-related dementia. Think about the RCT reasoning this way: if the HRT treatment makes no difference, people in the treated group would have had (or not had) dementia even if they'd been assigned to the control group. Some will get it, others won't (of course we can also consider degrees). Under the null hypothesis, any observed difference would be due to the accidental assignment to group T or C. So, if $\bar{x}_T - \bar{x}_C$ exceeds 0, it's just because more of the people who would have gotten dementia anyway happen to end up being assigned to the treated rather than the control group. Thanks to the random assignment, we can determine the probability of this occurring (under H_0). This is a particularly vivid illustration of a difference "due to chance" – where the chance is the way subjects were assigned to treatment. A statistical connection between the theoretical parameter Δ is created by dint of the design and execution of the experiment.[9]

Bayesians may find a home for randomized assignment (and possibly, double blindness) in the course of demonstrating "utmost good faith" (Senn 2007, p. 35). In the face of suspicious second parties: "Simple randomization is a method which by virtue of its very unpredictability affords the greatest degree of blinding. Some form of randomization is indispensable for any trial in which the issue of blinding is taken seriously..." (ibid., p. 70). Nevertheless, subjective Bayesians have generally concurred with Lindley and Novick that

---

[9] Stephen Stigler, in his clever *The Seven Pillars of Statistical Wisdom*, discusses some of the experiments performed by Peirce, who first defined randomization. In one, the goal was to test whether there's a threshold below which you can't discern the difference in weights between two objects. Psychologists had hypothesized that there was a minimal threshold "such that if the difference was below the threshold, termed the *just noticeable difference* (jnd), the two stimuli were indistinguishable ... [Peirce and Jastrow] showed this speculation was false" (Stigler 2016, p. 160). No matter how close in weight the objects were, the probability of a correct discernment of difference differed from ½. Another example of evidence for a "no-effect" null.

"[O]ne can do no better than . . . use an allocation which You think is unlikely to have important confounding effects" (Lindley and Novick 1981, p. 52).[10] Still, Berry and Kadane (1997) maintain that despite the "standard result that Bayesians need not randomize" (p. 813) there are scenarios where, because different actors have different subjective goals and beliefs, randomization is the optimal allocation. Say there's two treatments, 1 and 2, where each either shows the response of interest or does not. Dan, who will decide on whether the allocation should be randomized or not, is keen for the result to give a good estimate of the response rates over the whole population, whereas Phyllis, a doctor, believes one type of patient, say healthy ones, does better with treatment 1 than 2, and her goal is giving patients the best treatment (ibid., p. 818). "[I]f Dan has a positive probability that Phyllis, or whoever is allocating, is placing the patients on the two treatments unequally, then randomization is the preferred allocation scheme (optimal)" (ibid.). Presumably, the agent doesn't worry that he unconsciously biases his own study aimed at learning the success rate in the population. For non-subjective Bayesians, there may be an appeal to the fact that "with randomization, the posterior is much less sensitive to the prior. And I think most practical Bayesians would consider it valuable to increase the robustness of the posterior" (Wasserman 2013). An extensive discussion may be found in Gelman et al. (2004). Still, as I understand it, it's not the *direct* protection of error probabilities that drives the concern.

Randomization and the Philosophers

> It seems surprising that the value of randomisation should still be disputed at this stage, and of course it is not disputed by anybody in the business. There is, though, a body of philosophers who do dispute it. (Colquhoun 2011, p. 333)

It's a bit mortifying to hear Colquhoun allude to "subversion by philosophers of science" (p. 321). Philosophical arguments against randomization stem largely from influential Bayesian texts (e.g., Howson and Urbach 1993), "On the Bayesian view, randomization is optional, and the essential condition is for the comparison groups in a clinical trial to be adequately matched on factors believed to have prognostic significance" (Howson and Urbach 1993, p. 378). A criticism philosophers often raise is due to the possibility of unknown confounding factors that differentiate the treated from the control

---

[10] Lindley (1982, p. 439) argues that if practitioners would dismiss an allocation that appeared unsatisfactory "one might ask why randomize in the first place?" Just because we can fail to satisfy a statistical assumption, does not imply we shouldn't try to succeed, test if we've failed, and fix problems found.

groups (e.g., Worrall 2002, p. S324). As Stephen Senn explains (2013b, p. 1447), such "imbalance," which is expected, will not impugn the statistical significance computations under randomization. The analysis is of a ratio of the between group variability and the within group variability. The said imbalance between groups (the numerator) would also impinge on the within group variability (the denominator). Larger variability will at worst result in larger standard errors, wider confidence intervals, resulting in a non-statistically significant result. The relevance of the unknown factors "is bounded by outcome and if we have randomised, the variation within groups is related to the variation between in a way that can be described probabilistically by the Fisherian machinery" (ibid.). There's an observed difference between groups, and our question is how readily could the observed difference in outcome be generated by chance alone? Senn goes further.

It is not necessary for the groups to be balanced. In fact, the probability calculation applied to a clinical trial automatically *makes an allowance for the fact that groups will almost certainly be unbalanced*, and if one knew that they were balanced, then the calculation that is usually performed would not be correct. Every statistician knows that you should not analyse a matched pair's design as if it were a completely randomised design. (ibid., p. 1442)

The former randomly assigns a treatment to a pair that have been deliberately matched.

In clinical trials where subjects are assigned as they present themselves, you can't look over the groups for possibly relevant factors, but you can include them in your model. Suppose the sex of the patient is deemed relevant. According to Senn:

(1) If you have sex in the model the treatment estimate is corrected for sex whether or not the design is balanced; balancing makes it more efficient.
(2) Balancing for sex but not having it in the model does not give a valid inference.[11]

RCT4D

A different type of debate is cropping up in fields that are increasingly dabbling with using randomized controlled trials (RCTs) rather than a typical reliance on observational data and statistical modeling. The Poverty Action Lab at MIT led by Abhijit Banerjee and Esther Duflo (2011) is spearheading a major movement in development economics to employ RCTs to test the benefits of

---

[11] "If you refuse to choose at random between all possible designs in which sex is balanced I will cry fraud." (Senn, private communication; see also Senn 1994, p. 1721)

various aid programs for spurring economic growth and decreasing poverty, from bed nets and school uniforms, to micro-loans. For those advocating RCTs in development economics (abbreviated RCT4D), if you want to discover if school uniforms decrease teen pregnancy in Mumbai, you take k comparable schools and randomly assign uniforms to some and not to others; at the end of the study, differences in average results are observed. It is hoped thereby to repel criticisms from those who question if there are scientific foundations guiding aid-driven development.

Philosopher of science Nancy Cartwright allows that RCTs, if done very well, can show a program worked in a studied situation, but that's "a long way from establishing that it *will work* in a particular target" (2012, p. 299). A major concern is that what works in Kenya needn't work in Mumbai. In general, the results of RCTs apply to the experimental population and, unless that's a random sample of a given target population, the issue of extrapolating (or external validity) arises. That is true. Merely volunteering to be part of a trial may well be what distinguishes subjects from others.

The conflicting sides here are largely between those who advocate experimental testing of policies and those who think we can do better with econometric modeling coupled with theory. Opponents, such as Angus Deaton, think the attention should be "refocused toward the investigation of potentially generalizable mechanisms that explain why and in what contexts projects can be expected to work" (Deaton 2010, p. 426) via modeling and trial and error. But why not all of the above? Clearly RCTs limit what can be studied, so they can't be the only method. The "hierarchies of evidence" we agree should include explicit recognition of flaws and fallacies of extrapolation.

Giving the mother nutritional information improves child nourishment in city X, but not in city Y where the father does the shopping and the mother-in-law decides who eats what. Small classrooms improve learning in one case, but not if applying it means shutting down spaces for libraries and study facilities. I don't see how either the modelers or the randomistas (as they are sometimes called) can avoid needing to be attuned to such foibles. Shouldn't the kind of field trials described in Banerjee and Duflo (2011) reveal clues as to *why* what works in X won't work in Y? Perhaps one of the most valuable pieces of information emerges from talking with and interacting amongst the people involved.

Cartwright and Hardie worry that RCTs, being rule oriented, reduce or eliminate the use of necessary discretion and judgment:

If a rule such as 'follow the RCTs, and do so faithfully' were a good way of deciding about effectiveness, then certainly deliberation is second best (or worse) ... the orthodoxy, which is a rules system, discourages decision makers from thinking about their problems, because the aim of rules is to reduce or eliminate the use of discretion and judgment, . . . The aim of reducing discretion comes from a lack of trust in the ability of operatives to exercise discretion well . . . Thus, the orthodoxy not only discourage deliberation, as unnecessary since the rules are superior, but selects in favor of operatives who cannot deliberate. (Cartwright and Hardie 2012, pp. 158–9)

Do rules to prevent QRPs, conscious and unconscious, reflect a lack of trust? Absolutely, even those trying hard to get it right aren't immune to the tunnel vision of their tribe.

The truth is, performing and interpreting RCTs involve enormous amounts of discretion. One of the big problems with experiments in many fields is the way statistical-scientific gaps are filled in interpreting results. In development economics, negative results are hard to hide, but there's still plenty of latitude for post-data explanations. RCTs don't protect you from post-data hunting and snooping. One RCT gave evidence that free school uniforms decreased teenage pregnancy, by encouraging students to remain in school. Here, teen pregnancies serve as a proxy for contracting HIV/AIDS. However, combining uniforms with a curriculum on HIV/AIDS gave negative results.

In schools that had both the HIV/AIDS and the uniforms programs, girls were no less likely to become pregnant than those in the schools that had nothing. The HIV/AIDS education curriculum, instead of reducing sexual activity . . ., actually *undid* the positive effect of the [free uniforms]. (Banerjee and Duflo 2011, p. 115)

Several different theories are offered. Perhaps the AIDS curriculum, which stresses abstinence before marriage, encouraged marriages and thus pregnancies. Post-data explanations for insignificant results are as seductive here as elsewhere and aren't protected by an RCT without prespecified outcomes. If we are to evaluate aid programs, an accumulation of data on bungled implementation might be the best way to assess what works and why. Rather than scrap the trials, a call for explicit attention to how a program could fail in the new environment is needed. Researchers should also be open to finding that none of the existing models captures people's motivations. Sometimes those receiving aid might just resist being "helped," or being nudged to do what's "rational".

Randomization is no longer at the top of the evidence hierarchy. It' s been supplanted by systematic reviews or meta-analysis of all relevant RCTs. Here too, however, there are problems of selecting which studies to include and their differing quality. Still the need for meta-analysis has promoted "all trials," rather than hiding any in file-drawers, and with the emphasis by the Cochrane collaboration, are clearly not going away. Meta-analytic reviews have received some black eyes, but it's one of the central ways of combining results in frequentist statistics.[12] Our itinerary got too full to visit this important topic; you may catch it on a return tour.

Batch-Effect Driven Spurious Associations

> There is a relatively unknown problem with microarray experiments, in addition to the multiple testing problems [microarray] samples should be randomized over important sources of variation; otherwise p-values may be flawed. Until relatively recently, the microarray samples were not sent through assay equipment in random order. . . . Essentially all the microarray data pre-2010 is unreliable. (Young 2013)

The latest Big Data technologies are not immune from basic experimental design principles. We hear that a decade or more has been lost by failing to randomize microarrays. "Stop Ignoring Experimental Design (or my head will explode)" declares genomics researcher Christophe Lambert (2010). The result is spurious associations due to confounding "to the point, in fact, where real associations cannot be distinguished from experimental artifacts" (ibid., p. 1). Microarray analysis involves a great many steps, plating and processing, and washing and more; minute differences in entirely non-biological variables can easily swamp the difference of interest. Suppose a microarray, looking for genes differentially expressed between diseased and healthy tissue (cases and controls), processes the cases in one batch, say at a given lab on Monday, and the controls on Tuesday. The reported statistically significant differences may be swamped by artifacts – the tiny differences due to different technicians, different reagents, even ozone levels. A "batch" would be a set of microarrays processed in a relatively homogeneous way, say at a single lab on Monday. Batch effects are defined to be systematic non-biological variations between groups of samples (or batches) due to such experimental artifacts. A paper on genetic associations and longevity (Sebastiani et al. 2010) didn't live very long because it turned out the samples from centenarians were differently collected and processed than the control group of average people. The statistically significant difference disappeared when the samples were run on the same batch, showing the observed association to be an experimental artifact.

By randomly assigning the order of cases and controls, the spurious associations vanish! Ideally they also randomize over data collection techniques and balance experimental units to different batches, but "the case/control status is the most important variable to randomize" (Lambert 2010, p. 4). Then,

---

[12] See for example Ioannidis (2016). For applications, see Cumming (2012) and Senn (2007).

corrections due to site and data collection can be made later. But the reverse isn't true. As Fisher said, "To call in the statistician after the experiment is done .may be no more than asking him to perform a postmortem examination:. . . to say what the experiment died of" (Fisher 1938, p. 17). Nevertheless, in an attempt to fix the batch effect driven spurious associations,

[A] whole class of post-experiment statistical methods has emerged ... These methods ... represent a palliative, not a cure. ... the GWAS research community has too often accommodated bad experimental design with automated post-experiment cleanup. ... experimental designs for large-scale hypothesis testing have produced so many outliers that the field has made it standard practice to automate discarding outlying data. (Lambert and Black 2012, pp. 196–7)

By contrast, with proper design they find that post-experiment automatic filters are unneeded. In other words, the introduction of randomized design *frees them to deliberate* over the handful of extreme values to see if they are real or artifacts. (This contrasts with the worry raised by Cartwright and Hardie (2012).)

Souvenir T: Even Big Data Calls for Theory and Falsification

Historically, epidemiology has focused on minimizing Type II error (missing a relationship in the data), often ignoring multiple testing considerations, while traditional statistical study has focused on minimizing Type I error (incorrectly attributing a relationship in data better explained by random chance). When traditional epidemiology met the field of GWAS, a flurry of papers reported findings which eventually became viewed as nonreplicable. (Lambert and Black 2012, p. 199)

This is from Christophe Lambert and Laura Black's important paper "Learning from our GWAS Mistakes: From Experimental Design to Scientific Method"; it directly connects genome-wide association studies (GWAS) to philosophical themes from Meehl, Popper and falsification. In an attempt to staunch the non-replication, they explain, adjusted genome-wide thresholds of significance were required as well as replication in an independent sample (Section 4.6).

However, the intended goal is often thwarted by how this is carried out. "[R]esearchers commonly take forward, say, 20–40 nominally significant signals" that did not meet the stricter significance levels, "then run association tests for those signals in a second study, concluding that all the signals with a p-value ≤.05 have replicated (no Bonferroni adjustment). Frequently 1 or 2 associations replicate – which is also the number expected by random chance" (ibid.). Next these "replicated" cases are combined with the original data "to compute p-values considered genome-wide significant. This method has been propagated in publications, leading us to wonder if standard practice could become to publish random signals and tell a plausible biological story about the findings" (ibid.).

Instead of being satisfied with a post-data biological story to explain correlations, "[i]f journals were to insist that association studies also suggest possible experiments that could falsify a putative theory of causation based on association, the quality and durability of association studies could increase" (ibid., p. 201). At the very least, the severe tester argues, we should strive to falsify methods of inquiry and analysis. This might at least scotch the tendency Lambert and Black observe, for others to propagate a flawed methodology once seen in respected journals: "[W]ithout a clear falsifiable stance – one that has implications for the theory – associations do not necessarily contribute deeply to science" (ibid., p. 199).

Tour IV More Auditing: Objectivity and Model Checking

4.8 All Models Are False

> ... it does not seem helpful just to say that all models are wrong. The very word model implies simplification and idealization. ... The construction of idealized representations that capture important stable aspects of such systems is, however, a vital part of general scientific analysis. (Cox 1995, p. 456)

A popular slogan in statistics and elsewhere is "all Models are false!" Is this true? What can it mean to attribute a truth value to a model? Clearly what is meant involves some assertion or hypothesis about the model – that it correctly or incorrectly represents some phenomenon in some respect or to some degree. Such assertions clearly can be true. As Cox observes, "the very word model implies simplification and idealization." To declare, "all models are false" by dint of their being idealizations or approximations, is to stick us with one of those "all flesh is grass" trivializations (Section 4.1). So understood, it follows that all statistical models are false, but we have learned nothing about how statistical models may be used to infer true claims about problems of interest. Since the severe tester's goal in using approximate statistical models is largely to learn where they break down, their strict falsity is a given. Yet it does make her wonder why anyone would want to place a probability assignment on their truth, unless it was 0? Today's tour continues our journey into solving the problem of induction (Section 2.7).

Assigning a probability to either a substantive or a statistical model is very different from asserting it is approximately correct or adequate for solving a problem. The philosopher of science Peter Achinstein had hoped to discover that his scientific heroes, Isaac Newton and John Stuart Mill, were Bayesian probabilists, but he was disappointed; what he finds is enlightening:

Neither in their abstract formulations of inductive generalizations (Newton's rule 3; Mill's definition of 'induction') nor in their examples of particular inductions to general conclusions of the form 'all As are Bs' does the term 'probability' occur. Both write that from certain specific facts we can conclude general ones – not that we can conclude general propositions with probability, or that general propositions have a probability ... From the inductive premises we simply conclude that the generalization is true, or as Newton allows in rule 4, 'very nearly true,' by which he appears to mean not 'probably true' but 'approximately true' (as he does when he takes the orbits of the satellites of Jupiter to be circles rather than ellipses). (Achinstein 2010, p. 176)

There are two main ways the "all models are false" charge comes about:

1. The statistical inference refers to an idealized and partial representation of a theory or process.
2. The probability model, to which a statistical inference refers, is at most an idealized and partial representation of the actual data-generating source.

Neither of these facts precludes the use of these *false* models to find out true things, or to correctly solve problems. On the contrary, it would be impossible to learn about the world if we did not deliberately falsify and simplify.

Adequacy for a Problem. The statistician George Box, to whom the slogan "all models are wrong" is often attributed, goes on to add "But some are useful" (1979, p. 202). I'll go further still: all models are false, no useful models are true. Were a model so complex as to represent every detail of data "realistically," it wouldn't be useful for finding things out. Let's say a statistical model is useful by being adequate for a problem, meaning it may be used to find true or approximately true solutions. Statistical hypotheses may be seen as conjectured solutions to a problem. A statistical model is adequate for a problem of statistical inference (which is only a subset of uses of statistical models) if it enables controlling and assessing if purported solutions are well or poorly probed, and to what degree. Through approximate models, we learn about the "important stable aspects" or systematic patterns when we are in the context of phenomena that exhibit statistical variability. When I speak of ruling out mistaken interpretations of data, I include mistakes about theoretical and causal claims. If you're an anti-realist about science, you will interpret, or rather reinterpret, theoretical claims in terms of observable claims of some sort. One such anti-realist view we've seen is instrumentalism: unobservables including genes, particles, light bending may be regarded as at most instruments for finding out about observable regularities and predictions. Fortunately we won't have to engage the thorny problem of realism in science, we can remain agnostic. Neither my arguments, nor the error statistical philosophy in general, turn on whether one adopts one of the philosophies of realism or anti-realism. Today's versions of realism and anti-realism are quite frankly too hard to tell apart to be of importance to our goals. The most important thing is that both realists and non-realists require an account of statistical inference. Moreover, whatever one's view of scientific theories, a statistical analysis of problems of actual experiments involves abstraction and creative analogy.

Testing Assumptions is Crucial. You might hear it charged that frequentist methods presuppose the assumptions of their statistical models, which is puzzling because when it comes to testing assumptions it's to frequentist methods that researchers turn.

> It is crucial that any account of statistical inference provides a conceptual framework for this process of model criticism, ... the ability of the frequentist paradigm to offer a battery of simple significance tests for model checking and possible improvement is an important part of its ability to supply objective tools for learning. (Cox and Mayo 2010, p. 285)

Brad Efron is right to say the frequentist is the pessimist, who worries that "if anything can go wrong it will," while the Bayesian optimistically assumes if anything can go right it will (Efron 1998, p. 99). The frequentist error statistician is a worrywart, resigned to hoping things are half as good as intended. This also makes her an activist, deliberately reining in some portion of a problem so that it's sufficiently like one she knows how to check. Within these designated model checks, assumptions under test are intended to arise only as i-assumptions. They're assumptions for drawing out consequences, for possible falsification.

"In principle, the information in the data is split into two parts, one to assess the unknown parameters of interest and the other for model criticism" (Cox 2006a, p. 198). The number of successes in n Bernoulli trials, recall, is a *sufficient* statistic, and has a Binomial sampling distribution determined by θ, the probability of success on each trial (Section 3.3). If the model is appropriate then any permutation of the r successes in n trials has a known probability. Because this conditional distribution (X given s) is known, it serves to assess if the model is violated. If it shows statistical discordance, the model is disconfirmed or falsified. The key is to look at residuals: the difference between each observed value and what is expected under the model. (We illustrate with the runs test in Section 4.11.) It is also characteristic of error statistical methods to be relatively robust to violation.

Central Limit Theorem. In the presentation on justifying induction (Section 2.7), we heard Neyman stress how the empirical Law of Large Numbers (LLN) is in sync with the mathematical law in a number of "real random experiments." Supplementing the LLN is the Central Limit Theorem (CLT). It tells us that the mean $\overline{X}$ of n independent random variables, each X with mean μ, and finite non-zero σ^2, is approximately Normally distributed with its mean equal to μ and standard deviation $\sigma/\sqrt{n}$ – regardless of the underlying distribution of X. So long as n is reasonably large (say 40 or 50), and the underlying distribution is not too asymmetrical, the Normal distribution gives a good approximation, and is robust for many cases where IID is violated. The CLT tells us that $\overline{X}$ standardized is N(0,1). The finite non-zero variance isn't much of a restriction, and even this has been capable of being relaxed.

The CLT links a claim or question about a statistical hypothesis to claims about the relative frequencies that would be expected in applications (real or hypothetical) of the experiment. Owing to this link, we can use the sample mean to inquire about values of μ that are capable or incapable of bringing it about. Our standardized difference measures observed disagreement, and classifies those improbably far from hypothesized values. Thus, statistical models may be adequate for real random experiments, and hypotheses to this effect may pass with severity.

Exhibit (xii): Pest Control. Neyman (1952) immediately turns from the canonical examples of real random experiments – of coin tossing and roulette wheels – to illustrate how "the abstract theory of probability ... may be, and actually is, applied to solve problems of practice importance" such as pest control (p. 27)! Given the lack of human control here, he expects the mechanism to be complicated. The first attempt to model the variation in larvae hatched from moth eggs is way off.

[I]f we attempt to treat the distribution of larvae from the point of view of [the Poisson distribution], we would have to assume that each larva is placed on the field independently of the others. This basic assumption was flatly contradicted by the life of larvae as described by Dr. Beall. Larvae develop from eggs laid by moths. It is plausible to assume that, when a moth feels like laying eggs, it does not make any special choice between sections of a field planted with the same crop and reasonably uniform in other respects. (1952, pp. 34–5).

So it's plausible to suppose the Poisson distribution for the spots where moths lay their eggs. However, a data analysis made it "clear that a very serious divergence exists" between the actual distribution of larvae and the Poisson model (ibid., p. 34). Larvae expert, Dr. Beall, explains why: At each "sitting" a moth lays a whole batch of eggs and the number of eggs varies from one cluster to another. "After hatching ... the larvae begin to look for food and crawl around" but given their slow movement "if one larva is found, then it is likely that the plot will contain more than one from the same cluster" (ibid., p. 35). An independence assumption fails. (I omit many details; see Neyman 1952, Gillies 2001.)

The main thing is this: The misfit with the Poisson model leads Neyman to arrive at a completely novel distribution: he called it the type A distribution (a "contagious" distribution). Yet Neyman knows full well that even the type A distribution is strictly inadequate, and a far more complex distribution would be required for answering certain questions. He knows it's strictly false. Yet it suffices to show why the first attempt failed, and it's adequate to solving his immediate problem in pest control.

Souvenir U: Severity in Terms of Problem-Solving

The aim of inquiry is finding things out. To find things out we need to solve problems that arise due to limited, partial, noisy, and error-prone information. Statistical models are at best approximations of aspects of the data-generating process. Reasserting this fact is not informative about the case at hand. These models work because they need only capture rather coarse properties of the phenomena: the error probabilities of the test method are approximately and conservatively related to actual ones. A problem beset by variability is turned into one where the variability is known at least approximately. Far from wanting true (or even "truer") models, we need models whose deliberate falsity enables finding things out.

Our threadbare array of models and questions is just a starter home to grow the nooks and crannies between data and what you want to know (Souvenir E, Figure 2.1). In learning about the large-scale theories of sexy science, intermediate statistical models house two "would-be" claims. Let me explain. The theory of GTR does not directly say anything about an experiment we could perform. Splitting off some partial question, say about the deflection effect, we get a prediction about what *would be* expected were the deflection effect approximately equal to the Einstein value, 1.75". Raw data from actual experiments, cleaned and massaged, afford inferences about intermediate (astrometric) models; inferences as to what it would be like were we taking measurements at the limb of the sun. The two counterfactual inferences – from the theory down, and the data up – meet in the intermediate statistical models. We don't seek a probabilist assignment to a hypothesis or model. We want to know what the data say about a conjectured solution to a problem: What erroneous interpretations have been well ruled out? Which have not even been probed? The warrant for these claims is afforded by the method's capabilities to have informed us of mistaken interpretations. *Statistical methods are useful for testing solutions to problems when this capability/incapability is captured by the relative frequency with which the method avoids misinterpretations.*

If you want to avoid speaking of "truth" you can put the severity requirement in terms of solving a problem. A claim *H* asserts a proposed solution S to an inferential problem is adequate in some respects. It could be a model for prediction, or anything besides.

H: S is adequate for a problem

To reject *H* means "infer *S* is inadequate for a problem." If none of the possible outcomes lead to reject *H* even if *H* is false – the test is incapable of finding inadequacies in *S* – then "do not reject *H*" is BENT evidence that *H* is true. We move from no capability, to some, to high:

> If the test procedure (which generally alludes to a cluster of tests) very rarely rejects *H*, if *H* is true, then "reject *H*" provides evidence for falsifying *H* in the respect indicated.

You could say, a particular inadequacy is corroborated. It's still an inferential question: what's warranted to infer. We start, not with hypotheses, but questions and problems. We want to appraise hypothesized answers severely.

I'll meet you in the ship's library for a reenactment of George Box (1983) issuing "An Apology for Ecumenism in Statistics."

4.9 For Model-Checking, They Come Back to Significance Tests

> Why can't all criticism be done using Bayes posterior analysis . . .? The difficulty with this approach is that by supposing all possible sets of assumptions known *a priori*, it discredits the possibility of new discovery. But new discovery is, after all, the most important object of the scientific process. (George Box 1983, p. 73)

Why the apology for ecumenism? Unlike most Bayesians, Box does not view induction as probabilism in the form of probabilistic updating (posterior probabilism), or any form of probabilism. Rather, it requires critically testing whether a model M_i is "consonant" with data, and this, he argues, demands frequentist significance testing. Our ability "to find patterns in discrepancies $M_i - y_d$ between the data and what might be expected if some tentative model were true is of great importance in the search for explanations of data and of discrepant events" (Box 1983, p. 57). But the dangers of apophenia raise their head.

> However, some check is needed on [the brain's] pattern seeking ability, for common experience shows that some pattern or other can be seen in almost any set of data or facts. This is the object of diagnostic checks and tests of fit which, I will argue, require frequentist theory significance tests for their formal justification. (ibid.)

Once you have inductively arrived at an appropriate model, the move, on his view, "is entirely *deductive* and will be called *estimation*" (ibid., p. 56). The

302   Excursion 4: Objectivity and Auditing

deductive portion, he thinks, can be Bayesian, but the inductive portion requires frequentist significance tests, and statistical inference depends on an iteration between the two. Alluding to Box, Peter Huber avers: "Within orthodox Bayesian statistics, we cannot even address the question whether a model M_i, under consideration at stage i of the investigation, is consonant with the data y" (Huber 2011, p. 92). Box adds a non-Bayesian activity to his account.

A note on Box's slightly idiosyncratic use of deduction/induction: Frequentist significance testing is often called deductive, but for Box it's the inductive component. There's no confusion if we remember that Box is emphasizing that frequentist testing is the source of new ideas, it is the inductive achievement. It's in sync with our own view that inductive inference to claim C consists of trying and failing to falsify C with a stringent test: C should be well corroborated. In fact, the approach to misspecification (M-S) testing that melds seamlessly with the error statistical account has its roots in the diagnostic checking of Box and Jenkins (1976).

All You Need Is Bayes. Not

Box and Jenkins highlight the link between 'prove' and 'test': "A model is only capable of being 'proved' in the biblical sense of being put to the test" (ibid., p. 286). Box considers the possibility that model checking occurs as follows: One might imagine $A_1, A_2, \ldots, A_k$ being alternative assumptions and then computing $\Pr(A_i|y)$. Box denies this is plausible. To assume we start out with all models precludes the "something else we haven't thought of" so vital to science. Typically, Bayesians try to deal with this by computing a Bayesian catchall "everything else." Savage recommends reserving a low prior for the catchall (1962a), but Box worries that this may allow you to assign model M_i a high posterior probability *relative* to the other models considered. "In practice this would seem of little comfort" (Box 1983, pp. 73–4). For suppose of the three models under consideration the posteriors are 0.001, 0.001, 0.998, but unknown to the investigator a fourth model is a thousand times more probable than even the most probable one considered so far.

So he turns to frequentist tests for model checking. Is there any harm in snatching some cookies from the frequentist cookie jar? Not really. Does it violate the Likelihood Principle (LP)? Let's listen to Box:

The likelihood principle holds, of course, for the estimation aspect of inference in which the model is temporarily assumed true. However it is inapplicable to the criticism process in which the model is regarded as in doubt . . . In the criticism phase we are considering whether, given A, the sample y_d is likely to have occurred at all. To do this we *must* consider it in relation to the *other* samples that could have occurred but did not. (Box 1983, pp. 74–5)

Suppose you're about to use a statistical model, say n IID Normal trials for a primary inference about mean μ. Checking independence (I), identical distributed (ID), or the Normality assumption (N) are *secondary inferences* in relation to the primary one. In conducting secondary inferences, Box is saying, the LP must be violated, or simply doesn't apply. You can run a simple Fisherian significance test – the null asserting the model assumption A holds – and reject it if the observed result is statistically significantly far from what A predicts. A P-value (or its informal equivalent) is computed – a tail area – which requires considering outcomes other than the one observed.

Box gives the example of stopping rules. Stopping rules don't alter the posterior distribution, as we learned from the extreme example in Excursion 1 (Section 1.5). For a simple example, he considers four Bernoulli trials: $\langle S, S, F, S \rangle$. The same string could have come about if $n = 4$ was fixed in advance, or if the plan was to sample until the third success is observed. The latter are called negative Binomial trials, the former Binomial. The string enters the likelihood ratio the same way, $\binom{4}{3}\theta^3(1 - \theta)$ and $\binom{3}{2}\theta^3(1 - \theta)$ respectively: the only difference is the coefficients, which cancel. But the significance tester distinguishes them, because the sample space, and corresponding error probabilities, differ.[1] When it comes to model testing, Box contends, this LP violation is altogether reasonable, since "we are considering whether, given A, the sample is likely to have occurred at all" (ibid., p. 75).

This is interesting. Isn't it also our question at his estimation stage where the LP denies stopping rules matter? We don't know there's any genuine effect, or if a null is true. If we ignore the stopping rules, we may make it too easy to find one, even if it's absent. In the example of Section 1.5, we ensure erroneous rejection, violating "weak repeated sampling." A Boxian Bayesian, who retains the LP for primary statistical inference, still seems to owe us an explanation why we shouldn't echo Armitage (1962, p. 72) that "Thou shalt be misled" if your method hides optional stopping at the primary (Box's estimation) stage.

Another little puzzle arises in telling what's true about the LP: Is the LP violated or simply inapplicable in secondary testing of model assumptions. Consider Casella and R. Berger's text.

Most data analysts perform some sort of 'model checking' when analyzing a set of data . . . For example, it is common practice to examine *residuals* from a model, statistics that

---

[1] The sufficient statistic in the negative Binomial case is N, the number of trials until the fourth success. In the Binomial case, it is $\overline{X}$ (Cox and Mayo 2010, p. 286).

measure variation in the data not accounted for by the model ... (Of course such a practice directly violates the Likelihood Principle also.) Thus, *before* considering [the Likelihood Principle], we must be comfortable with the model. (Casella and R. Berger 2002, pp. 295–6)

For them, it appears, the LP is full out violated in model checking. I'm not sure how much turns on whether the LP is regarded as violated or merely inapplicable in testing assumptions; a question arises in either case. Say you have carried out Box's iterative moves between criticism and estimation, arrived at a model deemed adequate, and infer *H*: model M_i is adequate for modeling data x_0. My question is: How is this secondary inference qualified? Probabilists are supposed to qualify uncertain claims with probability (e.g., with posterior probabilities or comparisons of posteriors). What about this secondary inference to the adequacy/inadequacy of the model? For Boxians, it's admitted to be a non-Bayesian frequentist animal. Still a long-run performance justification wouldn't seem plausible. If you're going to accept the model as sufficiently adequate to build the primary inference, you'd want to say it had passed a severe test: that if it wasn't adequate for the primary inference, then you probably would have discovered this through the secondary model checking. However, if secondary inference is also a statistical inference, it looks like Casella and R. Berger, and Box, are right to consider the LP violated – as regards *that inference*. There's an appeal to outcomes other than the one observed.

Andrew Gelman's Bayesian approach can be considered an offshoot of Box's, but, unlike Box, he will avoid assigning a posterior probability to the primary inference. Indeed, he calls himself a falsificationist Bayesian, and is disgruntled that Bayesians don't test their models.

I vividly remember going from poster to poster at the 1991 Valencia meeting on Bayesian statistics ... not only were they not interested in checking the fit of the models, they considered such checks to be illegitimate. To them, any Bayesian model necessarily represented a subjective prior distribution and as such could never be tested. The idea of testing and p-values were held to be counter to the Bayesian philosophy. (Gelman 2011, pp. 68–9)

What he's describing is in sync with the classical subjective Bayesian: If "the Bayesian theory is about coherence, not about right or wrong" (Lindley 1976, p. 359), then what's to test? Lindley does distinguish a pre-data model choice:

Notice that the likelihood principle only applies to inference, i.e. to calculations once the data have been observed. Before then, e.g. in some aspects of model choice, in the design of experiments ..., a consideration of several possible data values is essential. (Lindley 2000, p. 310)

This he views as a decision based on maximizing an agent's expected utility. But wouldn't a correct assessment of utility depend on information on model adequacy?

Interestingly, there are a number of Bayesians who entertain the idea of a Bayesian P-value to check accordance of a model when there's no alternative in sight.[2] They accept the idea that significance tests and P-values are a good way, if not the only way, to assess the consonance between data and model. Yet perhaps they are only grinning and bearing it. As soon as alternative models are available, most would sooner engage in a Bayesian analysis, e.g., Bayes factors (Bayarri and Berger 2004).

But Gelman is a denizen of a tribe of Bayesians that rejects these traditional forms. "To me, Bayes factors correspond to a discrete view of the world, in which we must choose between models A, B, or C" (Gelman 2011, p. 74) or a weighted average of them as in Madigan and Raftery (1994). Nor will it be a posterior. "I do not trust Bayesian induction over the space of models because the posterior probability of a continuous-parameter model depends crucially on untestable aspects of its prior distribution" (Gelman 2011, p. 70). Instead, for Gelman, the priors/posteriors arise as an interim predictive device to draw out and test implications of a model. What is the status of the inference to the adequacy of the model? If neither probabilified nor Bayes ratioed, it can at least be well or poorly tested. In fact, he says, "This view corresponds closely to the error-statistics idea of Mayo (1996)" (ibid., p. 70). We'll try to extricate his approach in Excursion 6.

4.10 Bootstrap Resampling: My Sample Is a Mirror of the Universe

"My difficulty" with the Likelihood Principle (LP), declares Brad Efron (in a comment on Lindley), is that it "rules out many of our most useful data analytic tools without providing workable substitutes" (2000, p. 330) – notably, the method for which he is well known: bootstrap resampling (Efron 1979). Let's take a little detour to have a look around this hot topic. (I follow D. Freedman (2009), and A. Spanos (2019)).

We have a single IID sample of size 100 of the water temperatures soon after the accident $x_0 = \langle x_1, x_2, \ldots, x_{100} \rangle$. Can we say anything about its accuracy even if we couldn't take any more? Yes. We can lift ourselves up by the bootstraps with this single x_0 by treating it as its own population. Get the computer to take

---

[2] There is considerable discussion as to which involve pejorative "double use" of data, and which give adequate frequentist guarantees or calibrations (Ghosh et al. 2006, pp. 175–84; Bayarri and Berger 2004). But I find their rationale unclear.

a large number, say 10,000, independent samples from x_0 (with replacement), giving 10,000 *resamples*. Then reestimate the mean for each, giving 10,000 bootstrap means $\overline{X}_{b_1}, \overline{X}_{b_2}, \ldots, \overline{X}_{b_{10,000}}$. The frequency with which the bootstrapped means take different values approximates the sampling distribution of $\overline{X}$. It can be extended to medians, standard deviations, etc. "This is exactly the kind of calculation that is ruled out by the likelihood principle; it relies on hypothetical data sets different from the data that are actually observed and does so in a particularly flagrant way" (Efron 2000, p. 331). At its very core is the question: what would mean temperatures be like were we to have repeated the process many times? This lets us learn: How capable of producing our observed sample is a universe with mean temperature no higher than the temperature thought to endanger the ecosystem?

Averaging the 10,000 bootstrap means, we get the overall bootstrap sample mean, $\overline{X}_b$. If n is sufficiently large, the resampling distribution of $\overline{X}_b - \overline{x}$ mirrors the sampling distribution of $\overline{X} - \mu$, where μ is the mean of the population. We can use the sample deviation of $\overline{X}_b$ to approximate the standard error of $\overline{X}$.[3]

To illustrate with a tiny example, imagine that instead of 100 temperature measurements there are only 10: x_0: 150, 165, 151, 138, 148, 167, 164, 160, 136, 173, with sample mean $\overline{x} = 155.2$, and instead of 10,000 resamples, only 5. Since it's with replacement there can be duplicates.

x_0: 150, 165, 151, 138, 148, 167, 164, 160, 136, 173	$\overline{x} = 155.2$
Bootstrap resamples	**Bootstrap means**
x_{b_1}: 160, 136, 138, 165, 173, 165, 167, 148, 151, 167	157
x_{b_2}: 164, 136, 165, 167, 148, 138, 151, 160, 150, 151	153
x_{b_3}: 173, 138, 173, 160, 167, 167, 148, 138, 148, 165	157.7
x_{b_4}: 148, 138, 164, 167, 160, 150, 164, 167, 148, 173	157.9
x_{b_5}: 173, 136, 167, 138, 150, 160, 148, 164, 164, 148	154.8

Here are the rest of the bootstrap statistics:

Bootstrap overall mean: $\overline{x}^b = [157 + 153 + 157.7 + 157.9 + 154.8]/5 = 156.08$; Bootstrap variance: $[(157 - 156.08)^2 + (153 - 156.08)^2 + (157.7 - 156.08)^2 + (157.9 - 156.08)^2 + (154.8 - 156.08)^2]/4 = 4.477$; Bootstrap SE: $\sqrt{4.477} = 2.116$.

Note the difference between the mean of our observed sample and that of the overall bootstrap mean (the bias) is small: $(\overline{x} - \overline{x}^b) = 155.2 - 156.08 = -0.88$.

---

[3] The bootstrapped distribution is conditional on the observed x.

From our toy example, we could form the bootstrap 0.95 confidence interval: 156.08 ± 1.96 (2.116) approximately [152, 158]. You must now imagine we arrived at the interval via 10,000 samples, not 5. The observed mean just after the accident (155.2) exceeds 150 by around 2.5 SE, indicating our sample came from a population with $\theta > 150$. In fact, were $\theta \leq 152$, such large results would occur infrequently.

The non-parametric bootstrap works without relying on a theoretical probability distribution, at least when the sample is random, large enough, and has sufficiently many bootstraps. Statistical inference by non-parametrics still has assumptions, such as IID (although there are other variants). Many propose we do all statistics non-parametrically, and some newer texts advocate this. I'm all for it, because the underlying error statistical reasoning becomes especially clear. I concur, too, with the philosopher Jan Sprenger that philosophy of statistics should integrate "resampling methods into a unified scheme of data analysis and inductive inference" (Sprenger 2011, p. 74). That unified scheme is error statistical. (I'm not sure how it's in sync with his subjective Bayesianism.)

The philosophical significance of bootstrap resampling is twofold. (1) The relative frequency of different values of $\overline{X}_b$ sustains our error statistical argument: the probability model is a good way to approximate the empirical distribution analytically. Through a hypothetical – 'what it would be like' were this process repeated many times – we understand what produced the single observed sample. (2) By identifying exemplary cases where we manage to take approximately random samples, we can achieve inductive lift-off. It's through our deliberate data generation efforts, in other words, that we solve induction. I don't know if taking water samples is one such exemplar, but I'm using it just as an illustration. We may imagine ample checks of water sampling on bodies with known temperature show we're pretty good at taking random samples of water temperature. Thus we reason, it works when the mean temperature is unknown. Can supernal powers read my mind and interfere just in the cases of an unknown mean?

Nor is it necessary to deny altogether the existence of mysterious influences adverse to the validity of the inductive ... processes. So long as their influence were not too overwhelming, the wonderful self-correcting nature of the ampliative inference would enable us ... to detect and make allowance for them. (Peirce 2.749)

4.11   Misspecification (M-S) Testing in the Error Statistical Account

Induction – understood as severe testing – "not only corrects its conclusions, it even corrects its premises" (Peirce 3.575). In the land of statistical inference it does so by checking and correcting the assumptions underlying the inference. It's common to distinguish "model-based" and "design-based" statistical inference, but both involve assumptions. So let's speak of the adequacy of the model in both cases. It's to this auditing task that I now turn. Let's call violated statistical assumptions statistical model misspecifications. The term "misspecification" is often used to refer to a problem with a primary model, whereas for us it will always refer to the secondary problem of checking assumptions for probing a primary question (following A. Spanos). "Primary" is relative to the main inferential task: Once an adequate statistical model is at hand, an inquiry can grow to include many layers of primary questions.

Splitting things off piecemeal has payoffs. The key is for the relevant error probabilities to be sufficiently close to those calculated in probing the primary claim. Even if you have a theory, turning it into something statistically testable isn't straightforward. You can't simply add an error term at the end, such as $y =$ theory + error, particularly in the social sciences – although people often do. The trouble is that you can tinker with the error term to "fix" anomalies without the theory having been tested in the least. Aris Spanos, an econometrician, roundly criticizes this tendency to "the preeminence of theory in econometric modeling" (2010c, p. 202). This would be okay if you were only estimating quantities in a theory known to be true or adequate, but in fact, Spanos says, "mainstream economic theories have been invariably unreliable predictors of economic phenomena" (ibid., p. 203).

As always, different inquiry types have their own error repertoires that need to be mastered. Allow me to try to climb neatly through the vegetation of one of the more treacherous modeling fields: econometrics. Relying on seven figures from Mayo and Spanos (2004), I'll tell the story of a case Spanos presented to me in 2002.

Nonsense Regression

Suppose that in her attempt to find a way to understand and predict changes in the US population, an economist discovers an empirical relationship that appears to provide almost a "law-like" fit:

$$y_t = 167 + 2x_t + \hat{u}_t,$$

where y_t denotes the population of the USA (in millions), and x_t denotes a secret variable whose identity Spanos would not reveal until the end of the analysis. The subscript t is time. There are 33 annual data points for the period 1955–1989 ($t = 1$ is 1955, $t = 2$ is 1956, etc.) The data can be represented as 33 pairs $z_0 = \{(x_t, y_t), t = 1, 2, \ldots, 33\}$. The coefficients 167 and 2 come from the least squares fit, a purely mathematical operation.

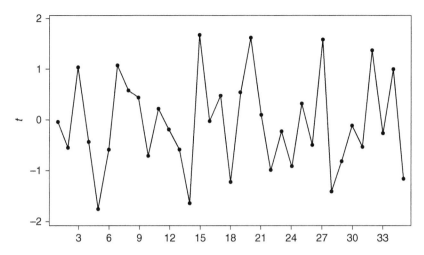

Figure 4.2 A typical realization of a NIID process.

This is an example of fitting a *Linear Regression Model* (LRM), which forms the backbone of most statistical models of interest:

$$M_0: \quad y_t = \beta_0 + \beta_1 x_t + u_t, \; t = 1, 2, \ldots, n.$$

The term $\beta_0 + \beta_1 x_t$ is viewed as the *systematic* component (and is the expected value of y_t), and $u_t = y_t - \beta_0 - \beta_1 x_t$ is the error or *non-systematic* component. The error u_t is a random variable assumed to be Normal, Independent, and Identically Distributed (NIID) with mean 0, variance σ^2. This is called Normal white noise. Figure 4.2 shows what NIID looks like.

A Primary Statistical Question: How Good a Predictor Is x_t? The empirical equation is intended to enable us to understand how y_t varies with x_t. Testing the statistical significance of the coefficients shows them to be highly significant: P-values are zero (0) to a third decimal, indicating a very strong relationship between the variables. The goodness-of-fit measure of how well this model "explains" the variability of y_t, $R^2 = 0.995$, an almost perfect fit (Figure 4.3). Everything looks hunky dory. Is it reliable? Only if the errors are approximately NIID with mean 0, variance σ^2.

The null hypotheses in M-S tests take the form

$$H_0: \text{the assumption(s) of statistical model } M \text{ hold for data } z,$$

as against not-H_0, which, strictly speaking, would include all of the ways one or more of its assumptions can fail. To rein in the testing, we consider specific departures with appropriate choices of test statistic d(y).

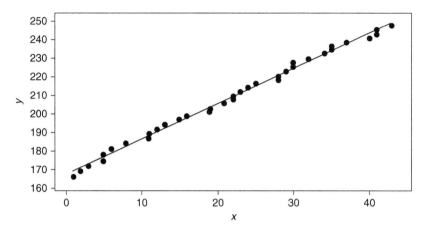

Figure 4.3 Fitted line plot, $y = 167.1 + 1.907x$.

Residuals Are the Key

Testing Randomness. The non-parametric *runs test* for IID is the test I showed Wes Salmon in relation to the Bernoulli model and justifying induction (Section 2.7). It falls under "omnibus" tests in Cox's taxonomy (Section 3.3). It can apply here by re-asking our question regarding the LRM. Look at the graph of the residuals (Figure 4.4), where the "hats" are the fitted values for the coefficients:

$$\{\hat{u}_t = y_t - \hat{\beta}_0 - \hat{\beta}_1 x_t, \ t = 1, 2, ..., n\}.$$

If the residuals do not fluctuate like pure noise, it's a sign the sample is not IID. Instead of the value of each residual, record whether the difference between successive observations is positive (+) or negative (−),

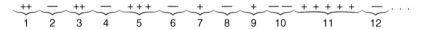

Each sequence of pluses only, or minuses only, is a *run*. We can calculate the probability of the number of runs just from the hypothesis that the assumption of randomness holds. It serves only as an argumentative (or i) assumption for the check. The expected number of runs, under randomness, is $(2n − 1)/3$, or in our case of $n = 35$ values, 23. Running out of letters, I'll use R again for the number of runs. The distribution of the test statistic, $\tau(y) = [R − E(R)]/\sqrt{\text{Var}(R)}$, under IID for $n \geq 20$, can be approximated by

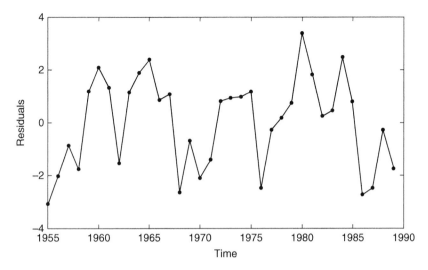

Figure 4.4 Plot of residuals over time.

$N(0, 1)$. We're actually testing H_0: $E(R) = (2n - 1)/3$ vs. H_1: $E(R) \neq (2n- 1)/3$, E is the expected value. We reject H_0 iff the observed R differs sufficiently (in either direction) from $E(R) = 23$. (SE = $\sqrt{(16n-29)/90}$.)

Our data yield 18 runs, around 2.4 SE units, giving a P-value of approximately 0.02. So 98% of the time, we'd expect an R closer to 23 if IID holds. Arguing from severity, the data indicate non-randomness. But rejecting the null only indicates a denial of IID: either independence is a problem or identically distributed is a problem. The test itself does not indicate whether the fault lies with the one or the other or both. More specific M-S testing enables addressing this Duhemian problem.

The Error in Fixing Error. A widely used parametric test for independence is the Durbin–Watson (D-W) test. Here, all the assumptions of the LRM are retained, except the one under test, independence, which is "relaxed." In particular, the original error term in M_0 is extended to allow for the possibility that the errors u_t are correlated with their own past, that is, *autocorrelated*,

$$u_t = \rho u_{t-1} + \varepsilon_t, \quad t = 1, 2, \ldots, n, \ldots$$

This is to propose a new overarching model, the *Autocorrelation-Corrected LRM*:

Proposed M_1: $\quad y_t = \beta_0 + \beta_1 x_t + u_t, \quad u_t = \rho u_{t-1} + \varepsilon_t, \quad t = 1, 2, \ldots, n, \ldots$

(Now it's ε_t that is assumed to be a Normal, white noise process.) The D-W test assesses whether or not $\rho = 0$, *assuming we are within model M_1*. One way to bring this out is to view the D-W test as actually considering the conjunctions:

$$H_0: \{M_1 \ \& \ \rho = 0\} \text{ vs. } H_1: \{M_1 \ \& \ \rho \neq 0\}.$$

With the data in our example, the D-W test statistic rejects the null hypothesis (at level 0.02), which is standardly taken as grounds to adopt H_1. This is a mistake. This move, to infer H_1, is warranted only if we are within M_1. True, if $\rho = 0$, we are back to the LRM, but $\rho \neq 0$ does not entail the particular violation of independence asserted in H_1. Notice we are in one of the "non-exhaustive" pigeonholes ("nested") of Cox's taxonomy. Because the assumptions of model M_1 have been retained in H_1, this check had *no chance* to uncover the various other forms of dependence that could have been responsible for $\rho \neq 0$. Thus any inference to H_1 lacks severity. The resulting model will *appear* to have corrected for autocorrelation even when it is statistically inadequate. If used for the "primary" statistical inferences, the actual error probabilities are much higher than the ones it is thought to license, and such inferences are unreliable at predicting values beyond the data used. "This is the kind of cure that kills the patient," Spanos warns. What should we do instead?

Probabilistic Reduction: Spanos

Spanos shows that any statistical model can be specified in terms of probabilistic assumptions from three broad categories: Distribution, Dependence, and Heterogeneity. In other words, a model emerges from selecting probabilistic assumptions from a menu of three groups: a choice of distribution; of type of dependence, if any; and a type of heterogeneity, i.e., how the generating mechanism remains the same or changes over the ordering of interest, such as time, space, or individuals. The LRM reflects just one of many ways of reducing the set of all possible models that could have given rise to the data $z_0 = \{(x_t, y_t), t = 1, \ldots, n\}$: Normal, Independent, Identically Distributed (NIID). Statistical inference need not be hamstrung by the neat and tidy cases. As a first step, we partition the set of all possible models coarsely:

	Distribution	Dependence	Heterogeneity
LRM	Normal	Independent	Identically distributed
Alternative (coarse partition)	Non-normal	Dependent	Non-IID

Since we are partitioning or reducing the space of models by means of the probabilistic assumptions, Spanos calls it the *Probabilistic Reduction* (PR)

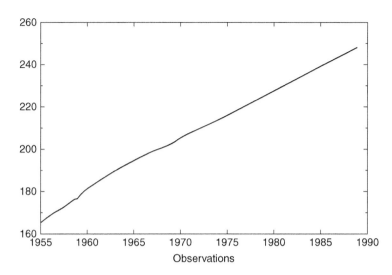

Figure 4.5 USA population (y) over time.

approach (first in Spanos 1986, 1999). The PR approach to M-S testing weaves together threads from Box–Jenkins, and what some dub the LSE (London School of Economics) tradition. Rather than give the assumptions by means of the error term, as is traditional, Spanos will specify them in terms of the observable random variables (x_t, y_t). This has several advantages. For one thing, it brings out hidden assumptions, notably assuming the parameters (β_0, β_1, σ^2) do not change with t. This is called *t-homogeneity* or *t*-invariance. Second, we can't directly probe errors given by the error term, but we can indirectly test them from the data.

We ask: What would be expected if each data series were to have come from a NIID process? Compare a typical realization of a NIID process (Figure 4.2) with the two series (t, x_t) and (t, y_t), in Figures 4.5 and 4.6, called *t*-plots.

Clearly, neither data series looks like the NIID of Figure 4.2. In each case the mean is increasing with time – there's a strong upward trend. Econometricians never use a short phrase when a long one will do, they call the trending mean: *mean heterogeneity*. The data can't be viewed as a realization of identically distributed random variables: ID is false. The very assumption of linear correlation between x and y is that x has a mean μ_x, and y has mean μ_y. If these are changing over the different samples, your estimate of correlation makes no sense.

We respecify, by adding terms of form t, t^2, . . ., to the model M_0. We don't know how far we'll have to go. We aren't inferring anything yet, just building a statistical model whose adequacy for the primary statistical inference will be

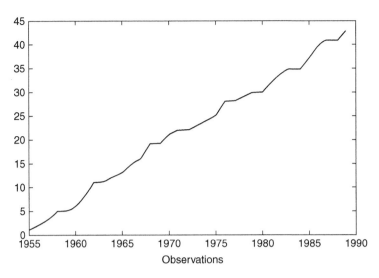

Figure 4.6 Secret variable (*x*) over time.

tested in its own right. With these data, adding *t* suffices to capture the trend, but in building the model you may include higher terms, allowing some to be found unnecessary later on. We can summarize our progress in detecting potential departures from the LRM model assumptions thus far:

	Distribution	Dependence	Heterogeneity
LRM	Normal	Independent	Identically distributed
Alternative	?	?	Mean heterogeneity

What about the independence assumption? We could check dependence if our data were ID and not obscured by the influence of the trending mean. We can "subtract out" the trending mean in a generic way to see what it would be like without it. Figures 4.7 and 4.8 show the *detrended* x_t and y_t. Reading data plots, and understanding how they connect to probabilistic assumptions, is a key feature of the PR approach.

The detrended data in both figures indicate, to a trained eye, positive dependence or "memory" in the form of cycles – this is called Markov dependence. So the independence assumption also looks problematic, and this explains the autocorrelation detected by the Durbin–Watson test and runs tests. As with trends, it comes in different orders, depending on how long the memory is found to be. It is modeled by adding terms called lags. To y_t add y_{t-1}, y_{t-2}, . . ., as many as needed. Likewise to x_t add x_{t-1}, x_{t-2}, . . . Our assessment so far, just on the basis of the graphical analysis is:

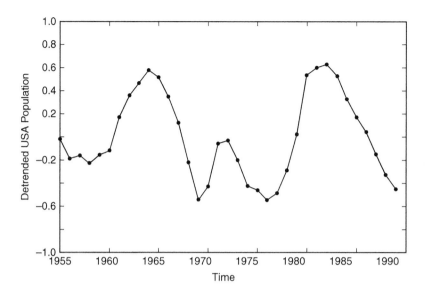

Figure 4.7 Detrended population data.

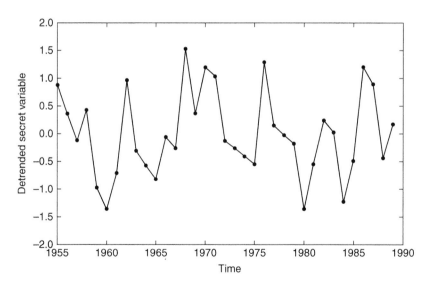

Figure 4.8 Detrended secret variable data.

	Distribution	Dependence	Heterogeneity
LRM	Normal	Independent	Identically distributed
Alternative	?	Markov	Mean heterogeneity

Finally, if we can see what the data $z_0 = \{(x_t, y_t), t = 1, 2, \ldots, 35\}$ would look like without the heterogeneity ("detrended") and without the dependence ("dememorized"), we could get some ideas about the appropriateness of the Normality assumption. We do this by subtracting them out "on paper" again (shown on graphs). The scatter-plot of (x_t, y_t) shows the elliptical pattern expected for Normality (though I haven't included a figure). We can organize our respecified model as an alternative to the LRM.

	Distribution	Dependence	Heterogeneity
LRM	Normal	Independent	Identically distributed
Alternative	Normal	Markov	Mean heterogeneity

While there are still several selections under each of the menu headings of Markov dependence and mean heterogeneity, the length of the Markov dependence (m), and the degree (ℓ) of the polynomial in t, I'm imagining we've carried out the subsequent rounds of the probing strategy.

The model derived by re-partitioning the set of all possible models, using the new reduction assumptions of Normality, Markov, and mean heterogeneity is called the Dynamic Linear Regression Model (DLRM). These data require one trend and two lags:

$$M_2: \ y_t = \beta_0 + \beta_1 x_t + ct \text{ [trending mean]} + (\gamma_1 y_{t-1} + \gamma_2 y_{t-2})$$
$$+ (\gamma_3 x_{t-1} + \gamma_4 x_{t-2}) \text{ [temporal lags]} + \varepsilon_t.$$

Back to the Primary Statistical Inference. Having established the statistical adequacy of the respecified model M_2 we are then licensed in making primary statistical inferences about the values of its parameters. In particular, does the secret variable help to predict the population of the USA (y_t)? No. A test of joint significance of the coefficients of (x_t, x_{t-1}, x_{t-2}), yields a P-value of 0.823 (using an F test). We cannot reject the hypothesis that they are all 0, indicating that x contributed nothing towards predicting or explaining y. The regression between x_t and y_t suggested by models M_0 and M_1 turns out to be a spurious or nonsense regression. Dropping the x variable from the model and reestimating the parameters we get what's called an *Autoregressive Model* of order 2 (for the 2 lags) with a trend

$y_t = 17 + 0.2t + 1.5y_{t-1} - 0.6y_{t-2} + \varepsilon_t.$

The Secret Variable Revealed. At this point, Spanos revealed that x_t was the number of pairs of shoes owned by his grandmother over the observation period! She lives in the mountains of Cyprus, and at last count continues to add to her shoe collection. You will say this is a quirky made-up example, sure. It serves as a "canonical exemplar" for a type of erroneous inference. Some of the best known spurious correlations can be explained by trending means. For live exhibits, check out an entire website by Tyler Vigen devoted to exposing them. I don't know who collects statistics on the correlation between death by getting tangled in bed sheets and the consumption of cheese, but it's exposed as nonsense by the trending means. An example from philosophy that is similarly scotched is the case of sea levels in Venice and the price of bread in Britain (Sober 2001), as shown by Spanos (2010d, p. 366). In some cases, x is a variable that theory suggests is doing real work; discovering the misspecification effectively falsifies the theory from which the statistical model is derived.

I've omitted many of the tests, parametric and non-parametric, single assumption and joint (several assumptions), used in a full application of the same ideas, and mentioned only the bare graphs for simplicity. As you add questions you might wish to pose, they become your new primary inferences. The first primary statistical inference might indicate an effect of a certain magnitude passes with severity, and then background information might enter to tell if it's substantively important. At yet another level, the question might be to test a new model with variables to account for the trending mean of an earlier stage, but this gets beyond our planned M-S testing itinerary. That won't stop us from picking up souvenirs.

Souvenir V: Two More Points on M-S Tests and an Overview of Excursion 4

M-S Tests versus Model Selection: Fit Is Not Enough. M-S tests are distinct from model selection techniques that are so popular. Model selection begins with a family of models to be ranked by one or another criterion. Perhaps the most surprising implication of statistical inadequacy is to call into question the most widely used criterion of model selection: the goodness-of-fit/prediction measures. Such criteria rely on the "smallness" of the residuals. Mathematical fit isn't the same as what's needed for statistical inference. The residuals can be "small" while systematically different from white noise.

Members of the model selection tribe view the problem differently. Model selection techniques reflect the worry of overfitting: that if you add enough factors (e.g., $n - 1$ for sample size n), the fit can be made as good as desired, even if the model is inadequate for future prediction. (In our examples the factors took the form of trends or lags.) Thus, model selection techniques make you pay a penalty for the number of factors. We share this concern – it's too easy to attain fit without arriving at an adequate model. The trouble is that it remains easy to jump through the model selector's hoops, and still not achieve model adequacy, in the sense of adequately capturing the systematic information in the data. The goodness-of-fit measures already assume the likelihood function, when that's what the M-S tester is probing.

Take the Akaike Information Criterion (AIC) developed by Akaike in the 1970s (Akaike 1973). (There are updated versions, but nothing in our discussion depends on this.) The best known defenders of this account in philosophy are Elliott Sober and Malcolm Forster (Sober 2008, Forster and Sober 1994). An influential text in ecology is by Burnham and Anderson (2002). Model selection begins with a family of models such as the LRM: $y_t = \beta_0 + \beta_1 x_t + u_t$. They ask: Do you get a better fit – smaller residual – if you add x^2_t? What about adding both x^2_t and x^3_t terms? And so on. Each time you add a factor, the fit improves, but Akaike kicks you in the shins and handicaps you by 1 for the additional parameter. The result is a preference ranking of models by AIC score.[4] For the granny shoe data above, the model that AIC likes best is

$$y_t = \beta_0 + \beta_1 x_t + \beta_2 x^2_t + \beta_3 x^3_t + u_t.$$

Moreover, it's selection is within this family. But we know that all these LRMs are statistically inadequate! As with other purely comparative measures, there's no falsification of models.

What if we start with the adequate model that the PR arrived at, the autoregressive model with a trend? In that case, the AIC ranks at the very top of the model with the wrong number of trends. That is, it ranks a statistically inadequate model higher than the statistically adequate one. Moreover, the Akaike method for ranking isn't assured of having decent error probabilities. When the Akaike ranking is translated into a N-P test comparing this pair of models, the Type I error probability is around 0.18, and

---

[4] For each y_t form the residual squared. The sum of the squared residuals:

$$\hat{\sigma}^2 = \frac{1}{n} \sum_{t=1}^{n} \left(y_t - \hat{\beta}_0 - \sum_{j=1}^{3} \hat{\beta}_j x^j_t \right)^2$$

gives an estimate of σ^2 for the model.

The AIC score for each contender in the case of the LRM, with sample size n, is $\log(\hat{\sigma}^2) + 2K/n$, where K is the number of parameters in model i. The models are then ranked with the smallest being preferred. The log-likelihood is the goodness-of-fit measure which is traded against simplicity, but if the statistical model is misspecified, one is using the wrong measure of fit.

For a comparison of the AIC using these data, and a number of related model-selection measures, see Spanos (2010a). None of these points change using the unbiased variant of AIC.

no warning of the laxity is given. As noted, model selection methods don't hone in on models outside the initial family. By contrast, building the model through our M-S approach is intended to accomplish both tasks – building and checking – in one fell swoop.

Leading proponents of AIC, Burnham and Anderson (2014, p. 627), are quite critical of error probabilities, declaring "P values are not proper evidence as they violate the likelihood principle (Royall 1997)." This tells us their own account forfeits control of error probabilities. "Burnham and Anderson (2002) in their textbook on likelihood methods for assessing models warn against data dredging . . . But there is nothing in the evidential measures recommended by Burnham and Anderson" to pick up on this (Dienes 2008, p. 144). See also Spanos (2014).

M-S Tests and Predesignation. Don't statistical M-S tests go against the error statistician's much-ballyhooed requirement that hypotheses be predesignated? The philosopher of science Rosenkrantz says yes:

[O]rthodox tests ... show how to test underlying assumptions of randomness, independence and stationarity, where none of these was the predesignated object of the test (the "tested hypothesis"). And yet, astoundingly in the face of all this, orthodox statisticians are one in their condemnation of "shopping for significance," picking out significant correlations in data post hoc, or "hunting for trends. . .". It is little wonder that orthodox tests tend to be highly ambivalent on the matter of predesignation. (Rosenkrantz 1977, 204–5)

Are we hoisted by our own petards? No. This is another case where failing to disentangle a rule's *raison d'être* leads to confusion. The aim of predesignation, as with the preference for novel data, is to avoid biasing selection effects in your primary statistical inference (see Tour III). The data are remodeled to ask a different question. Strictly speaking our model assumptions are predesignated as soon as we propose a given model for statistical inference. These are the pigeonholes in the PR menu. It has never been a matter of the time – of who knew what, when – but a matter of avoiding erroneous interpretations of the data at hand. M-S tests in the error statistical methodology are deliberately designed to be independent of (or orthogonal to) the primary question at hand. The model assumptions, singly or in groups, arise as argumentative assumptions, ready to be falsified by criticism. In many cases, the inference is as close to a deductive falsification as to be wished.

Parametric tests of assumptions may themselves have assumptions, which is why judicious combinations of varied tests are called upon to ensure their overall error probabilities. Order matters: Tests of the distribution, e.g.,

Normal, Binomial, or Poisson, assume IID, so one doesn't start there. The inference in the case of an M-S test of assumptions is not a statistical inference to a *generalization*: It's explaining given data, as with explaining a "known effect," only keeping to the statistical categories of distribution, independence/ dependence, and homogeneity/heterogeneity (Section 4.6). Rosenkrantz's concerns pertain to the kind of pejorative hunting for variables to include in a substantive model. That's always kept distinct from the task of M-S testing, including respecifying.

Our argument for a respecified model is a *convergent* argument: questionable conjectures along the way don't bring down the tower (section 1.2). Instead, problems ramify so that the specification finally deemed adequate has been sufficiently severely tested for the task at hand. The trends and perhaps the lags that are required to render the statistical model adequate generally cry out for a substantive explanation. It may well be that different statistical models are adequate for probing different questions.[5] Violated assumptions are responsible for a good deal of non-replication, and yet it has gone largely unattended in current replication research.

Take-away of Excursion 4. For a severe tester, a crucial part of a statistical method's objectivity (Tour I) is registering how test specifications such as sample size (Tour II) and biasing selection effects (Tour III) alter its error-probing capacities. Testing assumptions (Tour IV) is also crucial to auditing. If a probabilist measure such as a Bayes factor is taken as a gold standard for critiquing error statistical tests, significance levels and other error probabilities appear to overstate evidence – at least on certain choices of priors. From the perspective of the severe tester, it can be just the reverse. Preregistered reports are promoted to advance replication by blocking selective reporting. Thus there is a tension between preregistration and probabilist accounts that downplay error probabilities, that declare them only relevant for long runs, or tantamount to considering hidden intentions. Moreover, in the interest of promoting Bayes factors, researchers who most deserve censure are thrown a handy life preserver. Violating the LP, using the sampling distribution for inferences with the data at hand, and the importance of error probabilities form an interconnected web of severe testing. They are necessary for every one of the requirements for objectivity.

---

[5] When two different models capture the data adequately, they are called *reparameterizations* of each other.

Excursion 5 Power and Severity

Itinerary

Tour I Power: Pre-data and Post-data

> A salutary effect of power analysis is that it draws one forcibly to consider the magnitude of effects. In psychology, and especially in soft psychology, under the sway of the Fisherian scheme, there has been little consciousness of how big things are. (Cohen 1990, p. 1309)

So how would you use power to consider the magnitude of effects were you drawn forcibly to do so? In with your breakfast is an exercise to get us started on today's shore excursion.

> Suppose you are reading about a statistically significant result x (just at level α) from a one-sided test T+ of the mean of a Normal distribution with n IID samples, and known σ: H_0: $\mu \leq 0$ against H_1: $\mu > 0$.
>
> Underline the correct word, from the perspective of the (error statistical) philosophy, within which power is defined.

- If the test's power to detect μ' is very low (i.e., POW(μ') is low), then the statistically significant x is <u>poor</u>/good evidence that $\mu > \mu'$.
- Were POW(μ') reasonably high, the inference to $\mu > \mu'$ is <u>reasonably</u>/poorly warranted.

We've covered this reasoning in earlier travels (e.g., Section 4.3), but I want to launch our new tour from the power perspective. Assume the statistical test has passed an audit (for selection effects and underlying statistical assumptions) – you can't begin to analyze the logic if the premises are violated.

During our three tours on Power Peninsula, a partially uncharted territory, we'll be residing at local inns, not returning to the ship, so pack for overnights. We'll visit its museum, but mostly meet with different tribal members who talk about power – often critically. Power is one of the most abused notions in all of statistics, yet it's a favorite for those of us who care about magnitudes of discrepancies. Power is always defined in terms of a fixed cut-off, c_α, computed under a value of the parameter under test; since these vary, there is really a *power function*. If someone speaks of the power of a test *tout court*, you cannot make sense of it, without qualification. First defined in Section 3.1, the *power* of a test against μ' is the probability it would lead to rejecting H_0 when $\mu = \mu'$:

POW(T, μ') = Pr(d(X) $\geq c_\alpha$; $\mu = \mu'$), or Pr(test T rejects H_0; $\mu = \mu'$).

If it's clear what the test is, we just write POW(μ'). Power measures the capability of a test to detect μ' – where the detection is in the form of producing a $d \geq c_\alpha$. While power is computed at a point $\mu = \mu'$, we employ it to appraise claims of form $\mu > \mu'$ or $\mu < \mu'$.

Power is an ingredient in N-P tests, but even practitioners who declare they never set foot into N-P territory, but live only in the land of Fisherian significance tests, invoke power. This is all to the good, and they shouldn't fear that they are dabbling in an inconsistent hybrid.

Jacob Cohen's (1988) *Statistical Power Analysis for the Behavioral Sciences* is displayed at the Power Museum's permanent exhibition. Oddly, he makes some slips in the book's opening. On page 1 Cohen says: "The power of a statistical test is the probability it will yield statistically significant results." Also faulty is what he says on page 4: "The power of a statistical test of a null hypothesis is the probability that it will lead to the rejection of the null hypothesis, i.e., the probability that it will result in the conclusion that the phenomenon exists." Cohen means to add "computed under an alternative hypothesis," else the definitions are wrong. These snafus do not take away from Cohen's important tome on power analysis, yet I can't help wondering if these initial definitions play a bit of a role in the tendency to define power as 'the probability of a correct rejection,' which slips into erroneously viewing it as a posterior probability (unless qualified).

Although keeping to the fixed cut-off c_α is too coarse for the severe tester's tastes, it is important to keep to the given definition for understanding the statistical battles. We've already had sneak previews of "achieved sensitivity" or "attained power" [$\Pi(\gamma)$ = Pr(d(X) $\geq$ d(x_0); $\mu_0 + \gamma$)] by which members of Fisherian tribes are able to reason about discrepancies (Section 3.3). N-P accorded three roles to power: the first two are pre-data, for planning and comparing tests; the third is for interpretation post-data. It's the third that they don't announce very loudly, whereas that will be our main emphasis. Have a look at this museum label referring to a semi-famous passage by E. Pearson. Barnard (1950, p. 207) has just suggested that error probabilities of tests, like power, while fine for pre-data planning, should be replaced by other measures (likelihoods perhaps?) after the trial. What did Egon say in reply to George?

[I]f the planning is based on the consequences that will result from following a rule of statistical procedure, e.g., is based on a study of the power function of a test and then, having obtained our results, we do not follow the first rule but another, based on likelihoods, what is the meaning of the planning? (Pearson 1950, p. 228)

This is an interesting and, dare I say, powerful reply, but it doesn't quite answer George. By all means apply the rule you planned to, but there's still a legitimate question as to the relationship between the pre-data capability or performance measure, and post-data inference. The severe tester offers a view of this intimate relationship. In Tour II we'll be looking at interactive exhibits far outside the museum, including N-P post-data power analysis, retrospective power, and a notion I call shpower. Employing our understanding of power, scrutinizing a popular reinterpretation of tests as diagnostic tools will be straightforward. In Tour III we go a few levels deeper in disinterring the N-P vs. Fisher feuds. I suspect there is a correlation between those who took Fisher's side in the early disputes with Neyman and those leery of power. Oscar Kempthorne being interviewed by J. Leroy Folks (1995) said:

Well, a common thing said about [Fisher] was that he did not accept the idea of the power. But, of course, he must have. However, because Neyman had made such a point about power, Fisher couldn't bring himself to acknowledge it (p. 331).

However, since Fisherian tribe members have no problem with corresponding uses of sensitivity, P-value distributions, or CIs, they can come along on a severity analysis. There's more than one way to skin a cat, if one understands the relevant statistical principles. The issues surrounding power are subtle, and unraveling them will require great care, so bear with me. I will give you a money-back guarantee that by the end of the excursion you'll have a whole new view of power. Did I mention you'll have a chance to power the ship into port on this tour? Only kidding, however, you will get to show your stuff in a Cruise Severity Drill (Section 5.2).

5.1   Power Howlers, Trade-offs, and Benchmarks

In the Mountains out of Molehills (MM) Fallacy (Section 4.3), a rejection of H_0 just at level α with a larger sample size (higher power) is taken as evidence of a greater discrepancy from H_0 than with a smaller sample size (in tests otherwise the same). Power can be increased by increasing sample size, but also by computing it in relation to alternatives further and further from H_0. Some are careful to avoid the MM fallacy when the high power is due to large n, but then fall right into it when it is due to considering a very discrepant μ'. For our purposes, our one-sided T+ will do.

Mountains out of Molehills (MM) Fallacy (second form). Test T+: The fallacy of taking a just statistically significant difference at level α (i.e., $d(x_0) = d_\alpha$) as a better indication of a discrepancy μ' if the POW (μ') is high, than if POW(μ') is low.

Two Points Stephen Senn Correctly Dubs Nonsense and Ludicrous

Start with an extreme example: Suppose someone is testing H_0: the drug cures no one. An alternative H_1 is it cures nearly everyone. Clearly these are not the only possibilities. Say the test is practically guaranteed to reject H_0, if in fact H_1, the drug cures practically everyone. The test has high power to detect H_1. You wouldn't say that its rejecting H_0 is evidence of H_1. H_1 entails it's very probable you'll reject H_0; but rejecting H_0 doesn't warrant H_1. To think otherwise is to allow problematic statistical affirming the consequent – the basis for the MM fallacy (Section 2.1). This obvious point lets you zero in on some confusions about power.

Stephen Senn's contributions to statistical foundations are once again spot on. In drug development, it is typical to require a high power of 0.8 or 0.9 to detect effects deemed of clinical relevance. The clinically relevant discrepancy, as Senn sees it, is the discrepancy "one should not like to miss" (2007, p. 196). Senn labels this delta Δ. He is considering a difference between means, so the null hypothesis is typically 0. We'll apply severity to his example in Exhibit (iv) of this tour. Here the same points will be made with respect to our one-sided Normal test T+: H_0: $\mu \leq \mu_0$ vs. H_1: $\mu > \mu_0$, letting $\mu_0 = 0$, σ known. We may view Δ as the value of μ of clinical relevance. (Nothing changes in this discussion if it's estimated as s.) The test takes the form

Reject H_0 iff $Z \geq z_\alpha$ (Z is the standard Normal variate).

"Reject H_0" is the shorthand for "infer a statistically significant difference" at the level of the test. Though Z is the test statistic, it makes for a simpler presentation to use the cut-off for rejection in terms of $\bar{x}_\alpha$: Reject H_0 iff $\bar{X} \geq \bar{x}_\alpha = (\mu_0 + z_\alpha \sigma/\sqrt{n})$.

Let's abbreviate the alternative against which test T+ has 0.8 power by $\mu^{.8}$, when it's clear what test we're talking about. So POW($\mu^{.8}$) = 0.8, and let's suppose $\mu^{.8}$ is the clinically relevant difference Δ. Senn asks, what does $\mu^{.8}$ mean in relation to *what we are entitled to infer* when we obtain statistical significance? Can we say, upon rejecting the null hypothesis, that the treatment has a clinically relevant effect, i.e., $\mu \geq \mu^{.8}$ (or $\mu > \mu^{.8}$)?

"This is a surprisingly widespread piece of nonsense which has even made its way into one book on drug industry trials" (ibid., p. 201). The reason it is nonsense, Senn explains, is that $\mu^{.8}$ must be in excess of the cut-off for rejection, in particular, $\mu^{.8} = \bar{x}_\alpha + 0.85 \, \sigma_{\bar{x}}$ (where $\sigma_{\bar{x}} = \sigma/\sqrt{n}$). We know we are only entitled to infer μ exceeds the *lower bound* of the confidence interval at a reasonable level; whereas, $\mu^{.8}$ is actually the upper bound of a 0.8 (one-sided) confidence interval, formed having observed $\bar{x} = \bar{x}_\alpha$. All we are to infer, officially, from just reaching the cut-off $\bar{x}_\alpha$, is that $\mu > 0$.

Granted, as Senn admits, the test "lacks ambition" (ibid., p. 202), but with more data and with results surpassing the minimal cut-off, we may uncover a clinically relevant discrepancy. Why not just set up the test to enable the clinically relevant discrepancy to be inferred whenever the null is rejected?

$$H_0\colon \mu \le \Delta \text{ vs. } H_1\colon \mu > \Delta.$$

This requires redefining Δ. "It is no longer 'the difference we should not like to miss' but instead becomes 'the difference we should like to prove obtains'" (ibid.). Some call this the "clinically irrelevant difference" (ibid.). But then we can't also have high power to detect $H_1\colon \mu > \Delta$.

[I]f the true treatment difference is Δ, then the observed treatment difference will be less than Δ in approximately 50% of all trials. Therefore, the probability that it is less than the critical value must be greater than 50%. (ibid., p. 202)

Indeed, it will be approximately $1 - \alpha$. So the power – the probability the observed difference exceeds the critical value under H_1 – is, in this case, around α. The researcher is free to specify the null as $H_0\colon \mu \le \Delta$, but Senn argues against doing so, at least in drug testing, because "a nonsignificant result will often mean the end of the road for a treatment. It will be lost forever. However, a treatment which shows a 'significant' effect will be studied further" (ibid.). This goes beyond issues of interest now. The point is: Δ cannot be the value in H_0 and also the value against which we want 0.8 power to detect, i.e., $\mu^{.8}$.

If testing $H_0\colon \mu \le 0$ vs. $H_1\colon \mu > 0$, then a just α-significant result is poor evidence for $\mu \ge \mu^{.8}$ (or other alternative with high power). To think it's good evidence is *nonsense*. Senn's related point is that it is *ludicrous* to assume the effect is either 0 or a clinically relevant difference, as if we are testing

$$H_0\colon \mu = 0 \text{ vs. } H_1\colon \mu > \Delta.$$

"But where we are unsure whether a drug works or not, it would be ludicrous to maintain that it cannot have an effect which, while greater than nothing, is less than the clinically relevant difference" (ibid., p. 201). That is, it is ludicrous to cut out everything in between 0 and Δ. By the same token, it would seem odd to give a 0.5 prior probability to H_0, and the remaining 0.5 to H_1. We will have plenty of occasions to return to Senn's points about what's nonsensical and ludicrous.

Trade-offs and Benchmarks

Between H_0 and $\overline{x}_\alpha$ the power goes from α to 0.5. Keeping to our simple test T+ will amply reward us here.

a. *The power against H_0 is α.* We can use the power function to define the probability of a Type I error or the significance level of the test:

$$\text{POW}(\text{T+}, \mu_0) = \Pr(\overline{X} \geq \overline{x}_\alpha; \mu_0), \text{ where } \overline{x}_\alpha = \mu_0 + z_\alpha\, \sigma_{\overline{X}}$$

Standardizing $\overline{X}$, we get $Z = [(\mu_0 + z_\alpha\, \sigma_{\overline{X}}) - \mu_0]/\sigma_{\overline{X}}$.

The power at the null is $\Pr(Z \geq z_\alpha; \mu_0) = \alpha$.

It's the *low power* against H_0 that warrants taking a rejection as evidence that $\mu > \mu_0$. This is desirable: we infer an indication of discrepancy from H_0 because a null world would probably have resulted in a smaller difference than we observed.

b. *The power of T+ for $\mu_1 = \overline{x}_\alpha$ is 0.5.* In that case, $Z = 0$, and $\Pr(Z \geq 0) = 0.5$, so

$$\text{POW}(\text{T+}, \mu_1 = \overline{x}_\alpha) = 0.5.$$

The power only gets to be greater than 0.5 for alternatives that exceed the cut-off $\overline{x}_\alpha$, whatever it is. As noted, $\mu^{.8} = \overline{x}_\alpha + 0.85\, \sigma_{\overline{X}}$ since $\text{POW}(\text{T+}, \overline{x}_\alpha + 0.85\sigma_{\overline{X}}) = 0.8$. Tests ensuring 0.9 power are also often of interest: $\mu^{.9} = \overline{x}_\alpha + 1.28\sigma_{\overline{X}}$. We get these shortcuts:

Case 1: $\text{POW}(\text{T+}, \mu)$ for μ between H_0 and $\mu = \overline{x}_\alpha$:
If $\mu_1 = \overline{x}_\alpha - k\sigma_{\overline{X}}$ then $\text{POW}(\text{T+}, \mu_1)$ = area to the right of k under N(0,1) ($\leq$ 0.5).

Case 2: $\text{POW}(\text{T+}, \mu)$ for μ greater than $\overline{x}_\alpha$:
If $\mu_1 = \overline{x}_\alpha + k\sigma_{\overline{X}}$ then $\text{POW}(\text{T+}, \mu_1)$ = area to the right of $-k$ under N (0,1) (> 0.5).
Remember $\overline{x}_\alpha$ is $\mu_0 + z_\alpha\sigma_{\overline{X}}$.

Trade-offs Between the Type I and Type II Error Probability

We know that, for a given test, as the probability of a Type I error goes down the probability of a Type II error goes up (and power goes down). And as the probability of a Type II error goes down (and power goes up), the probability of a Type I error goes up, assuming we leave everything else the same. There's a trade-off between the two error probabilities. (No free lunch.) So if someone said: As the power increases, the probability of a Type I error *decreases*, they'd be saying, as the Type II error decreases, the probability of a Type I error decreases. That's the opposite of a trade-off! You'd know automatically they had made a mistake or were simply defining things in a way that differs from standard N-P statistical tests. Now you may say, "I don't care about Type I and II errors, I'm interested in inferring estimated effect sizes." I too want to infer magnitudes. But those will be ready to hand once we tell what's true about the existing concepts.

While $\mu^{.8}$ is obtained by adding 0.85 $\sigma_{\overline{X}}$ to $\overline{x}_\alpha$, in day-to-day rounding, if you're like me, you're more likely to remember the result of adding $1\sigma_{\overline{X}}$ to $\overline{x}_\alpha$. That takes us to a value of μ against which the test has 0.84 power, $\mu^{.84}$:

> The power of test T+ to detect an alternative that exceeds the cut-off $\overline{x}_\alpha$ by $1\sigma_{\overline{X}} = 0.84$.

In test T+ the range of possible values of $\overline{X}$ and μ are the same, so we are able to set μ values this way, without confusing the parameter and sample spaces.

Exhibit (i). Let test T+ ($\alpha = 0.025$) be H_0: $\mu = 0$ vs. H_1: $\mu \geq 0$, $\alpha = 0.025$, $n = 25$, $\sigma = 1$. Using the $2\sigma_{\overline{X}}$ cut-off: $\overline{x}_{0.025} = 2(1)/\sqrt{25} = 0.4$ (using 1.96 it's 3.92). Suppose you are instructed to decrease the Type I error probability α to 0.001 but it's impossible to get more samples. This requires the hurdle for rejection to be higher than in our original test. The new cut-off for test T+ will be $\overline{x}_{0.001}$. It must be 3 $\sigma_{\overline{X}}$ greater than 0 rather than only $2\sigma_{\overline{X}}$: $\overline{x}_{0.001} = 0 + 3(1)/\sqrt{25} = 0.6$. We decrease α (the Type I error probability) from 0.025 to 0.001 by moving the hurdle over to the right by $1\sigma_{\overline{X}}$ unit. But we've just made the power lower for any discrepancy or alternative. For what value of μ does this new test have 0.84 power?

POW(T+, $\alpha = 0.001$, $\mu^{.84} = ?$) = 0.84.

We know: $\mu^{.84} = 0.6 + (0.2) = 0.8$. So, POW(T+, $\alpha = 0.001$, $\mu = 0.8$) = 0.84. Decreasing the Type I error by moving the hurdle over to the right by $1\sigma_{\overline{X}}$ unit results in the alternative against which we have 0.84 power $\mu^{.84}$ also moving over to the right by $1\sigma_{\overline{X}}$ (Figure 5.1). We see the trade-off very neatly, at least in one direction.

Consider the discrepancy of $\mu = 0.6$ (Figure 5.2). The power to detect 0.6 in test T+ ($\alpha = 0.001$) is now only 0.5! In test T+ ($\alpha = 0.025$) it is 0.84. Test T+ ($\alpha = 0.001$) is less powerful than T+ ($\alpha = 0.025$).

Should you hear someone say that the higher the power, the higher the *hurdle* for rejection, you'd know they are confused or using terms in an

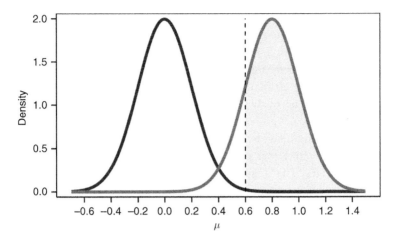

Figure 5.1 POW(T+, $\alpha = 0.001$, $\mu = 0.8$) = 0.84.

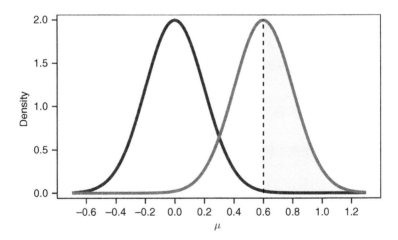

Figure 5.2 POW(T+, $\alpha = 0.001$, $\mu = 0.6$) = 0.5.

incorrect way. (The hurdle is how large the cut-off must be before rejecting at the given level.) Why then do Ziliak and McCloskey, popular critics of significance tests, announce: "refutations of the null are trivially easy to achieve if power is low enough or the sample is large enough" (2008a, p. 152)? Increasing sample size means increased power, so the second disjunct is correct. The first disjunct is not. One might be tempted to suppose they mean "power is high enough," but one would be mistaken. They mean what they wrote. Aris Spanos (2008a) points this out (in a review of their book), and I can't figure out why they dismiss such corrections as "a lot of technical smoke" (2008b, p. 166).

Ziliak and McCloskey Get Their Hurdles in a Twist

Still, their slippery slides are quite illuminating.

If the power of a test is low, say, 0.33, then the scientist will two times in three accept the null and mistakenly conclude that another hypothesis is false. If on the other hand the power of a test is high, say, 0.85 or higher, then the scientist can be reasonably confident that at minimum the null hypothesis (of, again, zero effect if that is the null chosen) is false and that therefore his rejection of it is highly probably correct. (Ziliak and McCloskey 2008a, p. 132–3)

With a wink and a nod, the first sentence isn't too bad, even though, at the very least, it is mandatory to specify a particular "another hypothesis," μ'. But what about the statement: if the power of a test is high, then a rejection of the null is probably correct?

We follow our rule of generous interpretation to try to see it as true. Let's allow the ";" in the first premise to be a conditional probability "|", using $\mu^{.84}$:

1. Pr(Test T+ rejects the null | $\mu^{.84}$) = 0.84.
2. Test T+ rejects the null hypothesis.
   Therefore, the rejection is correct with probability 0.84.

Oops. The premises are true, but the conclusion fallaciously transposes premise 1 to obtain conditional probability $Pr(\mu^{.84}$ | test T+ rejects the null) = 0.84. What I think they want to say, or at any rate what would be correct, is

$$Pr(\text{Test T+ } \textit{does not} \text{ reject the null hypothesis} \mid \mu^{.84}) = 0.16.$$

So the Type II error probability is 0.16. Looking at it this way, the flaw is in computing the complement of premise 1 by transposing (as we saw in the Higgs example, Section 3.8). Let's be clear about significance levels and hurdles. According to Ziliak and McCloskey:

It is the history of Fisher significance testing. One erects little "significance" hurdles, six inches tall, and makes a great show of leaping over them, . . . If a test does a good job of uncovering efficacy, then the test has high power and the hurdles are high not low. (ibid., p. 133)

They construe "little significance" as little hurdles! It explains how they wound up supposing high power translates into high hurdles. It's the opposite.

The higher the hurdle, the more difficult it is to reject, and the lower the power. High hurdles correspond to insensitive tests, like insensitive fire alarms. It might be that using "sensitivity" rather than power would make this abundantly clear. We may coin: The high power = high hurdle (for rejection) fallacy. A powerful test does give the null hypothesis a harder time in the sense that it's more probable that discrepancies from it are detected. That makes it easier for H_1. Z & M have their hurdles in a twist.

5.2   Cruise Severity Drill: How Tail Areas (Appear to) Exaggerate the Evidence

The most influential criticisms of statistical significance tests rest on highly plausible intuitions, at least from the perspective of a probabilist. We are about to visit a wonderfully instructive example from Steven Goodman. It combines the central skills gathered up from our journey, but with a surprising twist. As always, it's a canonical criticism – not limited to Goodman. He happens to give a much clearer exposition than most, and, on top of that, is frank about his philosophical standpoint. Let's listen:

To examine the inferential meaning of the p value, we need to review the concept of inductive evidence. An inductive measure assigns a number (a measure of support or credibility) to a hypothesis, given observed data. ... By this definition the p value is not an inductive measure of evidence, because it involves only one hypothesis and because it is based partially on unobserved data in the tail region.

To assess the quantitative impact of these philosophical issues, we need to turn to an inductive statistical measure: mathematical likelihood. (Goodman 1993, p. 490)

Well that settles things quickly. Influenced by Royall, Goodman has just listed the keynotes from the standpoint of "evidence is comparative likelihood" seen as far back as Excursion 1 (Tour II). Like the critics we visited in Excursion 4 (Sections 4.4 and 4.5), Goodman finds that the P-value exaggerates the evidence against a null hypothesis because the likelihood ratio (or Bayes factor) in favor of a chosen alternative is not as large as the P-value would suggest. He admits that one's assessment here will turn on philosophy. On Goodman's philosophy, it's the use of the tail area that deceitfully blows up the evidence against the null hypothesis. Now in Section 3.4, Jeffreys' tail area criticism, we saw that considering the tails makes it harder, not easier, to find evidence against a null. Goodman purports to show the opposite. That's the new twist.

Three Steps to the Argument

Goodman's context involves statistically significant results – he writes them as z values, as it is a case of Normal testing. We're not given the sample size or the precise test, but it won't matter for the key argument. He gives the two-sided α value, although "[w]e assume that we know the direction of the effect" (ibid., p. 491). I am not objecting. Even if we run a two-sided test, once we see the direction, it makes sense to look at the power of the relevant one-sided test, but double the α value. There are three steps. *First step*: form the likelihood ratio of the statistically significant outcome $z_{0.025}$ (i.e., 1.96) under the null hypothesis and some alternative (where Pr is the density):

$$\Pr(z_\alpha; H')/\Pr(z_\alpha; H_0).$$

But which alternative H'? Alternatives against which the test has high power, say 0.8, 0.84, or 0.9, are of interest, and he chooses $\mu^{.9}$. He writes this alternative as $\mu = \Delta_{0.05,\ 0.9}$, "the difference against which the hypothesis test has two-sided $\alpha = 0.05$ and one-sided $\beta = 0.10$ (power = 0.90)" (ibid., p. 496). We know from our benchmarks that $\mu^{.9}$ is approximately 1.28 standard errors from the one-sided cut-off: $(1.96 + 1.28)\sigma_{\overline{X}} = 3.24\sigma_{\overline{X}}$. The likelihood for alternative $\mu^{.9}$ is smaller than if one had used the maximum likely alternative (as in Johnson).[1]

I'll follow Goodman in computing the likelihood of the null over the alternative, although most of the authors we've considered do the reverse. Not that it matters so long as you keep them straight.

"Two likelihood ratios are compared, one for a 'precise' p value, e.g., $p = 0.03$, and one for … $p \le \alpha = 0.03$. The alternative hypothesis used here is the one against which the hypothesis test has 90 percent power (two-sided $\alpha = 0.05$) …" (ibid., pp. 490–1). The precise and imprecise P-values correspond to reporting $z = 1.96$ and $z \ge 1.96$, respectively.[2] The likelihood ratio for the precise P-value "corresponds to the ratio of heights of the two probability densities" (ibid., p. 491): Number this as (1):

(1) $\Pr(Z = 1.96; \mu_0)/\Pr(Z = 1.96; \mu^{.9}) = 0.058/0.176 = 0.33$.

---

[1] We saw this in Section 4.5. Goodman also shows the results for the maximum likely alternative.
[2] He writes the two descriptions as $p = \alpha$ vs. $p > \alpha$, but I think it's clearer using corresponding z values.

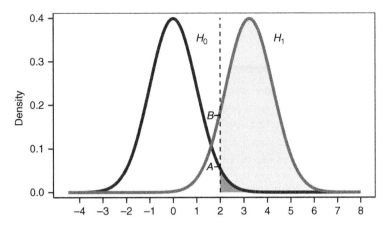

Figure 5.3 Comparing precise and imprecise P-values. The likelihood ratio is A/B, the ratio of the curve heights at the observed data. The likelihood ratio associated with the imprecise P-value ($p \leq \alpha$) is the ratio of the small darkly shaded area to the total shaded area (adapted from Goodman 1993, Figure 1 p. 492).

These are the ordinates not the tail areas of the Normal distribution.

That's the first step. The *second* step is to consider the likelihood ratio for the imprecise P-value, where the result is described coarsely as $\{z \geq 1.96\}$ rather than $z = 1.96$ (or equivalently $p \leq 0.025$ rather than $p = 0.025$):

(2) $\Pr(Z \geq 1.96; \mu_0)/ \Pr(Z \geq 1.96; \mu^{.9})$.

We see at once that the value of (2) is

$\alpha \ / \ \mathrm{POW}(\mu^{.9}) = 0.025/0.9 = 0.03$.

The comparative evidence for the null using (1) is considerably larger (0.33) than using (2) where it's only 0.03. Figure 5.3 shows what's being compared.

The difference Goodman wishes to highlight looks even more impressive if we flip the ratios in (1) and (2) to get the comparative evidence for the alternative compared to the null. We get $0.176/0.058 = 3.03$ using $z = 1.96$, and $0.9/0.025 = 36$ using $z \geq 1.96$. Either way you look at it, using the tail areas to compare support exaggerates the evidence against the null in favor of $\mu^{.9}$. Or so it seems.

Now for the *third* step: Assign priors of 0.5 to μ_0 and to $\mu^{.9}$:

With this alternative hypothesis $[\mu^{.9}]$, '$p \leq 0.05$' represents *11 times* (= 0.33/0.03) less evidence in support of the null hypothesis than does '$p = 0.05$.' Using Bayes' Theorem, with initial probabilities of 50 percent on both hypotheses (i.e., initial odds = 1), this means that after observing $p = 0.05$, the probability that the null hypothesis is true falls only to *25 percent* (= 0.33/(1 + 0.33)). When $p \leq 0.05$, the truth probability of the null hypothesis drops to 3 percent (= 0.03/(1 + 0.03)). (ibid., p. 491)

He concludes:

When we use the tail region to represent a result that is actually on the border, we misrepresent the evidence, making the case against the null hypothesis look much stronger than it actually is. (ibid.)

The posterior probabilities, he says, are assessments of credibility in the hypothesis. Join me for a break at the coffeehouse, called "Take Type II", where we'll engage a live exhibit.

Live Exhibit (ii): Drill Prompt. Using any relevant past travels, tours, and souvenirs, critically appraise his argument. Resist the impulse to simply ask, "where do significance tests recommend using error probabilities to form a likelihood ratio, and treat the result as a relative support measure?" Give him the most generous interpretation in order to see what's being claimed.

How do you grade the following test taker? Goodman actually runs two distinct issues together: The first issue contrasts a report of the observed P-values with a report of whether or not a predesignated cut-off is met (his "imprecise" P-value); the second issue is using likelihood ratios (as in (1)) as opposed to using tail areas. As I understand him, Goodman treats a report of the precise P-value as calling for a likelihood analysis as in (1), supplemented in the third step to get a posterior probability. But, even a reported P-value will lead to the use of tail areas in Fisherian statistics. This might not be a major point, but it is worth noting. For the error statistician, use of the tail area isn't to throw away the particular data and lump together all $z \geq z_\alpha$. It's to signal an interest in the probability the method would produce $z \geq z_\alpha$ under various hypotheses, to determine its capabilities (Translation Guide, Souvenir C). Goodman's hypothesis tester only reports $z \geq z_\alpha$ and so he portrays the Bayesian (or Likelihoodist) as forced to compute (2). The error statistical tester doesn't advocate this.

Let's have a look at (1): comparing the likelihoods of μ_0 and $\mu^{.9}$ given $z = 1.96$. Why look at an alternative so far away from μ_0? The value $\mu^{.9}$ gets a low likelihood, even given statistically significant z_α supplying a small denominator in (1). A classic issue with Bayes factors or likelihood ratios is the ease of finding little evidence against a null, and even evidence for it, by choosing an appropriately distant alternative. But $\mu^{.9}$ is a "typical choice" in epidemiology, Goodman notes (ibid., p. 491). Sure it's a discrepancy we want high power to detect. That's different from using it to form a comparative likelihood. As Senn remarks, it's "ludicrous" to consider just the null and $\mu^{.9}$, which scarcely exhaust the parameter space, and "nonsense" to take a single statistically significant result as even decent evidence for the discrepancy we should not like to miss. $\mu^{.9}$ is around 1.28 standard errors from the one-sided cut-off: $(1.96 + 1.28)\,\sigma_{\overline{X}} = 3.24\,\sigma_{\overline{X}}$. We know right away that any μ value in excess of the observed statistically significant z_{α} is one we cannot have good evidence for. We can only have decent evidence that μ exceeds values that would form the CI lower bound for a confidence (or severity) level at least 0.7, 0.8, 0.9, etc. That requires *subtracting* from the observed data, not adding to it. Were the data generated by $\mu^{.9}$, then 90% of the time we'd have gotten a larger difference than we observed. Goodman might reply that I am using error probability criteria to judge his Bayesian analysis; but his Bayesian analysis was to show the significance test exaggerates evidence against the test hypothesis H_0. To the tester, it's *his* analysis that's exaggerating by giving credibility 0.75 to $\mu^{.9}$. Perhaps it makes sense with the 0.5 prior, but does *that* make sense?

Assigning 0.75 to the alternative $\mu^{.9}$ (using the likelihood ratio in (1)) does not convey that this is terrible evidence for a discrepancy that large. Granted, the posterior for $\mu^{.9}$ would be even higher using the tail areas in (2), namely, 0.97 – and that is his real point, which I've yet to consider. He's right that using (2) in his Bayesian computation gives a whopping 0.97 posterior probability to $\mu^{.9}$ (instead of merely 0.75, as on the analysis he endorses). Yet a significance tester wouldn't compute (2), it's outside significance testing. Considering the tails makes it harder, not easier, to find evidence against a null – when properly used in a significance test.

He's right that using $\alpha/\text{POW}(\mu^{.9})$ as a likelihood ratio in computing a posterior probability for the point alternative $\mu^{.9}$ (using spiked priors of 0.5) gives some very strange answers. The tester isn't looking to compare point values, but to infer discrepancies. I'm doing the best I can to relate his criticism to what significance testers do.

The error statistician deservedly conveys the lack of stringency owed to any inference to $\mu^{.9}$. She finds the data give fairly good grounds that μ is less than $\mu^{.9}$ (confidence level 0.9 if one-sided, 0.8 if two-sided). Statistical quantities are like stew ingredients that you can jumble together into a farrago of computations, but there's a danger it produces an inference at odds with error statistical principles. For a significance tester, the alleged criticism falls apart; but it also leads me to question Goodman's posterior in (1).

What's Wrong with Using $(1 - \beta)/\alpha$ (or $\alpha/(1 - \beta)$) to Compare Evidence?
I think the test taker from last year's cruise did pretty well, don't you? But her "farrago" remark leads to a general point about the ills of using $(1 - \beta)/\alpha$ and $\alpha/(1 - \beta)$ to compare the evidence in favor of H_1 over H_0 or H_0 over H_1, respectively. Let's focus on the $(1 - \beta)/\alpha$ formulation. Benjamin and J. Berger (2016) call it the pre-data *rejection ratio*:

It is the probability of rejection when the alternative hypothesis is true, divided by the probability of rejection when the null hypothesis is true, i.e., the ratio of the power of the experiment to the Type I error of the experiment. The rejection ratio has a straightforward interpretation as quantifying the strength of evidence about the alternative hypothesis relative to the null hypothesis conveyed by the experimental result being statistically significant. (p. 1)

But does it? I say no (and J. Berger says he concurs[3]). Let's illustrate. Looking at Figure 5.3, we can select an alternative associated with power as high as we like by dragging the curve representing H_1 to the right.

Imagine pulling it even further than alternative $\mu^{.9}$. How about $\mu^{.999}$? If we consider the alternative $\mu = \bar{x}_{0.025} + 3\sigma_{\bar{X}}$ then $POW(T+, \mu_1)$ = area to the right of -3 under the Normal curve, which is a whopping 0.999. For some numbers, use our familiar T+: H_0: $\mu \leq 0$ vs. H_1: $\mu > 0$, $\alpha = 0.025$, $n = 25$, $\sigma = 1$, $\sigma_{\bar{X}} = 0.2$. So the cut-off $\bar{x}_{0.025} = \mu_0 + 1.96\,\sigma_{\bar{X}} = (0 + 1.96(0.2)) = 0.392$. Thus,

$$\mu^{.999} = 4.96\,\sigma_{\bar{X}} = 4.96(0.2) = 0.99 \cong 1.$$

Let the observed outcome just reach the cut-off to reject H_0, $z_0 = 0.392$. The rejection ratio is

$POW(T+, \mu = 1)/\alpha = 40$ (i.e., 0.999/0.025)!

Would even a Likelihoodist wish to say the strength of evidence for $\mu = 1$ is 40 times that of H_0? The data $\bar{x} = 0.392$ are even closer to 0 than to 1.

How then can it seem plausible, for comparativists, to compute a relative likelihood this way? We can view it through their eyes as follows: take $Z \geq 1.96$

---

[3] In private conversation.

as the lump outcome and reason along Likelihoodist lines. The probability is very high that $Z \geq 1.96$ under the assumption that $\mu = 1$:

$$\Pr(Z \geq 1.96; \mu^{0.999}) = 0.999.$$

The probability is low that $Z \geq 1.96$ under the assumption that $\mu = \mu_0 = 0$:

$$\Pr(Z \geq 1.96; \mu = \mu_0) = 0.025.$$

We've observed $z_0 = 1.96$ (so $Z \geq 1.96$).

Therefore, $\mu^{0.999}$ (i.e., 1) makes the result $Z \geq 1.96$ more probable than does $\mu = 0$.

Therefore, the result is better evidence that $\mu = 1$ than it is for $\mu = 0$. But this likelihood reasoning only holds for the specific value of z. Granted, Bayarri, Benjamin, Berger, and Sellke (2016) recommend the prerejection ratio before the data are in, and "the 'post-experimental rejection ratio' (or Bayes factor) when presenting their experimental results" (p. 91). The authors regard the pre-data rejection ratio as frequentist, but it turns out they're using Berger's "frequentist principle," which, you will recall, is in terms of error probability$_2$ (Section 3.6). A creation built on frequentist measures doesn't mean the result captures frequentist error statistical$_1$ reasoning. It might be a kind of Frequentstein entity!

Notably, power works in the opposite way. If there's a high probability you should have observed a larger difference than you did, assuming the data came from a world where $\mu = \mu_1$, then the data indicate you're *not* in a world where $\mu > \mu_1$.

If $\Pr(Z > z_0; \mu = \mu_1) = $ high, then $Z = z_0$ is strong evidence that $\mu \leq \mu_1$!

Rather than being evidence *for* μ_1, the just statistically significant result, or one that just misses, is evidence against μ being as high as μ_1. POW(μ_1) is not a measure of how well the data fit μ_1, but rather a measure of a test's capability to detect μ_1 by setting off the significance alarm (at size α). Having set off the alarm, you're not entitled to infer μ_1, but only discrepancies that have passed with severity (SIR). Else you're making mountains out of molehills.

5.3   Insignificant Results: Power Analysis and Severity

We're back at the Museum, and the display on power analysis. It is puzzling that many psychologists talk as if they're stuck with an account that says nothing about what may be inferred from negative results, when Cohen, the leader of power analysis, was toiling amongst them for years; and a central role for power, post-data, is interpreting non-significant results. The attention to power, of course, is a key feature of N-P tests, but apparently the prevalence of Fisherian tests in the social sciences, coupled, some speculate, with the difficulty in calculating power, resulted in power receiving short shrift. Cohen's work was to cure all that. Cohen supplied a multitude of tables (when tables were all we had) to encourage researchers to design tests with sufficient power to detect effects of interest. He bemoaned the fact that his efforts appeared to be falling on deaf ears. Even now, problems with power persist, and its use post-data is mired in controversy.

The focus, in the remainder of this Tour, is on negative (or non-statistically significant) results. Test T+ fails to reject the null when the test statistic fails to reach the cut-off point for rejection, i.e., $d(x_0) \leq c_\alpha$. A classic fallacy is to construe no evidence against H_0 as evidence of the correctness of H_0. A canonical example was in the list of slogans opening this book: Failing to find an increased risk is not evidence of no risk increase, if your test had little capability of detecting risks, even if present (as when you made your hurdles too high). The problem is the flip side of the fallacy of rejection: here the null hypothesis "survives" the test, but merely surviving can occur too frequently, even when there are discrepancies from H_0.

Power Analysis Follows Significance Test Reasoning

Early proponents of power analysis that I'm aware of include Cohen (1962), Gibbons and Pratt (1975), and Neyman (1955). It was the basis for my introducing severity, Mayo (1983). Both Neyman and Cohen make it clear that power analysis uses the same reasoning as does significance testing.[4] First Cohen:

[F]or a given hypothesis test, one defines a numerical value **i** (for *iota*) for the [population] ES (effect size), where **i** is so small that it is appropriate in the context to consider it negligible (trivial, inconsequential). Power $(1 - \beta)$ is then set at a high value, so that β is relatively small. When, additionally, α is specified, n can be found. Now, if the research is performed with this n and it results in nonsignificance, it is proper to conclude that the population ES is no more than **i**, i.e., that it is negligible . . . (Cohen 1988, p. 16; α, β substituted for his **a, b**).

Here Cohen imagines the researcher sets the size of a negligible discrepancy ahead of time – something not always available. Even where a negligible **i** may be specified, it's rare that the power to detect it is high. Two important points

---

[4] A key medical paper is Freiman et al. (1978).

can still be made: First, Cohen doesn't instruct you to infer there's no discrepancy from H_0, merely that it's "no more than **i**." Second, even if your test doesn't have high power to detect negligible **i**, you can infer the population discrepancy is less than whatever γ your test *does* have high power to detect. Some call this its *detectable discrepancy size*.

A little note on language. Cohen distinguishes the population ES and the observed ES, both in σ units. Keeping to Cohen's ES_s for "the effect size *in the sample*" (1988, p. 17) prevents a tendency to run them together. I continue to use "discrepancy" and "difference" for the population and observed differences, respectively, indicating the units being used.

Exhibit (iii): Ordinary Power Analysis. Now for how the inference from power analysis is akin to significance testing. Let $\mu^{1-\beta}$ be the alternative against which the null in T+ has high power, $1 - \beta$. Power analysis sanctions the inference that would accrue if we switched the null and alternative, yielding the one-sided test in the opposite direction, T−, we might call it. That is, T− tests $H_0: \mu \geq \mu^{1-\beta}$ vs. $H_1: \mu < \mu^{1-\beta}$ at the β level. The test rejects H_0 (at level β) when $\overline{X} \leq \mu_0 - z_\beta\, \sigma_{\overline{X}}$. Such a significant result would warrant inferring $\mu < \mu^{1-\beta}$ at level β. Using power analysis doesn't require making this switcheroo. The point is that there's essentially no new reasoning involved in power analysis, which is why members of the Fisherian tribe manage it without mentioning power.

> *Ordinary Power Analysis*: If data x are not statistically significantly different from H_0, and the power to detect discrepancy γ is high, then x indicates that the actual discrepancy is no greater than γ.

A simple example: Use $\mu^{.84}$ in test T+ ($H_0 : \mu \leq 0$ vs. $H_1; \mu > 0$, $\alpha = 0.025$, $n = 25$, $\sigma_{\overline{X}} = 0.2$) to create test T−. Test T+ has 0.84 power against $\mu^{.84} = 3\sigma_{\overline{X}} = 0.6$ (with our usual rounding). So, test T− is $H_0: \mu \geq 0.6$ vs. H_1: $\mu < 0.6$, and a result is statistically significantly *smaller* than 0.6 at level 0.16 whenever the sample mean $\overline{X} \leq 0.6 - 1\sigma_{\overline{X}} = 0.4$. To check, note that $\Pr(\overline{X} \leq 0.4; \mu = 0.6) = \Pr(Z \leq -1) = 0.16 = \beta$.

It will be useful to look at the two-sided alternative: test $T^\pm$. We'd combine the above one-sided test with a test of $H_0: \mu \geq -\mu^{1-\beta}$ vs. $H_1: \mu < -\mu^{1-\beta}$ at the β level. This will be to test $H_0: \mu \geq -0.6$ vs. $H_1: \mu < -0.6$, and find a result statistically significantly smaller than −0.6 at level 0.16 whenever the sample mean $\overline{X} \leq -0.8$ (i.e., $-0.6 - 1\sigma_{\overline{X}}$). If both nulls are rejected (at the 0.16 level), we infer $|\mu| < 0.6$ but the two-sided test has double the Type I error probability: 0.32.

How high a power should be regarded as high? How low as low? A power of 0.8 or 0.9 is common, we saw, in "clinically relevant" discrepancies. To anyone who complains that there's no way to draw a cut-off, note that we merely need to distinguish blatantly high from rather low values. Why have the probability of a Type II error exceed that of the Type I error? Some critics give Neyman and Pearson a hard time about this, but there's nothing in N-P tests to require it. Balance the errors as you like, N-P say. N-P recommend, based on tests in use, first to specify the test to reflect the Type I error as more serious than a Type II error. Second, choose a test that minimizes the Type II error probability, given the fixed Type I. In an example of testing a medical risk, Neyman says he places "a risk exists" as the test hypothesis since it's worse (for the consumer) to erroneously infer risk absence (1950, chapter V). Promoters of the precautionary principle are often surprised to learn this about N-P tests. However, there's never an automatic "accept/reject" in a scientific context.

Neyman Chides Carnap, Again

Neyman was an early power analyst? Yes, it's in his "The Problem of Inductive Inference" (1955) where we heard Neyman chide Carnap for ignoring the statistical model (Section 2.7). Neyman says:

I am concerned with the term 'degree of confirmation' introduced by Carnap . . . We have seen that the application of the locally best one-sided test to the data . . . failed to reject the hypothesis [that the n observations come from a source in which the null hypothesis is true]. The question is: does this result 'confirm' the hypothesis [that H_0 is true of the particular data set]? (ibid., p. 40)

The answer . . . depends very much on the exact meaning given to the words 'confirmation,' 'confidence,' etc. If one uses these words to describe one's intuitive feeling of confidence in the hypothesis tested H_0, then . . . the attitude described is dangerous . . . the chance of detecting the presence [of discrepancy from the null], when only [n] observations are available, is extremely slim, even if [the discrepancy is present]. Therefore, the failure of the test to reject H_0 cannot be reasonably considered as anything like a confirmation of H_0. The situation would have been radically different if the power function [corresponding to a discrepancy of interest] were, for example, greater than 0.95. (ibid., p. 41)

Ironically, Neyman also criticizes Fisher's move from a large P-value to inferring the null hypothesis as:

much too automatic [because] . . . large values of P may be obtained when the hypothesis tested is false to an important degree. Thus, . . . it is advisable to investigate . . . what is the probability (probability of error of the second kind) of obtaining a large value of P in cases when the [null is false . . . to a specified degree]. (1957a, p. 13)

Should this calculation show that the probability of detecting an appreciable error in the hypothesis tested was large, say 0.95 or greater, then and only then is the decision in favour of the hypothesis tested justifiable in the same sense as the decision against this hypothesis is justifiable when an appropriate test rejects it at a chosen level of significance. (1957b, pp. 16–17)

Typically, the hypothesis tested, $[H_0]$ in the N-P context, could be swapped with the alternative. Let's leave the museum where a leader of the severe testing tribe makes some comparisons.

Attained Power Π

So power analysis is in the spirit of severe testing. Still, power analysis is calculated relative to an outcome just missing the cut-off c_α. This corresponds to an observed difference whose P-value just exceeds α. This is, in effect, the worst case of a negative result. What if the actual outcome yields an even smaller difference (larger P-value)?

Consider test T+ ($\alpha = 0.025$) above. No one wants to turn pages so here it is: H_0: $\mu \leq 0$ vs. H_1: $\mu > 0$, $\alpha = 0.025$, $n = 25$, $\sigma = 1$, $\sigma_{\bar{x}} = 0.2$. So the cut-off $\bar{x}_{0.025} = \mu_0 + 1.96\,\sigma_{\bar{x}} = (0 + 1.96(0.2)) = 0.392$, or, with the $2\,\sigma_{\bar{x}}$ cut-off, $\bar{x}_{0.025} = 0.4$. Consider an arbitrary inference $\mu \leq 0.2$. We know POW(T+, $\mu = 0.2$) = 0.16 ($1\sigma_{\bar{x}}$ is subtracted from 0.4). A value of 0.16 is quite lousy power. It follows that no statistically insignificant result can warrant $\mu \leq 0.2$ for the power analyst. Power analysis only allows ruling out values as high as $\mu^{.8}$, $\mu^{.84}$, $\mu^{.9}$, and so on. The power of a test is fixed once and for all and doesn't change with the observed mean $\bar{x}$. Why consider every non-significant result as if it just barely missed the cut-off? Suppose, $\bar{x} = -0.2$. This is $2\sigma_{\bar{x}}$ lower than 0.2. Surely that should be taken into account? It is: 0.2 is the upper 0.975 confidence bound and SEV(T+, $\bar{x} = -0.2$, $\mu \leq 0.2$) = 0.975.[5]

What enables substituting the observed value of the test statistic, $d(x_0)$, is the counterfactual reasoning of severity:

> If, with high probability, the test would have resulted in a larger observed difference (a smaller P-value) than it did, if the discrepancy was as large as γ, then there's a good indication the discrepancy is no greater than γ, i.e., that $\mu \leq \mu_0 + \gamma$.

That is, if the *attained power* (att-power) of T+ against $\mu \leq \mu_0 + \gamma$ is very high, the inference to $\mu \leq \mu_0 + \gamma$ is warranted with severity. (As always, whether it's a mere indication or genuine evidence depends on whether it passes an audit.)

---

[5] To show, as crude power analysis does not, that SEV(T+, $\bar{x} = -0.2$, $\mu \leq 0.2$) = 0.975 when $\bar{x} = -0.2$: We standardize $\bar{x}$ to get $z = (-0.2 - 0.2)/0.2 = -2$ and so $\Pr(\bar{x} \geq -0.2; 0.2) = 0.975$.

If you elect to use the term attained power, you'll have to avoid confusing it with animals given similar names; I'll introduce you to them shortly.

Compare power analytic reasoning with severity (or att-power) reasoning from a negative or insignificant result from T+.

> *Power Analysis*: If $\Pr(d(X) \geq c_\alpha; \mu_1)$ = high and the result is not significant, then it's an indication or evidence that $\mu \leq \mu_1$.

> *Severity Analysis*: If $\Pr(d(X) \geq d(x_0); \mu_1)$ = high and the result is not significant, then it's an indication or evidence that $\mu \leq \mu_1$.

Severity replaces the predesignated cut-off c_α with the observed $d(x_0)$. Thus we obtain the same result if we choose to remain in the Fisherian tribe, as seen in the Frequentist Evidential Principle FEV(ii) (Section 3.3).

We still abide by the logic of power analysis, since if $\Pr(d(X) \geq d(x_0); \mu_1)$ = high, then $\Pr(d(X) \geq c_\alpha; \mu_1)$ = high, at least in a test with a sensible distance measure like T+. In other words, power analysis is conservative. It gives a sufficient but not a necessary condition for warranting an upper bound: $\mu \leq \mu_1$. But it can be way too conservative as we just saw.

> (1) $\Pr(d(X) \geq c_\alpha; \mu = \mu_0 + \gamma)$: Power to detect γ.

Ordinary power analysis requires (1) to be high (for non-significance to warrant $\mu \leq \mu_0 + \gamma$).

Just missing the cut-off c_α is the worst case. It is more informative to look at (2):

> (2) $\Pr(d(X) \geq d(x_0); \mu = \mu_0 + \gamma)$: Attained power ($\Pi(\gamma)$).

(1) can be low while (2) is high. The computation in (2) measures the severity (or degree of corroboration) for the inference $\mu \leq \mu_0 + \gamma$. The analysis with Cox kept to $\Pi(\gamma)$ (or "attained sensitivity"), keeping "power" out of it.

As an entrée to Exhibit (iv): Isn't severity just power? This is to compare apples and frogs. The power of a test to detect an alternative is an error probability of a method (one minus the probability of the corresponding Type II error). Power *analysis* is a way of using power to assess a statistical inference in the case of a negative result. Severity, by contrast, is always to assess a statistical inference. Severity is always in relation to a particular claim or inference C, from a test T and an outcome *x*. So with that out of the way, what if the question is put properly thus: If a result from test T+ is just statistically insignificant at level α, then is the test's power to detect μ_1 equal to the severity for inference C: $\mu > \mu_1$? The answer is no. It would be equal to the severity for inferring the *denial* of C! See Figure 5.4 comparing SEV$(\mu > \mu_1)$ and POW(μ_1).

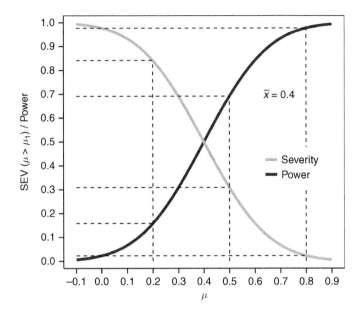

Figure 5.4 Severity for $(\mu > \mu_1)$ vs power (μ_1).

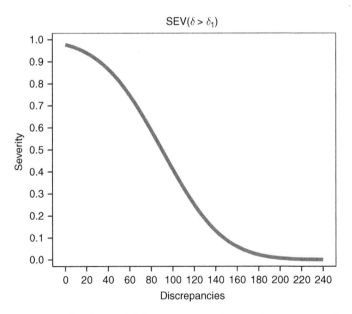

Figure 5.5 The observed difference is 90, each group has $n = 200$ patients, and the standard deviation is 450.

Exhibit (iv): Difference Between Means Illustration. I've been making use of Senn's points about the nonsensical and the ludicrous in motivating the severity assessment in relation to power. I want to show you severity for a difference between means, and fortuitously Senn (2019) alludes to severity in the latest edition of his textbook to make the same point. An example is a placebo-controlled trial in asthma where the amount of air a person can exhale in one second, the forced expiratory volume, is measured. "The clinically relevant difference is presumed to be 200 ml and the standard deviation 450 ml" (p. 197). It's a test of

$$H_0: \delta = \mu_1 - \mu_0 \le 0 \text{ vs. } H_1: \delta > 0.$$

He will use a one-sided test at the 2.5% level, using 200 patients in each group yielding the standard error (SE) for the difference between two means equal to $(450\sqrt{2}/\sqrt{n})$.[6] The test has over 0.99 power for detecting $\delta = 200$. The observed difference $d = (\overline{X} - \overline{Y}) = 90$, which is statistically significant at the 0.025 level (90/SE = 2), but it wouldn't warrant inferring $\delta > 200$, and we can see the severity for $\delta > 200$ is extremely low. The observed difference is statistically significantly different from H_0: in accord with (or in the direction of) H_1, so severity computes the probability of a worse fit with H_1 under $\delta = 200$:

$$\text{SEV}(\delta > 200) = \Pr(d < 90; \delta = 200).$$

While this is strictly a Student's t distribution, given the large sample size, we can use the Normal distribution:[7]

$$Z = \left\{ \frac{\sqrt{200}(90 - 200)}{450\sqrt{2}} \right\} = (90 - 200)/45 = -2.44.$$

$\text{SEV}(\delta > 200) =$ the area to the left of -2.44 (See Figure 5.5) which is 0.007,
    For a few examples, $\text{SEV}(\delta > 100) = 0.412$ the area to the left of $z = (90 - 100)/45 = -0.222$. $\text{SEV}(\delta > 50) = 0.813$ the area to the left of $z = (90 - 50)/45 = 0.889$. `

$\text{SEV}(\delta > 10) = 0.962$ the area to the left of $z = (90 - 10)/45 = 1.778$.

---

[6] From the general case, we get the case where $n_1 = n_2$,

$$\sigma_{M_1 - M_2} = \sqrt{\frac{\sigma_1^2}{n_1} + \frac{\sigma_2^2}{n_2}} = \sqrt{\frac{\sigma^2}{n} + \frac{\sigma^2}{n}} = \sqrt{\frac{2\sigma^2}{n}} = \frac{\sqrt{2}\sigma}{\sqrt{n}}$$

[7]

$$\frac{\sqrt{n}(\overline{X} - \overline{Y})}{s\sqrt{2}}$$

Senn gives another way to view the severity assessment of $\delta > \delta'$, namely "adopt $[\delta = \delta']$, as a null hypothesis and then turn the significance test machinery on it (2019). In the case of testing $\delta = 200$, the P-value would be 0.999. Scarcely evidence against it. We first visited this mathematical link in touring the connection between severity and confidence intervals in Section 3.8. As noted, the error statistician is loath to advocate modifying the null hypothesis because the point of a severity assessment is to supply a basis for interpreting tests that is absent in existing tests. Since significance tests aren't explicit about assessing discrepancies, and since the rationale for P-values is questioned in all the ways we've espied, it's best to supply a fresh rationale. I have offered the severity rationale as a basis for understanding, if not buying, error statistical reasoning. The severity computation might be seen as a rule of thumb to avoid misinterpretations; it could be arrived at through other means, including varying the null hypotheses. It's the idea of viewing statistical inference as severe testing that invites a non-trivial difference with probabilism.

5.4   Severity Interpretation of Tests: Severity Curves

We visit severity tribes who have prepared an overview that synthesizes nonsignificant results from Fisherian as well as ("do not reject") results from N-P tests. Following the minimal principle of severity:

(a) If data $d(x)$ are not statistically significantly different from H_0, but the capability of detecting discrepancy γ is low, then $d(x)$ is not good evidence that the actual discrepancy is less than γ.

What counts as a discrepancy "of interest" is a separate question, outside of statistics proper. You needn't know it to ask: What discrepancies, if they existed, would very probably have led your method to show a more significant result than you found? Upon finding this, you may infer that, at best, the test can rule out increases of that extent.

(b) If data $d(x)$ are not statistically significantly different from H_0, but the probability to detect discrepancy γ is high, then x is good evidence that the actual discrepancy is less than or equal to γ.

We are not changing the original null and alternative hypotheses! We're using the severe testing concept to interpret the negative results – the kind of scrutiny in which one might be interested, to follow Neyman, "when we are faced with . . . interpreting the results of an experiment planned and performed by someone else" (Neyman 1957b, p. 15). We want to know how well tested are claims of form $\mu \le \mu_1$, where $\mu_1 = (\mu_0 + \gamma)$, for some $\gamma \ge 0$.

Why object to applying the severity analysis by changing the null hypothesis, and doing a simple P-value computation? P-values, especially if plucked from thin air this way, are themselves in need of justification. That's a major goal of this journey. It's only by imagining we have either a best or good test or corresponding distance measure (let alone assuming we don't have to deal with lots of nuisance parameters) that substituting different null hypotheses works out.

Pre-data, we need a test with good error probabilities (as discussed in Section 3.2). That assures we avoid some worst case. Post-data we go further.

For a claim H to pass with severity requires not just that (S-1) the data accord with H, but also that (S-2) the test probably would have produced a worse fit, if H were false in specified ways. We often let the measure of accordance (in (S-1)) vary and train our critical focus on (S-2), but here it's a best test. Consider statistically insignificant results from test T+. The result "accords with" H_0, so we have (S-1), but we're wondering about (S-2): how probable is it that test T+ would have produced a result that accords *less* well with H_0 than x_0 does, were H_0 false? An equivalent but perhaps more natural phrase for "a result that accords *less* well with H_0" is "a result *more discordant*." Your choice.

Souvenir W: The Severity Interpretation of Negative Results (SIN) for Test T+

Applying our general abbreviation: SEV(test T+, outcome x, inference H), we get "the severity with which $\mu \leq \mu_1$ passes test T+, with data x_0":

$$\mathrm{SEV}(\text{T+}, \mathrm{d}(x_0), \mu \leq \mu_1),$$

where $\mu_1 = (\mu_0 + \gamma)$, for some $\gamma \geq 0$. If it's clear which test we're discussing, we use our abbreviation: $\mathrm{SEV}(\mu \leq \mu_1)$. We obtain a companion to the severity interpretation of rejection (SIR), Section 4.4, Souvenir R:

SIN (Severity Interpretation for Negative Results)

(a) If there is a very *low* probability that $\mathrm{d}(x_0)$ would have been larger than it is, even if $\mu > \mu_1$, then $\mu \leq \mu_1$ passes with *low* severity: $\mathrm{SEV}(\mu \leq \mu_1)$ is low.

(b) If there is a very *high* probability that $\mathrm{d}(x_0)$ would have been larger than it is, were $\mu > \mu_1$, then $\mu \leq \mu_1$ passes the test with *high* severity: $\mathrm{SEV}(\mu \leq \mu_1)$ is high.

To break it down, in the case of a statistically insignificant result:

$$\mathrm{SEV}(\mu \leq \mu_1) = \Pr(\mathrm{d}(X) > \mathrm{d}(x_0); \mu \leq \mu_1 \text{ false}).$$

We look at $\{d(X) > d(x_0)\}$ because severity directs us to consider a "worse fit" with the claim of interest. That $\mu \le \mu_1$ is false within our model means that $\mu > \mu_1$. Thus:

$$SEV(\mu \le \mu_1) = Pr(d(X) > d(x_0); \mu > \mu_1).$$

Now $\mu > \mu_1$ is a composite hypothesis, containing all the values in excess of μ_1. How can we compute it? As with power calculations, we evaluate severity at a point $\mu_1 = (\mu_0 + \gamma)$, for some $\gamma \ge 0$, because for values $\mu \ge \mu_1$ the severity increases. So we need only to compute

$$SEV(\mu \le \mu_1) > Pr(d(X) > d(x_0); \mu = \mu_1).$$

To compute SEV we compute $Pr(d(X) > d(x_0); \mu = \mu_1)$ for any μ_1 of interest. Swapping out the claims of interest (in significant and insignificant results), gives us a single criterion of a good test, severity.

Exhibit(v): Severity Curves. The severity tribes want to present severity using a standard Normal example, one where $\sigma_{\overline{X}} = 1$ (as in the water plant accident). For this illustration:

Test $T+ : H_0 : \mu \le 0$ vs. $H_1 : \mu > 0$, $\sigma = 10$, $n = 100, \sigma/\sqrt{n} = \sigma_{\overline{X}} = 1$.

If $\alpha = 0.025$, we reject H_0 iff $d(X) \ge c_{0.025} = 1.96$.

Suppose test $T+$ yields the statistically insignificant result $d(x_0) = 1.5$. Under the alternative $d(X)$ is $N(\delta, 1)$ where $\delta = (\mu - \mu_0)/\sigma_{\overline{X}}$.

Even without identifying a discrepancy of importance ahead of time, the severity associated with various inferences can be evaluated.

The severity curves (Figure 5.6) show $d(x_0) = 0.5, 1, 1.5$, and 1.96.

How severely does $\mu \le 0.5$ pass the test with $\overline{X} = 1.5$ $(d(x_0) = 1.5)$?

The easiest way to compute it is to go back to the observed $\overline{x}_0$, which would be 1.5.

$$SEV(\mu \le 0.5) = Pr(\overline{X} > 1.5; \mu = 0.5) = 0.16.$$

Here, $Z = [(1.5 - 0.5)/1] = 1$, and the area under the standard Normal distribution to the right of 1 is 0.16. Lousy. We can read it off the curve, looking at where the $d(x) = 1.5$ curve hits the bottom-most dotted line. The severity (vertical) axis hits 0.16, and the corresponding value on the μ axis is 0.5. This could be used more generally as a discrepancy axis, as I'll show.

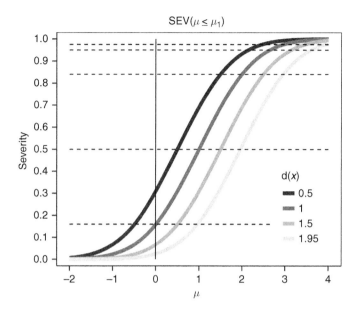

$$\text{SEV}(\mu \leq \mu_1)$$

Figure 5.6 Severity curves.

We can find some discrepancy from the null that this statistically insignificant result warrants ruling out at a reasonable level – one that very probably would have produced a more significant result than was observed. The value $d(x_0) = 1.5$ yields a severity of 0.84 for a discrepancy of 2.5. $\text{SEV}(\mu \leq 2.5) = 0.84$ with $d(x_0) = 1.5$. Compare this to $d(x) = 1.95$, failing to trigger the significance alarm. Now a larger upper bound is needed for severity 0.84, namely, $\mu \leq 2.95$. If we have discrepancies of interest, by setting a high power to detect them, we ensure – ahead of time – that any insignificant result entitles us to infer "it's not that high." Power against μ_1 evaluates the worst (i.e., lowest) severity for values $\mu \leq \mu_1$ for any outcome that leads to non-rejection. This test ensures any insignificant result entitles us to infer $\mu \leq 2.95$, call it $\mu \leq 3$. But we can determine discrepancies that pass with severity, post-data, without setting them at the outset. Compare four different outcomes:

$d(x_0) = 0.5$, $\text{SEV}(\mu \leq 1.5) = 0.84$; $d(x_0) = 1$, $\text{SEV}(\mu \leq 2) = 0.84$;

$d(x_0) = 1.5$, $\text{SEV}(\mu \leq 2.5) = 0.84$; $d(x_0) = 1.95$, $\text{SEV}(\mu \leq 2.95) = 0.84$.

In relation to test T+ (standard Normal): If you add $1\sigma_{\overline{X}}$ to $d(x_0)$, the result being μ_1, then $\text{SEV}(\mu \leq \mu_1) = 0.84$.[8]

We can also use severity curves to compare the severity for a given claim, say $\mu \leq 1.5$:

$d(x_0) = 0.5$, $\text{SEV}(\mu \leq 1.5) = 0.84$; $d(x_0) = 1$, $\text{SEV}(\mu \leq 1.5) = 0.7$;

$d(x_0) = 1.5$, $\text{SEV}(\mu \leq 1.5) = 0.5$; $d(x_0) = 1.95$, $\text{SEV}(\mu \leq 1.5) = 0.3$.

Low and high benchmarks convey what is and is not licensed, and suffice for avoiding fallacies of acceptance. We can deduce SIN from the case where T+ has led to a statistically significant result, SIR. In that case, the inference that passes the test is of form $\mu > \mu_1$, where $\mu_1 = \mu_0 + \gamma$. Because $(\mu > \mu_1)$ and $(\mu \leq \mu_1)$ partition the parameter space of μ, we get $\text{SEV}(\mu > \mu_1) = 1 - \text{SEV}(\mu \leq \mu_1)$.

The more devoted amongst you will want to improve and generalize my severity curves. Some of you are staying the night at Confidence Court Inn, others at Best Bet and Breakfast. We meet at the shared lounge, Calibration. Here's a souvenir of SIR, and SIN.

Souvenir X: Power and Severity Analysis

Let's record some highlights from Tour I:

First, ordinary power analysis versus severity analysis for Test T+:

Ordinary Power Analysis: If $\Pr(d(\mathbf{X}) \geq c_\alpha; \mu_1) = $ high and the result is not significant, then it's an indication or evidence that $\mu \leq \mu_1$.

Severity Analysis: If $\Pr(d(\mathbf{X}) \geq d(x_0); \mu_1) = $ high and the result is not significant, then it's an indication or evidence that $\mu \leq \mu_1$.

It can happen that claim $(\mu \leq \mu_1)$ is warranted by severity analysis but not by power analysis.

---

[8] • If you add $k\sigma_{\overline{X}}$ to $d(x_0)$, $k > 0$, the result being μ_1, then $\text{SEV}(\mu \leq \mu_1) = $ area to the right of $-k$ under the standard Normal (SEV > 0.5).
• If you subtract $k\sigma_{\overline{X}}$ from $d(x_0)$, the result being μ_1, then $\text{SEV}(\mu \leq \mu_1) = $ area to the right of k under the standard Normal (SEV $\leq$ 0.5).

For the general case of Test T+, you'd be adding or subtracting $k\sigma_{\overline{X}}$ to $(\mu_0 + d(x_0)\sigma_{\overline{X}})$. We know that adding $0.85\sigma_{\overline{X}}$, $1\,\sigma_{\overline{X}}$, and $1.28\sigma_{\overline{X}}$ to the cut-off for rejection in a test T+ results in μ values against which the test has 0.8, 0.84, and 0.9 power. If you treat the observed $\overline{x}$ as if it were being contemplated as the cut-off, and add $0.85\sigma_{\overline{X}}$, $1\sigma_{\overline{X}}$, and $1.28\sigma_{\overline{X}}$, you will arrive at μ_1 values such that $\text{SEV}(\mu \leq \mu_1) = 0.8$, 0.84, and 0.9, respectively. That's because severity goes in the same direction as power for non-rejection in T+. For familiar numbers of $\sigma_{\overline{X}}$'s added/subtracted to $\overline{x} = \mu_0 + d_0\sigma_{\overline{X}}$:

Claim	$(\mu \leq \overline{x} - 1\sigma_{\overline{X}})$	$(\mu \leq \overline{x})$	$(\mu \leq \overline{x} + 1\sigma_{\overline{X}})$	$(\mu \leq \overline{x} + 1.65\sigma_{\overline{X}})$	$(\mu \leq \overline{x} + 1.98\sigma_{\overline{X}})$
SEV	0.16	0.5	0.84	0.95	0.975

Now an overview of severity for test T+: Normal testing: H_0: $\mu \leq \mu_0$ vs. H_1: $\mu > \mu_0$ with σ known. The severity reinterpretation is set out using discrepancy parameter γ. We often use μ_1 where $\mu_1 = \mu_0 + \gamma$.

Reject H_0 (with x_0) licenses inferences of the form $\mu > [\mu_0 + \gamma]$, for some $\gamma \geq 0$, but with a warning as to $\mu \leq [\mu_0 + \kappa]$, for some $\kappa \geq 0$.

Non-reject H_0 (with x_0) licenses inferences of the form $\mu \leq [\mu_0 + \gamma]$, for some $\gamma \geq 0$, but with a warning as to values fairly well indicated $\mu > [\mu_0 + \kappa]$, for some $\kappa \geq 0$.

The severe tester reports the attained significance levels and at least two other benchmarks: claims warranted with severity, and ones that are poorly warranted.

Talking through SIN and SIR. Let $d_0 = d(x_0)$.

SIN *(Severity Interpretation for Negative Results)*

(a) *low*: If there is a very *low* probability that d_0 would have been larger than it is, even if $\mu > \mu_1$, then $\mu \leq \mu_1$ passes with *low* severity: SEV($\mu \leq \mu_1$) is low (i.e., your test wasn't very capable of detecting discrepancy μ_1 even if it existed, so when it's not detected, it's poor evidence of its absence).

(b) *high*: If there is a very *high* probability that d_0 would have been larger than it is, were $\mu > \mu_1$, then $\mu \leq \mu_1$ passes the test with *high* severity: SEV($\mu \leq \mu_1$) is high (i.e., your test was highly capable of detecting discrepancy μ_1 if it existed, so when it's not detected, it's a good indication of its absence).

SIR *(Severity Interpretation for Significant Results)*

If the significance level is small, it's indicative of some discrepancy from H_0, we're concerned about the magnitude:

(a) *low*: If there is a fairly high probability that d_0 would have been larger than it is, even if $\mu = \mu_1$, then d_0 is not a good indication $\mu > \mu_1$: SEV($\mu > \mu_1$) is low.[9]

(b) *high*: Here are two ways, choose your preferred:

  • (b-1) If there is a very high probability that d_0 would have been smaller than it is, if $\mu \leq \mu_1$, then when you observe so large a d_0, it indicates $\mu > \mu_1$: SEV($\mu > \mu_1$) is high.

---

[9] A good rule of thumb to ascertain if a claim C is warranted is to think of a statistical *modus tollens* argument, and find what would occur with high probability, were claim C false.

- (b-2) If there's a very low probability that so large a d_0 would have resulted, if μ were no greater than μ_1, then d_0 indicates $\mu > \mu_1$: SEV($\mu > \mu_1$) is high.[10]

---

[10] For a shorthand that covers both severity and FEV for Test T+ with small significance level (Section 3.1):

(FEV/SEV): If d(x_0) is not statistically significant, then $\mu \leq \bar{x} + k_\varepsilon \sigma/\sqrt{n}$ passes the test T+ with severity $(1 - \varepsilon)$

(FEV/SEV): If d(x_0) is statistically significant, then $\mu > \bar{x} - k_\varepsilon \sigma/\sqrt{n}$ passes test T+ with severity $(1 - \varepsilon)$,

where Pr(d(X) > k_ε) = ε (Mayo and Spanos (2006), Mayo and Cox (2006).)

Tour II How Not to Corrupt Power

5.5 Power Taboos, Retrospective Power, and Shpower

Let's visit some of the more populous tribes who take issue with power – by which we mean ordinary power – at least its post-data uses. Power Peninsula is often avoided due to various "keep out" warnings and prohibitions, or researchers come during planning, never to return. Why do some people consider it a waste of time, if not totally taboo, to compute power once we know the data? A degree of blame must go to N-P, who emphasized the planning role of power, and only occasionally mentioned its use in determining what gets "confirmed" post-data. After all, it's good to plan how large a boat we need for a philosophical excursion to the Lands of Overlapping Statistical Tribes, but once we've made it, it doesn't matter that the boat was rather small. Or so the critic of post-data power avers. A crucial disanalogy is that with statistics, we don't know that we've "made it there," when we arrive at a statistically significant result. The statistical significance alarm goes off, but you are not able to see the underlying discrepancy that generated the alarm you hear. The problem is to make the leap from the perceived alarm to an aspect of a process, deep below the visible ocean, responsible for its having been triggered. Then it is of considerable relevance to exploit information on the capability of your test procedure to result in alarms going off (perhaps with different decibels of loudness), due to varying values of the parameter of interest. There are also objections to power analysis with insignificant results.

Exhibit (vi): Non-significance + High Power Does Not Imply Support for the Null over the Alternative. Sander Greenland (2012) has a paper with this title. The first step is to understand the assertion, giving the most generous interpretation. It deals with non-significance, so our ears are perked for a fallacy of non-rejection. Second, we know that "high power" is an incomplete concept, so he clearly means high power against "the alternative." We have a handy example: alternative $\mu^{.84}$ in T+ (POW(T+, $\mu^{.84}$) = 0.84). Use the water plant case, T+: H_0: $\mu \leq 150$ vs. H_1: $\mu > 150$, $\sigma = 10$, $n = 100$. With $\alpha = 0.025$, $z_{0.025} = 1.96$, and the corresponding cut-off in terms of $\bar{x}_{0.025}$ is [150 + 1.96(10)/$\sqrt{100}$] = 151.96], $\mu^{.84} = 152.96$.

Now a title like this is supposed to signal a problem, a reason for those "keep out" signs. His point, in relation to this example, boils down to noting that an observed difference may not be statistically significant – $\bar{x}$ may fail to make it to the cut-off $\bar{x}_{0.025}$ – and yet be closer to $\mu^{.84}$ than to 0. This happens because the Type II error probability β (here, 0.16)[1] is greater than the Type I error probability (0.025).

For a quick computation let $\bar{x}_{0.025} = 152$ and $\mu^{.84} = 153$. Halfway between alternative 153 and the 150 null is 151.5. Any observed mean greater than 151.5 but less than the $\bar{x}_{0.025}$ cut-off, 152, will be an example of Greenland's phenomenon. An example would be those values that are closer to 153, the alternative against which the test has 0.84 power, than to 150 and thus, by a likelihood measure, support 153 more than 150 – even though POW($\mu = 153$) is high (0.84). Having established the phenomenon, your next question is: so what?

It *would* be problematic if power analysis took the insignificant result as evidence for $\mu = 150$ – maintaining compliance with the ecological stipulation – and I don't doubt some try to construe it as such, nor that Greenland has been put in the position of needing to correct them. Power analysis merely licenses $\mu \leq \mu^{.84}$ where 0.84 was chosen for "high power." Glance back at Souvenir X. So at least one of the "keep out" signs can be removed.

Shpower and Retrospective Power Analysis

It's unusual to hear books condemn an approach in a hush-hush sort of way without explaining what's so bad about it. This is the case with something called post hoc power analysis, practiced by some who live on the outskirts of Power Peninsula. Psst, don't go there. We hear "there's a sinister side to statistical power, ... I'm referring to post hoc power" (Cumming 2012, pp. 340–1), also called *observed* power and *retrospective* (retro) power. I will be calling it *shpower analysis*. It distorts the logic of ordinary power analysis (from insignificant results). The "post hoc" part comes in because it's based on the observed results. The trouble is that ordinary power analysis is also post-data. The criticisms are often wrongly taken to reject both.

Shpower evaluates power with respect to the hypothesis that the population effect size (discrepancy) equals the observed effect size, for example, that the parameter μ equals the observed mean. In T+ this would be to set $\mu = \bar{x}$. Conveniently, their examples use variations on test T+. We may define:

The Shpower of test T+: $\Pr(\bar{X} \geq \bar{x}_\alpha; \mu = \bar{x})$.

---

[1] That is, $\beta(\mu^{.84}) = \Pr(d < 0.4; \mu = 0.6) = \Pr(Z < -1) = 0.16$.

The thinking, presumably, is that, since we don't know the value of μ, we might use the observed $\bar{x}$ to estimate it, and then compute power in the usual way, except substituting the observed value. But a moment's thought shows the problem – at least for the purpose of using power analysis to interpret insignificant results. Why?

Since alternative μ is set equal to the observed $\bar{x}$, and $\bar{x}$ is given as statistically insignificant, we know we are in Case 1 from Section 5.1: the power can never exceed 0.5. In other words, since $\bar{x} < \bar{x}_\alpha$, the shpower = POW$(T+, \mu = \bar{x})$. But power analytic reasoning is all about finding an alternative against which the test has *high* capability to have rung the significance bell, were that the true parameter value – *high* power. Shpower is always "slim" (to echo Neyman) against such alternatives. Unsurprisingly, then, shpower analytical reasoning has been roundly criticized in the literature. But the critics think they're maligning power analytic reasoning.

Now we know the severe tester insists on using attained power Pr(d$(X) \geq$ d(x_0); μ') to evaluate severity, but when addressing the criticisms of power analysis, we have to stick to ordinary power:[2]

Ordinary power POW(μ'): Pr(d$(X) \geq c_\alpha$; $\mu')$

Shpower (aka post hoc or retro power): Pr(d$(X) \geq c_\alpha$; $\mu = \bar{x}$)

An article by Hoenig and Heisey (2001) ("The Abuse of Power") calls power analysis abusive. Is it? Aris Spanos and I say no (in a 2002 note on them), but the journal declined to publish it. Since then their slips have spread like kudzu through the literature.

Howlers of Shpower Analysis

Hoenig and Heisey notice that within the class of insignificant results, the more significant the observed $\bar{x}$ is, the higher the "observed power" against $\mu = \bar{x}$, until it reaches 0.5 (when $\bar{x}$ reaches $\bar{x}_\alpha$ and becomes significant). "That's backwards!" they howl. It is backwards if "observed power" is defined as shpower. Because, if you were to regard higher shpower as indicating better evidence for the null, you'd be saying the more statistically significant the observed difference (between $\bar{x}$ and μ_0), the more the evidence of the *absence of a discrepancy* from the null hypothesis μ_0. That *would* contradict the logic of tests.

---

[2] In deciphering existing discussions on ordinary power analysis, we can suppose that d(x_0) happens to be exactly at the cut-off for rejection, in discussing significant results; and just misses the cut-off for discussions on insignificant results in test T+. Then att-power for μ_1 equals ordinary power for μ_1.

Two fallacies are being committed here. The first we dealt with in discussing Greenland: namely, supposing that a negative result, with high power against μ_1, is evidence *for* the null rather than merely evidence that $\mu \leq \mu_1$. The more serious fallacy is that their "observed power" is shpower. Neither Cohen nor Neyman define power analysis this way. It is concluded that power analysis is paradoxical and inconsistent with P-value reasoning. You should really only conclude that shpower analytic reasoning is paradoxical. If you've redefined a concept and find that a principle that held with the original concept is contradicted, you should suspect your redefinition. It might have other uses, but there is no warrant to discredit the original notion.

The shpower computation is asking: What's the probability of getting $\overline{X} \geq \overline{x}_\alpha$, under $\mu = \overline{x}$? We still have that the larger the power (against $\mu = \overline{x}$), the better $\overline{x}$ indicates that $\mu \leq \overline{x}$ – as in ordinary power analysis – it's just that the indication is never more than 0.5. Other papers and even instructional manuals (Ellis 2010) assume shpower as what retrospective power analysis must mean, and ridicule it because "a nonsignificant result will almost always be associated with low statistical power" (p. 60). Not so. I'm afraid that observed power and retrospective power are all used in the literature to mean shpower. What about my use of severity? Severity will replace the cutoff for rejection with the observed value of the test statistic (i.e., $\Pr(d(X) \geq d(x_0); \mu_1)$), but not the parameter value μ. You might say, we don't know the value of μ_1. True, but that doesn't stop us from forming power or severity curves and interpreting results accordingly. Let's leave shpower and consider criticisms of ordinary power analysis. Again, pointing to Hoenig and Heisey's article (2001) is ubiquitous.

Anything Tests Can Do CIs Do Better

CIs do anything better than tests . . . No they don't, yes they do . . . *Annie Get Your Gun* is one of my favorite musicals, and while we've already seen the close connection between confidence limits and severity, they do not usurp tests. Hoenig and Heisey claim that power, by which they now mean ordinary power, is superfluous – once you have confidence intervals. We focused on CIs with a significant result (Section 4.3, Exhibit (vi)); our example now is a non-significant result. Let's admit right away that error statistical computations are interrelated, and if you have the correct principle directing you, you could get the severity computations by other means. The big deal is having the correct principle directing you, and this we'll see is what Hoenig and Heisey are missing.

Hoenig and Heisey consider an instance of our test T+: a one-sided Normal test of H_0: $\mu \leq 0$ vs. H_1: $\mu > 0$. The best way to address a criticism is to use the numbers given: "One might observe a sample mean $\overline{X} = 1.4$ with $\sigma_{\overline{X}} = 1$. Thus $Z = 1.4$ and $P = 0.08$, which is not significant at $\alpha = 0.05$" (ibid., p. 3). They don't tell us the sample size n, it could be that $\sigma = 5$ and $n = 25$, or any other combination to yield $(\sigma/\sqrt{n}) = 1$. Since the P-value is 0.08, $(\text{Pr}(Z > 1.4; \mu = 0) = 0.081)$, this is not significant at the 0.05 level (which requires $z = 1.65$), leading to a failure to reject H_0. They then point out that the power against $\mu = 3.29$ is high, 0.95 (i.e., $\text{Pr}(Z > 1.645; \mu = 3.29) = 0.95$).[3] Thus the power analyst would take the result as indicating $\mu < 3.29$. So what's the problem according to them?

They note that a 95% upper confidence bound on μ would be 3.05 (1.4 + 1.65), the implication being that it is more informative than what is given by the conservative power analysis. True, they get a tighter upper bound using the observed insignificant result, just as we do with severity. This they take to show that, "once we have constructed a confidence interval, power calculations yield no additional insights" (ibid., p. 4). Superfluous. There's one small problem: this is not the confidence interval that corresponds to test T+. The 95% confidence interval corresponding to test T+ is a one-sided interval: $\mu > \overline{x} - 1.65 \, \sigma_{\overline{X}}$ $(\mu > (1.4 - 1.65) = -0.25)$, not $\mu < 3.05$. That is, it corresponds to a one-sided lower bound, not an upper bound.

From the duality between CIs and tests (Section 3.7), as Hoenig and Heisey correctly state, "all values covered by the confidence interval could not be rejected" (ibid.). More specifically, the confidence interval contains the values that could not be rejected by the given test at the specified level of significance (Neyman 1937). But $\mu < 3.045$ does not give the set of values that T+ would fail to reject, were those values substituted for 0 in the null hypothesis of T+; there are plenty of μ values less than 3.045 that $\overline{X} = 1.4$ would reject, were they the ones tested, for instance, $\mu < -1$. The CI corresponding to test T+, namely, μ exceeds the lower confidence bound, doesn't help with the fallacy of insignificant results – the fallacy at which power analysis is aimed.

We don't deny it's useful to look at an upper bound (e.g., 3.05) to avoid the fallacy of non-rejection, just as it was to block fallacies of rejection (Section 4.3), but there needs to be a principled basis for this move, that's what severity gives us. Power analysis is a variation on the severity principle where $\overline{x} = \overline{x}_\alpha$. But Hoenig and Heisey are at pains to declare power

---

[3] Note: we are in "Case 2" where we've added 1.65 to the cut-off, meaning the power is the area to the right of -1.65 under the standard Normal curve (Section 5.1).

analysis superfluous! They plainly cannot have it both ways – they must either supplement confidence intervals with an adjunct along severity lines or be left with no way to avoid fallacies of insignificant results with the test they consider. Such an adjunct would require relinquishing their assertion: "It would be a mistake to conclude that the data refute any value within the confidence interval" (ibid., p. 4). The one-sided interval is [−0.245, infinity). We assume, of course, they don't literally mean "refute."

Now maybe they (or you) will say I'm being unfair, that one should always do a two-sided interval (corresponding to a two-sided test). But they are keen to argue that power analysis is superfluous for interpreting insignificant results from tests. Suppose we chuck tests and always do two-sided $1 - \alpha$ confidence intervals. We are still left with inadequacies already noted: First, the justification is purely performance: that the interval was obtained from a procedure with good long-run coverage; second, it relies on choosing a single confidence level and reporting, in effect, whether parameter values are inside or outside. Too dichotomous. Most importantly: The warrant for the confidence interval is just the one given by using attained power in a severity analysis. If this is right, it would make no sense for a confidence interval advocate to reject a severity analysis. You can see this revisiting Section 3.7 on capability and severity. (An example on bone sorting is Byrd 2008).

Inconclusive? Not only do we get an inferential construal of confidence intervals that differentiates the points within the interval rather than treating them all as on a par, we avoid a number of shortcomings of confidence intervals. Here's one: It is commonly taught that if a $1 - \alpha$ confidence interval contains both the null and a threshold value of interest, then only a diagnosis of "inconclusive" is warranted. While the inconclusive reading may be a reasonable rule of thumb in some cases, it forfeits distinctions that even ordinary significance levels and power analyses can reveal, if they are not limited to one fixed level. Ecologist Mark Burgman (2005, p. 341) shows how a confidence interval on the decline of threatened species reports the results as inconclusive, whereas a severity assessment shows non-trivial evidence of decline.

Go back to Hoenig and Heisey and $\overline{X} = 1.4$. Their two-sided 95% interval would be [−0.245, 3.04]. Suppose one were quite interested in a μ value in excess of 0.4. Both 0 and 0.4 are in the confidence interval. Are the results really uninformative about 0.4? Recognizing the test would fairly often (84% of the time) get such an insignificant result even if μ were as large as 0.4 should lead us to say no. Dichotomizing parameter values as rejected or not, as they do, turns the well-known arbitrariness in prespecifying confidence levels into an invidious distinction. Thus, we should deny Hoenig and Heisey's allegation that power analysis is "logically doomed" (p. 22), while endorsing a more nuanced use of both tests and intervals as in a severity assessment.

Our next exhibit looks at retrospective power in a different manner, and in relation, not to insignificant, but to significant results. It's not an objection to power analysis, but it appears to land us in a territory at odds with severity (as well as CIs and tests).

Exhibit (vii): Gelman and Carlin (2014) on Retrospective Power. They agree with the critiques of performing post-experiment power calculations (which are really shpower calculations), but consider "retrospective design analysis to be useful . . . in particular when apparently strong (statistically significant) evidence for nonnull effects has been found" (ibid., p. 2). They worry about "magnitude error," essentially our fallacy of making mountains out of molehills (MM). Unlike shpower, they don't compute power in relation to the observed effect size, but rather "on an effect size that is determined from literature review or other information external to the data at hand" (ibid.). They claim if you reach a just statistically significant result, yet the test had low power to detect a discrepancy from the null that is known from external sources to be correct, then the result "exaggerates" the magnitude of the discrepancy. In particular, when power gets much below 0.5, they say, statistically significant findings tend to be much larger in magnitude than true effect sizes. By contrast, "if the power is this high [.8], . . . overestimation of the magnitude of the effect will be small" (ibid., p. 3).

From the MM Fallacy, if $POW(\mu_1)$ is high then a just significant result is *poor* evidence that $\mu > \mu_1$; while if $POW(\mu_1)$ is low it's good evidence that $\mu > \mu_1$. Is their retrospective design analysis at odds with severity, P-values, and confidence intervals? Here's one way of making their assertion true using test T+: If you take the observed mean $\bar{x}_\alpha$ as the estimate of μ, and you happen to know the true value of μ is smaller than $\bar{x}_\alpha$ – between $\mu = \mu_0$ and $\mu = \bar{x}_\alpha$ (where the power ranges from α to 0.5.) – then obviously $\bar{x}_\alpha$ exceeds ("exaggerates") μ. Still I'm not sure this brings agreement.

Let's use our water plant accident testing $\mu \leq 150$ vs. $\mu > 150$ (with $\sigma = 10$, $\sigma/\sqrt{n} = 1$). The critical value for $\alpha = 0.025$ is $d_{0.025} = 1.96$, or $\bar{x}_{0.025} = 150 + 1.96(1) = 151.96$. You observe a *just* statistically significant result. You reject the null hypothesis and infer $\mu > 150$. Gelman and Carlin write:

[An] unbiased estimate will have 50% power if the true effect is 2 standard errors away from zero, it will have 17% power if the true effect is 1 standard error away from 0, and it will have 10% power if the true effect is 0.65 standard errors away from 0. (ibid., p. 4)

These correspond to μ =152, μ =151, and μ =150.65. It's odd to talk of an estimate having power; what they mean is that the test T+ has a power of 0.5 to detect a discrepancy 2 standard errors away from 150, and so on. The "unbiased estimate" here is the statistically significant $\bar{x}$. To check that we match their numbers, compute POW(μ = 152), POW(μ = 151), and POW(μ = 150.65)[4]:

(a) $\Pr(\bar{X} \geq 151.96;\ \mu = 152) = \Pr(Z \geq 0.04) = 0.51$;
(b) $\Pr(\bar{X} \geq 151.96;\ \mu = 151) = \Pr(Z \geq 0.96) = 0.17$;
(c) $\Pr(\bar{X} \geq 151.96;\ \mu = 150.65) = \Pr(Z \geq 1.31) = 0.1$.

They appear to be saying that there's better evidence for $\mu \geq 152$ than for $\mu \geq 151$ than for $\mu \geq 150.65$, since the power assessments go down. Nothing changes if we write >. Notice that the SEV computations for $\mu \geq 152$, $\mu \geq 151$, $\mu \geq 150.65$ are the complements of the corresponding powers 0.49, 0.83, 0.9. So the lower the power for μ_1 the stronger the evidence for $\mu > \mu_1$. Thus there's disagreement. But let's try to pursue their thinking.

Suppose we observe $\bar{x} = 152$. Say we have excellent reason to think it's too big. We're rather sure the mean temperature is no more than ~150.25 or 150.5, judging from previous cooling accidents, or perhaps from the fact that we don't see some drastic effects expected from water that hot. Thus 152 is an *overestimate*. The observed mean "exaggerates" what you know on good evidence to be the correct mean (< 150.5). No one can disagree with that, although they measure the exaggeration by a ratio.[5] Is this "power analytic" reasoning? No, but no matter. Some remarks:

First, the inferred estimate would not be 152 but rather the lower confidence bounds, say, $\mu > (152 - 2\sigma_{\bar{X}})$, i.e., $\mu > 150$ (for a 0.975 lower confidence bound). True, but suppose the lower bound at a reasonable confidence level is still at odds with what we assume is known. For example, a lower 0.93 bound is $\mu > 150.5$. What then? Then we simply have a conflict between what these data indicate and assumed background knowledge.

Second, do they really want to say that the statistically significant $\bar{x}$ fails to warrant $\mu \geq \mu_1$ for any μ_1 between 150 and 152 on grounds that the power in this range is low (going from 0.025 to 0.5)? If so, the result surely couldn't warrant values larger than 152. So it appears no values would be able to be inferred from the result.

---

[4] You can obtain these from the severity curves in Section 5.4.
[5] There are slight differences from their using a two-sided test, but we hardly add anything for the negative direction: For (a), $\Pr(\bar{X} < -2;\ \mu = 2) = \Pr(Z < -4) \simeq 0$. The severe tester would not compute power using both directions once she knew the result.

A way to make sense of their view is to construe it as saying the observed mean is so out of line with what's known that we suspect the assumptions of the test are questionable or invalid. Suppose you have considerable grounds for this suspicion: signs of cherry picking, multiple testing, artificiality of experiments, publication bias, and so forth – as are rife in both examples given in Gelman and Carlin's paper. *You have grounds to question the result* because you *question the reported error probabilities.* Indeed, no values can be inferred if the error probabilities are spurious, the severity is automatically low.

One reasons, if the assumptions are met, and the error probabilities approximately correct, then the statistically significant result *would* indicate $\mu > 150.5$, P-value 0.07, or severity level 0.93. But you happen to know that $\mu \leq 150.5$. Thus, that's grounds to question whether the assumptions are met. You suspect it would fail an audit. In that case put the blame where it belongs.[6]

Recall the (2010) study purporting to show genetic signatures of longevity (Section 4.3). Researchers found the observed differences suspiciously large, and sure enough, once reanalyzed, the data were found to suffer from the confounding of batch effects. When results seem out of whack with what's known, it's grounds to suspect the assumptions. That's how I propose to view Gelman and Carlin's argument; whether they concur is for them to decide.

5.6   Positive Predictive Value: Fine for Luggage

Many alarming articles about questionable statistics rely on alarmingly questionable statistics. Travelers on this cruise are already very familiar with the computations, because they stem from one or another of the "P-values exaggerate evidence" arguments in Sections 4.4, 4.5, and 5.2. They are given yet another new twist, which I will call the diagnostic screening (DS) criticism of significance tests. To understand how the DS criticism tests really took off, we should go back to a paper by John Ioannidis (2005):

Several methodologists have pointed out that the high rate of nonreplication (lack of confirmation) of research discoveries is a consequence of the convenient, yet ill-founded strategy of claiming conclusive research findings solely on the basis of a single study assessed by formal statistical significance, typically for a p-value less than 0.05. Research is not most appropriately represented and summarized by p-values, but, unfortunately, there is a widespread notion that medical research articles should

---

[6] The point can also be made out by increasing power by dint of sample size. If $n = 10,000$, $(\sigma/\sqrt{n}) = 0.1$. Test T+ ($n = 10,000$) rejects H_0 at the 0.025 level if $\overline{X} \geq 150.2$. A 95% confidence interval is [150, 150.4]. With $n = 100$, the just 0.025 significant result 152 corresponds to the interval [150, 154]. The latter is indicative of a larger discrepancy. Granted, sample size must be large enough for the statistical assumptions to pass an audit.

be interpreted based only on p-values. Research findings are defined here as any relationship reaching formal statistical significance, e.g., effective interventions, informative predictors, risk factors or associations.

It can be proven that most claimed research findings are false. (p. 0696)

First, do medical researchers claim to have "conclusive research findings" as soon as a single statistically significant result is spewed out of their statistical industrial complexes? Do they go straight to press? Ioannidis says that they do. Fisher's ghost is screaming. (He is not talking of merely identifying a possibly interesting result for further analysis.) However absurd such behavior sounds 80 years after Fisher exhorted us never to rely on "isolated results," let's suppose Ioannidis is right. But it gets worse. Even the single significant result is very often the result of the cherry picking and multiple testing we are all too familiar with:

... suppose investigators manipulate their design, analysis, and reporting so as to make more relationships cross the $p = 0.05$ threshold ... Such manipulation could be done, for example, with serendipitous inclusion or exclusion of certain patients or controls, post hoc subgroup analyses, investigation of genetic contrasts that were not originally specified ... Commercially available 'data mining' packages actually are proud of their ability to yield statistically significant results through data dredging. (ibid., p. 0699)

The DS criticism of tests shows that if

1. you publish upon getting a single P-value < 0.05,
2. you dichotomize tests into "up-down" outputs rather than report discrepancies and magnitudes of effect,
3. you data dredge and cherry pick and/or
4. there is a sufficiently low probability of genuine effects in your field, the notion of probability to be unpacked,

then the probability of true nulls among those rejected as statistically significant – a value we call the false finding rate (FFR)[7] – differs from and can be much greater than the Type I error set by the test.

However one chooses to measure "bad evidence, no test" (BENT) results, nobody is surprised that such bad behavior qualifies. For the severe tester, committing #3 alone is suspect, unless an adjustment to get proper error probabilities is achieved. Even if there's no cherry picking, and your test has a legitimate Type I error probability of 0.05, a critic will hold that the FFR can be much higher than 0.05, if you've randomly selected your null hypothesis from a group with a sufficiently high proportion of true "no

---

[7] Some call it the false discovery rate, but that was already defined by Benjamini and Hochberg in connection with the problem of multiple comparisons. (see Section 4.6).

effect" nulls. So is the criticism a matter of transposing probabilistic conditionals, only with the twist of trying to use "prevalence" for a prior? It is, but that doesn't suffice to dismiss the criticism. The critics argue that the quantity we should care about is the FFR, or its complement, the positive predictive value (PPV). Should we? Let's look at all this in detail.

Diagnostic Screening

In scrutinizing a statistical argument, particularly one that has so widely struck a nerve, we attempt the most generous interpretation. Still, if we are not to jumble up our freshly acquired clarity on statistical power, we need to use the proper terms for diagnostic screening, at least one model for it.[8]

We are all plagued by the TSA (Transportation Security Administration) screening in airports, although thankfully they have gotten rid of those whole body scanners in which all "your junk" is revealed to anonymous personnel. The latest test, we are told, would very rarely miss a dangerous item in your carry-on, and rarely trigger the alarm (+) for nothing. Yet most of the alarms are false alarms. That's because the dangerous items are relatively rare. On the other hand, sending positive (+) results for further scrutiny – usually in front of gloved representatives who have pulled you aside as they wave special wands and powder – ensures that, taken together, the false findings are quite rare. On the retest, they will usually discover you'd simply forgotten to remove that knife or box cutter from the last trip. Interestingly, the rarity of dangerous bags – that is, the low prevalence of D's (D for danger) – means we can be comforted in a negative result. So we'd often prefer not to lower the sensitivity, but control false positives relying on the follow-up retest given to any "+" result. (Mayo and Morey 2017.)

Positive Predictive Value (PPV) (1 – FFR). To get the (PPV) we are to apply Bayes' Rule using the given relative frequencies (or prevalences):

$$\text{PPV: } \Pr(D|+) = \frac{\Pr(+|D)\Pr(D)}{[\Pr(+|D)\Pr(D) + \Pr(+|\sim D)\Pr(\sim D)]} = \frac{1}{(1+B)}$$
$$B = \frac{\Pr(+|\sim D)\Pr(\sim D)}{\Pr(+|D)\Pr(D)}.$$

The *sensitivity* is the probability that a randomly selected item with D will be identified as "positive" (+):

$$\text{SENS: } \Pr(+|D).$$

---

[8] The screening model used here has also been criticized by many even for screening itself. See, for example, Dawid (1976).

The *specificity* is the probability a randomly selected item lacking D will be found negative (−):

SPEC: Pr(−|~D).

The *prevalence* is just the relative frequency of D in some population.

We run the test on the item (be it a person, a piece of luggage, or a hypothesis) and report either + or −. Instead of populations of carry-on bags and luggage, imagine an urn of null hypotheses, 50% of which are true. Randomly selecting a hypothesis, we run a test and output + (statistically significant) or − (non-significant). So our urn represents the proverbial "prior" probability of 50% true nulls.

The criticism turns on the PPV being too low. Even with Pr(D) = 0.5, with Pr(+|~D) = 0.05 and Pr(+|D)= 0.8, we still get a rather high PPV:

$$PPV = \frac{1}{\left[\frac{1+Pr(+|~D)}{Pr(+|D)}\right]}.$$

With Pr(D) = 0.5, all we need for a PPV greater than 0.5 is for Pr(+|~D) to be less than Pr(+|D). It suffices that the probability of ringing the alarm when we shouldn't is less than the probability of ringing it when we should. With a prevalence Pr(D) very small, e.g., <Pr(+|~D), we get a PPV < 0.5 even if we assume a maximal sensitivity Pr(+|D) of 1 (Van Belle 2008). In the field of diagnostics, it's scarcely worthless: there is still a boost from the prior prevalence.

Ioannidis rightly points out that many researchers are guilty of cherry picking and selection effects under his "bias" umbrella. The *actual* Pr(+|~D), with bias, is now the probability the "+" was generated by chance plus the probability it was generated by "bias." ~D plays the role of H_0. Even the lowest presumed bias, 0.10, changes a 0.05 into 0.14.

Actual Pr(+|~D):= "alleged" Pr(+| ~D) + Pr(−| ~D)(0.10) = (0.05) + (0.95)(0.10) = 0.14.

The PPV has now gone down to 0.85. Or consider if you're lucky enough to get a TSA official with 30% bias. Your "alleged" Pr(+| ~D) is again 0.05, but with 30% bias, the actual Pr(+|~D) = 0.05 + (0.95)(0.3) = 0.33. Table 5.1 lists some of the top (better) and bottom (worse) entries from Ioannidis' Table, keeping the notation of diagnostic tests. Some of the PPVs, especially for exploratory research with lots of data dredging, get very low PPVs.

Where do his bias adjustments come from? These are just guesses he puts forward. It would be interesting to see if they correlate with some of the better-

Table 5.1 Selected entries from Ioannidis (2005)

Pr(+|D)	PREV of D	Bias	Practical example	PPV
0.8	50%	0.10	Adequately powered RCT, little bias	0.85
0.95	67%	0.30	Confirmatory meta-analysis of good-quality RCTs	0.85
0.8	9%	0.3	Adequately powered exploratory epidemiological study	0.20
0.2	0.1%	0.8	Discovery-oriented exploratory research with massive testing	.001

known error adjustments, as with multiple testing. If so, maybe Ioannidis' bias assignments can be seen as giving another way to adjust error probabilities. The trouble is, the dredging can be so tortured in many cases that we'd be inclined to dismiss the study rather than give it a PPV number. (Perhaps confidence intervals around the PPV estimate should be given?)

Ioannidis will also adjust the prevalence according to the group that your research falls into, leading Goodman and Greenland (2007) to charge him with punishing the epidemiologist twice: by bias and low prevalence! I'm sympathetic with those who protest that rather than assume guilt (or innocence) by association (with a given field), it's better to see what crime was actually committed or avoided by the study at hand. Even bias violations are open to appeal, and may have been gotten around by other means. (No mention is given of failed statistical assumptions, which can quickly turn to mush the reported error probabilities, and preclude the substantive inference that is the actual output of research. Perhaps this could be added.) Others who mount the DS criticism allege that the problem holds even accepting the small α level and no bias.[9] Their gambit is to sufficiently lower the prevalence of D – which now stands for probability of a "true effect" – so that the PPV is low (e.g., Colquhoun 2014). Colquhoun's example retains Pr(+|~D) = 0.05, Pr(+|D)= 0.8, but shrinks the prevalence Pr(D) of true effects down to 10%. That is, 90% of the nulls in your research universe are true. This yields a PPV of 64%. The Pr (~D|+) is 0.36, much greater than Pr(+|~D) = 0.05.

So the DS criticism appears to go through with these computations. What about exporting the terms from significance tests into FFR or PPV assessments? We haven't said anything about treating ~D as H_0 in the DS criticism.

[9] Even without bias, it's expected that only 50% of statistically significant results will replicate as significantly on the next try, but such a probability is to be expected (Senn 2002). Senn regards such probabilities as irrelevant.

[$\alpha/(1-\beta)$] Again

Although we are keen to get away from coarse dichotomies, in the DS model of tests we are to consider just two possibilities: "no effect" and "real effect." The null hypothesis is treated as H_0: 0 effect ($\mu = 0$), while the alternative H_1: the discrepancy against which the test has power ($1 - \beta$). It is assumed the probability for finding any effect, regardless of size, is the same (Ioannidis 2005, p. 0696). Then [$\alpha/(1-\beta)$] is used as the likelihood ratio to compute the posterior of either H_0 or H_1 – a problematic move, as we know.

An example of one of their better tests might have H_1: $\mu = \mu^{.9}$ where $\mu^{.9}$ is the alternative against which the test has 0.9 power. But now the denial of the alternative H_1 does not yield the same null hypothesis used to obtain the Type I error probability of 0.05. Instead it would be high, nearly as high as 0.9. Likewise if the null is chosen to have low α, then its denial won't be one against which the test has high power (it will be close to α). Thus, the identification of "effect" and "no effect" with the hypotheses used to compute the Type I error probability and power are inconsistent with one another. The most plausible way to construe the DS argument is to assume the critics have in mind a test between a point null H_0, or a small interval around it, and a non-exhaustive alternative hypothesis $\mu = \mu_1$ against which there is a specified power such as 0.9. It is known that there are intermediate values of μ, but the inference will just compare two.

The DS critics will give a high PPV to alternatives with high power, which is often taken to be 0.8 or 0.9. We know the computation from Goodman (Section 5.2) that "the truth probability of the null hypothesis drops to 3 percent (= 0.03/(1 + 0.03))." The PPV for $\mu^{.9}$ is 0.97. We haven't escaped Senn's points about the nonsensical and the ludicrous, or making mountains out of molehills. To infer $\mu^{.9}$ based on $\alpha = 0.025$ (one-sided) is to be wrong 90% of the time. We'd expect a more significant result 90% of the time were $\mu^{.9}$ correct. I don't want to repeat what we've seen many times. Even using Goodman's "precise P-value" yields a high posterior. A DS critic could say: you compute error probabilities but we compute PPV, and our measure is better. So let's take a look at what the computation might mean.

In the typical illustrations it's the prevalence that causes the low PPV. But what is it? Colquhoun (2014) identifies Pr(D) with "the proportion of experiments we do over a lifetime in which there is a real effect" (p. 9). Ioannidis (2005) identifies it with "the number of 'true relationships' . . . among those tested in the field" (p. 0696). What's the relevant *reference class* for the prevalence Pr(D)? We scarcely have a list of all hypotheses to be tested in a field, much less do we know the proportion that are "true." With continuous parameters, it could be claimed there are infinitely many hypotheses;

individuating true ones could be done in multiple ways. Even limiting the considerations to discrete claims (effect/no effect), will quickly land us in quicksand. Classifying by study type makes sense, but any umbrella will house studies from different fields with different proportions of true claims.

One might aver that the PPV calculation is merely a heuristic to show the difference between α and FFR, or between $(1 - \alpha)$ and the PPV. It should always be kept in mind that even when a critic has performed a simulation, it is a simulation that assumes ingredients. If aspects of the calculation fail, then of what value is the heuristic? Furthermore, it is clear that the PPV calculation is intended to assess the results of actual tests. Even if we agreed on a reference class, say the proportion of true effects over your lifetime of testing is θ, this probability θ wouldn't be the probability that a selected effect is "true." It would not be a *frequentist* prior probability for the randomly selected hypothesis. We now turn to this.

Probabilistic Instantiation Fallacy. Suppose we did manage to do an experiment involving a random selection from an urn of null hypotheses, $100\theta\%$ assumed to be true. The outcome may be $X = 1$ or 0 according to whether the hypothesis we've selected is true. Even allowing it's known that the probability of $X = 1$ is 0.5, it does not follow that a specific hypothesis we might choose (say, your blood pressure drug is effective) has a frequentist probability of 0.5 of being true – any more than a particular 0.95 confidence interval estimate has a probability of 0.95. The issue, in this form, often arises in "base rate" criticisms (Mayo 1997a, 1997b, 2005b, 2010c, Spanos 2010b).

Is the PPV computation *relevant* to the very thing that working scientists want to assess: strength of the *evidence* for effects or their degree of corroboration?

Crud Factor. It is supposed in many fields of social and biological science that nearly everything is related to everything: "all nulls are false." Meehl dubbed this the crud factor. Meehl describes how he and David Lykken conducted a study of the crud factor in psychology in 1966. They used a University of Minnesota student questionnaire sent to 57,000 high school seniors, including family facts, attitudes toward school, leisure activities, educational plans, etc. Cross-tabulating variables including parents' occupation, education, siblings, birth order, family attitudes, sex, religious preferences, 22 leisure time activities, MCAT scores, etc., all 105 cross-tabulations were statistically significant at incredibly small levels.

These relationships are not, I repeat, Type I errors. They are facts about the world, and with $N = 57,000$ they are pretty stable. Some are theoretically easy to explain, others more difficult, others completely baffling. The 'easy' ones have multiple explanations, sometimes competing, usually not. Drawing theories from a pot and associating them whimsically with variable pairs would yield an impressive batch of H_0-refuting 'confirmations.' (Meehl 1990, p. 206)

He estimates the crud factor correlation at around 0.3 or 0.4.

So let's apply Ioannidis' analysis to two cases. In the first case, we've randomly selected a hypothesis from a social science urn with high crud factor. Even if I searched and cherry picked, perhaps looking for ones that correlate well with a theory I have in mind, statistical significance at the 0.05 level would still result in a fairly high prevalence of true claims (D's) among those found statistically significant. Since the test they passed lacked stringency, I wouldn't be able to demonstrate a genuine reproducible effect – in the manner that is understood in science. So nothing has been demonstrated about replicability or knowledge of real effects by dint of a high PPV.

You might say high prevalence could never happen with things like correlating genes and disease. But how can we count up the hypotheses? Should they include molecular biology, proteomics, stem cells, etc. Do we know what hypotheses will be conjectured next year? Why not combine fields for estimating prevalence? With a little effort, one could claim to have as high a prevalence as desired.

Now let's assume we are in one of those low prevalence situations. If I've done my homework and went beyond the one P-value before going into print, checked flaws, tested for violated assumptions, then even if I don't yet know the causal explanation, I may have a fairly good warrant for taking the effect as real. Having obeyed Fisher, I am in a good position to demonstrate the reality of the published finding. *Avoiding bias and premature publication is what's doing the work, not prevalence.*

There is a seductive blurring of rates of false positives over an imagined population, PPVs, on the one hand, with an assessment of what we know about reproducing any particular effect, on the other, and fans of the DS model fall into this equivocal talk. In other words, "positive predictive value," in this context, is a misnomer. The number isn't telling us how valuable the statistically significant result is for predicting the truth or reproducibility of *that effect*. Nor is it even assuring lots of the findings in the group will be reproducible over time. We want to look at how well tested the particular hypothesis of interest is. We might assess the prevalence with which hypotheses pass highly stringent tests, if false. Now look what's happened. We have come full circle to evaluating the severity of tests passed. *Prevalence has nothing to do with it.*

I am reminded of the story of Isaac. Not in the Bible, but in a discussion I had with Erich Lehmann in Princeton (when his wife was working at the Educational Testing Services). It coincided with a criticism by Colin Howson (1997a,b) to the effect that low prevalence (or "base rates") negates severity of test. Isaac is a high school student who has passed (+) a battery of tests for D: "college-readiness." It is given that Pr(+|~D) is 0.05, while Pr(+|D) ~1. But because he was randomly selected from Fewready town, where the prevalence of readiness is only 0.001, Pr(D|+) is still very low. Had Isaac been randomly selected from Manyready suburb with high (Pr(D), then Pr(D|+) is high. In fact Isaac, from Fewready town, would have to score quite a bit higher than if he had come from Manyready suburb for the same PPV. There is a real policy question here that officials disagree on. Should we demand higher test scores from students in Fewready town to ensure overall college-readiness amongst those accepted by college admissions boards? Or would that be a kind of reverse affirmative action?

We might go further and imagine Alex from Manyready scored lower than Isaac, maybe even cheated on just one or two questions. Even if their PPVs are equal, I submit that Isaac is in a better position to demonstrate his college readiness.[10]

The Dangers of the Diagnostic Screening Model for Science

What then can we infer is replicable? Claims that have passed with severity. If subsequent tests corroborate the severity assessment of an initial study, then it is replicated. But severity is not the goal of science. Lots of true but trivial claims are not the goal. Science seeks growth of knowledge and understanding. To take the diagnostic-screening model literally, by contrast, would point the other way: keep safe.

Large-scale evidence should be targeted for research questions where the pre-study probability is already considerably high, so that a significant research finding will lead to a post-test probability that would be considered quite definitive. (Ioannidis 2005, p. 0700)

Who would pursue seminal research that challenged the reigning biological paradigm, as did Prusiner, doggedly pursuing, over decades, the cause of mad cow and related diseases, and the discovery of prions? Would Eddington have gone to all the trouble of testing the deflection effect in Brazil? Newton was predicting fine. Replication is just a small step toward getting real effects. Lacking the knowledge of how to bring about an effect, and how to use it to change other known and checkable effects, your PPV may be swell but your science could be at a dead-end. To be clear: its advocates surely don't recommend the "keep safe" consequence, but addressing it is worthwhile to further emphasize the difference between good science and a good scorecard.

---

[10] Peter Achinstein and I have debated this on and off for years (Achinstein 2010; Mayo 1997a, 2005b, 2010c).

There are contexts in which the screening viewpoint is useful. Beyond diagnostic screening of disease, high-throughput testing of microarray data seeks to control the rates of genes worth following up. Nevertheless, we argue that the PPV does not quantify how well tested, warranted, or plausible a given scientific hypothesis is (including ones about genetic associations where a DS model is apt). I'm afraid the DS model has introduced confusion into the literature, by mixing up the probability of a Type I error (often called the "false positive rate") with the posterior probability given by the FFR: $\Pr(H_0|H_0$ is rejected). Equivocation is encouraged. In frequentist tests, reducing the Type II error probability results in *increasing* the Type I error probability: there is a trade-off. In the DS model, the trade-off disappears: reducing the Type II error rate also reduces the FFR.

Much of Ioannidis' work is replete with sagacious recommendations for better designs. My aim here was the limited one of analyzing the diagnostic screening model of tests. That it's the basis for popular reforms underscores the need for scrutiny.

Tour III Deconstructing the N-P versus Fisher Debates

> [Neyman and Pearson] began an influential collaboration initially designed primarily, it would seem, to clarify Fisher's writing. This led to their theory of testing hypotheses and to Neyman's development of confidence intervals, aiming to clarify Fisher's idea of fiducial intervals. As late as 1932 Fisher was writing to Neyman encouragingly about this work, but relations soured, notably when Fisher greatly disapproved of a paper of Neyman's on experimental design and no doubt partly because their being in the same building at University College London brought them too close to one another! (Cox 2006a, p. 195)

Who but David Cox could so expertly distill the nitty-gritty of a long story in short and crisp terms? It hits all the landmarks we want to visit in Tour III. Wearing error statistical spectacles gives a Rosetta Stone for a novel deconstruction of some of the best known artifacts. We begin with the most famous passage from Neyman and Pearson (1933), often taken as the essence of the N-P philosophy. We'll make three stops:

- First, we visit a local theater group performing "Les Miserables Citations";
- Next, I've planned a daytrip to Fisher's Fiducial Island for some little explored insights into our passage;
- Third, we'll get a look at how philosophers of statistics have deconstructed that same passage.

I am using "deconstruct" in the sense of "analyze or reduce to expose assumptions or reinterpret." A different sense goes along with "deconstructionism," whereby it's thought texts lack fixed meaning. I'm allergic to the relativistic philosophies associated with this secondary sense. Still, here we're dealing with methods about which advocates say the typical performance metaphor is just a heuristic, not an instruction for using the methods. In using them, they are to be given a subtle evidential reading. So it's fitting to speak of disinterring a new meaning.

5.7 Statistical Theatre: "Les Miserables Citations"

> We are inclined to think that as far as a particular hypothesis is concerned, no test based upon the theory of probability can by itself provide any valuable evidence of the truth or falsehood of that hypothesis.

But we may look at the purpose of tests from another view-point. Without hoping to know whether each separate hypothesis is true or false, we may search for rules to govern our behavior with regard to them, in following which we insure that, in the long run of experience, we shall not be too often wrong. (Neyman and Pearson 1933, pp. 141–2)

Neyman and Pearson wrote these paragraphs once upon a time when they were still in the midst of groping toward the basic concepts of tests – for example, "power" had yet to be coined. Yet they are invariably put forward as proof positive that N-P tests are relevant only for a crude long-run performance goal. I'm not dismissing the centrality of these passages, nor denying the 1933 paper records some of the crucial early developments. I am drawn to these passages because taken out of context, as they so often are, they have led to knee-jerk interpretations to which our famous duo would have objected. What was the real context of those passages? The paper opens, just five paragraphs earlier, with a discussion of two French probabilists – Joseph Bertrand, author of *Calculus of Probabilities* (1889), and Émile Borel, author of *Le Hasard* (1914)!

Neyman had attended Borel's lectures in Paris, and he returns to the Bertrand–Borel debate in no less than five different papers – one "an appreciation" for Egon Pearson when he died – and in recounting core influences on N-P theory to biographer Constance Reid. Erich Lehmann (1993a) wrote an entire paper on "The Bertrand-Borel Debate and the Origins of the Neyman Pearson Theory."[1] A deconstruction of the debate illuminates the inferential over the behavioristic construal of tests – somewhat surprisingly given the behavioristic-sounding passage to follow. We're in time for a matinee where the key characters are placed in an (imaginary) theater production. It's titled "Les Miserables Citations." (Lehmann's translation from the French is used where needed.)

The curtain opens with a young Neyman and Pearson (from 1933) standing mid-stage, lit by a spotlight. (All speaking parts are exact quotes; Neyman does the talking.)

NEYMAN AND PEARSON (N-P): Bertrand put into statistical form a variety of hypotheses, as for example the hypothesis that a given group of stars . . . form a 'system.' His method of attack, which is that in common use, consisted essentially in calculating the probability, P, that a certain character, x, of the observed facts would arise if the hypothesis tested were true. If P were very small, this would generally be considered as an indication that . . . H was probably false, and *vice*

---

[1] The pagination is from the Selected Works of E.L. Lehmann (2012).

versa. Bertrand expressed the pessimistic view that no test of this kind could give reliable results.

Borel, however, . . . considered that the method described could be applied with success provided that the character, x, of the observed facts were properly chosen – were, in fact, a character which he terms 'en quelque sorte remarquable'. (Neyman and Pearson 1933, p. 141).

The stage fades to black, then a spotlight shines on Bertrand, stage right.

BERTRAND: How can we decide on the unusual results that chance is incapable of producing? . . . The Pleiades appear closer to each other than one would naturally expect . . . In order to make the vague idea of closeness more precise, should we look for the smallest circle that contains the group? the largest of the angular distances? the sum of squares of all the distances? . . . Each of these quantities is smaller for the group of the Pleiades than seems plausible. Which of them should provide the measure of implausibility?

He turns to the audience, shaking his head.

The application of such calculations to questions of this kind is a delusion and an abuse. (Bertrand 1889, p. xvii; Lehmann 1993a, p. 963)

The stage fades to black, then a spotlight appears on Borel, stage left.

BOREL: The particular form that problems of causes often take . . . is the following: Is such and such a result due to chance or does it have a cause? It has often been observed how much this statement lacks in precision. Bertrand has strongly emphasized this point. But . . . to refuse to answer under the pretext that the answer cannot be absolutely precise, is to . . . misunderstand the essential nature of the application of mathematics. [Bertrand considers the Pleiades.] If one has observed a [precise angle between the stars] . . . in tenths of seconds . . . one would not think of asking to know the probability [of observing exactly this observed angle under chance] because one would never have asked that precise question before having measured the angle. . .

The question is whether one has the same reservations in the case in which one states that one of the angles of the triangle formed by three stars has 'une valeur remarquable' [a striking or noteworthy value], and is for example equal to the angle of the equilateral triangle. . .

Here is what one can say on this subject: One should carefully guard against the tendency to consider as striking an event that one has not specified *beforehand*, because the number of such events that may appear striking, from different points of view, is very substantial. (ibid., pp. 964–5)

The stage fades to black, then a spotlight beams on Neyman and Pearson mid-stage.

N-P: We appear to find disagreement here, but are inclined to think that . . . the two writers [Bertrand and Borel] are not really considering precisely the same problem. In general terms the problem is this: Is it possible that there are any efficient tests of hypotheses based upon the theory of probability, and if so, what is their nature? . . . What is the precise meaning of the words 'an efficient test of a hypothesis'?

[W]e may consider some specified hypothesis, as that concerning the group of stars, and look for a method which we should hope to tell us, *with regard to a particular group of stars*, whether they form a system, or are grouped 'by chance,' . . . their relative movements unrelated. (1933, p. 140; emphasis added)

If this were what is required of 'an efficient test,' we should agree with Bertrand in his pessimistic view. For however small be the probability that a particular grouping of a number of stars is due to 'chance,' does this in itself provide any evidence of another 'cause' for this grouping but 'chance'? . . . Indeed, if x is a continuous variable – as for example is the angular distance between two stars – then any value of x is a singularity of relative probability equal to zero. We are inclined to think that as far as a particular hypothesis is concerned, no test based upon the theory of probability can by itself provide any valuable evidence of the truth or falsehood of that hypothesis. But we may look at the purpose of tests from another view-point." (ibid., pp. 141–2)

Fade to black, spot on narrator mid-stage:

NARRATOR: We all know our famous lines are about to come. But let's linger on the "as far as a particular hypothesis is concerned." For any particular case, one may identify data dependent features x that would be highly improbable "under the particular hypothesis of chance." Every outcome would too-readily be considered statistically unusual. We must "carefully guard," Borel warns, "against the tendency to consider as striking an event that one has not specified *beforehand*." (Lehmann 1993a, p. 964.) If you are required to set the test's capabilities ahead of time, then you need to specify the type of falsity of H_0 – the test statistic – beforehand. An efficient test should capture Fisher's desire for tests sensitive to departures of interest. You should also wish to avoid tests that more probably find discrepancies when there are none than when present. Listen to Neyman's reflection on Borel's remarks much later on, in 1977.

Fade to black. Spotlight on an older Neyman, stage right. (He's in California, in the background there are palm trees, and Berkeley.)

NEYMAN: The question (what is an efficient test of a statistical hypothesis) is about an intelligible methodology for deciding whether the observed [difference] . . . contradicts the stochastic model . . .

[T]his question was the subject of a lively discussion by Borel and others. Borel was optimistic but insisted that: (a) the criterion to test a hypothesis (a 'statistical hypothesis') using some observations must be selected *not after the examination of the results of observation*, but before, and (b) this criterion should be a function of the observations 'en quelque sorte remarquable' [of a remarkable sort]. It is these remarks of Borel that served as an inspiration to Egon S. Pearson and myself in our effort to build a frequentist theory of testing hypotheses. (Neyman 1977, pp. 102–3)

Fade to black. Spotlight on an older Egon Pearson writing a letter to Neyman about the preprint Neyman sent of his 1977 paper. (The letter is unpublished, but I cite Lehmann 1993a.)

PEARSON: I remember that you produced this quotation [from Borel] when we began to get our [1933] paper into shape . . . The above stages [wherein he had been asking 'Why use that particular test statistic?'] led up to Borel's requirement of finding . . . a criterion which was "a function of the observations 'en quelque sorte remarquable'". Now my point is that you and I (perhaps my first leading) had ourselves reached the Borel requirement independently of Borel, because we were serious humane thinkers; Borel's expression neatly capped our own. (pp. 966–7)

Fade to black. End play.

Egon has the habit of leaving tantalizing claims unpacked, and this is no exception: What exactly is the Borel requirement he thinks they'd reached due to their being "serious humane thinkers"? I can well imagine turning this episode into something like Michael Frayn's expressionist play, *Copenhagen*, wherein a variety of alternative interpretations are entertained based on subsequent work and remarks. I don't say that a re-run would enjoy a long life on Broadway, but a small handful of us would relish it.

Inferential Rationales for Test Requirements

It's not hard to see that "as far as a particular" star grouping is concerned, we cannot expect a reliable inference to just any non-chance effect discovered in the data. The more specific the feature is to these particular observations, the more improbable. What's the probability of three hurricanes followed by two plane crashes? To cope with the fact that any sample is improbable in some respect, statistical methods do one of two things: appeal to prior probabilities or to error probabilities of a procedure. The former can check our tendency to find a more likely explanation H' than chance by an appropriately low prior weight to H'. The latter says, we need to consider the problem as of a *general* type. It's a general method, from a test statistic to some assertion about an alternative hypothesis, expressing the non-chance effect. Such assertions may be in error but we can control such erroneous interpretations.

Isn't this taken care of by Fisher's requirement that $\Pr(P < p_0; H_0) = P_0$ – that the test rarely rejects the null if true? It may be, in practice, Neyman and

Pearson thought, but only with certain conditions that were not explicitly codified by Fisher's simple significance tests. With just the null hypothesis, it is unwarranted to take low P-values as evidence for a specific "cause" or non-chance explanation. A statistical effect, even if genuine, *underdetermines* its explanation; several rivals can be erected post-data, but the ways they could be in error would not have been probed. The fallacy of rejection looms. Fisher (1935a, p. 187) is well aware that "the same data may contradict the hypothesis in any of a number of different ways," and that different corresponding tests would be used.

> The notion that different tests of significance are appropriate to test different features of the same null hypothesis presents no difficulty to workers engaged in practical experimentation. ... [T]he experimenter ... is aware of what observational discrepancy it is which interests him, and which he thinks may be statistically significant, before he enquires what test of significance, if any, is available appropriate to his needs. (ibid., p. 190)

Even if "an experienced experimenter" knows the appropriate test, this doesn't lessen the importance of N-P's interest in seeking to identify a statistical rationale for the choices made on informal grounds. There's legitimate concern about selecting the alternative that gives the more impressive P-value.

Here's Pearson writing with C. Chandra Sekar on testing if a sample has been drawn from a single Normal population:

> ... it is not possible to devise an efficient test if we only bring into the picture this single normal probability distribution with its two unknown parameters. We must also ask how sensitive the test is in detecting failure of the data to comply with the hypotheses tested, and to deal with this question effectively we must be able to specify the directions in which the hypothesis may fail. (Pearson and Chandra Sekar 1936, p. 121)

And while:

> It is sometimes held that the appropriate test can be chosen *after* examining the data. [but it will be hard to be unprejudiced at this point]. (ibid., p. 127)

Their position is:

> To base the choice of the test of a statistical hypothesis upon an inspection of the observations is a dangerous practice; a study of the configuration of a sample is almost certain to reveal some feature, or features, which are exceptions if the hypothesis is true ... By choosing the feature most unfavourable to H_0 out of a very large number of features examined it will usually be possible to find some reason for rejecting the hypothesis. It must be remembered, however, that the point now at issue will not be whether it is exceptional to find a given criterion with so unfavourable a value. We shall need to find an answer to the more difficult question. Is it exceptional that the most

unfavourable criterion of the n, say, examined should have as unfavourable a value as this? (ibid., p. 127)

In short, we'd have to adjust the attained P-value. In so doing, the goal is not behavioristic but avoiding glaring fallacies in the test at hand, fallacies we know all too well.

The statistician who does not know in advance with which type of alternative to H_0 he may be faced, is in the position of a carpenter who is summoned to a house to undertake a job of an unknown kind and is only able to take one tool with him! Which shall it be? Even if there is an 'omnibus' tool, it is likely to be far less sensitive at any particular job than a specialized one; but the specialized tool will be quite useless under the wrong conditions. (ibid., p. 126)

Neyman (1952) demonstrates that choosing the alternative post-data allows a result that leads to rejection in one test to yield non-rejection in another, despite both adhering to a fixed significance level. (Fisher concedes this as well.) If you are keen to ensure the test is capable of teaching about discrepancies of interest, you should prespecify an alternative hypothesis, where the null and alternative hypothesis exhaust the space, relative to a given question.

The Deconstruction So Far

If we accept the words, "an efficient test of the hypothesis H" to mean a statistical (methodological) falsification rule that controls the probabilities of erroneous interpretations of data, and ensures the rejection was *because* of the underlying cause (as modeled), then we agree with Borel that efficient tests are possible. This requires (i) a prespecified test criterion to avoid verification biases while ensuring power (efficiency), and (ii) consideration of alternative hypotheses to avoid fallacies of acceptance and rejection. We should steer away from isolated or particular curiosities to those that are tracking genuine effects. Fisher is to be credited, Pearson remarks, for his "emphasis on planning an experiment, which led naturally to the examination of the power function, both in choosing the size of sample so as to enable worthwhile results to be achieved, and in determining the most appropriate tests" (Pearson 1962, p. 277). If you're planning, you're prespecifying, perhaps, nowadays, by explicit preregistration.

"We agree also that not any character, x, whatever is equally suitable to be a basis for an efficient test (Neyman and Pearson 1933, p. 142)." The test "criterion should be a function of the observations," and the alternatives, such that there is a known statistical relationship between the characteristic of the data and the underlying distribution (Neyman 1977, p. 103). It must enable the error probabilities to be computed under the null and also under discrepancies from the null, despite any unknown parameters.

An exemplary characteristic of this sort is the remarkable properties offered by pivotal test statistics such as Z or T, whose distributions are known:

$$Z = \sqrt{n}(\overline{X} - \mu)/\sigma,$$

$$T = \sqrt{n}(\overline{X} - \mu)/s,$$

Z has the standard Normal distribution, and T the Student's t distribution, where σ is unknown and thus replaced by the estimator.

$$\frac{1}{n-1} \sum_{i=1}^{n} (x_i - \overline{x})^2$$

Consider the pivot Z. The probability $Z > 1.96$ is 0.025. But by pivoting, the $Z > 1.96$ is equivalent to

$$\mu < \overline{X} - 1.96 \, \sigma/\sqrt{n},$$

so it too has probability 0.025. Therefore, the procedure that asserts $\mu > \overline{X} - 1.96 \, \sigma/\sqrt{n}$ asserts correctly 95% of the time![2] We can make valid probabilistic claims about the method that holds post-data, *if interpreted correctly*. For the severe tester, these also inform about claims that are well and poorly tested (Section 3.7). This leads us on a side trip to Fisher's fiducial territory (Section 5.8), and the initial development of the behavioral performance idea. First, let's trace some of the ways our miserable citation has been interpreted by contemporaries.

The Miserable Passage in the Hands of Contemporaries

> Without hoping to know whether each separate hypothesis is true or false, we may search for rules to govern our behavior with regard to them, in following which we insure that, in the long run of experience, we shall not be too often wrong (Neyman and Pearson 1933, p. 142)

Ian Hacking. According to Ian Hacking (1965) this passage shows that Neyman and Pearson endorse something "more radical than anything I have mentioned so far" (p. 103). What they are saying is that "there is no alternative to certainty and ignorance" (p. 104). If probability only applies to the rule's long-run error control, Hacking is saying in 1965, it's not an account of inductive inference. This is precisely what he comes to deny in his 1980 "retraction" (Section 2.1 Exhibit (iii)), but here he's leading the posse in

---

[2] To get a fiducial distribution, the case has to be continuous.

philosophy toward the most prevalent view, even though it comes in different forms.

Isaac Levi. Isaac Levi (1980, p. 404), reacting to Hacking (1965) claims "Hacking misinterprets these remarks [in our passage] when he attributes to Neyman and Pearson the view that 'there is no alternative to certainty or ignorance.'" Even N-P allow intermediate standpoints when legitimate prior probabilities are available. Finding them to be so rarely available, N-P were led to methods whose validity would not depend on priors. Except for such cases, Levi concurs with the early Hacking, that N-P deny evidence altogether. According to Levi, N-P are "objectivist necessitarians" who stake out a rather robotic position: tests only serve as routine programs for "selecting policies rather than using such reports as evidence" (ibid., p. 408). While this might be desirable in certain contexts, Levi objects, "this does not entitle objectivist necessitarians to insist that rational agents should always assess benefits in terms of the long run nor to favor routinization over deliberation" (ibid.).

These construals by philosophers made sense in the context of seeking an inductive logic that would assign a degree of rational belief, support or confirmation to statistical hypotheses. N-P opposed such a view. Their attitude is, we're just giving examples to illustrate and capsulize a rationale to underwrite the tests chosen on intuitive grounds. Even in Neyman–Pearson (1928):

[T]he tests should only be regarded as tools which must be used with discretion and understanding, and not as instruments which in themselves give the final verdict. . . . we must not discard the original hypothesis until we have examined the alternative suggested, and have satisfied ourselves that it does involve a change in the real underlying factors in which we are interested . . . that the alternative hypothesis is not error in observation, error in record, variation due to some outside factor that it was believed had been controlled, or to any one of many causes . . . (p. 58)

In the 1933 paper, they explicitly distinguish their account from contexts where values enter other than controlling erroneous interpretations of data: "[I]t is possible that other conceptions of relative value may be introduced. But the problem is then no longer the simple one of discriminating between hypotheses" (1933, p. 148).

Howson and Urbach. Howson and Urbach interpret this same passage in yet another, radical manner. They regard it as "evident" that for Neyman and Pearson, acceptance and rejection of hypotheses is "the adoption of the same attitude towards them as one would take if one had an unqualified belief in their truth or falsehood" (1993, p. 204), putting up "his entire stock of worldly goods" upon a single statistically significant result (p. 203). Even on the strictest behavioristic formulation, "to accept a hypothesis H means only to decide to take action A rather than action B" (Neyman 1950, p. 259). It could be "decide" to declare the study unreplicable, publish a report, tell a patient to get another test, announce a genuine experimental effect, or whatever. A particular action always had to be spelled out, it was never to take any and all actions as if you had "unqualified belief."

Neyman, not Pearson, is deemed the performance-oriented one, but even he used conclude and decide interchangeably:

The analyses we performed led us to 'conclude' or 'decide' that the hypotheses tested could be rejected without excessive risk of error. In other words, after considering the probability of error (that is, after considering how frequently we would be in error if in conditions of our data we rejected the hypotheses tested), ... we *decided to act on the assumption* (or *concluded*) *that the two groups are not random* samples from the same population. (1976, 750–1; the emphasis is Neyman's)

What would make the reading closer to severity than performance is for the error probability to indicate what would/would not be a warranted cause of the observations. It's important, too, to recognize Neyman's view of inquiry: "A study of any serious substantive problem involves a sequence of incidents at which one is forced to pause and consider what to do next. In an effort to reduce the frequency of misdirected activities one uses statistical tests" (1976, p. 737). Rather than a series of unrelated tests, a single inquiry involves numerous piecemeal checks, and the error control promotes the "lift-off." Mistakes in one part ramify in others so as to check the overall inference. Even if Neyman wasn't consciously aware of the rationale behind tests picked out by these concerns, they still may be operative.

In his 1980 retraction, Hacking, following Peirce, denies there's a logic of statistical inference explaining it was a false analogy with deduction that led everyone to suppose the probability is to be assigned to the conclusion rather than to the overall method (Section 2.1). We should all be over that discredited view by now.

Elliott Sober. Our passage pops up in Elliott Sober, who combines the more disconcerting aspects of earlier interpretations. According to Sober (2008, p. 7), "Neyman and Pearson think of acceptance and rejection" as acts that should only "be regulated by prudential considerations, not by 'evidence,' which, for them, is a will o' the wisp ... There is no such thing as allowing 'evidence' to regulate what we believe. Rather, we must embrace a policy and stick to it." Sober regards this as akin to Pascal's theological argument for believing in God, declaring, "Pascal's concept of prudential acceptance lives on in frequentism" (ibid.).

I don't think it's plausible to read Neyman and Pearson, their theory or their applications, and come away with the view they are denying such a thing as evidence. Continuing in the paper critics love to cite: N-P 1933:

> We ask whether the variation in a certain character may be considered as following the normal law; . . . whether regression is linear; whether the variance in a number of samples differs significantly. . . . [W]e are not concerned with the exact value of particular parameters, but seek for information regarding the conditions and factors controlling the events. (ibid., p. 145)

Plainly, using data to obtain information regarding factors controlling events is indicative of using data as evidence. What goes under the banner of reliabilism in epistemology is scarcely different from what N-P offer for statistics: a means of arriving at a measurement through a procedure that is rarely wrong and, if wrong, is not far wrong, with mistakes likely discovered in later probes.

I could multiply ad nauseum similar readings of this passage. By the time statistician Robert Kass (2011, p. 8) gets to it, the construal is so hardened he doesn't even need to provide a reference:

> We now recognize Neyman and Pearson to have made permanent, important contributions to statistical inference through their introduction of hypothesis testing and confidence. From today's vantage point, however, their behavioral interpretation seems quaint, especially when represented by their famous dictum,

at which point our famous passage appears. I can see rejecting the extreme behavioristic view, but am not sure why Kass calls it "quaint" for an account to control error probabilities. I thought he (here and in Brown and Kass 2009) was at pains to insist on performance characteristics, declaring even Bayesians need them. I return to Kass in Excursion 6. At least he does not try to stick them with Pascal's wager!

Let's grant for the sake of argument that Neyman became a full blown behaviorist and thought the only justification for tests was low errors in the long run. Pearson absolutely disagreed. What's interesting is this. In the context of why Neyman regards the Bertrand–Borel debate as having "served as an inspiration to Egon S. Pearson and myself," the relevance of error probabilities is not hard to discern. Why report what would happen in repetitions were outcome x to be taken as indicating claim C? Because it's the way to design stringent tests and make probability claims pre-data that are highly informative post-data as to how well tested claims are.

5.8   Neyman's Performance and Fisher's Fiducial Probability

Many say fiducial probability was Fisher's biggest blunder; others suggest it still hasn't been understood. Most discussions avoid a side trip to the Fiducial Islands altogether, finding the surrounding brambles too thorny to negotiate. I now think this is a mistake, and it is a mistake at the heart of the consensus interpretation of the N-P vs. Fisher debate. We don't need to solve the problems of fiducial inference, fortunately, to avoid taking the words of the Fisher–Neyman dispute out of context. Although the Fiducial Islands are fraught with minefields, new bridges are being built connecting some of the islands to Power Peninsula and the general statistical mainland.

So what is fiducial inference? I begin with Cox's contemporary treatment, distilled from much controversy. The following passages swap his upper limit for the lower limit to keep to the example Fisher uses:

We take the simplest example, . . . the normal mean when the variance is known, but the considerations are fairly general. The lower limit

$$\bar{x} - z_c \sigma / \sqrt{n}$$

derived here from the probability statement

$$\Pr(\mu > \overline{X} - z_c \sigma / \sqrt{n}) = 1 - c$$

is a particular instance of a *hypothetical* long run of statements a proportion $1 - c$ of which will be true, . . . assuming our model is sound. We can, at least in principle, make such a statement for each c and thereby generate a collection of statements, sometimes called a *confidence distribution*. (Cox 2006a, p. 66; $\bar{x}$ for $\bar{y}$, $\overline{X}$ for $\overline{Y}$, and z_c for k_c^*)

Once $\bar{x}$ is observed, $\bar{x} - z_c \sigma / \sqrt{n}$ is what Fisher calls the *fiducial c percent limit* for μ. It is, of course, the *specific* $1-c$ lower confidence interval estimate $\hat{\mu}_{1-c}(\bar{x})$ (Section 3.7).

Here's Fisher in the earliest paper on fiducial inference in 1930. He sets $1 - c$ as 0.95. Starting from the significance test of a specific μ, he identifies the corresponding *95 percent value* $\bar{x}_{.05}$, such that in 95% of samples $\overline{X} < \bar{x}_{.05}$. In the normal testing example, $\bar{x}_{.05} = \mu + 1.65\sigma/\sqrt{n}$. Notice $\bar{x}_{.05}$ is the cut-off for a 0.05 one-sided test T+ (of $\mu \leq \mu_0$ vs. $\mu > \mu_0$).

[W]e have a relationship between the statistic $[\overline{X}]$ and the parameter μ, such that $[\bar{x}_{.05}]$ is the 95 per cent. value corresponding to a given μ, and this relationship implies the perfectly objective fact that in 5 per cent. of samples $[\overline{X} > \bar{x}_{.05}$. That is, $Pr(\overline{X} \leq \mu + 1.65\sigma/\sqrt{n}) = 0.95]$ (Fisher 1930, p. 533; substituting μ for θ and $\overline{X}$ for T.)

$\overline{X} > \bar{x}_{.05}$ occurs whenever $\mu < \overline{X} - 1.65\sigma/\sqrt{n}$ the *generic* $\hat{\mu}_{.95}(\overline{X})$. For a particular observed $\bar{x}$, $\bar{x} - 1.65\sigma/\sqrt{n}$ is the "fiducial 5 per cent. value of μ."

We may know as soon as $\overline{X}$ is calculated what is the fiducial 5 per cent. value of μ, and that the true value of μ will be less than this value in just 5 per cent. of trials. This then is a definite probability statement about the unknown parameter μ which is true irrespective of any assumption as to its *a priori* distribution. (ibid.)[3]

This seductively suggests $\mu < \hat{\mu}_{.95}(\overline{x})$ gets the probability 0.05 – a fallacious probabilistic instantiation.

However, there's a kosher probabilistic statement about $\overline{X}$, it's just not a probabilistic assignment to a parameter. Instead, a particular substitution is, to paraphrase Cox, "a particular instance of a hypothetical long run of statements 95% of which will be true." After all, Fisher was abundantly clear that the fiducial bound should not be regarded as an inverse inference to a posterior probability. We could only obtain an inverse inference by con-sidering μ to have been selected from a superpopulation of μ's, with known distribution. The posterior probability would then be a deductive inference and not properly inductive. In that case, says Fisher, we're not doing inverse or Bayesian inference.

In reality the statements with which we are concerned differ materially in logical content from inverse probability statements, and it is to distinguish them from these that we speak of the distribution derived as a *fiducial* frequency distribution, and of the working limits, at any required level of significance, . . . as the *fiducial limits* at this level. (Fisher 1936, p. 253)

So, what is being assigned the fiducial probability? It's the method of reaching claims to which the probability attaches. This is even clearer in his 1936 discussion where σ is unknown and must be estimated. Because $\overline{X}$ and S (using the Student's t pivot) are sufficient statistics "we may infer, without any use of probabilities *a priori*, a frequency distribution for μ which shall correspond with the aggregate of all such statements . . . to the effect that the probability μ is less than $\overline{x} - 2.145s/\sqrt{n}$ is exactly one in forty" (ibid., p. 253). This uses Student's t distribution with $n = 15$. It's plausible, at that point, to suppose Fisher means for $\overline{x}$ to be a random variable.

Suppose you're Neyman and Pearson working in the early 1930s aiming to clarify and justify Fisher's methods. 'I see what's going on,' we can imagine Neyman declaring. There's a method for outputting statements such as would take the general form

$$\mu > \overline{X} - 2.145 \, s/\sqrt{n}.$$

Some would be in error, others not. The method outputs statements with a probability (some might say a propensity) of 0.975 of being correct. "We may

---

[3] It's correct that $(\mu \leq \overline{X} - z_c \, \sigma/\sqrt{n})$ iff $(\overline{X} > \mu + z_c\sigma/\sqrt{n})$.

look at the purpose of tests from another viewpoint": probability ensures us of the performance of a method (it's methodological).

At the time, Neyman thought his development of confidence intervals (in 1930) was essentially the same as Fisher's fiducial intervals. There was evidence for this. Recall the historical side trip of Section 3.7. When Neyman gave a (1934) paper to the Royal Statistical Society discussing confidence intervals, seeking to generalize fiducial limits, he made it clear that the term confidence coefficient refers to "probability of our being right when applying a certain rule" for making statements set out in advance (p. 140). Much to Egon Pearson's relief, Fisher called Neyman's generalization "a wide and very handsome one," even though it didn't achieve the uniqueness Fisher had wanted (Fisher 1934c, p. 137). There was even a bit of a mutual admiration society, with Fisher saying "Dr Neyman did him too much honour" in crediting him for the revolutionary insight of Student's t pivotal, giving the credit to Student. Neyman (1934, p. 141) responds that of course in calling it Student's t he is crediting Student, but "this does not prevent me from recognizing and appreciating the work of Professor Fisher concerning the same distribution."

In terms of our famous passage, we may extract this reading: In struggling to extricate Fisher's fiducial limits, without slipping into fallacy, they are led to the N-P construal. Since fiducial probability was to apply to significance testing as well as estimation, it stands to reason that the performance notion would find its way into the N-P 1933 paper.[4] So the error probability applies to the method, but the question is whether it's intended to qualify a given inference, or only to express future long-run assurance (performance).

N-P and Fisher Dovetail: It's Interpretation, not Mathematics

David Cox shows that the Neyman–Pearson theory of tests and confidence intervals arrive at the same place as the Fisherian, even though in a sense they proceed in the opposite direction. Suppose that there is a full model covering both null and alternative possibilities. To establish a significance test, we need to have an appropriate test statistic $d(X)$ such that the larger the $d(X)$ the greater the discrepancy with the null hypothesis in the respect of interest. But it is also required that the probability distribution of $d(X)$ be known under the assumption of the null hypothesis. In focusing on the logic, we've mostly

---

[4] "[C]onsider that variation of the unknown parameter, μ, generates a continuum of hypotheses each of which might be regarded as a null hypothesis . . . [T]he data of the experiment, and the test of significance based upon them, have divided the continuum into two portions." One a region in which μ lies between the fixed fiducial limits, "is accepted by the test of significance, in the sense that values of μ within this region are not contradicted by the data at the level of significance chosen. The remainder. . . is rejected" (Fisher 1935a, p. 192).

considered just one unknown parameter, e.g., the mean of a Normal distribution. In most realistic cases there are additional parameters required to compute the P-value, sometimes called "nuisance" parameters λ, although they are just as legitimate as the parameter we happen to be interested in. We'd like to free the computation of the P-value from these other unknown parameters. This is the error statistician's way to ensure as far as possible that observed discordances may be blamed on discrepancies between the null and what's actually bringing about the data. We want to solve the classic Duhemian problems of falsification.

As Cox puts it, we want a test statistic with a distribution that is split off from the unknown nuisance parameters, which we can abbreviate as λ. The full parameter space Θ is partitioned into components $\Theta = (\psi, \lambda)$, such that the null hypothesis is that $\psi = \psi_0$, with λ an unknown nuisance parameter. Interest may focus on alternatives $\psi > \psi_0$. We do have information in the data about the unknown parameters, and the natural move is to estimate them using the data. The twin goals of computing the P-value, $\Pr(d > d_0; H_0)$, free of unknowns, and constructing tests that are appropriately sensitive, produce the same tests entailed by N-P theory, namely replacing the nuisance parameter by a sufficient statistic V. A statistic V, a sufficient statistic for nuisance parameter λ, means that the probability of the $d(X)$ conditional on the estimate V depends only on the parameter of interest ψ_0. So we are back to the simple situation with a null having just a single parameter ψ. This "largely determines the appropriate test statistic by the requirement of producing the most sensitive test possible with the data at hand" (Cox and Mayo 2010, p. 292). Cox calls this "conditioning for separation from nuisance parameters" (ibid.). I draw from Cox and Mayo (2010).

In the most familiar class of cases, this strategy for constructing appropriately sensitive or powerful tests, separate from nuisance parameters, produces the same tests as N-P theory. In fact, when statistic V is a special kind of sufficient statistic for nuisance parameter λ (called *complete*), there is no other way of achieving the N-P goal of an exactly α-level test that is fixed regardless of nuisance parameters – these are called *similar* tests.[5] Thus, replacing the nuisance parameter with a sufficient statistic "may be regarded as an outgrowth of the aim of calculating the relevant P-value independent of unknowns, or alternatively, as a byproduct of seeking to obtain most powerful

---

[5] The goal of exactly similar tests leads to tests that ensure

$$\Pr(d(X) \text{ is significant at level } \alpha | v; H_0) = \alpha,$$

where v is the value of the statistic V used to estimate the nuisance parameter. A good summary may be found in Lehmann (1981).

similar tests." These dual ways of generating tests reveal the underpinnings of a substantial part of standard, elementary statistical methods, including key problems about Binomial, Poisson, and Normal distributions, the method of least squares, and linear models.[6] (ibid., p. 293)

If you begin from the "three steps" in test generation described by E. Pearson in the opening to Section 3.2, rather than the later N-P–Wald approach, they're already starting from the same point. The only difference is in making the alternative explicit. Fisher (1934b) made the connection to the N-P (1933) result on uniformly most powerful tests:

... where a sufficient statistic exists, the likelihood, apart from a factor independent of the parameter to be estimated, is a function only of the parameter and the sufficient statistic, explains the principal result obtained by Neyman and Pearson in discussing the efficacy of tests of significance. Neyman and Pearson introduce the notion that any chosen test of a hypothesis H_0 is more powerful than any other equivalent test, with regard to an alternative hypothesis H_1, when it rejects H_0 in a set of samples having an assigned aggregate frequency ε when H_0 is true, and the greatest possible aggregate frequency when H_1 is true... (pp. 294–5)

It is inevitable, therefore, that if such a statistic exists it should uniquely define the contours best suited to discriminate among hypotheses differing only in respect of this parameter; ... When tests are considered only in relation to sets of hypotheses specified by one or more variable parameters, the efficacy of the tests can be treated directly as the problem of estimation of these parameters. Regard for what has been established in that theory, apart from the light it throws on the results already obtained by their own interesting line of approach, should also aid in treating the difficulties inherent in cases in which no sufficient statistics exists. (ibid., p. 296)

This article may be seen to mark the point after which Fisher's attitude changes because of the dust-up with Neyman.

Neyman and Pearson come to Fisher's Rescue

Neyman and Pearson entered the fray on Fisher's side as against the old guard (led by K. Pearson) regarding the key point of contention: showing statistical inference is possible without the sin of "inverse inference". Fisher denounced the *principle* of *indifference*: "We do not know the function ... specifying the super-population, but in view of our ignorance of the actual values of θ we may" take it that all values are equally probable (Fisher 1930, p. 531). "[B]ut

---

[6] Requiring exactly similar rejection regions, "precludes tests that merely satisfy the weaker requirement of being able to calculate P approximately, with only minimal dependence on nuisance parameters," which could be preferable especially when best tests are absent. (Ibid.)

however we might disguise it, the choice of this particular a priori distribution for the θ is just as arbitrary as any other. . ." (ibid.).

If, then, we follow writers like Boole, Venn, . . . in rejecting the inverse argument as devoid of foundation and incapable even of consistent application, how are we to avoid the staggering falsity of saying that however extensive our knowledge of the values of x . . . we know nothing and can know nothing about the values of θ? (ibid.)

When Fisher gave his paper in December 1934 ("The Logic of Inductive Inference"), the old guard were ready with talons drawn to attack his ideas, which challenged the overall philosophy of statistics they embraced. The opening thanks (by Arthur Bowley), which is typically a flowery, flattering affair, was couched in scathing, sarcastic terms (see Fisher 1935b, pp. 55–7). To Fisher's support came Egon Pearson and Jerzy Neyman. Neyman dismissed "Bowley's reaction to Fisher's critical review of the traditional view of statistics as an understandable attachment to old ideas (1935, p. 73)" (Spanos 2008b, p. 16). Fisher agreed: "However true it may be that Professor Bowley is left very much where he was, the quotations show at least that Dr. Neyman and myself have not been left in his company" (1935a, p. 77).

So What Happened in 1935?

A pivotal event was a paper Neyman gave in which he suggested a different way of analyzing one of Fisher's experimental designs. Then there was a meet-up in the hallway a few months later. Fisher stops by Neyman's office at University College, on his way to a meeting which was to decide on Neyman's reappointment in 1935:

And he said to me that he and I are in the same building . . . That, as I know, he has published a book – and that's *Statistical Methods for Research Workers* – and he is upstairs from me so he knows something about my lectures – that from time to time I mention his ideas, this and that – and that this would be quite appropriate if I were not here in the College but, say, in California . . . but if I am going to be at University College, then this is not acceptable to him. And then I said, 'Do you mean that if I am here, I should just lecture using your book?' And then he gave an affirmative answer. . . . And I said, 'Sorry, no. I cannot promise that.' And then he said, 'Well, if so, then from now on I shall oppose you in all my capacities.' And then he enumerated – member of the Royal Society and so forth. There were quite a few. Then he left. Banged the door. (Neyman in C. Reid 1998, p. 126)

Imagine if Neyman had replied: 'I'd be very pleased to use *Statistical Methods for Research Workers* in my class.' Or what if Fisher had said: 'Of course you'll want to use your own notes in your class, but I hope you will use a portion of my text when mentioning some of its key ideas.' Never mind. That was it. Fisher went on to a meeting wherein he attempted to get others to refuse Neyman a permanent position, but was unsuccessful. It wasn't just Fisher who seemed to need some anger management training, by the way. Erich Lehmann (in conversation and in 2011) points to a number of incidences wherein Neyman is the instigator of gratuitous ill-will. I find it hard to believe, however, that Fisher would have thrown Neyman's wooden models onto the floor.

One evening, late that spring, Neyman and Pearson returned to their department after dinner to do some work. Entering they were startled to find strewn on the floor the wooden models which Neyman had used to illustrate his talk . . . Both Neyman and Pearson always believed that the models were removed by Fisher in a fit of anger (C. Reid 124, noted in Lehmann 2011, p. 59).

Neyman left soon after to start the program at Berkeley (1939), and Fisher didn't remain long either, moving in 1943 to Cambridge and retiring in 1957 to Adelaide. I've already been disabusing you of the origins of the popular Fisher–N-P conflict (Souvenir L). In fact, it really only made an appearance long after the 1933 paper!

1955–6 Triad: Telling What's True About the Fisher–Neyman Conflict

If you want to get an idea of what transpired in the ensuing years, look at Fisher's charges and Neyman's and Pearson's responses 20 years later. This forms our triad: Fisher (1955), Pearson (1955), and Neyman (1956). Even at the height of mudslinging, Fisher said, "There is no difference to matter in the field of mathematical analysis . . . but in logical point of view" (1955, p. 70).

I owe to Professor Barnard . . . the penetrating observation that this difference in point of view originated when Neyman, thinking he was correcting and improving my own early work on tests of significance as a means to the 'improvement of natural knowledge,' in fact reinterpreted them in terms of that technological and commercial apparatus which is known as an acceptance procedure. . . . Russians are made familiar with the ideal that research in pure science can and should be geared to technological performance. (ibid., pp. 69–70)

Pearson's (1955) response: "To dispel the picture of the Russian technological bogey, I might recall how certain early ideas came into my head as I sat on a gate overlooking an experimental blackcurrant plot . . . !" (Pearson 1955, p. 204). He was "smitten" by an absence of logical justification for some of Fisher's tests, and turned to Neyman to help him solve the problem. This takes us to where we began with our miserable passages, leading them to pin down the required character for the test statistic, the need for the alternative and power considerations.

Until you disinter the underlying source of the problem – fiducial inference – the "he said/he said" appears to be all about something that it's not. The reason Neyman adopts a performance formulation, Fisher (1955) charges, is that he denies the soundness of fiducial inference. Fisher thinks Neyman is wrong because he "seems to claim that the statement (a) 'μ has a probability of 5 per cent. of exceeding $\overline{X}$' is a different statement from (b) '$\overline{X}$ has a probability of 5 per cent. of falling short of μ'" (p. 74, replacing θ and T with μ and $\overline{X}$). There's no problem about equating these two so long as $\overline{X}$ is a random variable. But watch what happens in the next sentence. According to Fisher, Neyman violates

... the principles of deductive logic [by accepting a] general symbolical statement such as

[1] $\Pr\{(\overline{x} - ts) < \mu < (\overline{x} - ts)\} = \alpha$,

as rigorously demonstrated, and yet, when numerical values are available for the statistics $\overline{x}$ and s, so that on substitution of these and use of the 5 per cent. value of t, the statement would read

[2] $\Pr\{92.99 < \mu < 93.01\} = 95$ per cent.,

to *deny* to this *numerical* statement any validity. This evidently is to deny the syllogistic process. (Fisher 1955, p. 75, in Neyman 1956, p. 291)

But the move from (1) to (2) is fallacious! Is Fisher committing this fallacious probabilistic instantiation (and still defending it in 1955)? I. J. Good describes how many felt, and still feel:

It seems almost inconceivable that Fisher should have made the error which he did in fact make. [That is why] ... so many people assumed for so long that the argument was correct. They lacked the *daring* to question it. (Good 1971a, p. 138).

Neyman (1956) declares himself at his wit's end in trying to convince Fisher of the inconsistencies in moving from (1) to (2). "Thus if X is a normal random variable with mean zero and an arbitrary variance greater than zero, then I expect" we may agree that $\Pr(X < 0) = 0.5$ (ibid., p. 292). But observing, say, $X = 1.7$ yields $\Pr(1.7 < 0) = 0.5$, which is clearly illicit. "It is doubtful whether the chaos and confusion now reigning in the field of fiducial argument were ever equaled in any other doctrine. The source of this confusion is the lack of realization that equation (1) does not imply (2)" (ibid., p. 293). It took the more complex example of Bartlett to demonstrate the problem: "Bartlett's revelation [1936, 1939] that the frequencies in repeated sampling [from the same or different populations] need not agree with Fisher's solution" in the case of a difference between two Normal means with different variances, "brought about an avalanche of rebuttals by Fisher and by Yates" (ibid., p. 292).[7] Some think it was only the collapse of Fisher's rebuttals that led Fisher to castigate N-P for assuming error probabilities and fiducial probabilities *ought* to agree, and begin to declare the idea "foreign to the development of tests of significance." As statistician Sandy Zabell (1992 p. 378) remarks, "such a statement is curiously inconsistent with Fisher's own earlier work" as in Fisher's (1934b) endorsement of UMP tests, and his initial attitude toward Neyman's confidence intervals. Because of Fisher's stubbornness "he engaged in a futile and unproductive battle with Neyman which had a largely destructive effect on the statistical profession" (ibid., p. 382).[8]

Fisher (1955) is spot on about one thing: When "Neyman denies the existence of inductive reasoning, he is merely expressing a verbal preference. For him 'reasoning' means what 'deductive reasoning' means to others" (p. 74). Nothing earth-shaking turns on the choice to dub every inference "an act of making an inference." Neyman, much like Popper, had a good reason for drawing a bright red line between the use of probability (for corroboration or probativeness) and the probabilists' use of confirmation: Fisher was blurring them.

... the early term I introduced to designate the process of adjusting our actions to observations is 'inductive behavior'. It was meant to contrast with the term 'inductive reasoning' which R. A. Fisher used in connection with his 'new measure of confidence or diffidence' represented by the likelihood function and with 'fiducial argument'. Both these concepts or principles are foreign to me. (Neyman 1977, p. 100)

The Fisher–Neyman dispute is pathological: there's no disinterring the truth of the matter. Perhaps Fisher altered his position out of professional antagonisms toward the new optimality revolution. Fisher's stubbornness on fiducial intervals seems to lead Neyman to amplify the performance construal louder and louder; whereas Fisher grew to renounce performance goals he himself had held when it was found that fiducial solutions disagreed with them. Perhaps inability to identify conditions wherein the error probabilities "rubbed off" – where there are no "recognizable subsets" with a different probability of success – led Fisher to move to a type of default Bayesian stance. That Neyman (with the contributions of Wald, and later Robbins) might have gone overboard in his behaviorism, to the extent that even Egon wanted to divorce him – ending his 1955 reply to Fisher with the claim that "inductive behavior" was

---

[7] In that case, "the test rejects a smaller proportion of such repeated samples than the proportion specified by the level of significance" (Fisher 1939, p. 173a). Prakash Gorroochurn (2016) has a masterful historical discussion.
[8] Buehler and Feddersen (1963) showed there were recognizable subsets even for the t test.

Neyman's field, not his – is a different matter. Ironically, Pearson shared Neyman's antipathy to "inferential theory" as Neyman (1962) defines it in the following:

In the present paper ... the term 'inferential theory' ... will be used to describe the attempts to solve the Bayes' problem with a reference to confidence, beliefs, etc., through some supplementation ... either a substitute *a priori* distribution [exemplified by the so called principle of insufficient reason] or a new measure of uncertainty [such as Fisher's fiducial probability] (p. 16).

Fisher may have started out seeing fiducial probability as both a frequency of correct claims in an aggregate, and a rational degree of belief (1930, p. 532), but the difficulties in satisfying uniqueness led him to give up the former. Fisher always showed inductive logic leanings, seeking a single rational belief assignment. N-P were allergic to the idea. In the N-P philosophy, if there is a difference in problems or questions asked, we expect differences in which solutions are warranted. This is in sync with the view of the severe tester. In this sense, she is closer to Fisher's viewing the posterior distribution to be an answer to a different problem from the fiducial limits, where we expect the sample to change (Fisher 1930, p. 535).

Bridges to Fiducial Island: Neymanian Interpretation of Fiducial Inference?

For a long time Fiducial Island really was an island, with work on it side-stepped. A notable exception is Donald Fraser. Fraser will have no truck with those who dismiss fiducial inference as Fisher's "biggest blunder." "What? We still have to do a little bit of thinking! Tough!" (Fraser 2011, p. 330). Now, however, bridges are being built, despite minefields. Numerous programs are developing confidence distributions (CDs), and the impenetrable thickets are being penetrated. The word "fiducial" is even bandied about in these circles.[9] Singh, Xie, and Strawderman (2007) say, "a CD is in fact Neymanian interpretation of Fisher's fiducial distribution" (p. 132).

"[A]ny approach that can build confidence intervals for all levels, regardless of whether they are exact or asymptotically justified, can potentially be unified under the confidence distribution framework" (Xie and Singh 2013, p. 5). Moreover, "as a frequentist procedure, the CD-based method can bypass [the] difficult task of jointly modelling [nuisance parameters] and focus directly on the parameter of interest" (p. 28). This turns on what we've been

---

[9] Efron predicts "that the old Fisher will have a very good 21st century. The world of applied statistics seems to need an effective compromise between Bayesian and frequentist ideas" (Efron 1998, p. 112).

calling the piecemeal nature of error statistics. "The idea that statistical problems do not have to be solved as one coherent whole is anathema to Bayesians but is liberating for frequentists" (Wasserman 2007, p. 261).

I'm not in a position to evaluate these new methods, or whether they lend themselves to a severity interpretation. The CD program does at least seem to show the wide landscape for which the necessary mathematical computations are attainable. While CDs do not supply the uniqueness that Fisher sought, given that a severity assessment is always relative to the question or problem of interest, this is no drawback. Nancy Reid claims the literature on the new frequentist–fiducial "fusions" isn't yet clear on matters of interpretation.[10] What is clear, is that the frequentist paradigm is undergoing the "historical process of development . . . which is and will always go on" of which Pearson spoke (1962, p. 394).

Back to the ship!

---

[10] The 4th Bayesian, Fiducial and Frequentist workshop (BFF4), May 2017. Other examples are Fraser and Reid (2002), Hannig (2009), Martin and Liu (2013), Schweder and Hjort (2016).

Excursion 6 (Probabilist) Foundations Lost, (Probative) Foundations Found

Itinerary

Tour I What Ever Happened to Bayesian Foundations?

> By and large, Statistics is a prosperous and happy country, but it is not a completely peaceful one. Two contending philosophical parties, the Bayesians and the frequentists, have been vying for supremacy over the past two-and-a-half centuries. . . . *Unlike most philosophical arguments, this one has important practical consequences.* The two philosophies represent competing visions of how science progresses. (Efron 2013, pp. 130; emphasis added)

Surveying the statistical landscape from a hot-air balloon this morning, a bird's-flight view of the past 100 of those 250 years unfolds before us. Except for the occasional whooshing sound of the balloon burner, it's quiet enough to actually hear some of the warring statistical tribes as well as peace offerings and reconciliations – at least with a special sound amplifier they've supplied. It's today's perspective I mainly want to show you from here. Arrayed before us is a most impressive smorgasbord of technical methods, as statistics expands over increasing territory. Many professional statisticians are eclecticists; foundational discussions are often in very much of a unificationist spirit. If you observe the territories undergoing recent statistical crises, you can see pockets, growing in number over the past decade or two, who are engaged in refighting old battles. Unsurprisingly, the methods most often used are the ones most often blamed for abuses. Perverse incentives, we hear, led to backsliding, to slothful, cronyist uses of significance tests that have been deplored for donkey's years. Big Data may have foisted statistics upon fields unfamiliar with the pitfalls stemming from vast numbers of correlations and multiple testing. A pow-wow of leading statisticians from different tribes was called by the American Statistical Association in 2015. We've seen the ASA 2016 Guide on how not to use *P*-values, but some of the "other approaches" also call for scrutiny:

> In view of the prevalent misuses of and misconceptions concerning *p*-values, some statisticians prefer to supplement or even replace *p*-values with other approaches. . . . confidence, credibility, or prediction intervals; Bayesian methods; . . . likelihood ratios or Bayes Factors; and other approaches such as decision-theoretic modeling and false discovery rates. (Wasserstein and Lazar 2016, p. 132)

Suppose you're appraising a recommendation that frequentist methods should or can be replaced by a Bayesian method. Your first question should be: Which type of Bayesian interpretation? The choices are basically three: subjectivist, default, or frequentist. The problem isn't just choosing amongst them but trying to pin down the multiple meanings being given to each! Classical subjective Bayesianism is home to a full-bodied statistical philosophy, but the most popular Bayesians live among rival tribes who favor one or another default or non-subjective prior probabilities. These are conventions chosen to ensure the data dominate the inference in some sense. By and large, these tribes do not see the growth of Bayesian methods as support for the classical subjective Bayesian philosophy, but rather as a set of technical tools that "work." Their leaders herald frequentist–Bayesian unifications as the way to serve multiple Gods. Zeus is throwing a thunderbolt!

Navigating the reforms requires a roadmap. In Tour I we'll visit the gallimaufry of very different notions of probability in current Bayesian discussions. Concerned that today's practice isn't captured by either traditional Bayesian or frequentist philosophies, new foundations are being sought – that's where we'll travel in Tour II.

Strange bedfellows: the classical subjective Bayesian and the classical frequentist tribes are at one in challenging non-subjective, default Bayesians. The small, but strong tribes of subjective Bayesians, we may imagine, ask them:

> How can you declare scientists want highly probable hypotheses (or comparatively highly probable ones) if your probabilities aren't measuring reasonable beliefs or plausibility (or the like)?

Frequentist error statisticians concur, but also, we may imagine, inquire:

> What's so good about high posterior probabilities if a method frequently assigns them to poorly tested claims?

Let's look back at Souvenir D, where Reid and Cox (2015, p. 295) press the weak repeated sampling requirement on non-frequentist assessments of uncertainty.

The role of calibration seems essential: even if an empirical frequency-based view of probability is not used directly as a basis for inference; it is unacceptable if a procedure yielding regions of high probability in the sense of representing uncertain knowledge would, if used repeatedly, give systematically misleading conclusions.

Frequentist performance is a necessary, though not a sufficient, condition for severe testing. Even those who deny an interest in performance might not want to run afoul of the minimal requirement for severity. The onus on those who declare what we really want in statistical inference are probabilities on hypotheses is to show, for existing ways of obtaining them, *why*?

Notice that the largest statistical territory is inhabited by practitioners who identify as eclecticists, using a toolbox of various and sundry methods. Some of the fastest growing counties of machine learners and data scientists point to spell checkers and self-driving cars that learn by conjecture and refutation algorithms, at times sidestepping probability models altogether. The year 2013 was dubbed *The International Year of Statistics* partly to underscore the importance of statistics to the Big Data revolution. The best AI algorithms appear to lack a human's grasp of deception based on common sense. That little skirmish is ongoing. Eclecticism gives all the more reason to clearly distinguish the meanings of numbers that stem from methods evaluating different things. This is especially so when it comes to promoting scientific integrity, reproducibility, and in the waves of methodological reforms from journals and reports. Efron has it right: "Unlike most philosophical arguments, this one has important practical consequences" (2013, p. 130). Let's land this balloon, we're heading back to the Museum of Statistics. If you've saved your stub from Excursion 1, it's free.

6.1 Bayesian Ways: From Classical to Default

Let's begin Excursion 6 on the museum floor devoted to classical, philosophical, subjective Bayesianism (which I'm not distinguishing from personalism). This will give us a thumbnail of the position that contemporary non-subjective Bayesians generally reject as a description of what they do. An excellent starting point that is not ancient history, and also has the advantage of contemporary responses, is Dennis Lindley's (2000) "Philosophy of Statistics." We merely click on the names, and authentic-looking figures light up and speak. Here's Lindley:

The suggestion here is that statistics is the study of uncertainty (Savage 1977): that statisticians are experts in handling uncertainty . . . (p. 294)

[Consider] any event, or proposition, which can either happen or not, be true or false. It is proposed to measure your uncertainty associated with the event . . . If you think that the event is just as uncertain as the random drawing of a red ball from an urn containing N balls, of which R are red, then the event has uncertainty R/N for you. (p. 295)

Historically, uncertainty has been associated with games of chance and gambling. Hence one way of measuring uncertainty is through the gambles that depend on it. (p. 297)

Consider before you an urn containing a known number N of balls that are as nearly identical as modern engineering can make them. Suppose that one ball is drawn at random from the urn . . . it is needful to define randomness. Imagine that the balls are numbered consecutively from 1 to N and suppose that, at no cost to you, you were offered a prize if ball 57 were drawn . . . [and] the same prize if ball 12 were drawn. If you are indifferent between the two propositions and, in extension, between any two numbers between 1 and N, then, for you, the ball is drawn at random. Notice that the definition of randomness is subjective; it depends on you. (p. 295)

It is immediate from [Bayes' Theorem] that the only contribution that the data make to inference is through the likelihood function for the observed x. This is the likelihood principle that values of x, other than that observed, play no role in inference. (pp. 309–10)

[U]nlike the frequency paradigm with its extensive collection of specialized methods, the coherent view provides a constructive method of formulating and solving any and every uncertainty problem of yours. (p. 333)[1]

This is so clear, clean, and neat. The severe tester, by contrast, doesn't object that "specialized" methods are required to apply formal statistics. Satisfying the requirements of severe testing demands it, and that's unity enough. But let's see what some of Lindley's critical responders said in 2000. Press the buttons under their names. I'll group by topic:

1. Subjectivity. *Peter Armitage*: "The great merit of the Fisherian revolution, apart from the sheer richness of the applicable methods, was the ability to summarize, and to draw conclusions from, experimental and observational data without reference to prior beliefs. An experimental scientist needs to report his or her findings, and to state a range of possible hypotheses with which these findings are consistent. The scientist will undoubtedly have prejudices and hunches, but the reporting of these should not be a primary aim of the investigation.... There were indeed important uncertainties, about possible biases . . . [and the] existence of confounding factors. But the way to deal with them was . . . by scrupulous argument rather than by assigning probabilities . . ." (ibid., pp. 319–20)

David Cox "It seems to be a fundamental assumption of the personalistic theory that all probabilities are comparable. Moreover, so far as I understand it, we are not allowed to attach measures of precision to probabilities. They are as they are . . . I understand Dennis Lindley's irritation at the cry 'where did the prior come from?' I hope that it is clear that my objection is rather different: why should I be interested in someone else's prior and why should anyone else be interested in mine? (ibid. p. 323) . . . [I]n my view the personalistic probability is virtually worthless for reasoned discussion unless it is based on

---

[1] "Frequency, however, is not adequate because there is ordinarily no repetition of parameters; they have unique unknown values . . . with the result that it has been necessary for them to develop incoherent concepts like confidence intervals." (p. 311) There are, however, repetitions of types of methods.

information, often directly or indirectly of a broadly frequentist kind. . . . For example, how often have very broadly comparable laboratory studies been misleading as regards human health? How distant are the laboratory studies from a direct process affecting health?" (ibid., p. 322)

2. Non-ampliative. *David Sprott*: "This paper relegates statistical and scientific inference to a branch (probability) of pure mathematics, where inferences are deductive statements of implication: if H_I then H_2. This can say nothing about whether there is reproducible objective empirical evidence for H_I or H_2, as is required by a scientific inference." (ibid., p. 331)

3. Science is Open-Ended. *John Nelder*: "Statistical science is not just about the study of uncertainty, but rather deals with inferences about scientific theories from uncertain data. . . . [Theories] are essentially open ended; at any time someone may come along and produce a new theory outside the current set. This contrasts with probability, where to calculate a specific probability it is necessary to have a bounded universe of possibilities over which the probabilities are defined. When there is intrinsic open-endedness it is not enough to have a residual class of all the theories that I have not thought of yet [the catchall]." (ibid., p. 324)

David Sprott: "Bayes's Theorem (1) requires that *all* possibilities $H_1, H_2, \ldots,$ H_k be specified in advance, along with their prior probabilities. Any new, hitherto unthought of hypothesis or concept H will necessarily have zero prior probability. From Bayes's Theorem, H will then always have zero posterior probability no matter how strong the empirical evidence in favour of H." (ibid., p. 331)

4. Likelihood Principle. *Brad Efron*: "The likelihood principle seems to be one of those ideas that is rigorously verifiable and yet wrong." (Efron 2000, p. 330)[2]

There are also supporters of course, notably, O' Hagan and Dawid, whose remarks we take up elsewhere. The fact that classical Bayesianism reduces statistical inference to probability theory – the very reason many take it as a respite from the chaos of frequentism – could also, Dawid observes, be thought to make it boring: "What is the principal distinction between Bayesian and classical statistics? It is that Bayesian statistics is fundamentally boring. There is so little to do: just specify the model and the prior, and turn the Bayesian handle." (ibid., p. 326). He's teasing I'm sure, but let's step back.

---

[2] I have argued (e.g., Mayo 2014b) the alleged verifications are circular. Efron, in private communication, said that he tried to argue against the result, but gave up; he was glad I did not.

The error statistician agrees with all these criticisms. In her view, statistics is collecting, modeling, and using data to make inferences about aspects of what produced them. Inferences, being error prone, are qualified by reports of the error probing capacities of the inferring method. There is a cluster of error types, real versus spurious effect, wrong magnitude for a parameter, violated statistical assumptions, and flaws in connecting formal statistical inference to substantive claims. It splits problems off piecemeal; there's no need for an exhaustive list of hypotheses that could explain data. Being able to *directly* pick up on gambits like cherry picking and optional stopping is essential for an account to be up to the epistemological task of determining if claims are poorly tested. While for Lindley this leads to incoherence (violations of the likelihood principle), for us it is the key to assessing if your tool is capable of deceptions. According to Efron: "The two philosophies, Bayesian and frequentist, are more orthogonal than antithetical" (Efron 2013, p. 145). Given the radical difference in goals between classical Bayesians and classical frequentists, he might be right. Vive la difference!

But Now Things Have Changed

What should we say now that the landscape has changed? That's what we'll explore in Excursion 6. We'll drop in on some sites we only visited briefly or passed up the first time around. We attempt to disinter the statistical philosophy practiced by the most popular of Bayesian tribes, those using non-subjective or default priors, picking up on Section 1.3, "The Current State of Play". Around 20 years ago, it began to be conceded that: "non-informative priors do not exist" (Bernardo 1997). In effect, they couldn't transcend the problems of "the principle of indifference" wherein lacking a reason to distinguish the probability of different values of θ is taken to render them all equally probable. The definitive review of default methods in statistics is Kass and Wasserman (1996). The default/non-subjective Bayesian focuses on priors that, in some sense, give heaviest weight to data. Impressive technical complexities notwithstanding, there's a multiplicity of incompatible ways to go about this job, none obviously superior. The problem is redolent of Carnap's problem of being faced with a continuum of inductive logics (Section 2.1). (A few are maximum entropy, invariance, maximizing the missing information, coverage matching.) Even for simple problems, recommended default Bayesian procedures differ.

If the proponents of this view thought their choice of a canonical prior were intellectually compelling, they would not feel attracted to a call for an internationally agreed convention on the subject, as have Berger and Bernardo (1992, p. 57) and Jeffreys (1955, p. 277). (Kadane 2011, p. 445–6)

No such convention has been held.

Default/non-subjective Bayesianism is often offered as a way to unify Bayesian and frequentist approaches.[3] It gives frequentist error statisticians a clearer and less contentious (re)entry into statistical foundations than when Bayesian "personalists" reigned (e.g., Lindley, Savage). At an earlier time, as Cox tells it, confronted with the position that "arguments for this personalistic theory were so persuasive that anything to any extent inconsistent with that theory should be discarded" (Cox 2006a, p. 196), frequentists might have felt alienated when it came to foundations. The discourse was snarky and divisive. Nowadays, Bayesians are more diffident. It's not that unusual to hear Bayesians admit that the older appeal to ideals of rationality were hyped. Listen to passages from Gelman (2011), Kass (2011), and Spiegelhalter (2004):

> Frequentists just took subjective Bayesians at their word and quite naturally concluded that Bayesians had achieved the goal of coherence only by abandoning scientific objectivity. Every time a prominent Bayesian published an article on the unsoundness of p-values, this became confirming evidence of the hypothesis that Bayesian inference operated in a subjective zone bounded by the prior distribution. (Gelman 2011, p. 71)

> [T]he introduction of prior distributions may not have been the central bothersome issue it was made out to be. Instead, it seems to me, the really troubling point for frequentists has been the Bayesian claim to a philosophical high ground, where compelling inferences could be delivered at negligible logical cost. (Kass 2011, p. 6)

> The general statistical community, who are not stupid, have justifiably found somewhat tiresome the tone of hectoring self-righteousness that has often come from the Bayesian lobby. Fortunately that period seems to be coming to a close, and with luck the time has come for the appropriate use of Bayesian thinking to be pragmatically established. (Spiegelhalter 2004, p. 172)

Bayesian empathy with objections to subjective foundations – "we feel your pain" – is a big deal, and still rather new to this traveler's ears. What's the new game all about? There's an important thread that needs to be woven into any answer. Not so long after the retreat from classical subjective Bayes, though it's impossible to give dates (early 2000s?), we saw the rise of irreproducible results and the statistical crisis in science. A new landscape of statistical conflict followed, but grew largely divorced from the older Bayesian–frequentist battles. "Younger readers . . . may not be fully aware of the passionate battles over Bayesian inference among statisticians in the last half of the twentieth century" (Gelman and Robert 2013, p. 1). Opening with Lindley's statistical philosophy

---

[3] Since we'll be talking a lot about default Bayesians in this tour, I'll use "default/non-subjective" lest I be seen as taking away an appealing name.

lets us launch into newer battles. Finding traditional Bayesian foundations ripped open, coupled with invitations for Bayesian panaceas to the reproducibility crisis, we are swept into a dizzying whirlpool where deeper and more enigmatic puzzles swirl. Do you still have that quicksand stick? (Section 3.6) Grab it and join me on some default, pragmatic, and eclectic Bayesian pathways.

(Note: Since we are discussing existing frequentist–Bayesian arguments, I'll usually use "frequentism" in this excursion, rather than our preferred error statistics.)

6.2   What Are Bayesian Priors? A Gallimaufry

The prevalent Bayesian position might be said to be: there are shortcomings or worse in standard frequentist methods, but classical subjective Bayesianism is, well, too subjective, so default Bayesianism should be used. Yet when you enter default Bayesian territory you'll need to juggle a plethora of competing meanings given to Bayesian priors, and consequently to posteriors.

To show you what I mean, look at a text by Ghosh, Delampady, and Samanta (2010): They say they will stress "objective" (default) priors, "because it still seems difficult to elicit fully subjective priors ... If a fully subjective prior is available we would indeed use it" (p. 36). Can we slip in and out of non-subjective and subjective priors so easily? Several contemporary Bayesian texts say yes. How should a default prior be construed? Ghosh et al. say that "it represents a shared belief or shared convention," while on the same page it is "to represent small or no information" (p. 30). Maybe it can be all three. The seminal points to keep in mind are spelled out by Bernardo:

> By definition, 'non-subjective' prior distributions are *not* intended to describe personal beliefs, and in most cases, they are *not even proper* probability distributions in that they often do not integrate [to] one. Technically they are *only* positive functions to be formally used in Bayes' theorem to obtain 'non-subjective posteriors' ... (Bernardo 1997, pp. 159–60)

Bernardo depicts them as a convention chosen "to make precise the type of prior knowledge which" for a given inference problem within a model "would make the data dominant" (ibid, p. 163). Can you just hear Fisher reply (as he did about washing out of priors), "we may well ask what [the prior] is doing in our reasoning at all" (1934b, p. 287). Bernardo might retort: They are merely formal tools "which, for a *given model*, are supposed to describe whatever the data 'have to say' about some *particular quantity*" (1997, p. 160). The desire for an inductive logic of probabilism is familiar to us. Statistician Christian Robert echoes this sentiment: "Having a prior attached to [a parameter θ] has nothing to do with 'reality,' it is a reference measure that is necessary for making probability statements" (2011, pp. 317–18). How then do we interpret the posterior, Cox asks? "If the prior is only a formal device and not to be interpreted as a probability, what interpretation is justified for the posterior as an adequate summary of information?" (2006a, p. 77)[4]

A Bayesian text by Gelman et al. (2014), to its credit, doesn't blithely assume that because probability works to express uncertainty about events in games of chance, we may assume it is relevant in making inferences about parameters. They aim to show the overall usefulness of the approach. What about meanings of priors?

We consider two basic interpretations that can be given to prior distributions. In the *population* interpretation, the prior distribution represents a population of possible parameter values, from which the θ of current interest has been drawn. In the more subjective *state of knowledge* interpretation, the guiding principle is that we must express our knowledge (and uncertainty) about θ as if its value could be thought of as a random realization from the prior distribution. (p. 34)

An example from Ghosh et al. (2010) lends itself to the "population interpretation," which to me sounds like a frequentist prior.

Exhibit (i): Blood Pressure and Historical Aside. "Let $X_1, X_2, \ldots, X_n$ be IID $N(\mu, \sigma^2)$ and assume for simplicity σ^2 is known. ... μ may be the expected reduction of blood pressure due to a new drug. You want to test H_0: $\mu \leq \mu_0$ vs. H_1: $\mu > \mu_0$, where μ_0 corresponds with a standard drug already in the market" (Ghosh et al. 2010, p. 34; their Example 2.4).

Here, μ can be viewed as a random variable that takes on values with different probabilities. The drug of interest may be regarded as a random selection from a population of drugs, each with its expected reductions in blood pressure, i.e., various values of μ. Neyman and Pearson would not have objected; here's a historical aside:

"I began as a quasi-Bayesian": Neyman

> I began as a quasi-Bayesian. My assumption was that the estimated parameter (just one!) is a particular value of a random variable having an unknown prior distribution. (Neyman 1977, p. 128)

Finding them so rarely available, he sought interval estimators with a probability of covering the true value being independent of the prior distribution.

---

[4] "When the parameter space is finite it [Bernardo reference priors] produces the maximum entropy prior of E. T. Jaynes and, for a one-dimensional parameter, the Jeffreys prior" (Cox 2006b, p. 6).

[My student Churchill Eisenhart in the 1930s] attended my lectures at the University College, London, and witnessed my introducing a prior distribution ... and then making efforts to produce an interval estimator, the properties of which would be independent of the prior. Once, Eisenhart's comment was that the whole theory would look nicer if it were built from the start without any reference to Bayesianism and priors. That remark proved inspiring. (ibid.)

Even the famous 1933 paper considered the Bayesian possibility. E. Pearson had been fully convinced by Fisher's non-Bayesian stance before Neyman, never mind the clash with his (Bayesian-leaning) father. It's one thing to forgo marriage with the woman you love because dad disapproves (as K. Pearson did); it's quite another to follow his view of probability (Section 3.2). Neyman was still exploring. He thought it "important to show that even if a statistician started from the point of view of inverse probabilities he would be led to the same" tests as those he and Pearson recommended (C. Reid 1998, p. 83). Neyman begged Pearson to sign on to a paper that included inverse probability solely for this purpose, but he would not. Pearson worried "they would find themselves involved in a disagreement with Fisher, who had come out decisively against [inverse probability]" (ibid., p. 84) and he never signed on. For more on this episode see C. Reid.

The kind of frequentist prior Neyman allowed were those in genetics. One might consider the probability a person is born with a trait as the effect of a combination of environmental and genetic factors that combine to produce the trait. In an example very like Exhibit (i), Neyman worries that we only know of a finite number of drugs, and we at best have estimates of their average pressure-lowering ability. However, Neyman (1977, p. 115) welcomes the "brilliant idea ... due to Herbert Robbins (1956)" launching "a novel chapter of frequentist mathematical statistics": Empirical Bayes Theory. There may be a sufficient stockpile of information of drugs (or, for that matter, black holes or pulsars) deemed similar to the one in question to arrive at frequentist priors, important for prediction. Some develop "enthusiastic priors" to be contrasted to "skeptical ones" in recommending policy (Spiegelhalter et al. 1994). The severe tester questions if even a fully warranted frequentist posterior gives a report of well-testedness, in and of itself. In any event, most cases aren't like this.

We sometimes hear: But the claim $\{\theta = \theta'\}$ and a claim $\{X = x\}$ are both statements, as if to say, if you can apply probability in one case, why not the other. There's a huge epistemic difference in assessing the probabilities of these different statements. There needs to be a well-defined model assigning probabilities to event statements – just what's missing when we are loath to assign probabilities to parameters. On the other hand, if we are limited to the legitimate frequentist priors of Neyman, there's no difference between what the Bayesian and frequentist can do, if they wanted to. Donald Fraser (2011) says only these frequentist priors should be called "objective" (p. 313) but, like Fisher, denies this is "Bayesian" inference, because it's just a deductive application of conditional probability, and "whether to include the prior becomes a modeling issue" (ibid., p. 302). But then it's not clear how much of the current Bayesian revolution is obviously Bayesian. Lindley himself famously said that there's "no one less Bayesian than an Empirical Bayesian . . . because he has to consider a sequence of similar problems" (1969, p. 421). Non-frequentist Bayesians switch the role of probability (compared to the frequentist) in a dramatic enough way to be a gestalt change of perspective.

A Dramatic Switch: Flipping the Role of Probability

"A Bayesian takes the view that all unknown quantities, namely the unknown parameter and data before observation, have a probability distribution" (Ghosh et al. 2010, p. 30). By contrast, frequentists don't assign probability to parameters (excepting the special cases noted), and data retain probabilities even after they are observed. This assertion, or close rewordings of it, while legion in Bayesian texts, is jarring to the frequentist ear because it flips the role of probability. Statisticians David Draper and David Madigan put it clearly:

When we reason in a frequentist way, . . . we view the data as random and the unknown as fixed. When we are thinking Bayesianly, we hold constant things we know, including the data values, . . . – the data are fixed and the unknowns are random. (1997, p. 18)

That's why, when the Higgs researchers spoke of the probability the results are mere statistical flukes, they appeared to be assigning probability to a hypothesis. There was nothing left as random, given the data – at least to a Bayesian. If known data x are given probability 1, we are led to the "old (or known) evidence" problem (Section 1.5) where no Bayes boost is forthcoming. Some further consequences will arise in this tour.[5]

Even where parameters are regarded as fixed, we may assign them probabilities to express uncertainty in them. Where do I get a probability on θ if fixed but unknown? The classic subjective way, we saw, is to find an event with known probability, and build a subjective prior by considering $\{\theta < \theta'\}$ for different values of parameter θ, now regarded as a random variable. If you locate an event E, with known frequentist probability k, such that you're indifferent to bets on $\{\theta < \theta'\}$ and E, then the former gets probability k. A non-

---

[5] The default Bayesian needn't give probability 1 to data, but it's unclear how they proceed with Bayes' Rule or other computations with a probability on the data and assumptions. Rejecting this possibility, Box and others use frequentist methods for model testing.

subjective/default approach can avoid this, and in some cases arrive at the same test as the frequentist in Exhibit (i) by setting a mathematically convenient conjugate, or an uninformative prior, say by viewing θ itself as Normally distributed $N(\eta, \tau^2)$. Instead of reporting the significance level of 0.05, this allows reporting that the posterior probability of H_0 is 0.05.

Pr($\theta = \theta_0$) = 0.95 is meaningless unless θ is a random variable. . . . this expression signifies that we are ready to bet that θ is equal to θ_0 with a 95/5 odds ratio, or, in other words, that the uncertainty about the value of θ is reduced to a 5% zone. (Robert 2007, p. 25; Pr for P)

Would we want to equate error probabilities to readiness to bet? As always it's most useful to look at cases of poor or weak evidence. Suppose you arrive at statistical significance of 0.2. We would be entitled to say we're as ready to bet on $\theta > \theta'$ as on the occurrence of an event with probability 0.8. I don't think we'd want to be so entitled. The default Bayesian replies, this just means the default prior doesn't reflect my beliefs. OK, but recall the question at the outset of this tour: *why assume we want a posterior probability on statistical hypotheses*, in any of the ways now available? The default Bayesian was to supply the (ideally) unique prior to use, not send us back to subjective priors.

The Bayesian treats the blood-pressure example very differently if the null is a point such as $\theta = 0$, whereas there's no difference for a frequentist. The spike and smear priors surveyed in Excursion 4 are common. Greenland and Poole (2013) suggest:

[A] null spike represents an assertion that, with prior probability q, we have background data that prove $\theta_t = 0$ with absolute certainty; $q = \frac{1}{2}$ thus represents a 50–50 bet that there is decisive information literally proving the null. [Otherwise a] spike at the null is an example of 'spinning knowledge out of ignorance.' (p. 66)

This is an interesting construal. Instead of how strongly you believe the null, it's how strongly you believe in a proof of it. That decisive information exists (their second clause) is weaker than actually having it (their first clause), but both are stronger than presuming they arise from a noncommittal "equipoise." Of course, the severe tester wants to know how strong the existing demonstration of H is, not how strong your belief in such a demonstration is.

Some subjective Bayesians would chafe at the idea of betting on scientific hypotheses or theoretical quantities. For one thing, it's hard to imagine people would be indifferent between a bet they know will be settled and one that is unlikely to be – as in the case of most scientific hypotheses. No one's going to put their money down now (unless they get interest). Still, cashing out Bayesian uncertainty with betting seems the most promising way to "operationalize it." Other types of scoring functions may be used, but still, there's a nagging feeling they leave us in the dark about what's really meant.

For both subjectivist and objectivist [default] Bayesians, probability models including both parameter priors and sampling models do not model the data-generating process, but rather represent plausibility or belief from a certain point of view. (Gelman and Hennig 2017, pp. 990–1)

Yet Gelman et al. (above) suggested expressing uncertainty as if a parameter's "value could be thought of as a random realization from the prior distribution" (2014, p. 34). If this is bending your brain, then you're getting it. Claims like it's "the knowledge [of fixed but unknown parameters] that Bayesians model as random" (Gelman and Robert 2013, p. 4) feel as if they ought to make perfect sense, but the more you think about them, the more they're liable to slip from grasp. For our purposes, let's understand claims that unknown quantities *have* probability distributions in terms of a person or persons who are doing the having – by assigning different degrees of belief (or other weights) to different parameter values.

The Probabilities of Events

Many Bayesian texts open with a focus on probabilities of simple events, or statements of events, like the "heads" on the toss of a coin. By focusing on probabilities of events which even frequentists condone, the reader may wonder what all the fuss is about. Problem is, the central point of contention between Bayesians and frequentists is whether to place probabilities on parameters in a statistical model. It isn't that frequentists don't assign probabilities to events, and any statistics based on them. It is that they recognize the need to infer a statistical model, and hypotheses about its parameters, in order to get those probabilities. How does that inference occur? It rests on probabilistic properties of a test method, which is very different from the deductive assignment of probability to hypotheses.

The severe tester uses probabilities assigned to events like {test T yields $d(X)$ $> d(x)$} to detach statistical inferences. She might argue: If $\Pr\{d(X) > d(x); H'\}$ is not very small, infer there's a poor indication of a discrepancy H'. Computing $\Pr\{d(X) > d(x); \theta\}$ for varying θ tells me the capability of the test to detect various discrepancies from a reference value of interest. This does not give a posterior probability to the hypothesis, but it allows making statistical inferences which are qualified by how well or poorly tested claims are.

True, when the frequentist assigns a probability to an event, it is seen as a general type, whereas a Bayesian can assign subjective probability to a unique event on November 8, 2016! Or so it is averred. But when they appeal to bets by reference to events with known probabilities, aren't they viewing it as a type? ("That's the kind of thing I'd bet 0.9 on.")

Cox points out that even subjectivists must think their probabilities have a frequentist interpretation. Consider n events/hypotheses:

... all judged by You to have the same probability p and not to be strongly dependent ... It follows from the Weak Law of Large Numbers obeyed by personalistic probability that Your belief that about a proportion p of the events are true has probability close to 1. (Cox 2006a, p. 79)

This suggests, Cox continues, that to elicit Your probability for H you try to find events or hypotheses that you judge for good reason to have the same probability as H, and then find out what proportion of this set is true. This proportion would yield Your subjective probability for H. Echoes of the screening model of tests (Section 5.6). Here the (hypothetical or actual) urn contains hypotheses that you thus far judge to be as probable as the H of interest. If the proportion of hypotheses in this urn that turned out true was, say, 80%, then H would get probability 0.8. It would be rare to know the truth rates of the hypotheses in this urn – would it be the proportion now assigned probability 1 by the subjectivist? Perhaps the proportion not yet falsified could be used.

Still, this would be a crazy way to actually go about evaluating evidence and hypotheses! But what if you considered H as if it were randomly selected from an urn of hypotheses that had passed severe tests, perhaps made up of claims in the same field. You check the relative frequency that are true or have held up so far. You'd still need to show why you're putting H in the high severity urn. In other words, you would have circled right back to the initial assignment of severity. All you'd be doing is reporting how often severely corroborated claims are true, or continue to solve their empirical problems (predicting or explaining). There would be nothing added by the imaginary urn.

A different attempt to assign a frequentist probability to a hypothesis H might try to consider how probable it is that the universe is such that H is true, considering fundamental laws, other worlds, multiverses, or what have you. One might consider the rarity of possible worlds that would have such a law. Even if we could somehow compute this, how would it be relevant to assessing hypotheses about this world? Here's C. S. Peirce:

[The present account] does not propose to look through all the possible universes, and say in what proportion of them a certain uniformity occurs; such a proceeding, were it possible, would be quite idle. The theory here presented only says how frequently, in this universe, the special form of induction or hypothesis would lead us right. The probability given by this theory is in every way different – in meaning, numerical value, and form – from that of those who would apply to ampliative inference the doctrine of inverse chances. (Peirce 2.748)

This objection, I take it, is different from trying to determine, on theoretical principles, how "fine tuned" this world would have to be for various parameters to be as we estimate them. Those pursuits, whose validity I'm in no position to judge, are aimed at deciding whether we should fiddle with theoretical assumptions so that this universe is not so "unnatural."

6.3 Unification or Schizophrenia: Bayesian Family Feuds

> COX: There's a lot of talk about what used to be called inverse probability and is now called Bayesian theory. That represents at least two extremely different approaches. How do you see the two? Do you see them as part of a single whole? Or as very different?
> MAYO: It's hard to give a single answer, because of a degree of schizophrenia among many Bayesians. On paper at least, the subjective Bayesian and the so-called default Bayesians . . . are wildly different. For the former the prior represents your beliefs apart from the data, . . . Default Bayesians, by contrast, look up 'reference' priors that do not represent beliefs and might not even be probabilities, . . . Yet in reality default Bayesians seem to want it both ways. (Cox and Mayo 2011, p. 104)

If you want to tell what's true about today's Bayesian debates, you should consider what they say in talking amongst themselves. I began to sense a shifting of sands in the foundations of statistics landscape with an invitation to comment on Jim Berger (2003). The trickle of discontent from family feuds issuing from Bayesian forums pulls back the curtain on how Bayesian–frequentist debates have metamorphosed. To show you what I mean, let's watch the proceedings of a conference at Carnegie Mellon, published in *Bayesian Analysis* (vol. 1, no. 3, 2006) in the museum library. Unlike J. Berger's (2003) attempted amalgam of Jeffreys, Neyman, and Fisher (Section 3.6), here it's Berger smoking the peace pipe, making "The Case for Objective Bayesianism" to his subjective compatriots (Section 4.1). The forum gives us a look at the inner sanctum, with Berger presenting a tough love approach: If we insist on subjectivity, we're out. "[T]hey come to statistics in large part because they wish it to provide objective validation of their science" (J. Berger 2006, p. 388).

Four Philosophical Positions

Admitting there is no unanimity as to either the definition or goal of "objective" (default) Bayesianism, Berger (2006, p. 386) outlines "four philosophical positions" that default Bayesianism might be seen to provide:

1. A complete coherent objective Bayesian methodology for learning from data.
2. The best method for objectively synthesizing and communicating the uncertainties that arise in a specific scenario, but is not necessarily coherent.
3. A convention we should adopt in scenarios in which a subjective analysis is not tenable.
4. A collection of ad hoc but useful methodologies for learning from data.

Berger regards (1) as unattainable; (2) as often attainable and should be done if possible, but concedes that often the best we can hope for is (3), or maybe (4). Lindley would have gone with (1).

Is a collection of ad hoc but useful methodologies good enough? There is a fascinating philosophical tension in Berger's work: while in his heart of hearts he holds "the (arguably correct) view that science should embrace subjective statistics", he realizes this "falls on deaf ears" (ibid., p. 388). When scientists demur: "I do not want to do a subjective analysis, and hence I will not use Bayesian methodology," Berger convincingly argues they can have it both ways (p. 389).

Among the advantages to adopting a default Bayesian methodology is avoiding a subjective elicitation of experts. Berger finds elicitation does not work out too well. Far from providing a route within which to describe background knowledge in terms of prior probabilities, he finds elicitation foibles are common even with statistically sophisticated practitioners. "[V]irtually never would different experts give prior distributions that even overlapped; there would be massive confusions over statistical definitions (e.g., what does a positive correlation mean?)" coupled with the difficulty of eliciting priors when, as is typical, "the expert has already seen the data" (ibid., p. 392). But if the prior is determined post-data, one wonders how it can be seen to reflect information independent of the data. I come back to this. In his own experience Berger found:

> ... for the many parameters for which there was data ... all of the expert time was used to assist model building. It was necessary to consider many different models, and expert insight was key to obtaining good models; there simply was no extra available expert time for prior elicitation. (ibid.)

He argues that the default choices have the advantage over trying to elicit a subjective prior:

> The problem is that, to elicit all features of a subjective prior $\pi(\theta)$, one must infinitely accurately specify a (typically) infinite number of things. In practice, only a modest number of (never fully accurate) subjective elicitations are possible, so practical

Bayesian analysis must somehow construct the entire prior distribution $\pi(\theta)$ from these elicitations. (ibid., p. 397)

A standard way to turn elicitations into full prior distributions is to use mathematically convenient priors (as with default priors). The trouble is this leads to Bayesian incoherence, in violation of the Likelihood Principle (LP). Why? Because "depending on the experiment designed to study θ, the subjective Bayesian following this 'prior completion' strategy would be constructing different priors for the same θ, clearly incoherent" (ibid.). Ironically, this LP violation is not directly driven by the need to compute the sampling distribution to obtain frequentist error probabilities: it is a way to try to capture a reasonably non-informative prior – as this is thought to depend on the experiment to be performed.[6] Any good error properties are touted as a nice bonus, not the deliberate aim, except for the special case of error probability matching priors.

Berger maintains that the default, at best, achieves "a readily understandable communication of the information in the observed data, as communicated through a statistical model, for any scientific question that is posed" (ibid., p. 388). We've seen that there's considerable latitude open to the default Bayesian – the source of arguments that P-values overstate the evidence. It's hard to view those spiked priors as merely conveying what the data say (especially when they use a two-sided test). Another issue is that it often distinguishes parameters of "interest" from additional "nuisance" parameters, each of which must be ordered. In Bernardo's system of reference priors, there are as many reference priors as possible parameters of interest (1997, p. 169). That's because what counts as data dominance (he calls it "maximizing the missing information") will differ for different parameters. Each ordering of parameters will yield different posteriors. Despite Berger's own misgivings in avoiding elicitation bias:

A common and reasonable practice is to develop subjective priors for the important parameters or quantities of interest in a problem, with the unimportant or 'nuisance' parameters being given objective priors. (ibid., p. 393)

Here again we see the default Bayesian inviting both types of priors: if you have information, put it in the elicitation; if not, keep it out and choose one of the

---

[6] One way to link the LP violation with Bayesian incoherence is to show that the posterior depends on the order of two independent experiments for the same parameter. We know the Binomial and Negative Binomial experiments have different sample spaces (Section 4.9), and yet are not distinguished on the LP. Default priors, however, are sample-space dependent. If the first experiment is Binomial and the second Negative Binomial, both for inferences about the probability of success on each trial, a different posterior results depending on the order that the default rule is applied. Excellent discussions are in Seidenfeld (1979) and Kass and Wasserman (1996, p. 1359). Note that some consider that coherence only concerns the assignment of the prior; a violation of Bayes' Rule is called a failure of Bayesian conditionalization.

conventional priors. You might say there's nothing really schizophrenic in this, even subjectivists argue that default priors are kosher as approximations to what they would have arrived at in cases of minimal information (O'Hagan, this forum). It's just faster. Should they be so sanguine? The tasks are quite different.

> [O]bjective priors can vary depending on the goal of the analysis for a given model. For instance, in a normal model, the reference prior will be different if inference is desired for the mean μ or if inference is desired for μ/σ. This, of course, does not happen with subjective Bayesianism. (Berger 2006, p. 394)

Trying to describe your beliefs is different from trying to make the data dominant relative to a given model and ordering of parameters. Subjectivists hold that prior beliefs in H shouldn't change according to the experiment to be performed. However, if they incorporate default priors, when required by complex problems, this changes. Since priors of different sorts are then combined in a posterior, how do you tell what's what? If nothing else, the simplicity that led Dawid to joke that Bayesianism is boring disappears.

Ironic and Bad Faith Bayesianism

A major impetus for developing default Bayesian methods, for Berger, is to combat what he calls "casual Bayesianism" or pseudo-Bayesianism.

> One of the mysteries of modern Bayesianism is the lip service that is often paid to subjective Bayesian analysis as opposed to objective Bayesian analysis, but then the practical analysis actually uses a very adhoc version of objective Bayes, including use of constant priors, vague proper priors, choosing priors to 'span' the range of the likelihood, and choosing priors with tuning parameters that are adjusted until the answer 'looks nice.' I call such analyses *pseudo-Bayes* because, while they utilize Bayesian machinery, they do not carry with them any of the guarantees of good performance that come with either true subjective analysis (with a very extensive elicitation effort) or (well-studied) objective Bayesian analysis. (Berger 2006, pp. 397–8)

Berger stops short of prohibiting casual Bayesianism, but warns that it "must be validated by some other route" (ibid.), left open. One thing to keep in mind: "good performance guarantees" mean disparate things to Bayesians and to frequentist error statisticians. Remember those subscripts. "In general reference priors have some good frequentist properties but except in one-dimensional problems it is unclear that they have any special merit in that regard" (Cox 2006b, p. 6). Judging from the ensuing discussion, Berger's concern here is with resulting improper posteriors that can remain hidden in the use of computer packages. Improper priors are often not problematic, but posteriors that are not probabilities (because they don't add to 1) are a disaster.

Interestingly, Lindley came to his subjective Bayesian stance after he was shown that conventional priors can lead to improper posteriors and thus to violations of probability theory (Dawid, Stone, and Zidek 1973). A remark that is especially puzzling or revealing, depending on your take:

> Too often I see people pretending to be subjectivists, and then using 'weakly informative' priors that the objective Bayesian community knows are terrible and will give ridiculous answers; subjectivism is then being used as a shield to hide ignorance . . . In my own more provocative moments, I claim that the only true subjectivists are the objective Bayesians, because they refuse to use subjectivism as a shield against criticism of sloppy pseudo-Bayesian practice. (Berger 2006, pp. 462–3)

How shall we deconstruct this fantastic piece of apparent doublespeak? I take him to mean that a subjectivist who properly recognizes her limits and biases and opts to be responsible for her priors would accept the constraints of default priors. A pseudo-Bayesian uses priors as if these really reflected properly elicited subjective judgments. In doing so, she (thinks that she) doesn't have to justify them – she claims that they reflect subjective judgments (and so who can argue with them?).

Although most Bayesians these days disavow classic subjective Bayesian foundations, even the most hard-nosed, "we're not squishy" Bayesians retain the view that a prior distribution is an important if not the best way to bring in background information. Here's Christian Robert:

> The importance of the prior distribution in a Bayesian statistical analysis is not at all that the parameter of interest θ can (or cannot) be perceived as generated from [prior distribution π] . . . but rather that the use of a prior distribution is the best way to summarize the available information (or even the lack of information) about this parameter. (Robert 2007, p. 10)

But is it? To suppose it is pulls in the opposite direction from the goal of the default prior which is to reflect just the data.

Grace and Amen Bayesians

> I edit an applied statistics journal. Perhaps one quarter of the papers employs Bayes' theorem, and most of these do *not* begin with genuine prior information. (Efron 2013, p. 134)

Stephen Senn wrote a paper "You Might Believe You Are a Bayesian But You Are Probably Wrong." More than a clever play on words, Senn's title highlights the common claim of researchers to have carried out a (subjective) Bayesian analysis when they have actually done something very different. They start and end with thanking the (subjective?) Bayesian account for housing all their uncertainties within prior probability distributions; in between, the analysis immediately turns to default priors, coupled with ordinary statistical modeling considerations that may well enter without being put in probabilistic form. "It is this sort of author who believes that he or she is Bayesian but in practice is wrong" (Senn 2011, p. 58). In one example Senn cites Lambert et al. (2005, p. 2402):

[T]he authors make various introductory statements about Bayesian inference. For example, 'In addition to the philosophical advantages of the Bayesian approach, the use of these methods has led to increasingly complex, but realistic, models being fitted,' and 'an advantage of the Bayesian approach is that the uncertainty in all parameter estimates is taken into account' . . . but whereas one can neither deny that more complex models are being fitted than had been the case until fairly recently, nor that the sort of investigations presented in this paper are of interest, these claims are clearly misleading. . . (Senn 2011, p. 62)

While the authors "considered thirteen different Bayesian approaches to the estimation of the so-called random effects variance in meta-analysis . . ." – techniques fully available to the frequentist, "[n]one of the thirteen prior distributions considered can possibly reflect what the authors believe about the random effect variance" (ibid., pp. 62–3).

Ironically, Senn says, a person who takes into account the specifics of the case in their statistical modeling is "being more Bayesian in the de Finetti sense" (ibid) than the default/non-subjective Bayesian. By focusing on how to dress the case into ill-fitting probabilistic clothing, Senn is insinuating, the Bayesians may miss context-dependent details solely because they were not framed probabilistically. Leo Breiman, an early leader in machine learning, needled Bayesians:

The Bayesian claim that priors are the only (or best) way to incorporate domain knowledge into the algorithms is simply not true. Domain knowledge is often incorporated into the structure of the method used. . . . In handwritten digit recognition, one of the most accurate algorithms uses nearest-neighbor classification with a distance that is locally invariant to things such as rotations, translations, and thickness. (Breiman 1997, p. 22)

Nor need context-dependent information of a repertoire of mistakes and pitfalls be cashed out in terms of priors. They'd surely be reflected in a post-data assessment of severity, which would be open to model builders from any camp.

Finally, the "the lip service that is often paid to subjective Bayesian analysis as opposed to objective Bayesian analysis," far from being the "modern mystery," Berger (2006, p. 397) dubs it, might reflect the degree of schizophrenia of default Bayesianism. After all, Berger[7] says that the default prior "is used to describe an individual's (or group's) 'degree of belief'" (ibid., p. 385), while ensuring the influence of subjective belief is minimal. Moreover, in using default priors, he maintains, you're getting closer to the subjective Bayesian ideal (absent a full elicitation). So there should be no surprise when a default Bayesian says she's being a good subjective Bayesian. The default Bayesians attain an aura of subjective foundations for philosophical appeal, and non-subjective foundations for scientific appeal. If you come face to face with a default posterior probability, you need to ask which default method was used, the ordering of parameters, the mixture of subjective and default priors and so on. Even a transparent description of all that may not help you appraise whether a high default posterior in H indicates warranted grounds for H.

6.4 What Happened to Updating by Bayes' Rule?

In striving to understand how today's Bayesians view their foundations, we find even some true-blue subjective Bayesians reject some principles thought to be important, such as Dutch book arguments. If it is agreed that we have degrees of belief in any and all propositions, then it is argued that if your beliefs do not conform to the probability calculus you are being incoherent. We can grant that if we had degrees of belief, and were required to take any bets on them, that, given we prefer not to lose, we do not agree to a series of bets that ensures losing. This is just a tautologous claim and entails nothing about degree of belief assignments. "That an agent ought not to accept a set of wagers according to which she loses come what may, if she would prefer not to lose, is a matter of deductive logic and not a property of beliefs" (Bacchus, Kyburg, and Thalos 1990, pp. 504–5).

The dynamic Dutch book argument was to show that the rational agent, upon learning some data E, would update by Bayes' Rule, else be guilty of irrationality. Confronted with counterexamples in which violating Bayes' Rule seems perfectly rational on intuitive grounds, many if not most Bayesian philosophers dismiss threats of being Dutch-booked as irrelevant. "It is the entirely rational claim that I may be induced to act irrationally that the dynamic Dutch book argument, absurdly, would condemn as incoherent" (Howson 1997a, p. 287). Howson declares it was absurd all along to consider it irrational to be induced to act irrationally. It's insisting on updating by Bayes' Rule that is irrational. "I am not inconsistent in planning … to entertain

---

[7] I've no objection to Berger's viewing "probability as a primitive concept" (p. 385). Theoretical concepts may arise in models and receive their explication through applications. It's problematic when the subsequent meanings shift, as happens with default probabilities.

a degree of belief [that is inconsistent with what I now hold], I have merely changed my mind" (ibid.). One thought the job of Bayesian updating was to show *how* to change one's mind reasonably.

Counterexamples to Bayes' Rule often take the following form: While an agent assigns probability 1 to event S at time t, i.e., $Pr(S) = 1$, he also believes that at some time in the future, say t', he may assign a low probability, say 0.1, to S, i.e., $Pr'(S) = 0.1$, where P' is the agent's belief function at later time t'.

> Let E be the assertion: $P'(S) = 0.1$.
> So at time t, $Pr(E) > 0$.
> But $Pr(S|E) = 1$ since $P(S) = 1$.

Now, Bayesian updating says:

> If $Pr(E) > 0$, then $Pr'(.) = Pr(. |E)$.
> But at t' we have, $Pr'(S) = 0.1$,

which contradicts $Pr'(S) = Pr(S| Pr'(S) = 0.1) = 1$ obtained by Bayesian updating. It is assumed, by the way, that learning E does not change any of the other degree of belief assignments held at t – never mind how one knows this.

The kind of example at the heart of this version of the counterexample was given by William Talbott (1991, p. 139). In one of his examples: S is "Mayo ate spaghetti at 6 p.m., April 6, 2016". $Pr(S) = 1$, where Pr is my degree of belief in S now (time t), and E is "$Pr'(S) = r$", where r is the proportion of times Mayo eats spaghetti (over an appropriate time period); say $r = 0.1$. As vivid as eating spaghetti is today, April 6, 2016, as Talbott explains, I believe, rationally, that next year at this time I will have forgotten, and will (rationally) turn to the relative frequency with which I eat spaghetti to obtain Pr'. Variations on the counterexample involve current beliefs about impairment at t' through alcohol or drugs. This is temporal incoherency.

It may seem surprising for a subjective Bayesian like Howson to reject Bayes' Rule: Typically, it's the subjectivist who recoils in finding the default/ non-subjective tribes living in conflict with it. Jon Williamson, a non-subjective Bayesian philosopher in a Carnap-maximum entropy mold,[8] identifies the problem in these examples as stemming from two sources of probabilistic information (Williamson 2010). Relative frequency information tells

---

[8] The noteworthy Carnapian part is his relativization to first order languages, rather than to statistical models.

us Pr′(S) = 0.1, but also, since this is known, Pr′(Pr′(S) = 0.1) = 1. Bayes' Rule holds, he allows, just when it holds. When there's a conflict with Bayes' Rule, default Bayesian "updating" takes place to reassign priors.

The position of Howson and Williamson is altogether plausible if one is forced with the given assignments. Ian Hacking, who is not a Bayesian, sympathizes, and blames universal Bayesianism. Universal Bayesianism, Hacking (1965, p. 223) remarks, forces Savage (Savage 1962, p. 16) to hold "if you come to doubt [e.g., the Normality of a sample], you must always have had a little doubt". To Hacking, it's plausible to be completely certain of something, betting the whole house and more, and later come to doubt it. All the more reason we should be loath to assign probability 1 to "known" data, while seeking a posterior probabilism.

Bayesian statisticians, at least of the default/non-subjective variety, follow suit, though for different reasons: "Betting incoherency thus seems to be too strong a condition to apply to communication of information" (J. Berger 2006, p. 395). Berger avers that even subjective Bayesianism is not coherent in practice, "except for trivial versions such as always estimate $\theta \in (0, \infty)$ by 17.35426 (a coherent rule, but not one particularly attractive in practice)" (pp. 396–7). His point appears to be that, while incoherence is part and parcel of default/non-subjective Bayesian accounts, in practice, idealizations lead the subjectivist to be incoherent as well. It gets worse: "in practice, subjective Bayesians will virtually always experience what could be called practical marginalization paradoxes" (p. 397), where posteriors don't sum to 1. If this is so, it's very hard to see how they can be happy using any kind of probability logic.

There are a great many complex twists and turns to the discussions of Dutch books; too many to do justice with a sample list.

Can You Change Your Bayesian Prior?

> As an exercise in *mathematics* [computing a posterior based on the clients prior probabilities] it is not superior to showing the client the data, eliciting a posterior distribution and then calculating the prior distribution; as an exercise in *inference* Bayesian updating does not appear to have greater claims than 'downdating' ... (Senn 2011, p. 59)

> If you could really express your uncertainty as a prior distribution, then you could just as well observe data and directly write your subjective posterior distribution, and there would be no need for statistical analysis at all. (Gelman 2011, p. 77)

Lindley's answer is that it would be a lot harder to be coherent that way, however, he's prepared to allow: "[I]f a prior leads to an unacceptable posterior then I modify it to cohere with properties that seem desirable in the inference" (Lindley 1971, p. 436). He resists saying the rejected prior was wrong though. Not wrong? No, just failing to cohere with desirable properties. I. J. Good (1971b) advocated his device of "imaginary results" whereby a subjective Bayesian would envisage all possible results in advance (p. 431) and choose a prior that she can live with regardless of results. Recognizing that his device is so difficult to apply that most are prepared to bend the rules, Good allowed "that it is possible after all to change a prior in the light of *actual* experimental results" (ibid.) – appealing to an informal, second-order rationality of "type II."

So can you change your Bayesian prior? I don't mean update it, but reject the one you had and replace it with another. I raised this question on my blog (June 18, 2015), hoping to learn what current practitioners think. Over 30 competing answers ensued (from over 100 comments), contributed by statisticians from different tribes. If the answer is yes you can, then how do they avoid the verification biases we are keen to block? Lindley seems to be saying it's quite open-ended. Cox says, "there is nothing intrinsically inconsistent in changing prior assessments" in the light of data, however the danger is that "even initially very surprising effects can post hoc be made to seem plausible" (Cox 2006b, p. 78). Berger had said elicitation typically takes place after "the expert has already seen the data" (2006, p. 392), a fact of life he understandably finds worrisome. If the prior is determined post-data, then it's not reflecting information *independent* of the data. All the work would have been done by the likelihoods, normalized to be in the form of a probability distribution or density. No wonder many look askance at changing priors based on the data.

[N]o conceivable possible constellation of results can cause you to wish to change your prior distribution. If it does, you had the wrong prior distribution and this prior distribution would therefore have been wrong even for cases that did not leave you wishing to change it. (Senn 2011, p. 63)

The prior, after all, is "like already having some data, but what statistical procedure would allow you to change your data?" (Senn 2015b). As with Good's appeal to type II rationality, Senn is saying this is tantamount to admitting "the informal has to come to the rescue of the formal" (Senn 2011, p. 58), which would otherwise permit counterintuitive results. He makes the interesting point that the post-data adjustment of priors could conceivably be taken account of in the posterior: "If you see there is a problem with the placeholder model and replace it, it may be that you can somehow reflect this 'sensible cheating' in your posterior probabilities" (Senn 2015b; see also Senn 2013a). I think Senn is applying a frequentist error statistical mindset to the Bayesian analysis, wherein the posterior might be qualified by an error statistical assessment. Bayesians would need a principle to this effect.

Dawid (2015) weighs in with his "prequential" approach. "In this approach the prior is constructed, not regarded as given in advance". Maybe the idea is that the subjective Bayesian is trying to represent her psychological states; the posterior from the data indicate her first stab failed to do so, so it makes sense to change it. The main thing for Dawid is to have a coherent package; his Bayesian starts over with a better prior and a new test. But it's not obvious how you block yourself from engineering the result you want. Gelman and Hennig say "priors in the subjectivist Bayesian conception are not open to falsification ... because by definition they have to be fixed before observation" (2017, p. 989). Now Howson's Bayesian changes his mind post-data, but admittedly this is not the same as falsification. In his comment to the blog discussion, Gelman (2015) says that "if some of the [posterior] inferences don't 'make sense,' this implies that you have additional information that has not been incorporated into the model" and it should be improved. But not making sense might just mean that more information would be necessary to get an answer, not that you rightfully have it. Shouldn't we worry that among the many ways you fix things, you choose one that protects or enhances a favored view, even if poorly probed? A reply might be that frequentists worry about data-dependent adjustments as well. There's one big difference.

In order for a methodological "worry" to be part of an inference account, it needs an explicit rationale, not generally found in contemporary Bayesianism – though Gelman is an exception. An error statistician changes her model in order to ensure the reported error probabilities are close to the actual ones (whether for performance or severe testing). There seem to be at least two situations where the default/non-subjective Bayesian may start over: The first, already noted, is when there's a conflict with Bayes' Rule. "Updating" takes place by going back to assign new prior probabilities using a chosen default prior. Philosophers Gaifman and Vasudevan (2012) describe it thus: "... the revision of an agent's [rational] subjective probabilities proceeds in fits and starts, with periods of conditionalization punctuated by abrupt alterations of the prior" (p. 170).[9] Why wouldn't this be taken as questioning the entire method of reaching a default prior assignment? Surely it relinquishes a key

---

[9] They have in mind maximal entropy methods advanced by Jaynes and also developed by Roger Rosenkrantz (1977). It is thought to work well in contexts where the current experiment may be seen as a typical instance of a known physical θ-generating process.

feature Bayesianism claims to provide: a method of accumulating and updating knowledge probabilistically. A second situation might be finding information that statistical assumptions are violated. But this brings up Duhemian problems as we'll see in Section 6.6.

The Bayesian Catchall

The key obstacle to probabilistic updating, and to viewing an evidential assessment at a given time in terms of a posterior probabilism, is the Bayesian catchall hypothesis. One is supposed to save some probability for a catchall hypothesis: "everything else," in case new hypotheses are introduced, which they certainly will be. Follow me to the gallery on the 1962 Savage Forum for a snippet from the discussion between Savage and Barnard (Savage 1962, pp. 79–84):

BARNARD: ... Professor Savage, as I understood him, said earlier that a difference between likelihoods and probabilities was that probabilities would normalize because they integrate to one, whereas likelihoods will not. Now probabilities integrate to one only if all possibilities are taken into account. This requires in its application to the probability of hypotheses that we should be in a position to enumerate all possible hypotheses which might explain a given set of data. Now I think it is just not true that we ever can enumerate all possible hypotheses. . . . If this is so we ought to allow that in addition to the hypotheses that we really consider we should allow something that we had not thought of yet, and of course as soon as we do this we lose the normalizing factor of the probability, and from that point of view probability has no advantage over likelihood. (p. 80)

SAVAGE: ... The list can, however, always be completed by tacking on a catchall 'something else.' ... In practice, the probability of a specified datum given 'something else' is likely to be particularly vague – an unpleasant reality. The probability of 'something else' is also meaningful of course, and usually, though perhaps poorly defined, it is definitely very small.

BARNARD: Professor Savage says in effect, 'add at the bottom of list H_1, H_2, ...'something else'.' But what is the probability that a penny comes up heads given the hypothesis 'something else.' We do not know.

Suppose a researcher makes the catchall probability small, as Savage recommends, and yet the true hypothesis is not in the set so far envisaged, call this set $\underline{H}$. Little by little, data might erode the probabilities in $\underline{H}$, but it could take a very long time until the catchall is probable enough so that a researcher begins to develop new theories. On the other hand, if a researcher suspects the existing hypothesis set $\underline{H}$ is inadequate, she might give the catchall a high prior. In that case, Barnard points out, it may be that none of the available hypotheses in $\underline{H}$ get a high posterior, even if one or more are adequate. Perhaps by suitably restricting the space ("small worlds") this can work, but the idea of inference as continually updating goes by the board.

The open-endedness of science is essential – as pointed out by Nelder and Sprott. The severe tester agrees. Posterior probabilism, with its single probability pie, is inimical to scientific discovery. Barnard's point at the Savage Forum was, why not settle for comparative likelihoods? I think he has a point, but for error control, that limited us to predesignated hypotheses. Nelder was a Likelihoodist and there's a lot of new work that goes beyond Royall's Likelihoodism – suitable for future journeys. The error statistician still seeks an account of severe testing, and it's hard to see that comparativism can ever give that. Despite science's open-endedness, hypotheses can pass tests with high severity. Accompanying reports of poorly tested claims point the way to novel theories. Remember Neyman's modeling the variation in larvae hatched from moth eggs (Section 4.8)? As Donald Gillies (2001) stresses, "Neyman did not consider any hypotheses other than that of the Poisson distribution" (p. 366) until it was refuted by statistical tests, which stimulated developing alternatives.

Yet it is difficult to see how all these changes in degrees of belief by Bayesian conditionalisation could have produced the solution to the problem, . . . The Bayesian mechanism seems capable of doing no more than change the statistician's degree of belief in particular values of λ [in the Poisson distribution]. (Gillies 2001, p. 367)

At the stage of inventing new models, Box had said, the Bayesian should call in frequentist tests. This is also how GTR and HEP scientists set out to extend their theories into new domains. In describing the goal of "efficient tests of hypotheses," Pearson said, if a researcher is going to have to abandon his hypothesis, he would like to do so quickly. The Bayesian, Gillies observes, might have to wait a very long time or never discover the problem (ibid., p. 368). By contrast, "The classical statisticians do not need to indulge in such toil. They can begin with any assumption (or conjecture) they like, provided only they obey the golden rule of testing it severely" (ibid., p. 376).

Souvenir Y: Axioms Are to Be Tested by You (Not Vice Versa)

Axiomatic Challenge. What do you say if you're confronted with a very authoritative-sounding challenge like this: To question classic subjective Bayesian tenets (e.g., your beliefs are captured by probability, must be betting coherent, and updated via Bayes' Rule) comes up against accepted mathematical axioms. First, recall a point from Section 2.1: You're free to use any formal deductive system, the issue will be soundness. Axioms can't run up against empirical claims: they are formal stipulations of a system that gets meaning, and thus truth value, by interpretations. Carefully cashed out, the axioms they have in mind subtly assume your beliefs are well represented by probability, and usually that belief change follows Bayes' Theorem. If this captures your intuitions, fine, but there's no non-circular proof of this.

Empirical Studies. We skipped over a wing of the museum that is at least worth mentioning: there have been empirical studies over many years that refute the claim that people are intuitive Bayesians: "we need not pursue this debate any further, for there is now overwhelming empirical evidence that no Bayesian model fits the thoughts or actions of real scientists" (Giere 1988, p. 149). The empirical studies refer to experiments conducted since the 1960s to assess how well people obey Bayes' Theorem. These experiments, such as those performed by Daniel Kahneman, Paul Slovic, and Amos Tversky (1982), reveal substantial deviations from the Bayesian model even in simple cases where the prior probabilities are given, and even with statistically sophisticated subjects. Some of the errors may result from terminology, such as the common understanding of probability as the likelihood. I don't know if anyone has debunked the famous "Linda paradox" this way, but given the data, it's more likely that Linda's a feminist and a bank teller than that she's a bank teller, in the technical sense of "likely." Gerd Gigerenzer (1991) gives a thorough analysis showing that rephrasing the most popular probability violations frequentistly has them disappear.

What is called in the heuristics and biases literature the "normative theory of probability" or the like is in fact a very narrow kind of neo-Bayesian view . . . (p. 86)

. . . Since "cognitive illusions" tend to disappear in frequency judgments, it is tempting to think of the intuitive statistics of the mind as frequentist statistics. (ibid., p. 104)

While interesting in their own right, I don't regard these studies as severe tests of whether Bayesian models are a good representation for scientific inference. Why? Because in these experiments the problem is set up to be one in which the task is calculating probabilities; the test-taker is right to assume they are answerable by probabilities.

Normative Epistemology. We have been querying the supposition that what we really want for statistical inference is a probabilism. What might appear as a direct way to represent beliefs may not at all be a direct way to use probability for a normative epistemology, to determine claims that are and are not evidentially warranted. An adequate account must be able to falsify claims statistically, and in so doing it's always from demonstrated effects to hypotheses, theories, or models. Neither a posterior probability nor a Bayes factor falsifies. Even to corroborate a real effect depends on falsifying "no effect" hypotheses. Granted, showing that you have a genuine effect is just a first step in the big picture of scientific inference. You need also to show you've correctly pinpointed causes, that you can triangulate with other ways of measuring the same quantity, and, more strongly still, that you understand a phenomenon well enough to exploit it to probe new domains. These abilities are what demarcate science and non-science (Section 2.3). Formal statistics hardly makes these assessments automatic, but we want piecemeal methods ready to serve these ends. If our language had kept to the root of probability, *probare*, to demonstrate or show how well you can put a claim to the test, and have it survive, we'd find it more natural to speak of claims being well probed rather than highly probable. Severity is not to be considered the goal of science or a sum-up of the growth of knowledge, but it has a crucial role in statistical inference.

Someone is bound to ask: Can a severity assessment be made to obey the probability axioms? If the severity for the statistical hypothesis H is high, then little problem arises in having a high degree of belief in H. But we know the axioms don't hold. Consider H: Humans will be cloned by 2030. Both H and $\sim H$ are poorly tested on current evidence. This always happens unless one of H, $\sim H$ is corroborated. Moreover, passing with low severity isn't akin to having a little bit of evidence but rather no evidence to speak of, or a poor test. What if we omitted cases of low severity due to failed audits (from violated assumptions or selection effects)? I still say no, but committed Bayesians might want to try. Since it would require the assessments to make use of sampling distributions and all that error statistics requires, it could at most be seen as a kind of probabilistic bookkeeping of inferences done in an entirely different way.

Nearly all tribes are becoming aware that today's practice isn't captured by tenets of classical probabilism. Even some subjective Bayesians, we saw, question updating by Bayes' Rule. Temporal incoherence can require a do-over. The most appealing aspects of non-subjective/default Bayesianism – a way to put in background information while allowing the data to dominate – are in tension with each other, and with updating. The gallimaufry of priors alone is an obstacle to scrutinizing the offerings. There are a few tribes where brand new foundations are being sought – that's our last port of call.

Tour II Pragmatic and Error Statistical Bayesians

6.5 Pragmatic Bayesians

> The protracted battle for the foundations of statistics, joined vociferously by Fisher, Jeffreys, Neyman, Savage and many disciples, has been deeply illuminating, but it has left statistics without a philosophy that matches contemporary attitudes. (Robert Kass 2011, p. 1)

Is there a philosophy that "matches contemporary attitudes"? Worried that "our textbook explanations have not caught up with the eclecticism of statistical practice," Robert Kass puts forward a statistical pragmatism "as a foundation for inference" (ibid.) as now practiced. It reflects the current disinclination to embrace subjective Bayes, skirts the rigid behavioral decision model of frequentist inference, and hangs on to a kind of frequentist calibration. "Subjectivism was never satisfying as a logical framework: an important purpose of the scientific enterprise is to go beyond personal decision-making" (ibid., p. 6). Nevertheless, Kass thinks "it became clear, especially from the arguments of Savage . . . the only solid foundation for Bayesianism is subjective" (ibid., p. 7). Statistical pragmatism pulls us out of that "solipsistic quagmire" (ibid.). Statistical pragmatism, says Kass, holds that confidence intervals, statistical significance, and posterior probability are all valuable tools. So long as we recognize that our statistical models exist in an abstract theoretical world, Kass avers, we can safely employ methods from either tribe, retaining unobjectionable common denominators: frequencies without long runs and Bayesianism without the subjectivity.

The advantage of the pragmatic framework is that it considers frequentist and Bayesian inference to be equally respectable and allows us to have a consistent interpretation, without feeling as if we must have split personalities in order to be competent statisticians. (ibid., p. 7)

Kass offers a valuable analysis of the conundrums of members of today's eclectic and non-subjective/default Bayesian tribes. We want to expose some hidden layers only visible after peeling away a more familiar mindset. Do we escape split personality? Or has he, perhaps inadvertently, explained the necessity for a degree of schizophrenia in current electic practice? That's where today's journey begins.

Kass and the Pragmatists

Bayes–frequentist agreement is typically closer with confidence intervals rather than tests. Kass (2011) uses a "paradigm case of confidence and posterior intervals for a Normal mean based on a sample of size n [49], with the standard deviation [1] being known" (ibid., pp. 3–4). Both the frequentist and the Bayesian arrive at the same interval, but he'll supply "mildly altered interpretations of frequentist and Bayesian inference" (ibid., p. 3).

We assume $X_1, X_2, \ldots, X_{49}$ are IID random variables from $N(\mu, 1)$. The 0.95 two-sided confidence interval estimator, we know is:

$$\mu = \overline{X} \pm Z_{0.025}(\sigma/\sqrt{n}).$$

Following Kass, take $Z_{0.025}$ to be 2 (instead of the more accurate 1.96):

$$\mu = (\overline{X} - 2/7, \overline{X} + 2/7).$$

We observe $\bar{x} = 10.2$. Plugging in $1/7 = 0.14, 2/7 = 0.28$, the particular interval estimate I is

$$(10.2 - 2/7, 10.2 + 2/7) = [9.92, 10.48].$$

He contrasts the usual frequentist interpretation with the pragmatic one he recommends:

FREQUENTIST INTERPRETATION... Under the assumptions above, if we were to draw infinitely many random samples from a $N(\mu, 1)$ distribution, 95% of the corresponding confidence intervals $(\overline{X} - 2/7, \overline{X} + 2/7)$ would cover μ. (ibid., p. 4)

PRAGMATIC INTERPRETION... If we were to draw a random sample according to the assumptions above, the resulting confidence interval $(\overline{X} - 2/7, \overline{X} + 2/7)$ would have probability 0.95 of covering μ. Because the random sample lives in the theoretical world, this is a theoretical statement. Nonetheless, substituting $\overline{X} = \bar{x}$ ... we obtain the interval I, and are able to draw useful conclusions as long as our theoretical world is aligned well with the real world that produced the data. (ibid.)

Kass's pragmatic Bayesian treatment runs parallel to his construal of the frequentist, except that the Bayesian also assumes a prior distribution for parameter μ. It is given its own mean μ_0 and variance τ^2.

BAYESIAN ASSUMPTIONS: Suppose $X_1, X_2, \ldots X_n$ form a random sample from a $N(\mu, 1)$ distribution and the prior distribution of μ is $N(\mu_0, \tau^2)$, with $\tau^2 \gg 1/49$ and $49\tau^2 \gg |\mu_0|$. (ibid.)

This conjugate prior might be invoked as what Berger called a "prior comple-tion strategy" or simply a choice of a prior enabling the data to sufficiently dominate. The results:

$$\text{Posterior mean: } \bar{\mu} = \frac{\tau^2}{1/49 + \tau^2} 10.2 + \frac{1/49}{1/49 + \tau^2} \mu_0,$$

$$\text{Posterior variance: } v = \left(49 + \frac{1}{\tau^2}\right)^{-1}.$$

Given the stipulations that $\tau^2 \gg 1/49$ and $49\tau^2 \gg |\mu_0|$, we get approximately the same interval as the frequentist, only it gets a posterior probability not just a coverage performance of 0.95.

BAYESIAN INTERPRETATION... Under the assumptions above, the probability that μ is in the interval I is 0.95.

PRAGMATIC INTERPRETATION... If the data were a random sample [as described and $\bar{x} = 10.2$], and if the assumptions above were to hold, then the probability that μ is in the interval I would be 0.95. This refers to a hypothetical value $\bar{x}$ of the random variable $\bar{X}$, and because $\bar{X}$ lives in the theoretical world the statement remains theoretical. Nonetheless, we are able to draw useful conclusions from the data as long as our theoretical world is aligned well with the real world that produced the data. (ibid., p. 4)

We get an agreement on numbers and we can cross over the Bayesian-frequentist bridge with aplomb. Assuming, that is, we don't notice the croco-diles in the water waiting to bite us!

Analysis. First off, the frequentist readily agrees with Kass that "long-run frequencies may be regarded as consequences of the Law of Large Numbers rather than as part of the definition of probability or confidence" (p. 2). The probability is methodological: the proportion of correct estimates produced in hypothetical applications is 0.95. The long run need only be long enough to see the pattern emerge and is hypothetical. One can simulate the statistical mechanism associated with the model to produce realizations of the process on a computer, today, not in a long run (Spanos 2013c).

According to Kass (2011), "the commonality between frequentist and Bayesian inferences is the use of theoretical assumptions ..." (p. 4). The prior becomes part of the model in the Bayesian case. For *both* cases, "When we use a statistical model to make a statistical inference we implicitly assert that the variation exhibited by data is captured reasonably well by the statistical model" (p. 2). Everything is left as a subjunctive conditional: if the "theoretical world is aligned well with the real world that produced the data"

(p. 4) then such and such follows. "Perhaps we might even say that most practitioners are subjunctivists" (p. 7).

Would either frequentists or Bayesians be content with this? The frequentist error statistician would not, because there's no detachment of an inference. It remains a subjunctive or 'would be' claim that is entirely deductive, not ampliative. Moreover, she insists on checking the adequacy of the model. I doubt the Bayesian would be satisfied with life as a subjunctivist either. The payoff for the extra complexity in positing a prior is the ability to detach the probability that μ is in (10.2 − 2/7, 10.2+ 2/7) is 0.95. Thus, locating the common frequentist/Bayesian ground in assuming the "theoretical world is aligned well with the real world that produced the data" doesn't get us very far, and it keeps key differences in goals under wraps. We can hear some Bayesian tribe members grumbling at the very assumption that they're modeling "the real world that produced the data" (p. 4).

A Deeper Analysis. We have developed enough muscle from our workouts to peel back some layers here. It only comes near the end of his paper, where Kass tells us what he says to his class (citing Brown and Kass 2009). I'll number his points, the highlight being (3):

(1) I explicitly distinguish the use of probability to describe variation and to express knowledge. . . . [These] are sometimes considered to involve two different kinds of probability . . . 'aleatory probability' and 'epistemic probability' [the latter in the sense of] quantified belief . . . .

(2) Bayesians merge these, applying the laws of probability to go from quantitative description to quantified belief (ibid., p. 5).

(3) But in every form of statistical inference aleatory probability is used, somehow, to make epistemic statements (ibid., pp. 5–6).

What do the frequentist and Bayesian say to (1)–(3)? I can well imagine all tribes mounting resistance. Let's assume for the moment the model–theory match entitles the detachment. Then we can revisit the frequentist and Bayesian inferences.

What does the frequentist do to get her epistemological claim according to Kass? Since the probability the estimator yields true estimates is 0.95 (performance), Kass's frequentist can assign probability 0.95 to the estimate. But this is a fallacy. Worse, it's to be understood as degree of belief (probabilism) rather than confidence (in the performance sense). If a frequentist is not to be robbed of a notion of statistical inference, "epistemic" couldn't be limited to posterior probabilism. An informal notion of probability can work, but that's still to rob probability from playing its intended frequentist role.

What about the pragmatic Bayesian? The pragmatic Bayesian infers 'I'm 95% sure that μ is in (10.2 − 2/7, 10.2 + 2/7).' The model, Kass says, gives the variability (here of μ), and "Bayesians merge these"(p. 5), variability with belief. That's what the theory–real-world match *means* for the Bayesian. One might rightly wonder if the permission to merge isn't tantamount to blessing the split personality we are to avoid.

Suppose you are a student sitting in Professor Kass's class. Probability as representing random variability is quite different from its expression as degree of belief, or uncertainty of knowledge, Professor Kass begins in (1). Nevertheless, Kass will tell you how to juggle them. If you're a pragmatic frequentist, you get your inferential claim by slyly misinterpreting the confidence coefficient as a (degree of belief) probability on the estimate. If you're a pragmatic Bayesian, however, you merge these. Aha, relief. But won't you wonder at the rationale? Kass might say 'in the cases where the numbers match, viewing probability as both variability and belief is unproblematic' for a Bayesian. What about where Bayesians and frequentist numbers don't match? Then "statistical pragmatism is agnostic" (p. 7). In such cases, Kass avers, "procedures should be judged according to their performance under . . . relevant real-world variation" (ibid.). But if the probabilistic assessment doesn't supply performance measures (as in the case of a mismatch), where do they come from?

Bayesians are likely to bristle at the idea of adjudicating disagreement by appeal to frequentist performance; whereas a frequentist like Fraser (2011) argues that it's misleading to even use "probability" if it doesn't have the performance sense of confidence. Christian Robert (commenting on Fraser) avers: "the Bayesian perspective on confidence (or credible) regions and statements does not claim 'correct coverage' from a frequentist viewpoint since it is articulated in terms of the parameters" (Robert 2011, p. 317). He suggests, "the chance identity occurring for location parameters is a coincidence" (ibid.), which raises doubts about a genuine consilience even in the case of frequentist matching. Even where they match, they mean different things.

Frequentist error probabilities relate to the sampling distribution, where we consider hypothetically different outcomes that could have occurred in investigating this one system of interest. The Bayesian allusion to frequentist 'matching' refers to the fixed data and considers frequencies over different systems (that could be investigated by a model like the one at hand). (Cox and Mayo 2010, p. 302)

These may be entirely different hypotheses, even in different fields, in reasoning about this particular *H*. A reply might be that, when there's matching, the frequentist uses error probability$_1$; the Bayesian uses error probability$_2$ (Section 3.6).

The frequentist may welcome adjudication by performance, so she feels she's doing well by Kass. But she still must carve out a kosher epistemological construal, keeping only to frequency. Now we get to the crux of our entire journey, hinted at way back in Excursion 1 (Souvenir D), "we may avoid the need for a different version of probability . . . by assessing the performance of proposed methods under hypothetical repetition" (Reid and Cox 2015, p. 295). But how? The common answer is for the error probability of the method to rub off on a particular inference in the form of confidence. I grant this is an evidential or epistemological use of probability.[1] Yet the severe tester isn't quite happy with it. It's still too performance oriented: We detach the inference, and the performance measure qualifies it by the trustworthiness of the proceeding. Our assessment of well-testedness is different. It's the sampling distribution of the given experiment that informs us of the capabilities of the method to have unearthed erroneous interpretations of data. From here, we reason, counterfactually, to what is well and poorly warranted. What justifies this reasoning? The severity requirements, weak and strong.[2]

We've often enough visited the severity interpretation of confidence intervals (Sections 3.7 and 4.3). Kass's 0.95 interval estimate is (9.92, 10.48). There's a good indication that $\mu > 9.92$ because if μ were 9.92 (or lower), we very frequently would have gotten a smaller value of $\overline{X}$ than we did. There's a poor indication that $\mu \geq 10.2$ because we'd frequently observe values even larger than we did in a world where $\mu < 10.2$, indeed, we'd expect this around 50% of the time. The evidence becomes indicative of $\mu < \mu'$ as we move from 10.2 toward the upper bound 10.48. What would occur in hypothetical repetitions, under various claims about parameters, is not for future assurance in using the rule, but to understand what caused the events on which this inference is based. If a method is incapable of reflecting changes in the data-generating source, its inferences are criticized on grounds of severity. It's possible that no parameter values are ruled out with reasonable severity, and we are left with the entire interval.

What then of Kass's attempt at peace? It may well describe a large tribe of "contemporary attitudes." As Larry Wasserman points out, if you ask people if a 95% CI means 95% of the time the method used gets it right, they will say yes. If you ask if 0.95 describes their belief in a resulting estimate, they may say yes too (2012c). But Kass was looking for respite from conceptual confusion.

---

[1] We know performance is necessary but not sufficient for severity, nor for confidence distributions or fiducial inference, but here we imagine we have got the *relevant* error probability.

[2] There's no need for the philosopher's appeal to things like closest possible worlds to use counter factuals either.

Maybe all Kass is saying is that in the special case of frequentist matching priors, the split personality scarcely registers, and no one will notice. A philosophy limited to frequentist matching results would very much restrict the Bayesian edifice. Notice we've been keeping to the simple examples on which nearly all statistics battles are based. Here matching is at least in principle possible. Fraser (2011, p. 299) shows that when we move away from examples on "location," as with mean μ, the matching goes by the board. We cannot be accused of looking to examples where the frequentist and Bayesian numbers necessarily diverge, no "gotcha" examples from our side. Nevertheless, they still diverge because of differences in meaning and goals.

Optional Stopping and Bayesian Intervals

Disagreement on numbers can also be due to different attitudes toward gambits that alter error probabilities but not likelihoods. Way back in Section 1.5, we illustrated a two-sided Normal test i.e., H_0: $\mu = 0$ vs. H_1: $\mu \neq 0$, $\sigma = 1$, with a rule that keeps sampling until H_0 is rejected. At the 1962 Savage Forum, Armitage needled Savage that the same thing happens with Bayesian methods:

> The departure of the mean by two standard errors corresponds to the ordinary five per cent level. It also corresponds to the null hypothesis being at the five per cent point of the posterior distribution. (Armitage 1962, p. 72)

The identical point can be framed in terms of the corresponding 95% confidence interval method. I follow Berger and Wolpert (1988, pp. 80–1).

Suppose a default/non-subjective Bayesian [they call him an "objective conditionalist"] states that with a fixed sample size n, he would use the interval

$$\mu = \bar{x} \pm 1.96\sigma/\sqrt{n}$$

He "would not interpret confidence in the frequency sense" but instead would "use a posterior Bayesian viewpoint with the non-informative prior density $\pi(\theta) = 1$, which leads to a $N(\bar{x}_n, 1/\sqrt{n})$ posterior" [given the variance is 1] (ibid., p. 80).

Consider the rule: keep sampling until the 95% confidence interval excludes 0.

Berger and Wolpert concede that the Bayesian "being bound to ignore the stopping rule will still use [the above interval] as his confidence interval, but this can *never* contain zero" (ibid., p. 81). The experimenter using this stopping rule "has thus succeeded in getting the [Bayesian] conditionalist to perceive that $\theta \neq 0$, and has done so honestly" (ibid.). This is so despite the Bayesian interval assigning a probability of 0.95 to the interval estimate. "The 'misleading,' however, is solely from a frequentist viewpoint, and will not be of concern to a conditionalist" (ibid.). Why are they unconcerned?

It's hard to pin down their response; they go on to other examples.[3] I take it they are unconcerned because they are not in the business of computing frequentist error probabilities. From their perspective, taking into account the stopping rule is tantamount to taking account of the experimenter's intentions (when to stop). Moreover, from the perspective of what the agent believes, Berger and Wolpert explain, he is not *really* being misled. They further suggest we should trust our intuitions about the Likelihood Principle in simple situations "rather than in extremely complex situations such as" with our stopping rule (ibid., p. 83).

As they surmise, this won't "satisfy a frequentist's violated intuition" (ibid.). It kills her linchpin for critically interpreting the data.[4] ·

It isn't that the stopping rule problem is such a big deal; but it's routinely given as an exemplar *in favor* of ignoring the sampling distribution, and Berger and Wolpert (1988) is the standard to which Bayesian texts refer. Ironically, as Roderick Little observes: "This example is cited as a counterexample by both Bayesians and frequentists! If we statisticians can't agree which theory this example is counter to, what is a clinician to make of this debate?" (Little 2006, p. 215). But by 2006, Berger embraces default priors that are model dependent, "leading to violations of basic principles, such as the likelihood principle and the stopping rule principle" (Berger 2006, p. 394). This is all part of the abandonment of Bayesian foundations. Still, he admits to having "trouble with saying that good frequentist coverage is a necessary requirement" (ibid., p. 463). Conditioning on the data, he says, is more important. "Since the calibrated Bayes inference is Bayesian, there are no penalties for peeking. . ." (Little 2006, p. 220). The disparate attitudes in default/non-subjective Bayesian texts are common – even in the same book.

For example, at the opening of Ghosh et al. (2010) we hear of the "stopping rule paradox in classical inference" (p. 38), happily avoided by the Bayesian. Then later on it's granted: "Inference based on objective [default] priors does violate the stopping rule principle, which is closely related to the likelihood principle" (p. 148). Arguably, any texts touting the stopping rule principle, while using default priors that violate it, should do likewise. Yet this doesn't bring conceptual clarity to their readers. Why are they upending their core principle?[5]

---

[3] They allow the possibility that the knowledge that optional stopping will be used alters their prior for 0. I take it they recognize this is at odds with the presumption that "optional stopping is no sin," and they don't press it. See Section 1.5 where we first took this up.

[4] In observing that "informative stopping rules occur only rarely in practice" (p. 90), Berger and Wolpert make the insightful point that disagreement on this is "due to the misconception that an informative stopping rule is one for which N carries information about θ."

[5] One explanation is in Bernardo's appeal to a decision theory that considers the sampling distribution in computing utilities.

Readers are assured that the violations of the Likelihood Principle are "minor," so the authors' hearts aren't in it. An error statistician wants *major* violations of a principle that denies the relevance of error probabilities. If the violation is merely an unfortunate quirk growing from the desire for a post-data probabilism, but with default priors, it may seem they gave up on those splendid foundations too readily. The non-subjective Bayesian might be seen as left in the worst of both frequentist and Bayesian worlds. They violate dearly held principles and relinquish the simplicity of classical Bayesianism, while coming up short on error statistical performance properties.

6.6 Error Statistical Bayesians: Falsificationist Bayesians

A final view, which can be seen as either more middle of the road or more extreme, is that of Andrew Gelman (writing both on his own and with others). It's the former for an error statistician, the latter for a Bayesian. I focus on what Gelman calls "falsificationist Bayesianism" as developed in Gelman and Shalizi (2013). You could see it as the outgrowth of an error-statistical-Bayesian pow-wow: Shalizi being an error statistician and Gelman a Bayesian. But it's consistent with Gelman's general (Bayesian) research program both before and after, and I don't think anyone can come away from it without recognizing how much statistical foundations are currently in flux. I begin by listing three striking points in their work (Mayo 2013b).

(1) Methodology is ineluctably bound up with philosophy. If nothing else, "strictures derived from philosophy can inhibit research progress" (Gelman and Shalizi 2013, p. 11), as when Bayesians are reluctant to test their models either because they assume they're subjective, or that checking involved non-Bayesian methods (Section 4.9).

(2) Bayesian methods need a new foundation. Although the subjective Bayesian philosophy, "strongly influenced by Savage (1954), is widespread and influential in the philosophy of science (especially in the form of Bayesian confirmation theory. . .)," and while many practitioners see the "rising use of Bayesian methods in applied statistical work" (ibid., p. 9) as supporting this Bayesian philosophy, the authors flatly declare that "most of the standard philosophy of Bayes is wrong" (p. 10, n. 2). While granting that "a statistical method can be useful even if its common philosophical justification is in error" (ibid.), their stance will rightly challenge many a Bayesian. This is especially so in considering their third thesis.

(3) The new foundation uses error statistical ideas. While at first professing that their "perspective is not new," but rather follows many other

statisticians in regarding "the value of Bayesian inference as an approach for obtaining statistical methods with good frequency properties"(p. 10), they admit they are "going beyond the evaluation of Bayesian methods based on their frequency properties – as recommended by Rubin (1984), Wasserman (2006), among others" (p. 21). "Indeed, crucial parts of Bayesian data analysis, such as model checking, can be understood as 'error probes' in Mayo's sense" (p. 10), which might be seen as using modern statistics to implement the Popperian criteria of severe tests.

Testing in Their Data-Analysis Cycle

Testing in their "data-analysis cycle" involves a "non-Bayesian checking of Bayesian models." This is akin to Box's eclecticism, except that their statistical analysis is used "not for computing the posterior probability that any particular model was true – we never actually did that" (p. 13), but rather "to fit rich enough models" and upon discerning that aspects of the model "did not fit our data", to build a more complex and better fitting model, which in turn called for alteration when faced with new data. They look to "pure significance testing" (p. 20) with just a null hypothesis, where no specific alternative models are considered. In testing for misspecifications, we saw, the null hypothesis asserts that a given model is adequate, and a relevant test statistic is sought whose distribution may be computed, at least under the null hypothesis (Section 4.9).

They describe their P-values as "generalizations of classical p-values, replacing point estimates of parameters θ with averages over the posterior distribution..." (p. 18). If a pivotal characteristic is available, their approach matches the usual significance test. The difference is that where the error statistician estimates the "nuisance" parameters, they supply them with priors. It may require a whole hierarchy of priors so that, once integrated out, you are left assigning probabilities to a distance measure $d(X)$. This allows complex modeling, and the distance measures aren't limited to those independent of unknowns, but if it seems hard to picture the distribution of μ, try to picture the distribution of a whole iteration of parameters. These will typically be default priors, with their various methods and meanings as we've seen. The approach is a variant on the "Bayesian P-value" research program, developed by Gelman, Meng, and Stern (1996) (Section 4.8). The role of the sampling distribution is played by what they call the *posterior predictive distribution*. The usual idea of the sampling distribution now refers to the probability of $d(X) > d(x)$ in future replications, averaging over all the priors. These are generally computed by simulation. Their P-value is a kind of error probability. Like the severe tester, I take it the concern is well-testedness, not ensuring good long-run performance decisions about misspecification. So, to put it in severe testing terms, if the model, which now includes a series of priors, was highly incapable of generating results so extreme, they infer a statistical indication of inconsistency. Some claim that, at least for large sample sizes, Gelman's approach leads essentially to "rediscovering" frequentist P-values (Ghosh et al. 2010, pp. 181–2; Bayarri and Berger 2004), which may well be the reason we so often agree. Moreover, as Gelman and Shalizi (2013, p. 18, n. 11) observe, all participants in the Bayesian P-value program implicitly "*disagree* with the standard inductive view" of Bayesianism – at least insofar as they are engaged in model checking (ibid.).

Non-significant results: They compute whether "the observed data set is the kind of thing that the fitted model produces with reasonably high probability" assuming the replicated data are of the same size and shape as $\mathbf{y}_0$, "generated under the assumption that the fitted model, prior and likelihood both, is true" (2013, pp. 17–18). If the Bayesian P-value is reasonably high, then the data are "unsurprising if the model is true" (ibid., p. 18). However, as the authors themselves note, "whether this is evidence *for* the usefulness of the model depends how likely it is to get such a high p-value when the model is false: the 'severity' of the test" (ibid.) associated with H: the adequacy of the combined model with prior. I'm not sure if they entertain alternatives needed for this purpose.

Significant results: A small P-value, on the other hand, is taken as evidence of incompatibility between model and data. The question that arises here is: what kind of incompatibility are we allowed to say this is evidence of? The particular choice of test statistic goes hand in hand with a type of discrepancy from the null. But this does not exhaust the possibilities. It would be fallacious to take a small P-value as directly giving evidence for a specific alternative that "explains" the effect – at least not without further work to pass the alternative with severity. (See Cox's taxonomy, Section 3.3.) Else there's a danger of a fallacy of rejection we're keen to avoid.

What lets the ordinary P-value work in M-S tests is that the null, a mere implicationary assumption, allows a low P-value to point the finger at the null – as a hypothesis about what generated this data. Can't they still see their null hypothesis as implicationary? Yes, but indicating a misfit doesn't pinpoint what's warranted to blame. The problem traces back to the split personality involved in interpreting priors. In most Bayesian model testing, it seems the prior is kept sufficiently vague (using default priors), so that the main work is finding flaws in the likelihood part of the model. They claim their methods provide ways for priors to be tested. To check if something is satisfying its role, we had better be clear on what its intended role is. Gelman and Shalizi tell us what a prior need not be: It need not be a default prior (p. 19), nor need it represent a statistician's beliefs. They suggest the model combines the prior and the likelihood "each of which represents some compromise among scientific knowledge, mathematical convenience, and computational tractability" (p. 20). It may be "a regularization device," (p. 19) to smooth the likelihood, making fitted models less sensitive to details of the data. So if the prior fails the goodness-of-fit test, it could mean it represented false beliefs, or that it was not so convenient after all, or . . .? Duhemian problems loom large; there are all kinds of things one might consider changing to make it all fit.

There is no difficulty with the prior serving different functions, so long as its particular role is pinned down for the given case at hand.[6] Since Gelman regards the test as error statistical, it might work to use the problem solving variation on severe testing (Souvenir U), the problem, I surmise, being one of prediction. For prediction, it might be that the difference between M-S testing and Gelman's tests is more of a technical issue (about the best way to deal with nuisance parameters), and less a matter of foundations, on which we often seem to agree. I don't want to downplay differences and others are better equipped to locate them.[7]

What's most provocative and welcome is that by moving away from probabilisms (posteriors and Bayes factors) and inviting error statistical falsification, we alter the conception of which uses of probability are direct, and which indirect.[8] In Gelman and Hennig (2017), "falsificationist Bayesianism" is described as:

a philosophy that openly deviates from both objectivist and subjectivist Bayesianism, integrating Bayesian methodology with an interpretation of probability that can be seen as frequentist in a wide sense and with an error statistical approach to testing assumptions . . . (p. 991)

In their view, falsification requires something other than probabilism of any type. "Plausibility and belief models can be modified by data in ways that are specified *a priori*, but they cannot be falsified by data" (p. 991). Actually any Bayesian (or even Likelihoodist) account can become falsificationist, indirectly, by adding a falsification rule – provided it has satisfactory error probabilities. But Gelman and Hennig are right that subjective and default

---

[6] The error statistical account would suggest first checking the likelihood portion of the model, after which they could turn to the prior.

[7] Note, for example, that for a given parameter θ, one has presumably only selected a single θ, not the n samples of our usual M-S test. It's not clear why we should expect it to produce typical outcomes. I owe this point to Christian Hennig.

[8] A co-developer of posterior predictive checks, Xiao-Li Meng, is a leader of the "Bayes–Fiducial–Frequentist" movement.

Bayesianism, in current formulations, do not falsify, although they can undergo prior redos or shifts. The Bayesian probabilist regards error probabilities as indirect because they seek a posterior; for the Bayesian falsificationist, like the severe tester, the shoe is on the other foot.

Souvenir Z: Understanding Tribal Warfare

We began this tour asking: Is there an overarching philosophy that "matches contemporary attitudes"? More important is changing attitudes. Not to encourage a switch of tribes, or even a tribal truce, but something more modest and actually achievable: to understand and get beyond the tribal warfare. To understand them, at minimum, requires grasping how the goals of probabilism differ from those of probativeness. This leads to a way of changing contemporary attitudes that is bolder and more challenging. Snapshots from the error statistical lens let you see how frequentist methods supply tools for controlling and assessing how well or poorly warranted claims are. All of the links, from data generation to modeling, to statistical inference and from there to substantive research claims, fall into place within this statistical philosophy. If this is close to being a useful way to interpret a cluster of methods, then the change in contemporary attitudes is radical: it has never been explicitly unveiled. Our journey was restricted to simple examples because those are the ones fought over in decades of statistical battles. Much more work is needed. Those grappling with applied problems are best suited to develop these ideas, and see where they may lead. I never promised, when you bought your ticket for this passage, to go beyond showing that viewing statistics as severe testing will let you get beyond the statistics wars.

6.7   Farewell Keepsake

Despite the eclecticism of statistical practice, conflicting views about the roles of probability and the nature of statistical inference – holdovers from long-standing frequentist–Bayesian battles – still simmer below the surface of today's debates. Reluctance to reopen wounds from old battles has allowed them to fester. To assume all we need is an agreement on numbers – even if they're measuring different things – leads to statistical schizophrenia. Rival conceptions of the nature of statistical inference show up unannounced in the problems of scientific integrity, irreproducibility, and questionable research practices, and in proposed methodological reforms. If you don't understand the assumptions behind proposed reforms, their ramifications for statistical practice remain hidden from you.

Rival standards reflect a tension between using probability (a) to constrain the probability that a method avoids erroneously interpreting data in a series of applications (*performance*), and (b) to assign degrees of support, confirmation, or plausibility to hypotheses (*probabilism*). We set sail on our journey with an informal tool for telling what's true about statistical inference: If little if anything has been done to rule out flaws in taking data as evidence for a claim, then that claim has not passed a *severe test*. From this minimal severe-testing requirement, we develop a statistical philosophy that goes beyond probabilism and performance. The goals of the severe tester (*probativism*) arise in contexts sufficiently different from those of probabilism that you are free to hold both, for distinct aims (Section 1.2). For statistical inference in science, it is severity we seek. A claim passes with severity only to the extent that it is subjected to, and passes, a test that it probably would have failed, if false. Viewing statistical inference as severe testing alters long-held conceptions of what's required for an adequate account of statistical inference in science. In this view, a *normative statistical epistemology* – an account of what's warranted to infer – must be:

- directly altered by biasing selection effects
- able to falsify claims statistically
- able to test statistical model assumptions
- able to block inferences that violate minimal severity

These overlapping and interrelated requirements are disinterred over the course of our travels. This final keepsake collects a cluster of familiar criticisms of error statistical methods. They are not intended to replace the detailed arguments, pro and con, within; here we cut to the chase, generally keeping to the language of critics. Given our conception of evidence, we retain testing language even when the statistical inference is an estimation, prediction, or proposed answer to a question. The concept of severe testing is sufficiently general to apply to any of the methods now in use. It follows that a variety of statistical methods can serve to advance the severity goal, and that they can, in principle, find their foundations in an error statistical philosophy. However, each requires supplements and reformulations to be relevant to real-world learning. Good science does not turn on adopting any formal tool, and yet the statistics wars often focus on whether to use one type of test (or estimation, or model selection) or another. Meta-researchers charged with instigating reforms do not agree, but the foundational basis for the disagreement is left unattended. It is no wonder some see the statistics wars as proxy wars between competing tribe leaders, each keen to advance one or another tool, rather than about how to do better science. Leading minds are drawn into inconsequential battles, e.g., whether to use a pre-specified cut-off of 0.025 or 0.0025 – when in fact good inference is not about cut-offs altogether but about a series of small-scale steps in collecting, modeling and analyzing data that work together to find things out. Still, we need to get beyond the statistics wars in their present form. By viewing a contentious battle in terms of a difference in goals – finding highly probable versus highly well probed hypotheses – readers can see why leaders of rival tribes often talk past each other. To be clear, the standpoints underlying the following criticisms are open to debate; we're far from claiming to do away with them. What should be done away with is rehearsing the same criticisms ad nauseum. Only then can we hear the voices of those calling for an honest standpoint about responsible science.

1. NHST Licenses Abuses. First, there's the cluster of criticisms directed at an abusive NHST animal: NHSTs infer from a single P-value below an arbitrary cut-off to evidence for a research claim, and they encourage P-hacking, fishing, and other selection effects. The reply: this ignores crucial requirements set by Fisher and other founders: isolated significant results are poor evidence of a genuine effect and statistical significance doesn't warrant substantive, (e.g., causal) inferences. Moreover, selective reporting invalidates error probabilities. Some argue significance tests are un-Popperian because the higher the sample size, the easier to infer one's research hypothesis. It's true that with a sufficiently high sample size any discrepancy from a null hypothesis has a high probability of being detected, but statistical significance does not license inferring a research claim H. Unless H's errors have been well probed by merely finding a small P-value, H passes an extremely insevere test. No mountains out of molehills (Sections 4.3 and 5.1). Enlightened users of statistical tests have rejected the cookbook, dichotomous NHST, long lampooned: such criticisms are behind the times. When well-intentioned aims of replication research are linked to these retreads, it only hurts the cause. One doesn't need a sharp dichotomy to identify rather lousy tests – a main goal for a severe tester. Granted, policy-making contexts may require cut-offs, as do behavioristic setups. But in those contexts, a test's error probabilities measure overall error control, and are not generally used to assess well-testedness. Even there, users need not fall into the NHST traps (Section 2.5). While attention to banning terms is the least productive aspect of the statistics wars, since NHST is not used by Fisher or N-P, let's give the caricature its due and drop the NHST acronym; "statistical tests" or "error statistical tests" will do. Simple significance tests are a small part of a conglomeration of error statistical methods.

2. Against Error Probabilities: Inference Should Obey the LP. A common criticism is that error statistical methods use error probabilities post-data. Facets of the same argument take the form of criticizing methods that take account of outcomes other than the one observed, the sampling distribution, the sample space, and researcher "intentions" in optional stopping. It will also be charged that they violate the Likelihood Principle (LP), and are incoherent

(Sections 1.5, 4.6, and 6.6). From the perspective of a logic of induction, considering what other outputs might have resulted seems irrelevant. If there's anything we learn from the consequences of biasing selection effects it is that such logics come up short: data do not speak for themselves. To regard the sampling distribution irrelevant is to render error probabilities irrelevant, and error probability control is necessary (though not sufficient) for severe testing. The problem with cherry picking, hunting for significance, and a host of biasing selection effects – the main source of handwringing behind the statistics crisis in science – is they wreak havoc with a method's error probabilities. It becomes easy to arrive at findings that have not been severely tested. Ask yourself: what bothers you when cherry pickers selectively report favorable findings, and then claim to have good evidence of an effect? You're not concerned that making a habit out of this would yield poor long-run performance. What bothers you, and rightly so, is they haven't done a good job in ruling out spurious findings in the case at hand. The severity requirement explains this evidential standpoint. You can't count on being rescued by the implausibility of cherry-picked claims. It's essential to be able to say that, a claim is plausible but horribly tested by these data.

There is a tension between popular calls for preregistration – arguably, one of the most promising ways to boost replication – and accounts that downplay error probabilities (Souvenir G, Section 4.6). The critical reader of a registered report, post-data, looks at the probability that one or another hypothesis, stopping point, choice of grouping variables, and so on, could have led to a false positive–in effect, she looks at the sampling distribution even without a formal error probability computation. We obtain a rationale never made clear by users of significance tests or confidence intervals as to the relevance of error probabilities in the case at hand. Ironically, those who promote methodologies that reject error probabilities are forced to beat around the bush rather than directly upbraid researchers for committing QRPs that damage error probabilities. They give the guilty party a life raft (Section 4.6). If you're in the market for a method that directly registers flexibilities, p-hacking, outcome-switching and all the rest, then you want one that picks up on a method's error probing capacities.

Granted the rejection of error probabilities is often tied to presupposing they only serve for behavioristic or performance goals. The severe tester breaks out of the behavioristic prison from which this charge arises. Error probabilities are used to assess and control how severely tested claims are.

3. **Fisher and N-P Form an Inconsistent Hybrid.** We debunk a related charge: that Fisherian and N-P methods are an incompatible hybrid and

should be kept segregated. They offer distinct tools under the overarching umbrella of error statistics, along with other methods that employ a method's sampling distribution for inference (confidence intervals, N-P and Fisherian tests, resampling, randomization). While in some quarters the incompatibilist charge is viewed as merely calling attention to the very real, and generally pathological, in-fighting between Fisher and Neyman, it's not innocuous (Section 5.8). Incompatibilist or segregationist positions keep people adhering to caricatures of both approaches where Fisherians can't use power, and N-P testers can't use P-values. Some charge that Fisherian P-values are not error probabilities because Fisher wanted an evidential, not a performance, interpretation. In fact, N-P and Fisher used P-values in both ways. Reporting the actual P-value is recommended by N-P, Lehmann, and others (Section 3.5). It is a post-data error probability. To paraphrase Cox and Hinkley (1974, p. 66), it's the probability we'd mistakenly report evidence against H_0 were we to regard the data as just decisive for issuing such a report. The charge that N-P tests preclude distinguishing highly significant from just significant results is challenged on historical, statistical, and philosophical grounds. Most importantly, even if their founders were die-hard behaviorists it doesn't stop us from giving them an inferential construal (Section 3.3). Personality labels should be dropped. It's time we took responsibility for interpreting tests. It's the methods, stupid.

The most consequential variant under the banner of "P-values aren't error probabilities" goes further, and redefines error probabilities to refer to one or another variant on a posterior probability of hypotheses. It is no wonder the disputants so often talk past each other. Once we pull back the curtain on this equivocal use of "error probability," with the help of subscripts, it is apparent that all these arguments must be revisited (Sections 3.5 and 3.6). Even where it may be argued the critics haven't left the frequentist station, the train takes us to probabilities on hypotheses – requiring priors.

4. P-values Overstate Evidence Against the Null Hypothesis. This is a very common charge (Sections 4.5 and 4.6). What's often meant is that the P-value can be smaller than a posterior probability on a point null hypothesis H_0, based on a lump prior (often 0.5) on H_0. Why take the context that leads to the criticism – one where a point null value has a high prior probability – as typical? It has been questioned by Bayesians and frequentists alike (some even say all such nulls are false). Moreover, P-values can also agree with the posterior: in short, there is wide latitude for Bayesian assignments. Other variations on the criticism judge P-values on the standard of a comparative probabilism (Bayes factors, likelihood ratios). We do not discount any of these criticisms simply because they hinge on taking probabilist measures as an appropriate standard for judging an error probability. The critics think their standard is relevant, so we go with them as far as possible, until inseverity hits. It does hit. A small P-value appears to exaggerate the evidence from the standpoint of probabilism, while from that of performance or severity, it can be the other way around (Sections 4.4, 4.5, 5.2, and 5.6). Without unearthing the presuppositions of rival tribes, users may operate with inconsistent recommendations. Finally, that the charge it is too easy to obtain small P-values is belied by how difficult it is to replicate low P-values – particularly with preregistration (replication paradox): the problem isn't P-values but selective reporting and other abuses (Section 4.6).

5. Inference Should Be Comparative. Statistical significance tests, it may be charged, are not real accounts of evidence because they are not comparative. A comparative assessment takes the form of hypothesis H_1 is comparatively better supported, believed (or otherwise favored) than H_0. Comparativist accounts do not say there's evidence against one hypothesis, nor for the other: neither may be warranted by the data. Nor do they statistically falsify a model or claim as required by a normative epistemology. So why be a comparativist? Comparativism is an appealing way to avoid the dreaded catchall factor required by a posterior probabilism: all hypotheses that could explain the data (Section 6.4).

What of the problems that are thought to confront the non-comparativist who is not a probabilist but a tester? (The criticism is usually limited to Fisherian tests, but N-P tests aren't comparative in the sense being used here.) The problems fall under fallacies of rejection (Section 2.5). Notably, it is assumed a Fisherian test permits an inference from reject H_0 to an alternative H_1 further from the data than is H_0. Such an inference is barred as having low severity (Section 4.3). We do not deny the informativeness of a comparativist measure within an overall severe testing rationale.[9] We agree with Fisher in denying there's just one way to use probability in statistical inquiry (he used likelihoods, P-values, and fiducial intervals). Our point is that the criticism of significance tests for not being comparativist is based on a straw man. Actually, it's their ability to test accordance with a single model or hypothesis that makes simple significance tests so valuable for testing assumptions, leading to (6).

6. Accounts Should Test Model Assumptions. Statistical tests are sometimes criticized as assuming the correctness of their statistical models. In fact, the battery of diagnostic and M-S tests are error statistical. When it comes to testing model assumptions – an important part of auditing – it's to significance

---

[9] We would need predesignation of hypotheses (and/or other restrictions) if there is to be error control.

tests, or the analogous graphical analysis, to which people turn (Sections 4.9, 4.11, and 6.7). Sampling distributions are the key. Also under the heading of auditing are: checking for violated assumptions in linking statistical to substantive inferences, and illicit error probabilities due to biasing selection effects. Reports about what has been poorly audited, far from admissions of weakness, should become the most interesting parts of research reports, at least if done in the severity spirit. They are what afford building a cumulative repertoire of errors, pointing to rival theories to probe. Even domains that lack full-blown theories have theories of mistakes and fallibilities. These suffice to falsify inquiries or even entire measurement procedures, long assumed valid.

7. Inference Should Report Effect Sizes. Pre-data error probabilities and P-values do not report effect sizes or discrepancies – their major weakness. We avoid this criticism by interpreting statistically significant results, or "reject H_0," in terms of indications of a discrepancy γ from H_0. In test T+: [Normal testing: $H_0: \mu \leq \mu_0$ vs. $H_1: \mu > \mu_0$], reject H_0 licenses inferences of the form: $\mu > [\mu_0 + \gamma]$; non-reject H_0, to inferences of the form: $\mu \leq [\mu_0 + \gamma]$. A report of discrepancies poorly warranted is also given (Section 3.1). The severity assessment takes account of the particular outcome x_0 (Souvenir W). In some cases, a qualitative assessment suffices, for instance, that there's no real effect.

The desire for an effect size interpretation is behind a family feud among frequentists, urging that tests be replaced by confidence intervals (CIs). In fact there's a duality between CIs and tests: the parameter values within the $(1 - \alpha)$ CI are those that are not rejectable by the corresponding test at level α (Section 3.7). Severity seamlessly connects tests and CIs. A core idea is arguing from the capabilities of methods to what may be inferred, much as we argue from the capabilities of a key to open a door to the shape of the key's teeth.[10] In statistical contexts, a method's capabilities are represented by its probabilities of avoiding erroneous interpretations of data (Section 2.7).

The "CIs only" battlers have encouraged the use of CIs as supplements to tests, which is good; but there have been casualties. They often promulgate the perception that the only alternative to standard CIs is the abusive NHST animal, with cookbook, binary thinking. The most vociferous among critics in group (1) may well be waging a proxy war for replacing tests with CIs. Viewing statistical inference as severe testing leads to improvements that the CI advocate should welcome (Sections 3.7, 4.3, and 5.5): (a) instead of a fixed confidence level, usually 95%, several levels are needed, as with confidence distributions CDs. (b) We move away from the dichotomy of parameter

---

[10] I allude to a pin and tumbler lock.

values being inside or outside a CI estimate; points within a CI correspond to distinct claims, and get different severity assignments. (c) CIs receive an inferential rather than a performance "coverage probability" justification. (d) Fallacies and chestnuts of confidence intervals (vacuous intervals) are avoided.

8. Inference Should Provide Posterior Probabilities, final degrees of support, belief, probability. Wars often revolve around assuming what is wanted is a posterior probabilism of some sort. Informally, we might say we want probable claims. But when we examine each of the ways this could be attained with formal probablity, the desirability for science evanesces. The onus is on those who hold that what we want are probabilities on hypotheses to show, for existing ways of obtaining them, why.[11] (Note that it's not provided by comparative accounts, Bayes factors, likelihood ratios or model selections.) The most prevalent Bayesian accounts are default/non-subjective, but there is no agreement on suitable priors. The priors are mere formal devices for obtaining a posterior so that the data dominate in some sense. The main assets of the Bayesian picture – a coherent way to represent and update beliefs – go by the board.

Error statistical methods, deemed indirect for probabilism, are direct for severe probing and falsification. Severe testers do not view scientists as seeking highly probable hypotheses, but learning which interpretations of data are well and poorly tested. Of course we want well-warranted claims, but arriving at them does not presuppose a single probability pie with its requirements of exhaustiveness: science must be open ended. We want methods that efficiently find falsity, not ones that are based on updating values for parameters in an existing model. We want to infer local variants of theories piecemeal, falsify others, and be free to launch a probe of any hypothesis which we can subject to severe testing. If other ways to falsify satisfy error statistical requirements, then they are happily in sync with us.

9. Severe Testing Is Not All You Do in Inquiry. Agreed. I have used a neutral word "warranted" to mean justified, adding "with severity" when appropriate. There's a distinctive twist that goes with severely warranting claims – some prefer to say "beliefs," and you could substitute throughout if you wish. It is this twist that makes it possible to have your probabilist cake, and probativism too – each for distinct contexts. The severe testing assessment is not measuring how strong your belief in H is but how well you can show why H ought to be believed. It is relevant when the aim is to know *why* claims pass (or fail) the tests they do. View the error statistical notions as

---

[11] Some will use a *P*-value as a degree of inconsistency with a null hypothesis.

a picturesque representation of the real life, flesh and blood, capability or incapability to put to rest reasonable skeptical challenges. It's in the spirit of Fisher's requiring you know how to bring about results that would rarely fail to corroborate *H*. It's not merely knowing, but *showing* you're prepared, or would be, to tackle skeptical challenges. I'm using "you" but it could be a group or a machine. It's just not all that you do in inquiry. I admitted at the outset that we do not always want to find things out. If your goal is belief probabilism, or you're in a context where the aim is to assign direct probabilities to events (a deductive task), then you are better off recognizing the differences than trying to unify or reconcile. Let me be clear, severe testing isn't reserved for cases of strong evidence; it is operative at every stage of inquiry, but even more so in early stages – where skepticism is greatest. The severity demand is what we naturally want as consumers of statistics, namely, grounds that reports would very probably have revealed flaws of relevance when they're present. To pass tests with severity gives strong evidence, yes, but most of the time it's to learn that much less than was thought or hoped has passed. Showing (with severity!) that a study was poorly run is important in its own right, even if done semi-formally. Better still is to pinpoint a flaw that's been overlooked.

Our journey has taken you far beyond the hackneyed statistical battles that make up much of today's statistics wars. I've chosen to focus on some of them in your final "keepsake" because, if you have to refight them, you can begin from the places we've reached. These criticisms can no longer be blithely put forward as having weight without wrestling with the underlying presuppositions and challenges about evidence and inference. You might say that the criticisms have force against garden-variety treatments of error statistical methods, that I've changed things by adding an explicit severe testing philosophy. I'll happily concede this, but that is the whole reason for taking this journey. You needn't accept this statistical philosophy to use it to peel back the layers of the statistics wars; you will then be beyond them. It's time.

Live (Final) Exhibit. What Does the Severe Tester Say About Positions 1–9? What do you say?

Souvenirs

References

Achinstein, P. (2000). 'Why Philosophical Theories of Evidence Are (And Ought to Be) Ignored by Scientists', in Howard, D. (ed.), *Proceedings of the 1998 Biennial Meetings of the Philosophy of Science Association, Philosophy of Science* 67, S180–92.

Achinstein, P. (2001). *The Book of Evidence*. Oxford: Oxford University Press.

Achinstein, P. (2010). 'Mill's Sins or Mayo's Errors?' In Mayo, D. and Spanos, A. (eds.), pp. 170–88.

Akaike, H. (1973). 'Information Theory and an Extension of the Maximum Likelihood Principle', in Petrov B. and Csaki, F. (eds.), *2nd International Symposium on Information Theory*. Akademia Kiado, Budapest, pp. 267–81.

American Statistical Association (2017). *Recommendations to Funding Agencies for Supporting Reproducible Research*. amstat.org/asa/News/ASA-Develops-Reproducible-Research-Recommendations.aspx.

Armitage, P. (1961). 'Contribution to the Discussion in Smith, C.A.B., "Consistency in Statistical Inference and Decision"', *Journal of the Royal Statistical Society: Series B (Methodological)* 23, 1–37.

Armitage, P. (1962). 'Contribution to Discussion', in Savage, L. J. (ed.), pp. 62–103.

Armitage, P. (1975). *Sequential Medical Trials*, 2nd edn. New York: Wiley.

Armitage, P. (2000). 'Comments on the Paper by Lindley', *Journal of the Royal Statistical Society: Series D* 49(3), 319–20.

ATLAS Collaboration (2012a). 'Latest Results from ATLAS Higgs Search', Press statement, *ATLAS Updates*, July 4, 2012. http://atlas.cern/updates/press-statement/latest-results-atlas-higgs-search.

ATLAS Collaboration (2012b). 'Observations of a New Particle in the Search for the Standard Model Higgs Boson with the Atlas Detector at the LHC', *Physics Letters B* 716(2012), 1–29.

ATLAS Collaboration (2012c). 'Updated ATLAS Results on the Signal Strength of the Higgs-like Boson for Decays into WW and Heavy Fermion Final States', ATLAS-CONF-2012-162. *ATLAS Note*, November 14, 2012. http://cds.cern.ch/record/1494183/files/ATLAS-CONF-2012-162.pdf.

Bacchus, F., Kyburg H., and Thalos, M. (1990). 'Against Conditionalization', *Synthese* 85(3), 475–506.

Baggerly, K. and Coombes, K. (2009). 'Deriving Chemosensitivity from Cell Lines: Forensic Bioinformatics and Reproducible Research in High-throughput Biology', *Annals of Applied Statistics* 3(4), 1309–34.

Bailar, J. (1991). 'Scientific Inferences and Environmental Health Problems', *Chance* 4 (2), 27–38.

Bakan, D. (1970). 'The Test of Significance in Psychological Research', in Morrison, D. and Henkel, R. (eds.), pp. 231–51.

Baker, M. (2016). '1,500 Scientists Lift the Lid on Reproducibility', *Nature* 533, 452–4.

Banerjee, A. and Duflo, E. (2011). *Poor Economics: A Radical Rethinking of the Way to Fight Global Poverty*, 1st edn. New York: PublicAffairs.

Barnard, G. A. (1950). 'On the Fisher-Behrens Test', *Biometrika* 37(3/4), 203–7.

Barnard, G. (1962). 'Contribution to Discussion', in Savage, L. J. (ed.), pp. 62–103.

Barnard, G. (1971). 'Scientific Inferences and Day to Day Decisions', in Godambe, V. and Sprott, D. (eds.), pp. 289–300.

Barnard, G. (1972). 'The Logic of Statistical Inference (Review of "The Logic of Statistical Inference" by Ian Hacking)', *British Journal for the Philosophy of Science* 23(2), 123–32.

Barnard, G. (1985). *A Coherent View of Statistical Inference*, Statistics Technical Report Series. Department of Statistics & Actuarial Science, University of Waterloo, Canada.

Bartlett, M. (1936). 'The Information Available in Small Samples', *Proceedings of the Cambridge Philosophical Society* 32, 560–6.

Bartlett, M. (1939). 'Complete Simultaneous Fiducial Distributions', *Annals of Mathematical Statistics*, 10, 129–38.

Bartlett, T. (2012a). 'Daniel Kahneman Sees "Train-Wreck Looming" for Social Psychology', *Chronicle of Higher Education*, online (10/4/2012).

Bartlett, T. (2012b). 'The Researcher Behind the Ovulation Voting Study Responds', *Chronicle of Higher Education* online (10/28/2012).

Bayarri, M. and Berger, J. (2004). 'The Interplay between Bayesian and Frequentist Analysis', *Statistical Science* 19, 58–80.

Bayarri, M., Benjamin, D., Berger, J., and Sellke, T. (2016). 'Rejection Odds and Rejection Ratios: A Proposal for Statistical Practice in Testing Hypotheses', *Journal of Mathematical Psychology*, 72, 90–103.

Bem, D. (2011). 'Feeling the Future: Experimental Evidence for Anomalous Retroactive Influences on Cognition and Affect', *Journal of Personality and Social Psychology* 100(3), 407–425.

Bem, D., Utts, J., and Johnson, W. (2011). 'Must Psychologists Change the Way They Analyze Their Data?' *Journal of Personality and Social Psychology*, 101(4), 716–719.

Benjamin, D. and Berger, J. (2016). 'Comment: A Simple Alternative to P-values on Wasserstein, R. and Lazar, N. 2016, "The ASA's Statement on p-Values: Context, Process, and Purpose"', *The American Statistician* 70(2) (supplemental materials).

Benjamin, D., Berger, J., Johannesson, M., et al. (2017). 'Redefine Statistical Significance', *Nature Human Behaviour* 2, 6–10.

Benjamini, Y. (2008). 'Comment: Microarrays, Empirical Bayes and the Two-Groups Model', *Statistical Science* 23(1), 23–8.

Benjamini, Y. and Hochberg, Y. (1995). 'Controlling the False Discovery Rate: A Practical and Powerful Approach to Multiple Testing', *Journal of the Royal Statistical Society, B* 57, 289–300.

Berger, J. (2003). 'Could Fisher, Jeffreys and Neyman Have Agreed on Testing?' and 'Rejoinder', *Statistical Science* 18(1), 1–12; 28–32.

Berger, J. (2006). 'The Case for Objective Bayesian Analysis' and 'Rejoinder', *Bayesian Analysis* 1(3), 385–402; 457–64.

Berger, J. (2008). 'A Comparison of Testing Methodologies', in *Proceedings of the PHYSTAT-LHC Workshop on Statistical Issues for LHC Physics*, June 2008, CERN 2008-001, pp. 8–19.

Berger, J. and Bernardo, J. (1992). 'On the Development of Reference Priors', in Bernardo, J., Berger, J., Dawid, A. and Smith A. (eds.), *Bayesian Statistics Volume 4*, Oxford: Oxford University Press, pp. 35–60.

Berger, J. and Sellke, T. (1987). 'Testing a Point Null Hypothesis: The Irreconcilability of P Values and Evidence (with Discussion and Rejoinder)', *Journal of the American Statistical Association* 82(397), 112–22; 135–9.

Berger, J. and Wolpert, R. (1988). *The Likelihood Principle*, 2nd edn., Vol. 6. Lecture Notes-Monograph Series. Hayward, CA: Institute of Mathematical Statistics.

Berger, R. (2014). 'Comment on S. Senn's post: "Blood Simple?" The complicated and controversial world of bioequivalence', Guest Blogpost on Errorstatistics.com (7/31/2014).

Berger, R. and Hsu, J. (1996). 'Bioequivalence Trials, Intersection-union Tests and Equivalence Confidence Sets', *Statistical Science* 11(4), 283–302.

Bernardo, J. (1997). 'Non-informative Priors Do Not Exist: A Discussion', *Journal of Statistical Planning and Inference* 65, 159–89.

Bernardo, J. (2008). 'Comment on Article by Gelman', *Bayesian Analysis* 3(3), 451–4.

Bernardo, J. (2010). 'Integrated Objective Bayesian Estimation and Hypothesis Testing' (with discussion), *Bayesian Statistics* 9, 1–68.

Berry, S. and Kadane, J. (1997). 'Optimal Bayesian Randomization', *Journal of the Royal Statistical Society: Series B* 59(4), 813–19.

Bertrand, J. ([1889]/ 1907). *Calcul des Probabilités*. Paris: Gauthier-Villars.

Birnbaum, A. (1962). 'On the Foundations of Statistical Inference', in Kotz, S. and Johnson, N. (eds.), *Breakthroughs in Statistics*, 1, Springer Series in Statistics, New York: Springer-Verlag, pp. 478–581. (First published with discussion in *Journal of the American Statistical Association* 57(298), 269–326.)

Birnbaum, A. (1969). 'Concepts of Statistical Evidence', in Morgenbesser, S., Suppes, P., and White, M. (eds.), *Philosophy, Science, and Method: Essays in Honor of Ernest Nagel*, New York: St. Martin's Press, pp. 112–43.

Birnbaum, A. (1970). 'Statistical Methods in Scientific Inference' (letter to the Editor), *Nature* 225(5237), 1033.

Birnbaum, A. (1977). 'The Neyman-Pearson Theory as Decision Theory, and as Inference Theory; with a Criticism of the Lindley-Savage Argument for Bayesian Theory', *Synthese* 36(1), 19–49.

Bogen, J. and Woodward, J. (1988). 'Saving the Phenomena', *Philosophical Review* 97 (3), 303–52.

Borel, E. ([1914]/ 1948). *Le Hasard*. Paris: Alcan.

Bowley, A. (1934). 'Discussion and Commentary' pp. 131–3 in Neyman 1934.

Box, G. (1979). 'Robustness in the Strategy of Scientific Model Building', in Launer, R. and Wilkinson, G. (eds.), *Robustness in Statistics*, New York: Academic Press, 201–36.

Box, G. (1983). 'An Apology for Ecumenism in Statistics', in Box, G., Leonard, T., and Wu, D. (eds.), *Scientific Inference, Data Analysis, and Robustness*, New York: Academic Press, 51–84.

Box, G. and Jenkins, G. (1976). *Time Series Analysis: Forecasting and Control*. San Francisco: Holden-Day.

Box, J. (1978). *R. A. Fisher: The Life of a Scientist*. New York: John Wiley.

Breiman, L. (1997). 'No Bayesians in Foxholes', part of 'Banter on Bayes: Debating the Usefulness of Bayesian Approaches to Solving Practical Problems', hosted by Hearst, M., *IEEE Expert: Intelligent Systems and Their Applications* 12(6), 21–4.

Brown, E. N. and Kass, R. E. (2009). 'What is Statistics?' (with discussion), *The American Statistician* 63, 105–23.

Buchen, L. (2009). 'May 29, 1919: A Major Eclipse, Relatively Speaking', *Wired* online (5/29/2009).

Buehler, R. J. and Feddersen, A. P. (1963). 'Note on a Conditional Property of Student's t', *The Annals of Mathematical Statistics* 34(3), 1098–100.

Burgman, M. (2005). *Risk and Decision for Conservation and Environmental Management*. Cambridge: Cambridge University Press.

Burnham, K. and Anderson, D. (2002). *Model Selection and Multimodal Inference: A Practical Information-Theoretic Approach*. New York: Springer-Verlag.

Burnham, K. and Anderson, D. (2014). 'P values Are Only an Index to Evidence: 20th- vs. 21st-century Statistical Science', *Ecology* 95(3), 627–30.

Byrd, J. (2018). 'Models and Methods for Osteometric Sorting', in Adams, B. and Byrd, J. (eds.), *Recovery, Analysis, and Identification of Commingled Human Remains*, Totowa, NJ: Humana Press, pp. 199–220.

Carlin, B. and Louis, T. (2008). *Bayesian Methods for Data Analysis*, 3rd edn. Boca Raton, FL: Chapman and Hall/CRC.

Carnap, R. (1962). *Logical Foundations of Probability*, 2nd edn. Chicago, IL: University of Chicago Press.

Cartlidge, E. (2016). 'Theorizing About the LHC's 750 GeV Bump', Physicsworld.com (4/19/2016).

Cartwright, N. (2012). 'RCTs, Evidence, and Predicting Policy Effectiveness', in Kincaid, H., *The Oxford Handbook of Philosophy of Social Science*, Oxford: Oxford University Press.

Cartwright, N. and Hardie, J. (2012). *Evidence-Based Policy: A Practical Guide to Doing It Better*. Oxford: Oxford University Press.

Casella, G. and Berger, R. (1987a). 'Reconciling Bayesian and Frequentist Evidence in the One-sided Testing Problem', *Journal of the American Statistical Association* 82 (397), 106–11.

Casella, G. and Berger, R. (1987b). 'Comment on Testing Precise Hypotheses by J. O. Berger and M. Delampady', *Statistical Science* 2(3), 344–7.

Casella, G. and Berger, R. (2002). *Statistical Inference*, 2nd edn. Belmont, CA: Duxbury Press.

Castelvecchi, D. and Gibney, E. (2016). 'Hints of New LHC Particle Get Slightly Stronger', *Nature* online (3/18/2016).

Center for Open Science, University of Virginia (2015). *Reproducibility Project*: cos.io/pr/2015-09-29.

Chalmers, A. (2010). 'Can Scientific Theories Be Warranted?', in Mayo, D. and Spanos, A. (eds.), pp. 58–72.

Cherkassky, V. (2012). 'Vladimir Cherkassky Responds on Foundations of Simplicity', Guest Blogpost on Errorstatistics.com (7/6/2012).

CMS Experiment (2012). 'Observation of a New Particle with a Mass of 125 GeV', Press statement, *Compact Muon Solenoid Experiment at CERN* [blog], July 4, 2012. http://cms.web.cern.ch/news/observation-new-particle-mass-125-gev.

The Cochrane Collaboration: cochrane.org.

Cohen, J. (1962). 'The Statistical Power of Abnormal-Social Psychological Research: A Review', *Journal of Abnormal and Social Psychology* 65, 145–53.

Cohen, J. (1988). *Statistical Power Analysis for the Behavioral Sciences*, 2nd edn. Hillsdale, NJ: Erlbaum.

Cohen, J. (1990). 'Things I Have Learned (So Far)', *American Psychologist* 45(12), 1304–12.

Colquhoun, D. (2011). 'In Praise of Randomisation: The Importance of Causality in Medicine and Its Subversion by Philosophers of Science', *Proceedings of the British Academy* 171, 321–40.

Colquhoun, D. (2014). 'An Investigation of the False Discovery Rate and the Misinterpretation of P-values', *Royal Society Open Science* 1(3), 140216 (16 pages).

Committee on Science, Technology, Law Policy and Global Affairs, Committee on the Development of the Third Edition of the Reference Manual on Scientific Evidence, and Federal Judicial Center (2011). *Reference Manual on Scientific Evidence*, 3rd edn. Washington, D.C.: National Academy Press.

COMpare Team (2015). 'Letter to NEJM (Dan Longo, Deputy Editor)' on 11/ 25/2015 (compare-trials.org).

Coombes, K., Wang J., and Baggerly, K. (2007). 'Microarrays: Retracing Steps', *Nature Medicine* 13(11), 1276–7.

Cousins, R. (2017). 'The Jeffreys-Lindley Paradox and Discovery Criteria in High Energy Physics', *Synthese* 194, 395–432.

Cox, D. (1958). 'Some Problems Connected with Statistical Inference', *Annals of Mathematical Statistics* 29(2), 357–72.

Cox, D. (1977). 'The Role of Significance Tests' (with Discussion), *Scandinavian Journal of Statistics* 4, 49–70.

Cox, D. (1978). 'Foundations of Statistical Inference: The Case for Eclecticism', *Australian Journal of Statistics* 20(1), 43–59.

Cox, D. (1982). 'Statistical Significance Tests', *British Journal of Clinical Pharmacology* 14, 325–331.

Cox, D. (1995). 'Comment on "Model Uncertainty, Data Mining and Statistical Inference by C. Chatfield"', *Journal of the Royal Statistical Society: Series A* 158, 419–66.

Cox, D. (2000). 'Comments on the Paper by Lindley', *Journal of the Royal Statistical Society: Series D* 49(3), 321–4.

Cox, D. (2006a). *Principles of Statistical Inference*. Cambridge: Cambridge University Press.

Cox, D. (2006b). 'Frequentist and Bayesian Statistics: A Critique' (keynote address), in *Statistical Problems in Particle Physics, Astrophysics and Cosmology*, Lyons, L. and Müge Karagöz Ünel (eds.), London: Imperial College Press, pp. 3–6.

Cox, D. and Hinkley, D. (1974). *Theoretical Statistics*. London: Chapman and Hall.

Cox, D. and Mayo, D. (2010). 'Objectivity and Conditionality in Frequentist Inference', in Mayo, D and Spanos, A. (eds.), pp. 276–304.

Cox, D. and Mayo, D. (2011). 'A Statistical Scientist Meets a Philosopher of Science: A Conversation between Sir David Cox and Deborah Mayo', in *Rationality, Markets and Morals (RMM)* 2, 103–14.

Crupi, V. and Tentori, K. (2010). 'Irrelevant Conjunction: Statement and Solution of a New Paradox', *Philosophy of Science* 77, 1–13.

Cumming, G. (2012). *Understanding the New Statistics: Effect Sizes, Confidence Intervals, and Meta-analysis.* New York: Routledge.

Dawid, A. (1976). 'Properties of Diagnostic Data Distributions', *Biometrics* 32(3), 647–58.

Dawid, A. (2000). 'Comment on a Paper by Lindley', *Journal of the Royal Statistical Society: Series D* 49(3), 325–6.

Dawid, A. (2015). 'Comment' on 'Can You Change Your Bayesian Prior?', Blogpost on Errorstatistics.com (6/18/2015).

Dawid, A., Stone, M., and Zidek, J. (1973). 'Marginalization Paradoxes in Bayesian and Structural Inference', *Journal of the Royal Statistical Society: Series B* 35(2), 189–233.

Deaton, A. (2010). 'Instruments, Randomization, and Learning about Development', *Journal of Economic Literature* 48(2), 424–55.

de Finetti, B. (1972). *Probability, Induction and Statistics: The Art of Guessing.* New York: John Wiley & Sons.

de Finetti, B. (1974). *Theory of Probability: A Critical Introductory Treatment*, 2 volumes. New York: John Wiley & Sons.

Diaconis, P. (1978). 'Statistical Problems in ESP Research', *Science* 201(4351), 131–6.

Dienes, Z. (2008). *Understanding Psychology as a Science: An Introduction to Scientific and Statistical Inference.* Basingstoke, UK: Palgrave Macmillan.

Dominus, S. (2017). 'When the Revolution Came for Amy Cuddy', *The New York Times Magazine*, New York Times Company (October 18, 2017).

Doudna, J. and Charpentier, E. (2014). 'The New Frontier of Genome Engineering with CRISPR-Cas9', *Science* 346(6213), 1077 (1–9).

Draper, D. and Madigan, D. (1997). 'The Scientific Value of Bayesian Statistical Methods', part of 'Banter on Bayes: Debating the Usefulness of Bayesian Approaches to Solving Practical Problems,' hosted by Hearst, M., *IEEE Intelligent Systems and Their Applications* 12(6), 18–20.

Duhem, P. (1954). *The Aim and Structure of Physical Theory.* New Jersey: Princeton University Press.

Dupré, J. (1993). *The Disorder of Things.* Cambridge, MA: Harvard University Press.

Durante, K., Rae, A., and Griskevicius, V. (2013). 'The Fluctuating Female Vote: Politics, Religion, and the Ovulatory Cycle', *Psychological Science* 24(6), 1007–16.

Dyson, E., Eddington, A., and Davidson, C. (1920). 'A Determination of the Deflection of Light by the Sun's Gravitational Field, from Observations Made at the Total Eclipse of May 29, 1919', *Philosophical Transactions of the Royal Society of London,* A220, 291–333.

Earman, J. (1992). *Bayes or Bust?: A Critical Examination of Bayesian Confirmation Theory.* Cambridge, MA: MIT Press.

Earman, J. and Glymour, C. (1980). 'Relativity and Eclipses: The British Eclipse Expeditions of 1919 and Their Predecessors', *Historical Studies in the Physical Sciences* 11(1), 49–85.

Eddington, A. ([1920]1987). *Space, Time and Gravitation: An Outline of the General Relativity Theory*, Cambridge Science Classics Series. Cambridge: Cambridge University Press.

Edwards, A. F. (1972). *Likelihood*. Cambridge: Cambridge University Press.

Edwards, W., Lindman, H., and Savage, L. (E, L, & S) (1963). 'Bayesian Statistical Inference for Psychological Research', *Psychological Review* 70(3), 193–242.

Efron, B. (1979). 'Bootstrap Methods: Another Look at the Jackknife', *The Annals of Statistics* 7(1), 1–26.

Efron, B. (1986). 'Why Isn't Everyone a Bayesian?', *The American Statistician* 40(1), 1–5.

Efron, B. (1998). 'R. A. Fisher in the 21st Century' and 'Rejoinder', *Statistical Science* 13 (2), 95–114; 121–2.

Efron, B. (2000). 'Comments on the Paper by Lindley', *Journal of the Royal Statistical Society: Series D* 49(3), 330–1.

Efron, B. (2013). 'A 250-Year Argument: Belief, Behavior, and the Bootstrap', *Bulletin of the American Mathematical Society* 50(1), 126–46.

Ellis, P. (2010). *Essential Guide to Effect Sizes: Statistical Power, Meta-Analysis, and the Interpretation of Research Results*. Cambridge: Cambridge University Press.

Feynman, R. (1974). 'Cargo Cult Science,' Caltech Commencement Speech. Reprinted in *Surely You're Joking Mr. Feynman!: Adventures of a Curious Character* (1985), New York: Norton & Company, pp. 382–91.

Fisher, R. A. (1926). 'The Arrangement of Field Experiments', *Journal of Ministry of Agriculture* 33, 503–13. Reprinted in *R. A. Fisher: Contributions to Mathematical Statistics* (1950), Paper 17, New York: John Wiley & Sons.

Fisher, R. A. (1930). 'Inverse Probability', *Mathematical Proceedings of the Cambridge Philosophical Society* 26(4), 528–35.

Fisher, R. A. (1934a). *Statistical Methods for Research Workers*. Edinburgh: Oliver and Boyd. (First published in 1925.) Reprinted in Fisher 1990.

Fisher, R. A. (1934b). 'Two New Properties of Mathematical Likelihood', *Proceedings of the Royal Society of London. Series A* 144 (852), 285–307.

Fisher. R. A. (1934c). 'Discussion and Commentary' pp. 137–8 in Neyman 1934.

Fisher, R. A. (1935a). *The Design of Experiments*, 1st edn., Edinburgh: Oliver and Boyd. Reprinted in Fisher 1990.

Fisher, R. A. (1935b). 'The Logic of Inductive Inference', *Journal of the Royal Statistical Society* 98(1), 39–82.

Fisher, R. A. (1936), 'Uncertain Inference', *Proceedings of the American Academy of Arts and Sciences* 71, 248–58.

Fisher, R. A. (1938). 'Presidential Address', *Sankhyā: The Indian Journal of Statistics* 4 (1), 14–17.

Fisher, R. A. (1939). 'The Comparison of Samples with Possibly Unequal Variances', *Annals of Eugenics* IX (pt II), 174–80. Reprinted in Fisher, R. (1950) ('Author's Note' 35.173a-35.173b); 35.174-35.180.

Fisher, R. A. (1950). *Contributions to Mathematical Statistics*. New York: John Wiley & Sons.

Fisher, R. A. (1955). 'Statistical Methods and Scientific Induction', *Journal of the Royal Statistical Society: Series B* 17(1), 69–78.

Fisher, R. A. (1956). *Statistical Methods and Scientific Inference.* Edinburgh: Oliver and Boyd. Reprinted in Fisher 1990.

Fisher, R. A. (1990). *Statistical Methods, Experimental Design, and Scientific Inference,* (ed.), Bennett, J. H. Oxford: Oxford University Press.

Fitelson, B. (2002). 'Putting the Irrelevance Back into the Problem of Irrelevant Conjunction', *Philosophy of Science* 69(4), 611–22.

Folks, J. (1995). 'A Conversation with Oscar Kempthorne', *Statistical Science* 10(4), 321–36.

Forster, M. and Sober, E. (1994). 'How to Tell When Simpler, More Unified, or Less Ad Hoc Theories Will Provide More Accurate Predictions', *The British Journal for the Philosophy of Science* 45(1), 1–35.

Fraser, D. (1961). 'On Fiducial Inference', *Annals of Mathematical Statistics* 32(3), 661–676.

Fraser, D. (1998). '[R. A. Fisher in the 21st Century]: Comment', *Statistical Science* 13 (2), 118–120.

Fraser, D. (2011). 'Is Bayes Posterior Just Quick and Dirty Confidence?' and 'Rejoinder', *Statistical Science* 26(3), 299–316; 329–31.

Fraser, D. (2014). 'Why Does Statistics Have Two Theories?', in Lin, X., Banks, D., Genest, C., Molenberghs, G., Scott, D. and Wang, J.-L. (eds.), *Past, Present and Future of Statistical Science,* 237–252. CRC Press, Boca Raton, FL.

Fraser, D. and Reid, N. (2002). 'Strong Matching of Frequentist and Bayesian Parametric Inference', *Journal of Statistical Planning and Inference* 103, 263–85.

Freedman, D. (2009). *Statistical Models: Theory and Practice,* 2nd edn. Cambridge: Cambridge University Press.

Freiman, J. A., Chalmers, T. C., Smith, H., and Kuebler, R. R. (1978). 'The Importance of Beta, the Type II Error and Sample Size in the Design and Interpretation of the Randomized Control Trial: Survey of 71 Negative Trials', *The New England Journal of Medicine* 299(13), 690–4.

Gaifman, H. and Vasudevan, A. (2012). 'Deceptive Updating and Minimal Information Methods', *Synthese* 187(1), 147–78.

Gelman, A. (2011). 'Induction and Deduction in Bayesian Data Analysis', *Rationality, Markets and Morals* (*RMM*) 2, 67–78.

Gelman, A. (2012). 'Ethics and the Statistical Use of Prior Information', *Chance Magazine* 25(4), 52–4.

Gelman, A. (2013). 'P Values and Statistical Practice', *Epidemiology* 24(1), 69–72.

Gelman, A. (2015). 'Comment on "Can You Change Your Bayesian Prior (ii)"'. Blogpost on Errorstatistics.com (6/18/2015).

Gelman, A. and Carlin, J. (2014). 'Beyond Power Calculations: Assessing Type S (Sign) and Type M (Magnitude) Errors', *Perspectives on Psychological Science* 9, 641–51.

Gelman, A., Carlin, J., Stern, H., et al. (2004). *Bayesian Data Analysis,* 2nd edn. Boca Raton, FL: Chapman & Hall/CRC.

Gelman, A. and Hennig, C. (2017). 'Beyond Subjective and Objective in Statistics', *Journal of the Royal Statistical Society: Series A* 180(4), 967–1033.

Gelman, A. and Loken, E. (2014). 'The Statistical Crisis in Science', *American Scientist* 2, 460–5.

Gelman, A., Meng, X-L., and Stern, H. (1996). 'Posterior Predictive Assessment of Model Fitness Via Realized Discrepancies' (with Discussion), *Statistica Sinica* 6, 733–807.

Gelman, A. and Robert, C. (2013). '"Not Only Defended but Also Applied": The Perceived Absurdity of Bayesian Inference' and 'Rejoinder', *The American Statistician* 67(1), 1–5; 15–6.

Gelman, A. and Shalizi, C. (2013). 'Philosophy and the Practice of Bayesian Statistics' and 'Rejoinder', *British Journal of Mathematical and Statistical Psychology* 66(1), 8–38; 76–80.

Gibbons, J. D. and Pratt, J. W. (1975). 'P-values: Interpretation and Methodology', *The American Statistician* 29(1), 20–5.

Giere, R. N. (1969). 'Bayesian Statistics and Biased Procedures', *Synthese* 20(3), 371–87.

Giere, R. N. (1988). *Explaining Science*. Chicago: University of Chicago Press.

Gigerenzer, G. (1991). 'How to Make Cognitive Illusions Disappear: Beyond "Heuristics and Biases"', *European Review of Social Psychology* 2, 83–115.

Gigerenzer, G. (2002). *Adaptive Thinking: Rationality in the Real World*. Oxford: Oxford University Press.

Gigerenzer, G. (2004). 'Mindless Statistics', *Journal of Socio-Economics* 33(5), 587–606.

Gigerenzer, G. and Marewski, J. (2015). 'Surrogate Science: The Idol of a Universal Method for Scientific Inference', *Journal of Management* 41(2), 421–40.

Gigerenzer, G., Swijtink, Z., Porter, T., et al. (1989). *Empire of Chance: How Probability Changed Science and Everyday Life*. Cambridge: Cambridge University Press.

Gill, R. (2014). 'Who Ya Gonna Call for Statistical Fraudbusting? R. A. Fisher, P-values, and Error Statistics (Again)', Guest Blogpost on Errorstatistics.com (5/10/2014).

Gillies, D. (2000). *Philosophical Theories of Probability*. New York: Routledge.

Gillies, D. (2001). 'Bayesianism and the Fixity of the Theoretical Framework', in Corfield, D. and Williamson, J. (eds.), *Foundations of Bayesianism* 24, The Netherlands: Springer, pp. 363–79.

Ghosh, J., Delampady, M., and Samanta, T. (2010). *An Introduction to Bayesian Analysis: Theory and Methods*. New York: Springer.

Glymour, C. (1980). *Theory and Evidence*. Princeton: Princeton University Press.

Glymour, C. (2010). 'Explanation and Truth', in Mayo, D. and Spanos, A. (eds.), pp. 331–50.

Godambe, V. and Sprott, D. (eds.) (1971). *Foundations of Statistical Inference*. Toronto: Holt, Rinehart and Winston of Canada.

Goldacre, B. (2008). *Bad Science*. London: HarperCollins Publishers.

Goldacre, B. (2012). *Bad Pharma: How Drug Companies Mislead Doctors and Harm Patients*. London: Fourth Estate.

Goldacre, B. (2016). 'Make Journals Report Clinical Trials Properly', *Nature* 530(7588), 7.

Goldman, A. (1986). *Epistemology and Cognition*. Cambridge, MA: Harvard University Press.

Goldman, A. (1999). *Knowledge in a Social World*. Oxford: Oxford University Press.

Good, I. J. (1971a). 'The Probabilistic Explication of Information, Evidence, Surprise, Causality, Explanation, and Utility' and 'Reply', in Godambe, V. and Sprott, D. (eds.), pp. 108–22, 131–41.

Good, I. J. (1971b). 'Commentary on D. J. Bartholomew' in Godambe, V. and Sprott, D. (eds.), p. 431.

Good, I. J. (1976). 'The Bayesian Influence, or How to Sweep Subjectivism Under the Carpet', in Harper, W. and Hooker, C. (eds.), pp. 25–174.

Good, I. J. (1983). *Good Thinking: The Foundations of Probability and Its Applications.* Minneapolis, MN: University of Minnesota Press.

Goodman, S. (1992). 'A Comment on Replication, P-values and Evidence', *Statistics in Medicine* 11(7), 875–9.

Goodman, S. (1993). 'P-values, Hypothesis Tests, and Likelihood: Implications for Epidemiology of a Neglected Historical Debate', *American Journal of Epidemiology* 137(5), 485–96.

Goodman, S. (1999). 'Toward Evidence-Based Medical Statistics. 2: The Bayes Factor', *Annals of Internal Medicine*, 130(12), 1005–13.

Goodman, S. and Greenland S. (2007). 'Assessing the Unreliability of the Medical Literature: A Response to "Why Most Published Research Findings Are False"', Johns Hopkins University, Department of Biostatistics Working Papers. Working Paper 135, pp. 1–25.

Gopnik, A. (2009). August 11 Interview in *The Edge*, https://edge.org/conversation/amazing-babies.

Gorroochurn, P. (2016). *Classic Topics on the History of Modern Mathematical Statistics: From Laplace to More Recent Times*, Hoboken, NJ: Wiley.

Greenland, S. (2012). 'Nonsignificance Plus High Power Does Not Imply Support for the Null Over the Alternative', *Annals of Epidemiology* 22, 364–8.

Greenland, S. and Poole, C. (2013). 'Living with P Values: Resurrecting a Bayesian Perspective on Frequentist Statistics' and 'Rejoinder: Living with Statistics in Observational Research', *Epidemiology* 24(1), 62–8; 73–8.

Greenland, S., Senn, S., Rothman, K., et al. (2016). 'Statistical Tests, P values, Confidence Intervals, and Power: A Guide to Misinterpretations', *European Journal of Epidemiology* 31(4), 337–50.

Gurney, J., Mueller, B., Davis S., et al. (1996). 'Childhood Brain Tumor Occurrence in Relation to Residential Power Line Configurations, Electric Heating Sources, and Electric Appliance Use', *American Journal of Epidemiology* 143, 120–8.

Hacking, I. (1965). *Logic of Statistical Inference.* Cambridge: Cambridge University Press.

Hacking, I. (1972). 'Review: Likelihood', *British Journal for the Philosophy of Science* 23 (2), 132–7.

Hacking, I. (1980). 'The Theory of Probable Inference: Neyman, Peirce and Braithwaite', in Mellor, D. (ed.), *Science, Belief and Behavior: Essays in Honour of R. B. Braithwaite*, Cambridge: Cambridge University Press, pp. 141–60.

Haidt, J. and Iyer, R. (2016). 'How to Get Beyond Our Tribal Politics', *Wall Street Journal* (11/10/2016).

Haig, B. (2016). 'Tests of Statistical Significance Made Sound', *Educational and Psychological Measurement* 77(3) 489–506.

Hand, D. (2014). *The Improbability Principle: Why Coincidences, Miracles, and Rare Events Happen Every Day*, 1st edn. New York: Scientific American/ Farrar, Straus and Giroux.

Hannig, J. (2009). 'On Generalized Fiducial Inference', *Statistica Sinica* 19, 491–544.

Harlow, H. (1958). 'The Nature of Love', *American Psychologist* 13, pp. 673–85.

Harper, W. and Hooker, C. (eds.) (1976). *Foundations of Probability Theory, Statistical Inference and Statistical Theories of Science*, Volume II. Boston, MA: D. Reidel.

Hawthorne, J. and Fitelson, B. (2004). 'Re-Solving Irrelevant Conjunction with Probabilistic Independence', *Philosophy of Science* 71, 505–14.

Hempel, C. G. (1945). 'Studies in the Logic of Confirmation (I.)', *Mind* 54(213), 1–26.

Hendry, D. (2011). 'Empirical Economic Model Discovery and Theory Evaluation', *Rationality, Markets and Morals (RMM)* 2, 115–45.

Hitchcock, C. and Sober, E. (2004). 'Prediction versus accommodation and the risk of overfitting', *The British Journal for the Philosophy of Science* 55(1), 1–34.

Hoenig, J. and Heisey, D. (2001). 'The Abuse of Power: The Pervasive Fallacy of Power Calculations in Data Analysis', *The American Statistician* 55(1), 1–6.

Howson, C. (1997a). 'A Logic of Induction', *Philosophy of Science* 64(2), 268–90.

Howson, C. (1997b). 'Error Probabilities in Error', *Philosophy of Science* 64, Supplemental Issue PSA 1996: Symposia Papers. Edited by L. Darden (1996). S185–S194.

Howson, C. (2017). 'Putting on the Garber Style? Better Not', *Philosophy of Science* 84 (4), 659–76.

Howson, C. and Urbach, P. (1993). *Scientific Reasoning: The Bayesian Approach*. La Salle, IL: Open Court.

Hubbard, R., and Bayarri, M. J. (2003). 'Confusion Over Measures of Evidence versus Errors' and 'Rejoinder', *The American Statistician* 57(3), 171–8; 181–2.

Huber, P. J. (2011). *Data Analysis: What Can Be Learned from the Past 50 Years?*, New York: Wiley.

Huff, D. (1954). *How to Lie with Statistics*, 1st edn. New York: W. W. Norton & Company.

Hume, D. (1739). *A Treatise of Human Nature*. BiblioBazaar.

Hurlbert, S. and Lombardi, C. (2009). 'Final Collapse of the Neyman-Pearson Decision Theoretic Framework and Rise of the NeoFisherian', *Annales Zoologici Fennici* 46, 311–49.

Ioannidis, J. (2005). 'Why Most Published Research Findings are False', *PLoS Medicine* 2(8), 0696–0701.

Ioannidis, J. (2016). 'The Mass Production of Redundant, Misleading, and Conflicted Systematic Reviews and Meta-analyses', *Milbank Quarterly* 94(3), 485–514.

Irony, T. and Singpurwalla, N. (1997). 'Non-informative Priors Do not Exist: A Dialogue with José M. Bernardo', *Journal of Statistical Planning and Inference* 65(1), 159–77.

Jefferys, W. and Berger, J. (1992). 'Ockham's Razor and Bayesian Analysis', *American Scientist* 80, 64–72.

Jeffreys, H. (1919). 'Contribution to Discussion on the Theory of Relativity', and 'On the Crucial Test of Einstein's Theory of Gravitation', *Monthly Notices of the Royal Astronomical Society* 80, 96–118; 138–54.

Jeffreys, H. ([1939]/ 1961). *Theory of Probability*. Oxford: Oxford University Press.

Jeffreys, H. (1955). 'The Present Position in Probability Theory', *The British Journal for the Philosophy of Science* 5, 275–89.

Johnson, V. (2013a). 'Revised Standards of Statistical Evidence', *Proceedings of the National Academy of Sciences (PNAS)* 110(48), 19313–17.

Johnson, V. (2013b). 'Uniformly Most Powerful Bayesian Tests', *The Annals of Statistics* 41(4), 1716–41.

Kadane, J. (2006). 'Is "Objective Bayesian Analysis" Objective, Bayesian, or Wise? (Comment on Articles by Berger and by Goldstein)', *Bayesian Analysis* 1(3), 433–6.

Kadane, J. (2008). 'Comment on Article by Gelman', *Bayesian Analysis* 3(3), 455–8.

Kadane, J. (2011). *Principles of Uncertainty*. Boca Raton, FL: Chapman and Hall/CRC.

Kadane, J. (2016). 'Beyond Hypothesis Testing', *Entropy* 18(5), article 199, 1–5.

Kahneman, D. (2012). 'A proposal to deal with questions about priming effects' email. Link to letter in Bartlett 2012a.

Kahneman, D. (2014). 'A New Etiquette for Replication', *Social Psychology* 45(4), 299–311.

Kahneman, D., Slovic, P., and Tversky, A. (1982). *Judgment under Uncertainty: Heuristics and Biases*. New York: Cambridge University Press.

Kaku, M. (2005). *Einstein's Cosmos: How Albert Einstein's Vision Transformed Our Understanding of Space and Time (Great Discoveries)*. New York: W. W. Norton & Company.

Kalbfleisch, J. and Sprott, D. (1976). 'On Tests of Significance', in Harper, W. and Hooker, C., pp. 259–72.

Kass, R. (1998). '[R. A. Fisher in the 21st Century]: Comment.' *Statistical Science* 13(2), 115–16.

Kass, R. (2011). 'Statistical Inference: The Big Picture (with discussion and rejoinder)', *Statistical Science* 26(1), 1–20.

Kass, R. and Wasserman, L. (1996). 'The Selection of Prior Distributions by Formal Rules', *Journal of the American Statistical Association* 91, 1343–70.

Kaye, D. and Freedman, D. (2011). 'Reference Guide on Statistics', in *Reference Manual on Scientific Evidence*, 3rd edn. pp. 83–178.

Kempthorne, O. (1976). 'Statistics and the Philosophers', in Harper, W. and Hooker, C. (eds.), 273–314.

Kennefick, D. (2009). 'Testing Relativity from the 1919 Eclipse: A Question of Bias', *Physics Today* 62(3), 37–42.

Kerridge, D. (1963). 'Bounds for the Frequency of Misleading Bayes Inferences', *The Annals of Mathematical Statistics* 34(3), 1109–10.

Keynes, J. (1921). *A Treatise on Probability*. London: MacMillan and Co.

Kheifets, L., Sussman, S., and Preston-Martin, S. (1999). 'Childhood Brain Tumors and Residential Electromagnetic Fields (EMF)', *Reviews of Environmental Contamination and Toxicology* 159, 111–29.

Kish, L. (1970). 'Some Statistical Problems in Research Design', in Morrison, D. and Henkel, R. (eds.), pp. 127–41. (First published 1959, *American Sociological Review* 24(3), 328).

Kruschke, J. K. and Liddell, T. M. (2017). 'The Bayesian New Statistics: Hypothesis Testing, Estimation, Meta-analysis, and Power Analysis from a Bayesian Perspective', *Psychonomic Bulletin & Review*, 1–29.

Kuhn, T. (1970). 'Logic of Discovery or Psychology of Research?', in Lakatos, I. and Musgrave, A. (eds.), pp. 1–23.

Kyburg, H. (1992). 'The Scope of Bayesian Reasoning', *PSA: Proceedings of the Biennial Meeting of the Philosophy of Science Association 1992*, 139–52.

Kyburg, H. (2003). 'Probability as a Guide to Life', in Kyburg, H. E. and Thalos, M. (eds.), pp. 135–52.

Kyburg, H. and Thalos, M. (eds.) (2003). *Probability Is the Very Guide of Life: The Philosophical Uses of Chance*. Chicago, IL: Open Court.

Lad. F. (2006). 'Objective Bayesian Statistics . . . Do You Buy It? Should We Sell It? (Comment on Articles by Berger and by Goldstein)', *Bayesian Analysis* 1(3), 441–4.

Lakatos, I. (1970). 'Falsification and the Methodology of Scientific Research Programmes', in Lakatos, I. and Musgrave, A. (eds.), pp. 91–138.

Lakatos, I. (1978). *The Methodology of Scientific Research Programmes*. Cambridge: Cambridge University Press.

Lakatos, I. and Musgrave, A. (eds.) (1970). *Criticism and the Growth of Knowledge*. Cambridge: Cambridge University Press.

Lakens, D. (2017). 'Equivalence Tests: A Practical Primer for t Tests, Correlations, and Meta-analyses', *Social Psychological & Personality Science* 8(4), 355–62.

Lakens, D., et al. (2018). 'Justify your Alpha', *Nature Human Behavior* 2, 168–71.

Lambert, C. (2010). 'Stop Ignoring Experimental Design (or My Head Will Explode)', Blogpost on GoldenHelix.com (9/29/2010).

Lambert, C. and Black, L. (2012). 'Learning From Our GWAS Mistakes: From Experimental Design to Scientific Method', *Biostatistics* 13(2), 195–203.

Lambert, P., Sutton, A., Burton, P., Abrams, K., and Jones, D. (2005). 'How Vague is Vague? A Simulation Study of the Impact of the Use of Vague Prior Distributions in MCMC Using WinBUGS', *Statistics in Medicine* 24, 2401–28.

Laudan, L. (1978). *Progress and Its Problems*. Berkeley, CA: University of California Press.

Laudan, L. (1983). 'The Demise of the Demarcation Problem', in R. S. Cohen and L. Laudan (eds.), *Physics, Philosophy and Psychoanalysis*. Dordrecht, The Netherlands: D. Reidel, pp. 111–27.

Laudan, L. (1996). *Beyond Positivism and Relativism: Theory, Method, and Evidence*. Boulder, CL: Westview Press.

Laudan, L. (1997). 'How About Bust? Factoring Explanatory Power Back into Theory Evaluation', *Philosophy of Science* 64, 303–16.

Leek, J. (2016). 'Statistical Vitriol', Blogpost on SimplyStatistics.com (09/29/16).

Lehmann, E. (1981). 'An Interpretation of Completeness and Basu's Theorem', *Journal of the American Statistical Association* 76(374), 335–40.

Lehmann, E. (1986). *Testing Statistical Hypotheses*, 2nd edn. New York: Wiley.

Lehmann, E. (1988). 'Jerzy Neyman, 1894–1981', Technical Report No. 155, May 1988.

Lehmann, E. (1990). 'Model Specification: The Views of Fisher and Neyman, and Later Developments', *Statistical Science* 5(2), 160–168.

Lehmann, E. (1993a). 'The Bertrand-Borel Debate and the Origins of the Neyman-Pearson Theory', in Ghosh, J., Mitra, S., Parthasarathy, K. and Prak Ma Rao, L. (eds.), *Statistics and Probability: A Raghu Raj Bahadur Festschrift*, New Delhi: Wiley Eastern, 371–80. Reprinted in Lehmann 2012, pp. 965–74.

Lehmann, E. (1993b). 'The Fisher, Neyman-Pearson Theories of Testing Hypotheses: One Theory or Two?', *Journal of the American Statistical Association* 88 (424), 1242–9.

Lehmann, E. (2011). *Fisher, Neyman, and the Creation of Classical Statistics*, 1st edn. New York: Springer.

Lehmann, E. (2012). *Selected Works of E. L. Lehmann*, Rojo, J. (ed.). New York: Springer.

Lehmann, E. and Romano, J. (2005). *Testing Statistical Hypotheses*, 3rd edn. New York: Springer.

Letzter, R. (2016). 'Scientists Are Furious after a Famous Psychologist Accused Her Peers of "Methodological Terrorism"', *Business Insider* (9/22/2016), businessinsider.com/susan-fiske-methodological-terrorism-2016-9.

Levelt Committee, Noort Committee, Drenth Committee (2012). 'Flawed Science: The Fraudulent Research Practices of Social Psychologist Diederik Stapel', *Stapel Investigation: Joint Tilburg/Groningen/Amsterdam investigation of the publications by Mr. Stapel* (www.commissielevelt.nl/).

Levi, I. (1980). *The Enterprise of Knowledge: An Essay on Knowledge, Credal Probability, and Change*. Cambridge, MA: MIT Press.

Lindemann, F. (1919). 'Contribution to "Discussion on the Theory of Relativity"', *Monthly Notices of the Royal Astronomical Society* 80, 114.

Lindley, D. (1957). 'A Statistical Paradox', *Biometrika* 44, 187–92.

Lindley, D. (1969). 'Discussion of Compound Decisions and Empirical Bayes, J. B. Copas', *Journal of the Royal Statistical Society: Series B* 31, 397–425.

Lindley, D. (1971). 'The Estimation of Many Parameters', in Godambe, V. and Sprott, D. (eds.), pp. 435–55.

Lindley, D. (1976). 'Bayesian Statistics', in Harper, W. and Hooker, C. (eds.), pp. 353–62.

Lindley, D. (1982). 'The Role of Randomization in Inference', *PSA: Proceedings of the Biennial Meeting of the Philosophy of Science Association, 1982*(2), 431–46.

Lindley, D. (2000). 'The Philosophy of Statistics' (with Discussion), *Journal of the Royal Statistical Society: Series D* 49(3), 293–337.

Lindley, D. and Novick, M. (1981). 'The Role of Exchangeability in Inference', *Annals of Statistics* 9(1), 45–58.

Little, R. (2006). 'Calibrated Bayes: A Bayes/Frequentist Roadmap', *The American Statistician* 60(3), 213–23.

Lodge, O. (1919). 'Contribution to "Discussion on the Theory of Relativity"', *Monthly Notices of the Royal Astronomical Society* 80, 106–9.

Logan, B. (2012). 'Jackie Mason Review', *The Guardian* (2/21/2012).

Longino, H. (2002). *The Fate of Knowledge*. Princeton, NJ: Princeton University Press.

Madigan, D. and Raftery, A. (1994). 'Model Selection and Accounting for Model Uncertainty in Graphical Models Using Occam's Window', *Journal of the American Statistical Association* 89(428), 1535–46.

Maher, P. (2004). 'Bayesianism and Irrelevant Conjunction', *Philosophy of Science* 71, 515–20.

Marcus, G. (2018). 'Deep Learning: A Critical Appraisal', arXiv: 1801.00631 preprint, 1–27.

Martin, R. and Liu, C. (2013). 'Inferential Models: A Framework for Prior-free Posterior Probabilistic Inference', *Journal of the American Statistical Association* 108, 301–13.

Mayo, D. (1980). 'The Philosophical Relevance of Statistics', *PSA: Proceedings of the Biennial Meeting of the Philosophy of Science Association, 1980*, 97–109.

Mayo, D. (1983). 'An Objective Theory of Statistical Testing', *Synthese* 57(3), 297–340.

Mayo, D. (1988). 'Toward a More Objective Understanding of the Evidence of Carcinogenic Risk', *PSA: Proceedings of the Biennial Meeting of the Philosophy of Science Association 1988*, 2, 489–503.

Mayo, D. (1991). 'Novel Evidence and Severe Tests', *Philosophy of Science* 58(4), 523–52.

Mayo, D. (1996). *Error and the Growth of Experimental Knowledge*. Chicago: University of Chicago Press.

Mayo, D. (1997a). 'Duhem's Problem, the Bayesian Way, and Error Statistics, or "What's Belief Got to Do with It?"' and 'Response to Howson and Laudan', *Philosophy of Science* 64(2), 222–44, 323–33.

Mayo, D. (1997b). 'Severe Tests, Arguing From Error, and Methodological Underdetermination', *Philosophical Studies* 86 (3), 243–66.

Mayo, D. (2003a). 'Severe Testing as a Guide for Inductive Learning', in Kyburg, H. E. and Thalos, M. (eds.), pp. 89–117.

Mayo, D. (2003b). 'Could Fisher, Jeffreys and Neyman Have Agreed on Testing? Commentary on J. Berger's Fisher Address', *Statistical Science* 18, 19–24.

Mayo, D. (2004). 'An Error-statistical Philosophy of Evidence' and 'Rejoinder', in Taper, M. and Lele, S. (eds.), pp. 79–97, 101–18.

Mayo, D. (2005a). 'Peircean Induction and the Error-Correcting Thesis', *Transactions of the Charles S. Peirce Society: A Quarterly Journal in American Philosophy* 41(2), 299–319.

Mayo, D. (2005b). 'Evidence as Passing Severe Tests: Highly Probable versus Highly Probed Hypotheses', in Achinstein, P. (ed.), *Scientific Evidence*, Johns Hopkins, pp. 95–127.

Mayo, D. (2008). 'How to Discount Double-Counting When It Counts: Some Clarifications', *The British Journal for the Philosophy of Science* 59, 857–79.

Mayo, D. (2010a). 'Learning from Error, Severe Testing, and the Growth of Theoretical Knowledge', in Mayo, D. and Spanos, A. (eds.), pp. 28–57.

Mayo, D. (2010b). 'An Error in the Argument from Conditionality and Sufficiency to the Likelihood Principle', in Mayo, D. and Spanos, A. (eds.), pp. 305–14.

Mayo, D. (2010c). 'Sins of the Epistemic Probabilist: Exchanges with Peter Achinstein', in Mayo, D. and Spanos, A. (eds.), pp. 189–201.

Mayo, D. (2010d). 'An Ad Hoc Save of a Theory of Adhocness?: Exchanges with John Worrall', in Mayo, D. and Spanos, A. (eds.), pp. 155–169.

Mayo, D. (2010e). 'Learning from Error: The Theoretical Significance of Experimental Knowledge', *The Modern Schoolman*. Guest editor, Kent Staley. 87(3/4), (March/May 2010). Experimental and Theoretical Knowledge, The Ninth Henle Conference in the History of Philosophy, 191–217.

Mayo, D. (2011). 'Statistical Science and Philosophy of Science: Where Do/Should They Meet in 2011 (and Beyond)?', *Rationality, Markets and Morals (RMM)* 2, 79–102.

Mayo, D. (2012a). 'Statistical Science Meets Philosophy of Science Part 2: Shallow Versus Deep Explorations', *Rationality, Markets and Morals (RMM)* 3, 71–107.

Mayo, D. (2012b). 'Fallacy of Rejection and the Fallacy of Nouvelle Cuisine', Blogpost on Errorstatistics.com (4/4/2012).

Mayo, D. (2013a). 'Presented Version: On the Birnbaum Argument for the Strong Likelihood Principle', *JSM Proceedings*, Section on Bayesian Statistical Science. Alexandria, VA: American Statistical Association, (2013), 440–53.

Mayo, D. (2013b). 'The Error-Statistical Philosophy and the Practice of Bayesian Statistics: Comments on Gelman and Shalizi: "Philosophy and the Practice of

Bayesian Statistics"', *British Journal of Mathematical and Statistical Psychology* 66 (1), 57–64.

Mayo, D. (2014a). 'Learning from Error: How Experiment Gets a Life (of Its Own)', in *Error and Uncertainty in Scientific Practice*, Boumans, M., Hon, G. and Petersen, A. (eds.), London: Pickering and Chatto, pp. 57–77.

Mayo, D. (2014b). 'On the Birnbaum Argument for the Strong Likelihood Principle' (with discussion), *Statistical Science* 29(2), 227–39; 261–6.

Mayo, D. (2015). 'Can You Change Your Bayesian Prior? (and discussion)', Blogpost on ErrorStatistics.com (6/18/2015).

Mayo, D. (2016). 'Don't Throw Out the Error Control Baby with the Bad Statistics Bathwater: A Commentary on Wasserstein, R. L. and Lazar, N. A. 2016, "The ASA's Statement on p-Values: Context, Process, and Purpose"', *The American Statistician* 70(2) (supplemental materials).

Mayo, D. and Cox, D. (2006). 'Frequentist Statistics as a Theory of Inductive Inference', in Rojo, J. (ed.), *Optimality: The Second Erich L. Lehmann Symposium*, Lecture Notes-Monograph series, Institute of Mathematical Statistics (IMS), 49, pp. 77–97. (Reprinted 2010 in Mayo, D. and Spanos, A. (eds.), pp. 247–75.)

Mayo, D. and Kruse, M. (2001). 'Principles of Inference and Their Consequences', in Cornfield, D. and Williamson, J. (eds.), *Foundations of Bayesianism*. Dordrecht, The Netherlands: Kluwer Academic Publishers, pp. 381–403.

Mayo, D. and Morey, R. (2017). 'A Poor Prognosis for the Diagnostic Screening Critique of Statistical Tests', OSF Preprint.

Mayo, D. and Spanos, A. (2004). 'Methodology in Practice: Statistical Misspecification Testing', *Philosophy of Science: Symposia 2002* 71, 1007–25.

Mayo, D. and Spanos, A. (2006). 'Severe Testing as a Basic Concept in a Neyman–Pearson Philosophy of Induction', *British Journal for the Philosophy of Science* 57(2), 323–57.

Mayo, D. and Spanos, A. (eds.) (2010). *Error and Inference: Recent Exchanges on Experimental Reasoning, Reliability, and the Objectivity and Rationality of Science*. Cambridge: Cambridge University Press.

Mayo, D. and Spanos, A. (2011). 'Error Statistics', in Bandyopadhyay, P. and Forster, M. (eds.), *Philosophy of Statistics, 7, Handbook of the Philosophy of Science*, Amsterdam: Elsevier, pp. 153–98.

Meehl, P. (1978). 'Theoretical Risks and Tabular Asterisks: Sir Karl, Sir Ronald, and the Slow Progress of Soft Psychology', *Journal of Consulting and Clinical Psychology* 46, 806–34.

Meehl, P. (1990). 'Why Summaries of Research on Psychological Theories Are Often Uninterpretable', *Psychological Reports* 66(1), 195–244.

Meehl, P. and Waller, N. (2002). 'The Path Analysis Controversy: A New Statistical Approach to Strong Appraisal of Verisimilitude', *Psychological Methods* 7(3), 283–300.

Mill, J. S. (1888). *A System of Logic*, 8th edn. New York: Harper and Brothers.

Michell, J. (2008). 'Is Psychometrics Pathological Science?' *Measurement* 6, 7–24.

Miller, J. (2008). *Naturalism & Objectivity: Methods and Meta-methods* (dissertation), Virginia Tech.

Mignard, F. and Klioner, S. (2009). 'Gaia: Relativistic Modelling and Testing', *Proceedings of the International Astronomical Union* 5(S261), 306–14.

Morrison, D. and Henkel, R. (eds.) (1970). *The Significance Test Controversy: A Reader*. Chicago, IL: Aldine De Gruyter.

Munafò, M., Nosek, B., Bishop, D., et al. (2017). 'A Manifesto for Reproducible Science', *Nature Human Behaviour* 1 (art 0021), 1–9.

Musgrave, A. (1974). 'Logical versus Historical Theories of Confirmation', *The British Journal for the Philosophy of Science* 25(1), 1–23.

Musgrave, A. (2010). 'Critical Rationalism, Explanation and Severe Tests', in Mayo, D. and Spanos, A. (eds.), pp. 88–112.

The National Women's Health Network (NWHN) (2002). *The Truth About Hormone Replacement Therapy: How to Break Free from the Medical Myths of Menopause*. Roseville, CA: Random House, Prima Publishing.

Nelder, J. (2000). 'Commentary on Lindley's "The Philosophy of Statistics"', *Journal of the Royal Statistical Society: Series D* 49(3), 324–5.

Nevins, J. and Potti, A. (2015). 'Nevins and Potti Respond to Perez's Questions and Worries', *The Cancer Letter* (January 9, 2015), cancerletter.com/articles/20150109_10/.

Neyman, J. (1934). 'On the Two Different Aspects of the Representative Method: The Method of Stratified Sampling and the Method of Purposive Selection', *The Journal of the Royal Statistical Society* 97(4), 558–625. Reprinted 1967 *Early Statistical Papers of J. Neyman*, 98–141.

Neyman, J. (1937). 'Outline of a Theory of Statistical Estimation Based on the Classical Theory of Probability', *Philosophical Transactions of the Royal Society of London Series A* 236(767), 333–80. Reprinted 1967 in *Early Statistical Papers of J. Neyman*, 250–90.

Neyman, J. (1941). 'Fiducial Argument and the Theory of Confidence Intervals', *Biometrika* 32(2), 128–50. Reprinted 1967 in *Early Statistical Papers of J. Neyman*: 375–94.

Neyman, J. (1950). *First Course in Probability and Statistics*. New York: Henry Holt.

Neyman, J. (1952). *Lectures and Conferences on Mathematical Statistics and Probability*, 2nd edn. Washington, DC: Graduate School of U.S. Department of Agriculture.

Neyman, J. (1955). 'The Problem of Inductive Inference', *Communications on Pure and Applied Mathematics* 8(1), 13–46.

Neyman, J. (1956). 'Note on an Article by Sir Ronald Fisher', *Journal of the Royal Statistical Society, Series B (Methodological)* 18(2), 288–94.

Neyman, J. (1957a). '"Inductive Behavior" as a Basic Concept of Philosophy of Science', *Revue de l'Institut International de Statistique/Review of the International Statistical Institute* 25(1/3), 7–22.

Neyman, J. (1957b). 'The Use of the Concept of Power in Agricultural Experimentation', *Journal of the Indian Society of Agricultural Statistics* IX(1), 9–17.

Neyman, J. (1962). 'Two Breakthroughs in the Theory of Statistical Decision Making', *Revue De l'Institut International De Statistique / Review of the International Statistical Institute*, 30(1),11–27.

Neyman, J. (1967). *Early Statistical Papers of J. Neyman*. Berkeley: University of California Press.

Neyman, J. (1976). 'Tests of Statistical Hypotheses and Their Use in Studies of Natural Phenomena', *Communications in Statistics: Theory and Methods* 5(8), 737–51.

Neyman, J. (1977). 'Frequentist Probability and Frequentist Statistics', *Synthese* 36(1), 97–131.

Neyman, J. (1981). 'Egon S. Pearson (August 11, 1895–June 12, 1980). An Appreciation', *The Annals of Statistics* 9(1), 1–2.

Neyman, J. and Pearson, E. (1928). 'On the Use and Interpretation of Certain Test Criteria for Purposes of Statistical Inference: Part I', *Biometrika* 20A(1/2), 175–240. Reprinted in *Joint Statistical Papers*, 1–66.

Neyman, J. and Pearson, E. (1933). 'On the Problem of the Most Efficient Tests of Statistical Hypotheses', *Philosophical Transactions of the Royal Society of London Series A* 231, 289–337. Reprinted in *Joint Statistical Papers*, 140–85.

Neyman, J. and Pearson, E. (1936). 'Contributions to the Theory of Testing Statistical Hypotheses', *Statistical Research Memoirs* 1, 1–37. Reprinted in *Joint Statistical Papers*, 203–39.

Neyman, J. and Pearson, E. (1967). *Joint Statistical Papers of J. Neyman and E. S. Pearson*. Berkeley, CA: University of California Press.

O'Hagan, T. (2012). 'Higgs Boson: Digest and Discussion', August 20, 2012. https://www.scribd.com/document/220900274/Higgs-Boson.

O'Neil, C. (2016). *Weapons of Math Destruction: How Big Data Increases Inequality and Threatens Democracy*, New York: Penguin Books

Open Science Collaboration (2015). 'Estimating the Reproducibility of Psychological Science', *Science* 349(6251), 943–51.

Overbye, D. (2013). 'Chasing the Higgs', *New York Times* (March 5, 2013).

Pearson, E. (1947). 'The Choice of Statistical Tests Illustrated on the Interpretation of Data Classed in a 2 × 2 Table', *Biometrika* 34 (1/2), 139–167. Reprinted 1966 in *The Selected Papers of E. S. Pearson*, pp. 169–200.

Pearson, E. (1950). 'On Questions Raised by the Combination of Tests Based on Discontinuous Distributions', *Biometrika*, 37(3/4),383–98. Reprinted 1966 in *The Selected Papers of E. S. Pearson*, pp. 217–32.

Pearson, E. (1955). 'Statistical Concepts in Their Relation to Reality', *Journal of the Royal Statistical Society Series B* 17(2), 204–7.

Pearson, E. (1962). 'Some Thoughts on Statistical Inference', *The Annals of Mathematical Statistics* 33(2), 394–403. Reprinted 1966 in *The Selected Papers of E. S. Pearson*, pp. 276–83.

Pearson, E. (1966). *The Selected Papers of E. S. Pearson*. Berkeley, CA: University of California Press.

Pearson, E. (1970). 'The Neyman Pearson Story: 1926–34', in Pearson, E. and Kendall, M. (eds.), *Studies in History of Statistics and Probability, I*. London: Charles Griffin & Co., 455–77.

Pearson, E. and Chandra Sekar, C. (1936). 'The Efficiency of Statistical Tools and a Criterion for the Rejection of Outlying Observations', *Biometrika* 28 (3/4), 308–20. Reprinted 1966 in *The Selected Papers of E. S. Pearson*, pp. 118–30.

Pearson, E. and Neyman, J. (1930). 'On the Problem of Two Samples,' *Bulletin of the Academy of Polish Sciences*, 73–96. Reprinted 1966 in *Joint Statistical Papers*, 99–115.

Peirce, C. S. (1931–35). *Collected Papers*, Volumes 1–6. Hartsthorne, C. and Weiss, P. (eds.), Cambridge, MA: Harvard University Press.

Perez, B. (2015). 'Research Concerns, The Med. Student's Memo', *Cancer Letter*, 1/9/ 2015.

Pigliucci, M. (2010). *Nonsense on Stilts: How to Tell Science from Bunk*. Chicago, IL: University of Chicago Press.

Pigliucci, M. (2013). 'The Demarcation Problem: A (Belated) Response to Laudan', in Pigliucci, M. and Boudry, M. (eds.), *Philosophy of Pseudoscience: Reconsidering the Demarcation Problem*, Chicago: University of Chicago Press, pp. 9–28.

Popper, K. (1959). *The Logic of Scientific Discovery*. London, New York: Routledge.

Popper, K. (1962). *Conjectures and Refutations: The Growth of Scientific Knowledge*. New York: Basic Books.

Popper, K. (1983). *Realism and the Aim of Science*. Totowa, NJ: Rowman and Littlefield.

Potti, A., Dressman H. K., Bild, A., et al. (2006). 'Genomic Signatures to Guide the Use of Chemotherapeutics', *Nature Medicine* 12(11), 1294–300.

Potti, A. and Nevins, J. (2007). 'Potti et al. Reply', *Nature Medicine* 13(11), 1277–8.

Pratt, J. (1961). 'Review of Testing Statistical Hypotheses by E. L. Lehmann', *Journal of the American Statistical Association* 56, 163–7.

Pratt, J. (1965). 'Bayesian Interpretation of Standard Inference Statements', *Journal of the Royal Statistical Society: Series B* 27(2), 169–203.

Pratt, J., Raiffa, H., and Schlaifer, R. (1995). *Introduction to Statistical Decision Theory*. Cambridge, MA: MIT Press.

Prusiner Labs, University of California, press release, San Francisco (2004, July 30). Prion Finding Offers Insight into Spontaneous Protein Diseases. *ScienceDaily*.

Prusiner, S. (1982). 'Novel Proteinaceous Infectious Particles Cause Scrapie', *Science* 216(4542), 136–44.

Prusiner, S. (1997). 'Stanley B. Prusiner: Biographical', Nobelprize.org, Nobel Media AB 2014, online November 21, 2017 (www.nobelprize.org/nobel_prizes/medi cine/laureates/1997/prusiner-bio.html).

Prusiner, S. (2014). *Madness and Memory: The Discovery of Prions: A New Biological Principle of Disease*. New Haven, CT: Yale University Press.

Ratliff, K. A. and Oishi, S. (2013). 'Gender Differences in Implicit Self-Esteem Following a Romantic Partner's Success or Failure', *Journal of Personality and Social Psychology* 105(4), 688–702.

Reich, E. (2012). 'Flaws Found in Faster than Light Neutrino Measurement: Two Possible Sources of Error Uncovered', *Nature* online (2/22/2012).

Reid, C. (1998). *Neyman*. New York: Springer Science & Business Media.

Reid, N. (2003). 'Could Fisher, Jeffreys and Neyman Have Agreed on Testing? Commentary on J. Berger's Fisher Address', *Statistical Science* 18, 27.

Reid, N. and Cox. D. (2015). 'On Some Principles of Statistical Inference', *International Statistical Review* 83(2), 293–308.

Robert, C. (2007). *The Bayesian Choice: From Decision-Theoretic Foundations to Computational Implementation*. New York: Springer-Verlag.

Robert, C. (2011). 'Discussion of "Is Bayes Posterior Just Quick and Dirty Confidence?" by D. A. S. Fraser', *Statistical Science* 26(3), 317–18.

Robbins, H. (1956). 'An Empirical Bayes' Approach to Statistics', in *Proceedings Third Berkeley Symposium on Mathematical Statistics and Probability, Vol. 1*, Berkeley, CA: University of California Press, pp. 157–64.

Rosenkrantz, R. (1977). *Inference, Method and Decision: Towards a Bayesian Philosophy of Science*. Dordrecht, The Netherlands: D. Reidel.

Rosenthal, R. and Gaito, J. (1963). 'The Interpretation of Levels of Significance by Psychological Researchers', *Journal of Psychology* 55(1), 33–8.

Rosenthal, R. and Rubin, D. (1994). 'The Counternull Value of an Effect Size: A New Statistic', *Psychological Science* 5(6), 329–34.

Rothman, K. (1990). 'No Adjustments Are Needed for Multiple Comparisons', *Epidemiology* 1(1), 43–6.

Rothman, K., Lash, T., and Schachtman, N. (2013). *Brief to United States Supreme Court of Amici Curiae in Support of Petitioner*. W. Scott Harkonen v. United States, No. 13–180, (9th Cir., filed September 4, 2013)

Royall, R. (1997). *Statistical Evidence: A Likelihood Paradigm*. Boca Raton. FL: Chapman and Hall, CRC Press.

Royall, R. (2004). 'The Likelihood Paradigm for Statistical Evidence' and 'Rejoinder', in Taper, M. and Lele, S. (eds.) *The Nature of Scientific Evidence*, Chicago: University of Chicago Press, pp. 119–138; 145–151.

Rubin, D. (1984). 'Bayesianly Justifiable and Relevant Frequency Calculations for the Applied Statistician', *The Annals of Statistics* 12(4), 1151–72.

Salmon, W. (1966). *The Foundations of Scientific Inference*. Pittsburgh, PA: University of Pittsburgh Press.

Salmon, W. (1988). 'Dynamic Rationality: Propensity, Probability, and Credence', in Fetzer, J. (ed.), *Probability and Causality: Essays in Honor of Wesley C. Salmon*, Dordrecht, The Netherlands: D. Reidel, pp. 3–40.

Savage, L. J. (1954). *The Foundations of Statistics*. New York: Wiley.

Savage, L. J. (1961). 'The Foundations of Statistics Reconsidered', *Proceedings of the Fourth Berkeley Symposium on Mathematical Statistics and Probability* 1, Berkeley: University of California Press, 575–86.

Savage, L. J. (ed.) (1962). *The Foundations of Statistical Inference: A Discussion*. London: Methuen.

Savage, L. J. (1964). 'The Foundations of Statistics Reconsidered,' in Kyburg, H. E. and Smolker, H. (eds.), *Studies in Subjective Probability*, New York: Wiley, pp. 173–188. (Published originally 1961.)

Schnall, S., Benton, J., and Harvey, S. (2008). 'With a Clean Conscience: Cleanliness Reduces the Severity of Moral Judgments', *Psychological Science* 19(12), 1219–22.

Schweder, T. and Hjort, N. (2016). *Confidence, Likelihood, Probability, Statistical Inference with Confidence Distributions*. Cambridge: Cambridge University Press.

Sebastiani, P., Solovieff, N., Puca, A., et al. (2010). 'Genetic Signatures of Exceptional Longevity in Humans', *Science* (epub. July 1, 2010). (Retracted: *Science* July 22, 2011, 333(6041), 404.)

Seidenfeld, T. (1979). 'Why I Am Not an Objective Bayesian; Some Reflections Prompted by Rosenkrantz', *Theory and Decision* 11(4), 413–40.

Sellke, T., Bayarri, M., and Berger, J. (2001). 'Calibration of ρ Values for Testing Precise Null Hypotheses', *The American Statistician* 55(1), 62–71.

Selvin, H. (1970). 'A Critique of Tests of Significance in Survey Research' in Morrison, D. and Henkel, R. (eds.), pp. 94–106.

Senn, S. (1994a). 'Testing for Baseline Balance in Clinical Trials', *Statistics in Medicine* 13, 1715–26.

Senn, S. (1994b). 'Fisher's Game with the Devil', *Statistics in Medicine* 13(3), 217–30.

Senn, S. (2001a). 'Statistical Issues in Bioequivalence', *Statistics in Medicine* 20(17–18), 2785–2799.

Senn, S. (2001b). 'Two Cheers for P-values?' *Journal of Epidemiology and Biostatistics* 6 (2), 193–204.

Senn, S. (2002). 'A Comment on Replication, P-values and Evidence', S. N.Goodman, *Statistics in Medicine* 1992; 11:875-879', *Statistics in Medicine* 21(16), 2437–44.

Senn, S. (2007). *Statistical Issues in Drug Development*, 2nd edn. Chichester, UK: Wiley Interscience.

Senn, S. (2008). 'Comment on an Article by Gelman', *Bayesian Analysis* 3(3), 459–62.

Senn, S. (2011). 'You May Believe You Are a Bayesian But You Are Probably Wrong', *Rationality, Markets and Morals (RMM)* 2, 48–66.

Senn, S. (2013a). 'Comment on Gelman and Shalizi', *British Journal of Mathematical and Statistical Psychology* 66, 65–7.

Senn, S. (2013b). 'Seven Myths of Randomisation in Clinical Trials', *Statistics in Medicine* 32(9), 1439–50.

Senn, S. (2014). 'Blood Simple? The Complicated and Controversial World of Bioequivalence', Guest Blogpost on Errorstatistics.com (6/5/2014).

Senn, S. (2015a). 'Double Jeopardy?: Judge Jeffreys Upholds the Law', Guest Blogpost on Errorstatistics.com (5/9/2015).

Senn, S. (2015b). 'Comment' on Blogpost 'Can You Change Your Bayesian Prior?' on Errorstatistics.com (6/18/2015).

Senn, S. (2019). *Statistical Issues in Drug Development*, 3rd edn. Chichester, UK: Wiley Interscience.

Sewell, W. (1952). 'Infant Training and the Personality of the Child', *American Journal of Sociology* 58, 150–9.

Shaffer, J. (1995). 'Multiple Hypothesis-Testing', *Annual Review of Psychology* 46(1), 561–84.

Silberstein, L. (1919). 'Contribution to "Joint Eclipse Meeting of the Royal Society and the Royal Astronomical Society"', *The Observatory* 42, 389–98.

Silver, N. (2017). 'There Really Was a Liberal Media Bubble', on FiveThirtyEight.com (3/10/2017).

Simmons, J., Nelson, L., and Simonsohn, U. (2011). 'False-Positive Psychology: Undisclosed Flexibility in Data Collection and Analysis Allow Presenting Anything as Significant', *Psychological Science* 22(11), 1359–66.

Simmons, J., Nelson, L., and Simonsohn, U. (2012). 'A 21 word solution', *Dialogue: The Official Newsletter of the Society for Personality and Social Psychology* 26(2), 4–7.

Simonsohn, U. (2013). 'Just Post It: The Lesson from Two Cases of Fabricated Data Detected by Statistics Alone', *Psychological Science* 24(10), 1875–88.

Simonsohn, U., Nelson, L., and Simmons, J. (2014). 'P-Curve: A Key to the File-Drawer', *Journal of Experimental Psychology: General* 143(2), 534–47.

Singh, K., Xie, M., and Strawderman, W. (2007). 'Confidence Distribution (CD)–Distribution Estimator of a Parameter', IMS Lecture Notes–Monograph Series, Volume 54, *Complex Datasets and Inverse Problems: Tomography, Networks and Beyond*, pp. 132–50.

Skyrms, B. (1986). *Choice and Chance: An Introduction to Inductive Logic*, 3rd edn. Belmont, CA: Wadsworth.

Sober, E. (2001). 'Venetian Sea Levels, British Bread Prices, and the Principle of the Common Cause', *The British Journal for the Philosophy of Science* 52(2), 331–46.

Sober, E. (2008). *Evidence and Evolution: The Logic behind the Science*. Cambridge: Cambridge University Press.

Spanos, A. (1986). *Statistical Foundations of Econometric Modeling*. Cambridge: Cambridge University Press.

Spanos, A. (1999). *Probability Theory and Statistical Inference: Econometric Modeling with Observational Data*. Cambridge: Cambridge University Press.

Spanos, A. (2000). 'Revisiting Data Mining: "Hunting" with or without a License', *Journal of Economic Methodology* 7(2), 231–64.

Spanos, A. (2007). 'Curve Fitting, the Reliability of Inductive Inference, and the Error-Statistical Approach', *Philosophy of Science* 74(5), 1046–66.

Spanos, A. (2008a). 'Review of S. T. Ziliak and D. N. McCloskey's *The Cult of Statistical Significance*', *Erasmus Journal for Philosophy and Economics* 1(1), 154–64.

Spanos, A. (2008b). 'Statistics and Economics', in Durlauf, S. and Blume, L. (eds.), *The New Palgrave Dictionary of Economics*, 2nd edn., London: Palgrave Macmillan, pp. 1057–97.

Spanos, A. (2010a). 'Akaike-type Criteria and the Reliability of Inference: Model Selection Versus Statistical Model Specification', *Journal of Econometrics* 158(2), 204–20.

Spanos, A. (2010b). 'Is Frequentist Testing Vulnerable to the Base-Rate Fallacy?', *Philosophy of Science* 77(4), 565–83.

Spanos, A. (2010c). 'Theory Testing in Economics and the Error-Statistical Perspective', in Mayo, D. and Spanos, A. (eds.), pp. 202–46.

Spanos, A. (2010d). 'Graphical Causal Modeling and Error Statistics: Exchanges with Clark Glymour', in Mayo, D. and Spanos, A. (eds.), 364–75.

Spanos, A. (2011a). 'Revisiting the Welch Uniform Model: A Case for Conditional Inference?', *Advances and Applications in Statistical Science* 5, 33–52.

Spanos, A. (2011b). 'Foundational Issues in Statistical Modeling: Statistical Model Specification and Validation', *Rationality, Markets and Morals (RMM)* 2, 146–78.

Spanos, A. (2012). 'Revisiting the Berger Location Model: Fallacious Confidence Interval or a Rigged Example?', *Statistical Methodology*, 9, 555–61.

Spanos, A. (2013a). 'R. A. Fisher: How an Outsider Revolutionized Statistics', on Error Statistics blog (2/17/13). errorstatistics.com/2013/02/17/r-a-fisher-how-an-outsider-revolutionized-statistics/.

Spanos, A. (2013b). 'Who Should Be Afraid of the Jeffreys-Lindley Paradox?', *Philosophy of Science* 80, 73–93.

Spanos, A. (2013c). 'A Frequentist Interpretation of Probability for Model-based Inductive Inference', *Synthese* 190(9), 1555–85.

Spanos, A. (2014). 'Recurring Controversies about P values and Confidence Intervals Revisited', *Ecology* 95(3), 645–51.

Spanos, A. (2019). *Probability Theory and Statistical Inference: Empirical Modeling with Observational Data.* Cambridge: Cambridge University Press.

Spanos, A. and Mayo, D. (2015). 'Error Statistical Modeling and Inference: Where Methodology Meets Ontology', *Synthese* 192(11), 3533–55.

Spiegelhalter, D. (2004). 'Incorporating Bayesian Ideas into Health-Care Evaluation', *Statistical Science* 19(1), 156–74.

Spiegelhalter, D. (2012). 'Explaining 5 Sigma for the Higgs: How Well Did They Do?', Blogpost on Understandinguncertainty.org (8/7/2012).

Spiegelhalter, D., Freedman, L., and Parmar, M. (1994). 'Bayesian Approaches to Randomized Trials', *Journal of the Royal Statistical Society, Series A* 157(3), 357–416.

Sprenger, J. (2011). 'Science without (Parametric) Models: The Case of Bootstrap Resampling', *Synthese* 180(1), 65–76.

Sprott, D. (2000). 'Comments on the Paper by Lindley', *Journal of the Royal Statistical Society, Series D* 49(3), 331–2.

Staley, K. (2014). *An Introduction to the Philosophy of Science.* Cambridge: Cambridge University Press.

Staly, K. (2017). 'Pragmatic Warrant for Frequentist Statistical Practice: The Case of High Energy Physics', *Synthese* 194(2),355–76

Staley, K. and Cobb, A. (2011). 'Internalist and Externalist Aspects of Justification in Scientific Inquiry', *Synthese* 182(3), 475–92.

Stapel, D. (2014). *Faking Science: A True Story of Academic Fraud.* Translated by Brown, N. from the original 2012 Dutch *Ontsporing* (Derailment), http://nick .brown.free.fr/stapel.

Steegen, S., Tuerlinckx, F., Gelman, A., and Vanpaemel, W. (2016). 'Increasing Transparency Through a Multiverse Analysis', *Perspectives on Psychological Science* 11(5), 702–12.

Stevens, S. (1946). 'On the Theory of Scales of Measurement', *Science* 103, 677–80.

Stigler, S. (2016). *The Seven Pillars of Statistical Wisdom*, Cambridge, MA: Harvard University Press.

Stone, M. (1997). 'Discussion of Papers by Dempster and Aitkin', *Statistics and Computing* 7, 263–4.

Strassler, M. (2013a). 'CMS Sees No Excess in Higgs Decays to Photons', Blogpost on *Of Particular Significance* (profmattstrassler.com) (3/14/2013).

Strassler, M. (2013b). 'A Second Higgs Particle', Blogpost on *Of Particular Significance* (profmattstrassler.com) (7/2/2013).

Sugden, R. (2005). 'Experiments as Exhibits and Experiments as Tests', *Journal of Economic Methodology* 12(2), 291–302.

Suppes, P. (1969). 'Models of Data', in *Studies in the Methodology and Foundations of Science*, Dordrecht, The Netherlands: D. Reidel, pp. 24–35.

Talbott, W. (1991). 'Two Principles of Bayesian Epistemology', *Philosophical Studies: An International Journal for Philosophy in the Analytic Tradition* 62(2), 135–50.

Taleb, N. (2013). 'Beware the Big Errors of "Big Data"', *WIRED/Opinion.* Blogpost on Wired.com (2/8/2013).

Taleb, N. (2018). *Skin in the Game: The Thrills and Logic of Risk Taking.* London: Penguin Books.

Thaler, R. H. (2013). 'Breadwinning Wives and Nervous Husbands', *New York Times* (June 2, 2013: 3(L)).

van Belle, G. (2008). *Statistical Rules of Thumb*, 2nd edn. Hoboken, NJ: John Wiley and Sons.

Vigen, T. (2015). 'Tangled Bedsheets & Consumption of Cheese', in *Spurious Correlations*, New York: Hyperion (tylervigen.com/spurious-correlations).

von Mises, R. (1957). *Probability, Statistics and Truth.* New York: Dover.

Wagenmakers, E.-J. (2007). 'A Practical Solution to the Pervasive Problems of P values', *Psychonomic Bulletin & Review* 14(5), 779–804.

Wagenmakers, E.-J. and Grünwald, P. (2006). 'A Bayesian Perspective on Hypothesis Testing: A Comment on Killeen (2005)', *Psychological Science* 17(7), 641–2.

Wagenmakers, E.-J., Wetzels, R., Borsboom, D., and van der Maas, H. (2011). 'Why Psychologists Must Change the Way They Analyze Their Data: The Case of Psi: Comment on Bem (2011)', *Journal of Personality and Social Psychology* 100, 426–32.

Wasserman, L. (2006). 'Frequentist Bayes is Objective', *Bayesian Analysis*, 1, 451–6.

Wasserman, L. (2007). 'Why Isn't Everyone a Bayesian?', in Morris, C. and Tibshirani, R. (eds.), *The Science of Bradley Efron*, New York: Springer, pp. 260–1.

Wasserman, L. (2008). 'Comment on an Article by Gelman', *Bayesian Analysis* 3(3), 463–6.

Wasserman, L. (2012a). 'The Higgs Boson and the P-value Police', Blogpost on normal deviate.wordpress.com (7/11/2012).

Wasserman, L. (2012b). 'What is Bayesian/Frequentist Inference?', Blogpost on normal deviate.wordpress.com (11/7/2012).

Wasserman, L. (2012c). 'Nate Silver is a Frequentist: Review of "The Signal and the Noise"', Blogpost on normaldeviate.wordpress.com (12/4/2012).

Wasserman, L. (2013). 'The Value of Adding Randomness', Blogpost on normaldeviate .wordpress.com (6/9/2013).

Wasserstein, R. and Lazar, N. (2016). 'The ASA's Statement on P-values: Context, Process and Purpose', (and supplemental materials), *The American Statistician* 70(2), 129–33.

Wellek, S. (2010). *Testing Statistical Hypotheses of Equivalence and Noninferiority*, 2nd edn. Boca Raton, FL: Chapman and Hall, CRC Press.

Wertheimer, N. and Leeper E. (1979). 'Electrical Wiring Configurations and Childhood Cancer', *American Journal of Epidemiology* 109, 273–84.

Westfall, P. and Young, S. (1993). *Resampling-Based Multiple Testing: Examples and Methods for P-Value Adjustment.* New York: Wiley.

Wiki How to Do Anything (2017). 'How to Get Out of Quicksand' (wikihow.com/Get-out-of-Quicksand).

Wilks, S. (1938). 'The Large-Sample Distribution of the Likelihood Ratio for Testing Composite Hypotheses', *Annals of Mathematical Statistics* 9, 60–2.

Wilks, S. (1962). *Mathematical Statistics.* New York: John Wiley & Sons.

Will, C. M. (1986). *Was Einstein Right?: Putting General Relativity to the Test*, 1st edn. New York: Basic Books.

Will, C. M. (1993). *Theory and Experiment in Gravitational Physics.* Cambridge: Cambridge University Press.

Williamson, J. (2010). *In Defence of Objective Bayesianism.* Oxford: Oxford University Press.

Wilson, R. A. (1971). *Feminine Forever*, 3rd edn. New York: Pocket Books. (First published 1968 by M. Evans & Company.)

Woodward, J. (2000). 'Data, Phenomena, and Reliability', *Philosophy of Science 67,* S163–S179.

Worrall, J. (1978). 'Research Programmes, Empirical Support, and the Duhem Problem: Replies to Criticism', in Radnitzky, G. and Andersson, G. (eds.), *Progress and Rationality in Science,* Dordrecht, The Netherlands: D. Reidel, pp. 321–38.

Worrall, J. (1989). 'Fresnel, Poisson and the White Spot: The Role of Successful Predictions in the Acceptance of Scientific Theories', in Gooding, D., Pinch, T. and Schaffer, S. (eds.), *The Uses of Experiment: Studies in the Natural Sciences,* Cambridge: Cambridge University Press, pp. 135–57.

Worrall, J. (2002). 'What Evidence in Evidence-Based Medicine?', *Philosophy of Science* 69(S3), S316–S330.

Worrall, J. (2010). 'Error, Tests, and Theory Confirmation', in Mayo, D. and Spanos, A. (eds.), pp. 125–54.

Xie, M. and Singh, K. (2013). 'Confidence Distribution, the Frequentist Distribution Estimator of a Parameter: A Review', *International Statistical Review* 81(1), 3–39.

Xu, F. and Garcia, V. (2008). 'From the Cover: Intuitive Statistics by 8-month-old Infants', *Proceedings of the National Academy of Sciences* 105(13), 5012–15.

Young, S. (2013). 'Better P-values Through Randomization in Microarrays', Guest Blogpost on Errorstatistics.com (6/19/2013).

Zabell, S. L. (1992). 'R. A. Fisher and Fiducial Argument', *Statistical Science* 7(3), 369–87.

Ziliak, S. and McCloskey, D. (2008a). *The Cult of Statistical Significance: How the Standard Error Costs Us Jobs, Justice, and Lives.* Ann Arbor, MI: University of Michigan Press.

Ziliak, S. and McCloskey, D. (2008b). 'Science is Judgment, Not Only Calculation: A Reply to Aris Spanos's review of "The Cult of Statistical Significance"', *Erasmus Journal for Philosophy and Economics* 1(1), 165–70.

Index

severity (cont.)
  attained power, 342–3
  and confidence levels, 193
  and difference between two means, 345–6
  disobeys the probability axioms, 423
  and explanatory content/informativeness,
    79–80, 237
  and Fisherian tribes, 146
  function (SEV), 143
  and large-scale theories, 128, 162
  in meta-methodology, 9, 32
  and Popperian corroboration, 72, 75, 87
  vs. power analysis, 343
  and replicability, 370
  and sensitivity, 152
  when not calculable, 200
severity curves, 348–9, 360n4
severity interpretation of negative results
    (SIN), 143–5, 152, 212, 343, 346–7, 351;
    see also severity
severity interpretation of rejection (SIR), 143,
    265–6, 351
severity requirement/principle, 5, 92, 125, 258
  and biasing selection effects, 92
  and error control, 269
  and failed replication, 158, 266
  to block fallacies of rejection, 144, 357
  as heuristic tool, 12, 264
  informal, 109
  from low *P*-value, 209
  as minimal principle of evidence, 5, 396
  for non-significance (Higgs), 212
  in terms of solving a problem, 300
  vs. fit measures, 72
  weak and strong, 22, 108
  strong, 14
Sewell, W., 280
sexy science: severe testing in large-scale
    theories, 121, 163, 300
Shaffer, J., 275
Shalizi, C., 27, 432, 434
Sharpe, G., 137
shpower (retrospective power analysis), 354–6
  howlers of, 355–6
  vs. severity, 356; *see also* power analysis.
significance levels
  as predesignated, 137
  attained vs. predesignated, 173–5, 177;
    see also *P*-values
significance tests
  vs. comparativism, 35

Cox definition of, 93
criticisms of, 93–5, 438; *see also* chestnuts
  and howlers of tests
fallacies of rejection/non-rejection
falsifying alternatives in, 159
Fisherian (pure/simple), 132, 150
in Higgs, 202
roles for model testing and discovery,
  298–304; *see also* M-S tests
simple or point hypotheses, 33
test T+, 144; *see also* Neyman and Pearson
  (N-P) Tests
Silberstein, L., 127
Silver, N., 232–3
similar tests, 385, 386n6
Simmons, J., 43, 237, 270
Simonsohn, U., 43, 237, 270, 284, 285
Singh, K., 391
skin off your nose, 273
Skyrms, B., 62, 73
Slovic, P., 422
Smeesters, D., 284
Smith, C., 47
Smith, H., 339n4
Sober, E., 35–6, 47, 92n3, 242, 317–18, 380–1
Spanos, A., 120, 133–4, 139, 146n5, 200, 254–5,
  305, 308, 312–13, 317–19, 331, 352n10,
  355, 367, 387, 426
Spiegelhalter, D., 204–5, 401, 404
spike and smear priors, 239, 248, 250–1, 259,
  336
  Bayesian justification for, 251
  coffee shop, 257
  criticisms of, 252n4, 256, 259, 406, 440
  cult of the holy, 252
  severe tester on, 257–8
spongiform diseases, 81, *see also* kuru
Sprenger, J., 307
Sprott, D., 180, 399, 421
spurious associations, 3
  batch effects, 293
  in longevity study, 293, 362
  population and number of shoes, 308, 317
  sea level and price of bread, 317
Staley, K., 203, 235–6
standard model (SM) physics, 203, 206, 214, 215
Stapel, D., 78, 97, 100, 276
statistical battles, current state of play, 11–12,
  23–8, 395–7, 400–2, 444
  proxy, 437; *see also* getting beyond statistics
  wars